Regeneration of Complex Capital Goods

Joerg R. Seume · Berend Denkena · Philipp Gilge
Editors

Regeneration of Complex Capital Goods

Contributions to the Final Symposium
of the Collaborative Research Center 871

 Springer

Editors
Joerg R. Seume
Institute of Turbomachinery and Fluid
Dynamics
Leibniz University Hannover
Garbsen, Germany

Berend Denkena
Institute of Production Engineering
and Machine Tools
Leibniz University Hannover
Garbsen, Germany

Philipp Gilge
Institute of Turbomachinery and Fluid
Dynamics
Formerly Leibniz University Hannover
Garbsen, Germany

ISBN 978-3-031-51394-7   ISBN 978-3-031-51395-4 (eBook)
https://doi.org/10.1007/978-3-031-51395-4

This Springer imprint is published by the registered company Springer Nature Switzerland AG
The registered company address is: Gewerbestrasse 11, 6330 Cham, Switzerland

If disposing of this product, please recycle the paper.

Contents

Introduction

Joerg R. Seume

1 Preface

This volume on the results of the research of the Collaborative Research Center 871 (CRC 871) contains selected papers presented at the Final Symposium of the CRC 871, which took place from March 31 through April 1, 2022 at Leibniz University Hannover, Germany. The chapters herein are a compilation of articles presented by the sub-projects at the symposium. The CRC 871 "Regeneration of Complex Capital Goods" was funded by the Deutsche Forschungsgemeinschaft (DFG, German Research Foundation)—SFB 871/3—119193472. All participants of the CRC 871 would like to express their gratitude to the DFG for the funding. Overall, the book provides readers with the state of the art of Maintenance, Repair, and Overhaul (MRO) in different engineering and economic disciplines with a special focus on the regeneration of civil aircraft engines.

2 Introduction Collaborative Research Centre 871: Regeneration of Complex Capital Goods

From 2010 through 2022, the CRC 871 contributed to developing a scientific basis for maintaining complex capital goods. The main goal of the CRC 871 was to develop new and innovative methods for restoring or even improving the functional properties of capital goods in order to reuse as many of the worn components as possible. The functional benefit of regenerated components and of the entire capital good are assessed through model-based simulations. Rule-based decisions are used to select

J. R. Seume (✉)
Institute of Turbomachinery and Fluid Dynamics, Leibniz University Hannover, An der Universitaet 1, 30823 Garbsen, Germany
e-mail: seume@tfd.uni-hannover.de

© The Author(s) 2025
J. R. Seume et al. (eds.), *Regeneration of Complex Capital Goods*,
https://doi.org/10.1007/978-3-031-51395-4_1

the optimum regeneration path in order to achieve the maximum benefit for the customer.

In the CRC 871, jet engines were chosen as an application example because of the great challenge provided by their high complexity. The methods and procedures developed can also be transferred to other complex capital goods such as wind turbines, railway vehicles, and heavy-duty gas turbines. The objective of the CRC 871 required a cooperation of scientists from engineering, including the areas of product development and product design, industrial and production engineering, as well as business administration. Overall, twelve institutes from three faculties of Gottfried Wilhelm Leibniz University Hannover (LUH), the Laser Zentrum Hannover (LZH), and one institute of the Technical University of Braunschweig participated in the CRC 871. One researcher was associated as he moved to the Technical University Dresden.

2.1 Motivation and Objectives

The maintenance and servicing of capital goods such as aircraft engines, wind turbines, or rail vehicles contributes significantly to the operating cost. To reduce this share and save expensive resources, maintenance processes and repair procedures need to become more efficient. In state of the art regeneration processes, when the CRC 871 started in 2010, skilled workers carried out maintenance work and repairs on the basis of prescribed guidelines, see Fig. 1. The individual experience played an essential role, which can lead to poor reproducibility of decisions and a great variability of repair results.

The objective of the CRC 871 is to maintain and, if possible, improve the functionality of complex capital goods. A breakdown of the main objectives of the SFB 871 is presented below:

- Early identification of the components to be regenerated and the possible regeneration steps
- Continuous consideration of successive findings in the planning of the ongoing regeneration process
- Improved processes of regeneration and their integration in regeneration paths
- Model-based prediction of the functional properties of complex capital goods and regeneration-related expenses
- Based on the model-based predictions, rule-based decisions about the regeneration steps in order to maximize the benefit for the customer
- Integration of the model-based prediction of production-related expenses and functional benefits at the level of the entire capital goods

To achieve these subobjectives, the CRC 871 developed the scientific basis for an innovative approach: The example of aircraft engines was used to develop a combined virtual and real repair process. This novel approach used classic methods formerly reserved for product and manufacturing development and transferred them

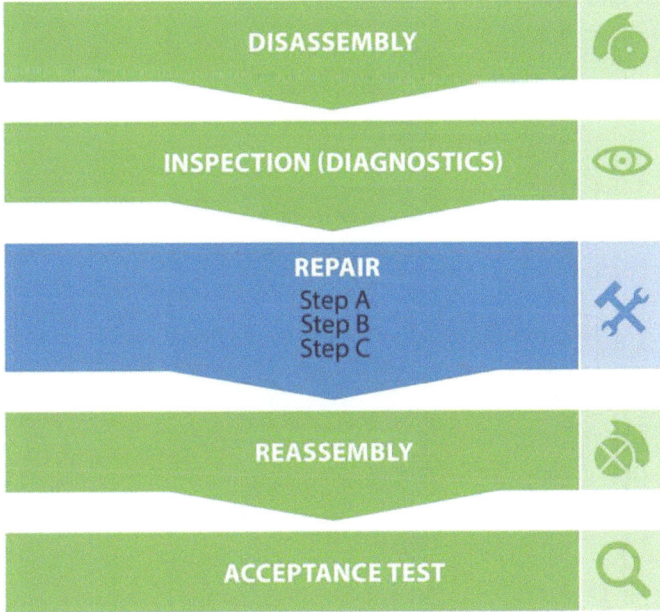

Fig. 1 Example of the current repair process of a civil aircraft engine

to maintenance (Fig. 2): Before selecting a regeneration path and the subsequent repair in a virtual process with the aid of a digital twin for all regeneration paths, the production-related cost and the functional benefit was evaluated and a decision for the most efficient regeneration path was derived from this rule-based assessment. The selection of the most efficient regeneration path depends on the customer business model and thus may differ by the customer.

2.2   Research Program

The aircraft engine was chosen directly at the beginning of the CRC 871 in the year 2010 as the central research object for the joint research. Aircraft engines are very complex machines where all involved disciplines reach the limits of physical understanding and design methods. This applies to both, the individual components and sub-assemblies as well as to the overall system behavior. Therefore, aircraft engines are perfect for the interdisciplinary research of regeneration in the CRC 871. In the first funding period, the CRC focused on a turbine blade and developed an innovative regeneration process. In the second funding period, the developed methods were transferred to an assembly using the example of a compressor blisk. In the third funding period, the research was focused on the overall system, i.e. the aircraft engine, reaching maximum complexity (Fig. 3).

Fig. 2 New repair process
for civil aircraft engines
based on the CRC 871

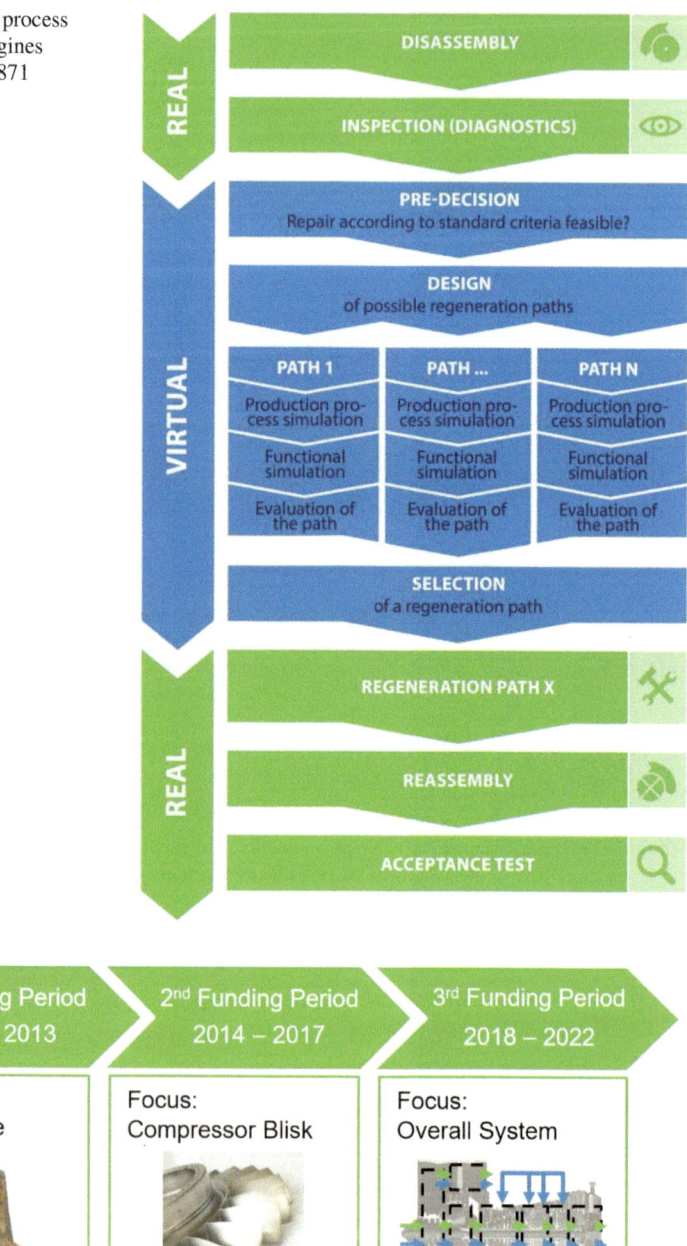

Fig. 3 Funding periods and its research focus

The sub-projects of the CRC 871 are grouped into four project areas. These project areas and sub-projects are depicted in Fig. 4 and were derived from the need to conduct research on the regeneration of civil aircraft engines.

Project area A "Inspection and Diagnostics" aimed to improve the volume and quality of information from the inspection, which is necessary for planning the regeneration process. The sub-projects of this project area developed methods for an earlier, cheaper, more comprhensive, and faster detection of the condition of the engine or of single components.

In project area B "Interaction of the Production Process with Functional Product Properties", the objective was to determine and to consider the complex interaction between production processes and functional properties of the components. To this end, the sub-projects simulated effects of the production process on the components and the overall system. The results of these simulations are used to determine

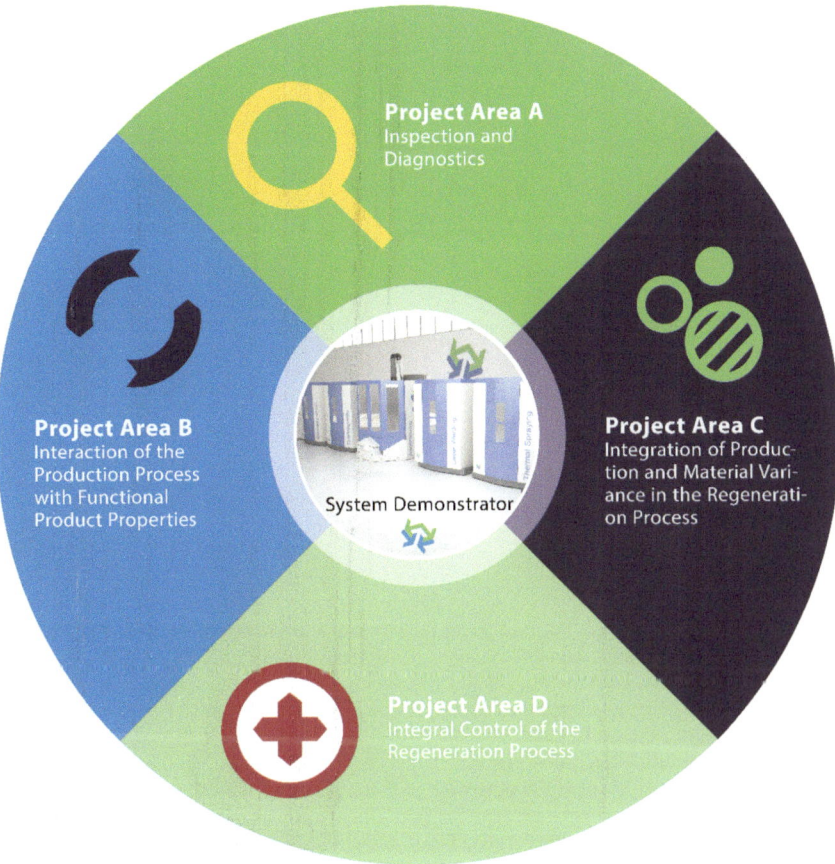

Fig. 4 Project areas of the CRC 871

the influence on the functional properties like the component life expectancy, fuel consumption, or vibration behavior.

The focus of project area C "Variance of Production and Material Properties in Regeneration" was on the variances which occur due to wear during operation and due to repair. The objective was to handle these variances and to estimate the influence on the functional properties.

In project area D "Integral Control of the Regeneration Process", the results from the other project areas were used for the integral control of each regeneration step in order to build an integrated process. An additional objective was to integrate the simulations of the functional properties of the components, which made it possible to evaluate the functional properties of the overall system, such that the complete regeneration process could be controlled.

In addition to the project areas, research was also carried out on bringing the CRC's subjects into the curricula of primary and secondary schools by an additional, dedicated sub-project during the last funding period. Furthermore, the practical applicability of the research results was shown by a system demonstrator.

Inspection and Condition Assessment (Project Area A)

Non-destructive Characterization of Coating and Material Conditions of Heavily Stressed Turbine Components

Maximilian K.-B. Weiss, Sebastian Barton, and Hans Jürgen Maier

Abstract This study presents non-destructive testing techniques for the fast assessment and detailed characterization of turbine blades. Different inspection techniques were developed to realize a flexible and adaptive inspection sequence. In addition to an initial inspection, the advantages of this inspection system with regard to quality assurance measures are also discussed. In this context, the focus is on regeneration. Furthermore, fatigue tests with combined non-destructive testing techniques are presented and the benefit of this new approach for in-situ monitoring of the microstructural evolution is shown.

Keywords Non-Destructive Testing · Eddy Current · Induction Thermography · Fatigue Testing · Maintenance Repair and Overhaul

1 Introduction

The requirements regarding performance and reliability of modern aircraft engines place high demands on the integrity of severely stressed engine components such as the turbine, compressor and fan blades and their condition assessment within the scope of inspections and maintenance measures. In order to be able to detect and assess the need for repair at an early stage and to consider it in further planning, a fast and flexible initial assessment of engine components is becoming increasingly important.

In particular, the turbine blades of the first stage after the combustion chamber are subject to high thermal, mechanical and chemical loads during operation, which leads to changes in the microstructure of the high-temperature materials used, and thus also in the safety-relevant material properties with increasing operating time (Bräunling 2015). In order to withstand the severe environmental conditions, the turbine blades are equipped with a coating system that reduces the effects of thermal

M. K.-B. Weiss (✉) · S. Barton · H. J. Maier
Institut für Werkstoffkunde (Materials Science), Leibniz University Hannover, An der Universitaet 2, 30823 Garbsen, Germany
e-mail: weiss@iw.uni-hannover.de

© The Author(s) 2025
J. R. Seume et al. (eds.), *Regeneration of Complex Capital Goods*,
https://doi.org/10.1007/978-3-031-51395-4_2

and corrosive loads. Due to the specific structure of such a turbine blade, appropriate testing technology is necessary for valid testing. Within this subproject of the CRC 871, a systematic investigation of the corresponding structural changes was carried out and the causes and effects on the component properties was analyzed. Within the context of the CRC 871, several new repair processes for engine components have been developed. Non-destructive defect testing (e.g. crack detection) and material characterization of the repaired areas were carried out. The focus was on quality assurance of the regeneration measures in the regeneration path. Thus, the material and component condition were evaluated depending on the selected repair processes as the new repair processes also affect the resulting material and component properties in the repair-welded, repair-brazed and coated areas.

The basic structure of a turbine blade shown in Fig. 1 can be divided into three different materials and functions. The substrate, which ultimately has to withstand the mechanical, thermal and chemical stresses is usually made of a nickel (Ni)- or cobalt (Co)-based superalloys. The degradation resulting from chemical attack is counteracted with the help of a corrosion protection layer, which is applied directly onto the base material. The standard are MCrAlY coating systems. The external thermal barrier coating (TBC) consists of a ceramic, often yttrium-stabilised zirconium oxide (YSZ), which significantly reduces the temperature of the base material due to its low thermal conductivity (Kumar et al. 2015; Reed 2006).

In addition to the structure of a turbine blade, knowledge of typical types of damage is necessary in order to adapt the testing techniques in a targeted manner and to be able to evaluate the quality of the test results. The damage that occurs in the course of operation can be assigned to geometry, such as foreign object damage (FOD) or burnings, to structure, for example in the form of cracks, or to the material itself in form of corrosion (Carter 2005). Figure 2 illustrates typical damage types of an operated turbine blade. These defects affect the functionality of the blade and can

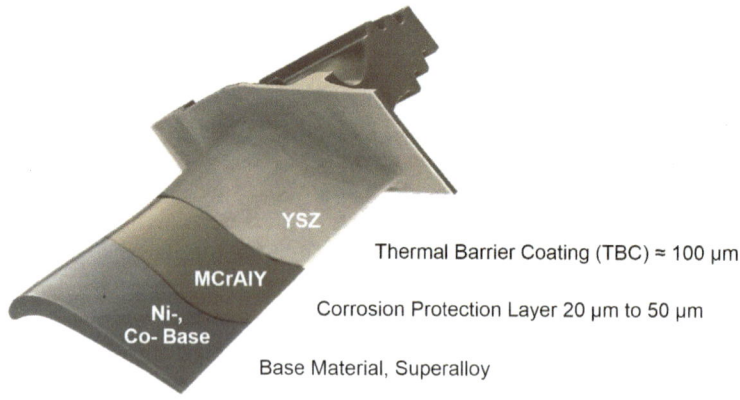

Fig. 1 Structure of a turbine blade consisting of base material, corrosion protection layer and thermal barrier coating; the coating thicknesses are given in μm

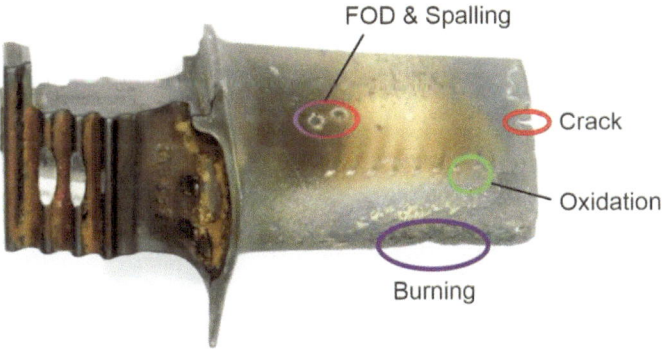

Fig. 2 Turbine blade showing typical damages due to operational loads: the individual damage types have been assigned to the categories geometry (purple), structure (red) and material (green)

lead to final failure if regeneration measures are not taken. Therefore, early detection of defects and evaluation of their impact on reliability is essential.

2 Objective

In order to carry out a targeted regeneration of individual components, it is essential to have detailed information about the location and extent of defects. This will be achieved, in particular, with the aim of flexible initial assessment, which represents a core element of this subproject. The special properties of a high-pressure turbine blade require a combination of different non-destructive testing (NDT) methods to fully characterize the condition of the coating system and the base material itself. By combining NDT methods, the inspection resources can be used such that high quality of the inspection results with the lowest possible inspection effort. This adaptive inspection sequence is to be used after the disassembly of an engine to allow a targeted alignment of the regeneration paths. With regard to highly stressed turbine blades, the characterization of thin layer and edge zone conditions plays a particularly important role. However, an intact microstructure of the base material, which can be negatively influenced mainly by high thermal and corrosive loads, is crucial for component integrity. Furthermore, the detected defects must be classified.

The flexible initial assessment, illustrated in Fig. 3, shows one approach how such a system can be designed. In the first process step an evaluation of the condition of the base material is performed. For this purpose, a harmonic analysis of eddy currents (HA-EC) is carried out with a sensor in integral design. This test is intended to evaluate microstructural degradation so that the base material can subsequently be assessed as operational or not. If the condition assessment of the base material leads to a negative result, no further testing resources need to be used, as this criterion leads to a reject. The next step is to check the blade for cracks, spalling and delamination of

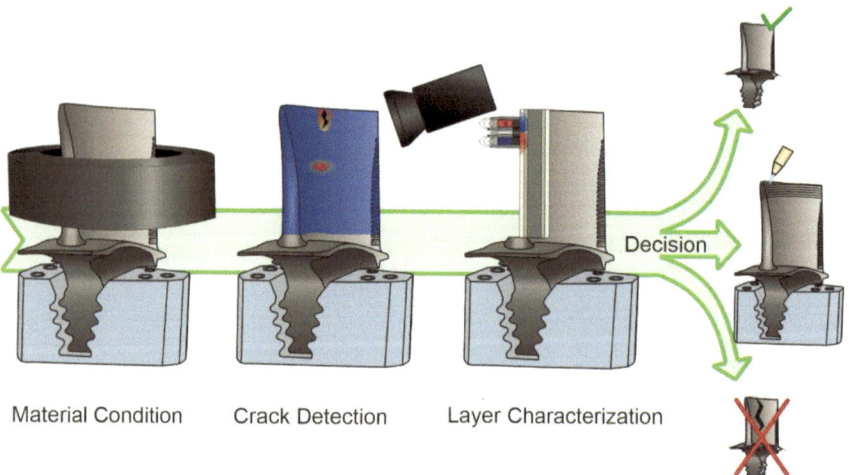

Fig. 3 Schematic illustrating the flexible initial assessment process chain, see main text for details

the coating system using active high-frequency thermography (HF-Thermography). These defects can be regenerated depending on the location and size of the defect. As a final test to ensure a holistic view of the blade condition, a characterization of the coating system needs to be performed. Here, high-frequency eddy current (HF-EC) test is used. The high frequencies ensure that the eddy currents mainly propagate in the subsurface zone, and thus the condition of the coating system can be evaluated. These steps can be carried out without having to strip the coating of the turbine blade. If no relevant microstructural change or damage can be detected in all three steps, the turbine blade can still be used without repair measures.

However, the flexible initial assessment developed is not only intended to allow targeted regeneration, but also to evaluate the regeneration measures themselves and to check the condition of the regenerated component. Here, it is important to reliably detect any process faults, such as lack of fusion after a welding process. With this approach, a process chain is available at the end that can also be used in quality assurance (Melchert et al. 2021; Rocha et al. 2018; Wang 2013).

A further working hypothesis is that the non-destructive in situ recording and characterization of the damage development of regenerated specimens in comparison to reference specimens under thermal and cyclic loading provide relevant information about the regeneration-related changes in the component properties, especially under cyclic thermomechanical loading. This allows evaluation of the repair methods used in connection with the functional benefit of the regenerated components, which results in a significant gain in knowledge for the evaluation of the repair measures.

By combining the non-destructive testing techniques with fatigue tests, a statement on the expected remaining service life can also be made. For this purpose, eddy current techniques, thermographic and acoustic methods were used in situ on a fatigue test rig. In addition to a residual life estimation, this combination also generated evaluation criteria for the HA-EC technique.

3   Material and Methods

Harmonic Analysis of Eddy Currents

The flexible initial assessment consists of three different testing techniques. HA-EC was used for material characterization. In the course of preliminary investigations, it was shown that the use of nickel-based materials at high temperatures in a sulphurous atmosphere leads to significant changes in the magnetic property profile due to the formation of ferromagnetic phases. This indicates a local change in the microstructure of the material, which is paramagnetic in its initial state. The eddy current sensors can be designed for integral testing, as shown in Fig. 3, or for local testing in scanning technology. To realize a problem-adapted testing, various sensors were developed and used in the subproject.

Thus, the magnetic inductive test method of harmonic analysis of eddy current signals was used in the present study as it can be applied to actual components. The harmonic analysis of eddy current signals is a non-destructive test method, which is suitable for the sensitive evaluation of magnetic material properties. The measuring principle of this non-destructive testing technique is shown in Fig. 4. A coil carrying an alternating current (AC) generates an alternating electromagnetic field (excitation signal), which induces eddy currents in an electrically conductive test specimen. In a (locally) ferromagnetic specimen, this also causes substantial remagnetizing processes. The resulting eddy currents and the remagnitization in the material generate a secondary magnetic field, which is superimposed on the primary field. These magnetic fields then induce a voltage in a measuring coil. If the sample has ferromagnetic or ferrimagnetic properties, the sinusoidal wave form of the voltage–time profile obtained for paramagnetic substances is now superimposed by so-called higher harmonics, which are due to the distortion of the magnetic field due to the nonlinear relationship between field strength and flux density. Using a Fast-Fourier-Transformation, the measured signal can be spectrally decomposed and the amplitude of the fundamental (1st harmonic) wave as well as the 3rd and 5th harmonic can be obtained. The eddy current test in combination with harmonic analysis is a well-established method for characterizing the properties of steel materials (Maaß et al. 2004, Mercier et al. 2006; Stegemann et al. 1998). In addition to the values for the amplitudes of the harmonics, this method can also be used to determine the Curie temperature of the examined area. The Curie temperature describes the temperature at which a material changes between ferromagnetic and paramagnetic behaviour. If the material is in the paramagnetic state, there is no distortion of the signal, and the higher harmonics vanish. This can be exploited to determine the Curie temperature. The Curie temperature is not a constant but depends on the loading history of the material. If the material is exposed to increased temperatures, more oxidised volume is formed. This is reflected in an increase in the Curie temperature, so that a statement

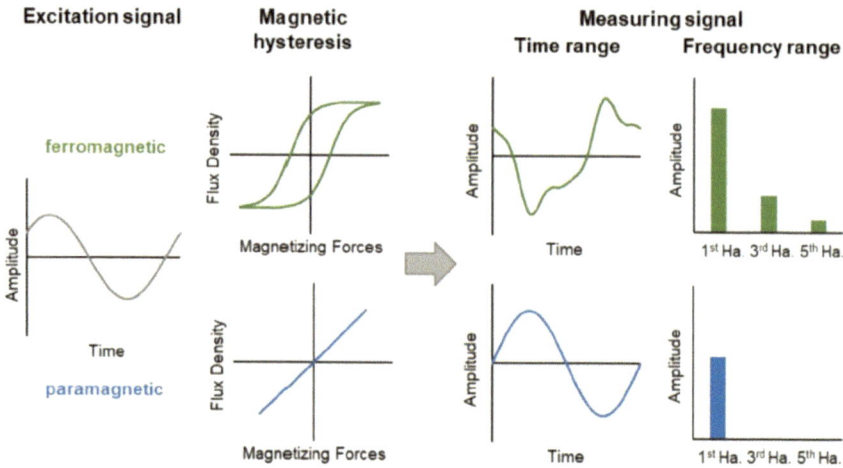

Fig. 4 Influence of ferromagnetic (top) and paramagnetic (bottom) materials on eddy current signals (Barton et al. 2021)

about the material's oxidation condition can be made with this method (Barton 2022, Barton et al. 2021, Delaunay et al. 2000, Garat et al. 2005, Jia et al. 2014).

3.1 Induction Thermography

Thermographic test methods can be divided into passive and active methods. In active methods, it is not the natural thermal radiation of the test object that is decisive, but an actively generated temperature change. Active methods are particularly suitable for detecting detection of cracks and other thin and sharp-edged defects in electrically conductive materials. Different from other optical methods that use the visible light spectrum, induction thermography is not based on effects such as reflection and shadowing, but uses heat generation and heat flow resulting from actively introduced electromagnetic energy in an examined area. The strong eddy current field created during inductive excitation causes increased heat release in areas of disturbed flux due to the local increase in current density. The increase in current density and the corresponding temperature rise take place mainly at sharp edges and other narrow geometric shape features of the test object, as well as in areas with defects and inhomogeneities in the material. The defect information resulting from the temperature distribution on the sample surface can be obtained and visualised using a thermal imaging system. The fundamental parameter that determines the applicability of this method and its sensitivity for defect detection is the eddy current penetration depth, which depends on the magnetic permeability and electrical conductivity of the material (Netzelmann et al. 2016, Schloblom et al. 2016). The thin-walled turbine blades made of nickel-based superalloys with low electrical conductivity ($\sigma \approx 0.6$ MS/m) and low magnetic permeability ($\mu r \approx 1$) generally require high-frequency excitation frequencies for damage detection. Assuming that not only the base material of the turbine blade but also the thin protective layer system of the blade is to be tested, excitation frequencies in the megahertz range are necessary to effectively generate strong eddy currents in the subsurface zone of the test object. The usual transistorised high-frequency induction generators operate in the kilohertz range. By contrast, the setup used was equipped with a high-performance tube-type generator that excites at frequencies between 0.1 and 3.5 MHz. An external microcontroller controlled the induction generator in pulsed mode. The locally introduced short-time high-energy pulses improve the dynamics and sharpness in the image of the temperature field and simultaneously reduce the global heating of the component. A series of successive excitation pulses can be used to generate a thermal response in the component, resulting in a specific temperature–time profile for each pixel. The best results were obtained with excitation times between 50 and 100 ms at 50% duty cycle (Schlobohm et al. 2016). When testing metallic components, long excitation times lead to temperature equalisation in the excited area and ultimately in the entire component due to the thermal conductivity of the material. As a result, the differences in the temperature field quickly disappear. Consequently, long excitation times reduce the contrast in the image between irregularities and the background and the measurement sensitivity becomes lower (Reimche et al. 2008, 2009). With pulsed mode excitation, the local relative temperature change was recorded. The thermal response was analysed and filtered with a Fast Fourier Transform (FFT) based algorithm at each pixel of the recorded image sequence. The advantage of this approach is that the local emissivity variations of the sample surface, e.g. due to impurities, can be compensated.

Figure 5 shows the schematic structure of the inductive high-frequency thermography system. The synchronisation of the individual components by the control system in particular is of major importance here. Induction thermography is suitable for real-time detection of cracks and similar defects with high optical resolution at different scales. By using changeable lenses with different focal lengths, it is possible to vary the magnification and the inspection times, i.e. coarse-fast/fine-slow. Furthermore, the inspection can be executed in automatic mode. For the inspection of the specimens, the thermography system was equipped with a macro lens that allows a lateral resolution of 15 µm per pixel and a free working distance of 300 mm. The presented high-frequency induction thermography technique allows distinguishing between cracks in the bulk and non-critical surface scratches, thus avoiding false defect indications. Tests on TBC-coated samples have shown that high-frequency induction thermography can be used not only to detect cracks in the substrate and in the coating material, but also to detect changes in the thickness of the thermal barrier coating and detachments between the coating and the substrate (Schlobohm et al. 2017).

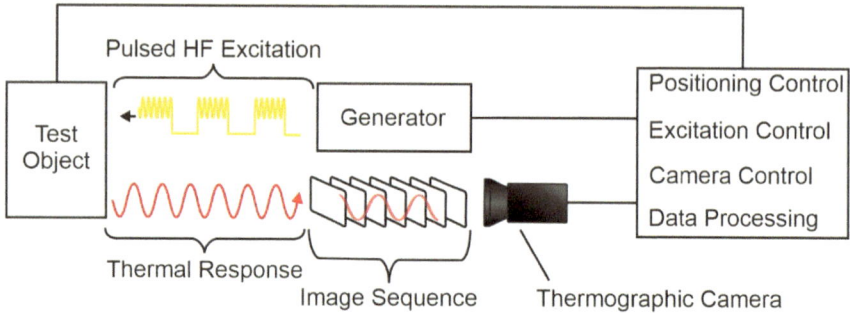

Fig. 5 Schematic diagram of the HF thermographic system; see main text for details

3.2 *High-Frequency Eddy Current Testing*

During the operating life of a turbine blade, the chemical composition of the aluminium-based corrosion protection layer (PtAl, MCrAlY) changes due to oxidation at high temperatures. In particular, aluminium oxides form in the highly stressed areas of the turbine blade and the content of metallic aluminium decreases significantly—the corrosion protection layer loses its protective function and must be regenerated (Bräunling 2015; Schlobohm et al. 2017). Maintenance, repair and overhaul (MRO) companies are permitted to repair a turbine blade coating system only two or three times. Therefore, there is a great interest in non-destructive characterization of the coating condition before the repair process. Since the electrical conductivity of the corrosion protection layer also decreases with the reduction of the metallic aluminium, the electrical conductivity can be used as an indicator of the condition of the corrosion protection layer. As will be shown below, the eddy current technique is a suitable test method for a quick and non-destructive assessment of the coating condition. Modern eddy current test systems use an eddy current sensor to generate an alternating magnetic field with a characteristic frequency in the low single-digit MHz range. Depending on the test frequency, the eddy currents can be induced at different depths of the test object. These eddy currents generate a secondary magnetic field with a characteristic phase shift and field strength and lead to an equally characteristic, sinus-shaped voltage induced in the sensor's measuring coil. The phase shift and amplitude of this measurement signal are mainly based on the electrical conductivity in dia- and para-magnetic materials and can therefore be used to estimate the layer condition (Delauny et al. 2000). However, due to the high penetration depth of the eddy current, conventional eddy current testing systems are not suitable for characterizing the coating state of the very thin corrosion protection layer separately from the base material (Bräunling 2015; Cosack 2009).

Within the framework of the CRC 871 a new HF-EC test system that operates in the frequency range up to 100 MHz with custom-build and miniaturised sensors, as shown in Fig. 6 was developed. With these sensors, it is possible to test in the gas path without having to dismantle the individual blades. Due to the high test frequency,

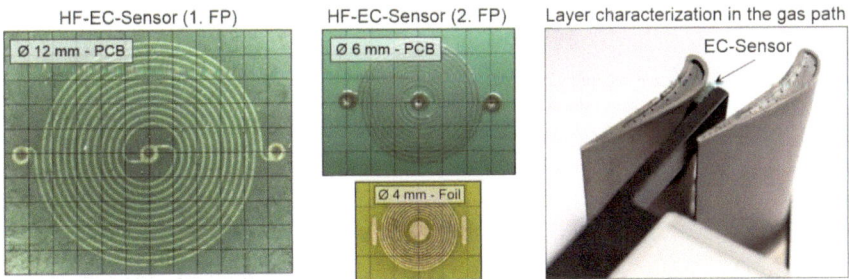

Fig. 6 Miniaturised HF-EC sensors of the 1st and 2nd funding period and application of the sensors in the gas path

a typical penetration depth of < 100 μm can be realized, allowing a separate and non-destructive characterization of the turbine blade coating condition (Frackowiak et al. 2018; Bruchwald et al. 2016).

3.3 Fatigue Tests with Combined NDT

As a result of the application of new repair processes, the mechanical-technological characteristic values of the components in the repair-welded, repair-brazed and coated areas change. In order to determine the characteristic material properties in the area of the repair points and to evaluate the repair methods used, a test rig that allow non-destructive in situ recording and characterization of the damage evolution in the repaired area under thermal and cyclic stress was developed, cf. Fig. 7. The system is capable of applying temperature cycles and mechanical load cycles to a sample. The sample is heated via an induction coil and can be cooled down as required by an active air cooling system.

Since the test rig is capable of heating the samples to over 1000 °C by induction, additional measures had to be implemented to protect the test equipment. The sensor for harmonic analysis of eddy currents was installed on a positioning unit so in cooling phases it could be moved until it contacted the specimen surface. Since the sensor only records values ones the temperature is below the Curie temperature, a fixed installation of the sensor does not offer any benefit especially in high-temperature phases. The acoustic emission used for crack detection, on the other hand, depends on a continuous recording of measured values. It was not possible to attach the sensor directly to the specimen surface or the grips of the testing machine, as the electromagnetic compatibility between inductive heating and acoustic emission sensor was not given. Due to the inductive heating in the frequency range of 200 kHz, the AE sensor was massively disturbed without shielding. Another challenge was the high temperatures in the immediate vicinity of the sample. The temperature acting on the sensor could be reduced to tolerable values with the help of a sound conductor. At the same time, this also reduced the negative effect of the electromagnetic field

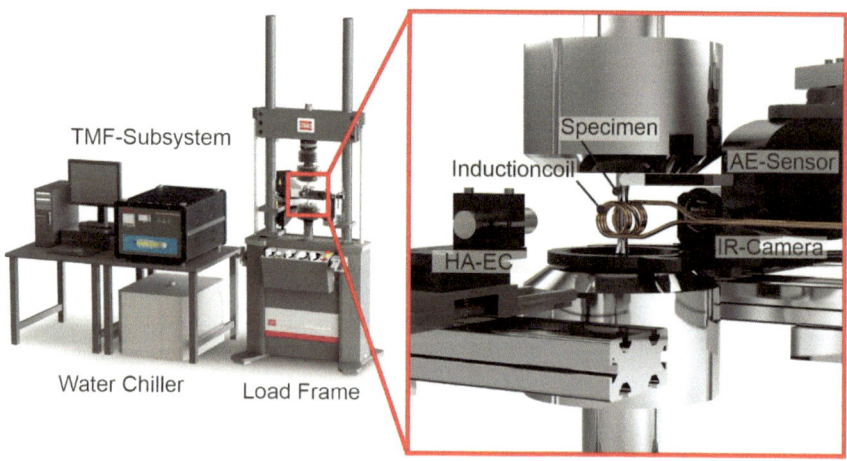

Fig. 7 Fatigue test system with integrated NDT: The detailed view shows the eddy current sensor (HA-EC) in waiting position, the acoustic emission (AE) sensor in the shielding housing including sound conductor with contact to the round specimen and in the background the thermographic camera (IR-Camera)

of the induction coil on the sensor. Still, a complex shielding of the sensor against the intensive electromagnetic field was necessary. The thermographic camera was permanently installed. Recordings were performed by internal timer or by an external trigger of the test system.

4 Results

4.1 Harmonic Analysis of Eddy Currents

The results of the fast damage detection with an integral HA-EC sensor are shown in Fig. 8 (right). Nineteen blades of the same type were examined using this technique. Fourteen blades showed a high third harmonic amplitude reading, which correlates with significant changes in the microstructure, caused by corrosion, as the material does not have ferromagnetic properties at room temperature in the non-degraded state. The amplitudes here show large differences, which indicate a different volume of altered material. Five blades show no amplitude (blade number: 4, 6, 7, 9, 11), so that no structural change was indicated at room temperature.

In addition to integral testing, it is also possible to carry out local surface testing using the same method. Only different sensors are needed for this purpose. The inspection times are increased in this case, but with this option, the damaged areas can be displayed in relation to the location. In order to reveal the blade areas damaged by high temperature oxidation, a local HA-EC sensor was used, cf. Fig. 9. The green

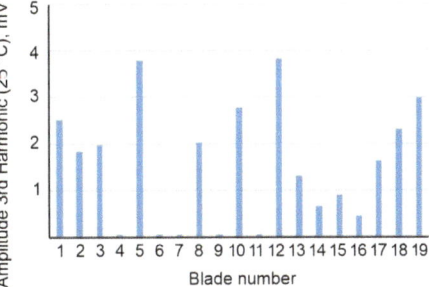

Fig. 8 Fast damage detection. Left: experiment set-up consisting of sensor, turbine blade and gripper at the end effector of an axial robot; right: measured amplitudes of the 3rd harmonic

circle in the left figure corresponds to the measuring spot size of the sensor. Especially in the area of the leading edge, the blade shows an increased amplitude of the 3rd harmonic. The remaining measuring points show clear external damage in the form of spalling. However, the base material is not affected by oxidation at these points (Barton 2022).

The Fig. 9 on the right visualises the test result of a complete inspection to identify oxidised areas in the base material. As expected, the leading edge is most affected by oxidation according to the measurements. The oxidised portion decreases in the direction of the blade root. In order to only determine the oxidised volume, but also about the intensity of the oxidation present, the Curie temperature must also be taken into account in addition to the amplitude of the 3rd harmonic. Figure 10 shows a clear linear correlation between the Curie temperature and the minimum chromium content. As the chromium content decreases, the Curie temperature increases. The chromium content is in itself directly connected to the oxidation progress, as the

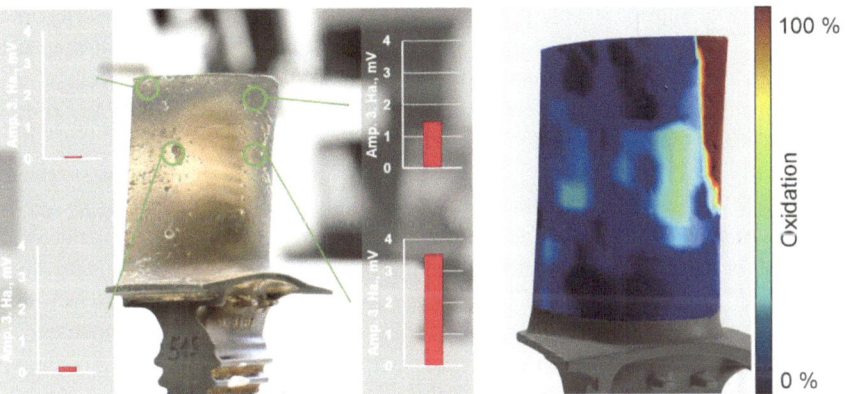

Fig. 9 A local HA-EC sensor was used to determine the blade areas affected by high temperature oxidation. Left: Amplitudes of the 3rd harmonic assigned to individual measuring points. Right. Testing of an entire blade surface and subsequent transfer of the measured values to a 3D mapping

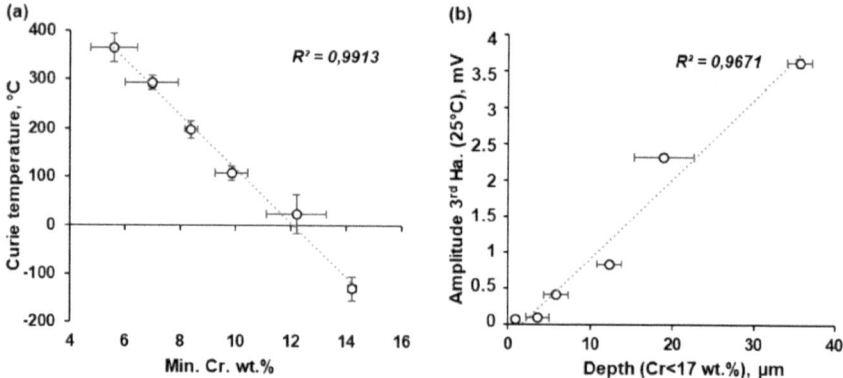

Fig. 10 Curie temperature versus minimum chromium content (**a**) and amplitude of the 3rd harmonic as function of the depth of the chromium-depleted region (**b**) (Barton et al. 2021)

chromium atoms diffuse out of the subsurface zone and form chromium oxides on the surface. Figure 10b demonstrates the good correlation the 3rd harmonic in relation to the chromium depleted volume (Barton et al. 2021).

4.2 Induction Thermography

The HF induction thermography is able to detect delamination of the coating system and cracks in the base material below the coating. Figure 11 shows the crack detection using the example of a turbine blade with an artificial crack. The crack can be seen in Fig. 11, left. Prior to inspection, the blade was coated with a thermal insulation layer. After the coating process, the crack is not longer visible. No additional brazing material was applied at this point, i.e. the crack was still present underneath the coating. The blade was then inspected for cracks and delamination in the coated state. The test results are visualised in Fig. 11 on the right with the help of a 3D mapping. The result shows a clear heating at the position of the crack. As expected, the heating is only visible at the crack tip.

Clearly, the artificially generated crack is different in its dimension and shape from real cracks. Therefore, in further tests, blades with cracks formed during service, which are significantly finer and smaller, were also tested. Since the system can also be used for quality assurance, it was also important to ensure that any defects present after the MRO process, such as lack of fusion in the welding process, can be reliably detected. For this purpose, a series of turbine blades was examined, which had real cracks in the tip area. These cracks were closed with a high-temperature brazing process and then a new coating system was applied to the blades.

Figure 12 shows a blade after the regeneration process, the previously clearly visible crack is no longer visible here and seems to be filled by the brazing material. Yet, the superimposed data from the thermographic test clearly revealed that there

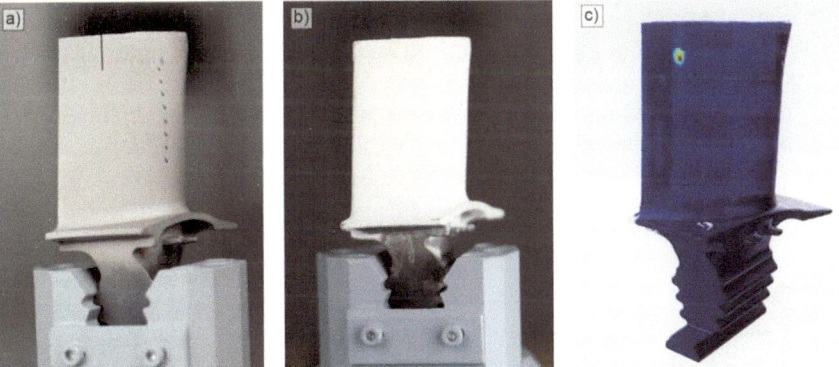

Fig. 11 Stripped turbine blade with artificial crack (**a**), which was subsequently coated with a new corrosion protection layer and thermal barrier coating (**b**) and then tested for cracks using HF thermography (**c**)

is still a crack-like defect present. To verify this, the area was metallographically analysed. The blade was cut at the expected position of the crack and then examined under a light microscope. The inset to Fig. 12 (right) shows that the crack present prior to the repair has not been completely closed. At the pressure side, the crack is closed. This defect is therefore classified as an internal defect.

High-Frequency Eddy Current Testing

The high-frequency eddy current technique with test frequencies in the MHz range allows separate characterization of the thin coatings and testing for defects in the coating system and the edge zone of turbine blades. The eddy current technique

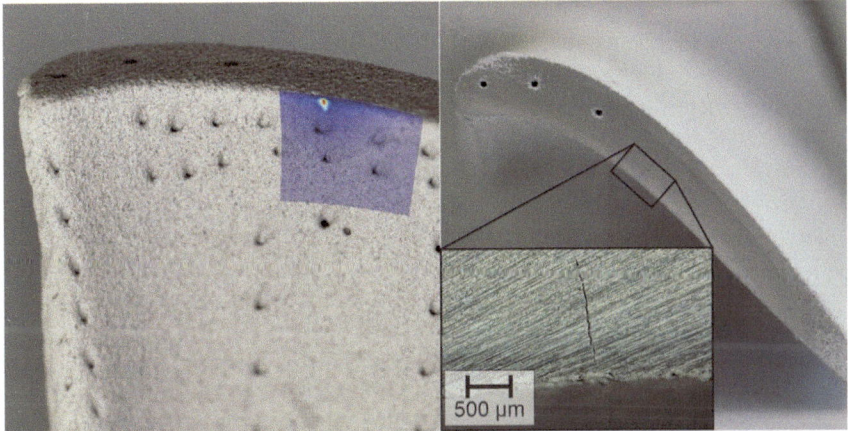

Fig. 12 Optical images of a turbine blade with a crack in the tip area that required a regeneration process. Left: superimposed data from HF thermography test revealing the crack; Right: subsequent metallographic analysis of the tip area

has its main advantage in determining the TBC thickness and assessing the corrosion protection potential of the underlying corrosion protection layer (Frackowiak et al. 2018).

The experimental tests were carried out on first stage turbine blades made of the nickel-based superalloy René 142 with a PtAl corrosion protection coating and YSZ as the TBC. Figure 13 shows the test result for the individual measuring points of the turbine blade shown. The measurement result is within the tolerance range, which was defined by measuring two reference blades coated with a ceramic non-conductive TBC with a thickness between 0–200 μm. One of these reference blades had an unstressed PtAl layer, while the other completely lacked such a layer. The phase position of the eddy current signal was chosen in such a way that the relevant geometry and the lift-off effect were mainly represented by the real part of the measured values in the complex plane. In order to be able to specifically evaluate and represent the occurring influence of the lift-off and geometry effect, the sensor was slowly moved towards the sample. This leads to the characteristic profiles of the working points in the complex plane. The magnitude of the lift-off effect corresponds to the thickness of the non-conductive TBC. When the sensor is in full contact, the lift-off effect is minimal and corresponds to the layer thickness at that position. The degradation of the metallic aluminium in the anti- corrosion coating due to oxidation and formation of the non-conductive ceramic Al3O2 during turbine operation leads to changes in the electrical conductivity of the coating over time. These changes are represented by the shift of the measured values mainly along the imaginary axis of the complex plane. The electrical conductivity of the corrosion protection layer is thus a characteristic property that can be used to assess the condition of this part of the coating system using the HF-EC measurement technique (Frackowiak et al. 2018; Reimche et al. 2013).

Fatigue Tests Combined with in Situ NDT

The developed fatigue test rig with combined NDT allows correlation of the NDT data with the fatigue curves. In general, tests were carried out with virgin smooth specimens, as well as tests with notched specimens and specimens regenerated by other subprojects.

The use of acoustic analysis to detect cracks proved to be challenging due to the inductive heating used. In isothermal fatigue tests at room temperature, i.e. without the use of inductive heating, the system can be coupled to the specimen by means of sound conductors to detect crack signatures. In the case of thermo-mechanical loading, the system was at least temporarily disturbed in all tests to such an extent that no reliable in situ crack detection could be realized.

By contrast, thermographic tracking of the surface could be reliably implemented. Still, it is still difficult to monitor the crack growth. On one hand, this is due to the duration of fatigue tests in the low-cycle fatigue (LCF) range, and thus the very large amounts of data that would be generated by a complete thermographic observation. In addition, it is not possible to monitor the entire surface of the sample with a single thermographic camera; this would require at least three cameras, and even then, part of the surface area would still be obscured by the induction coil.

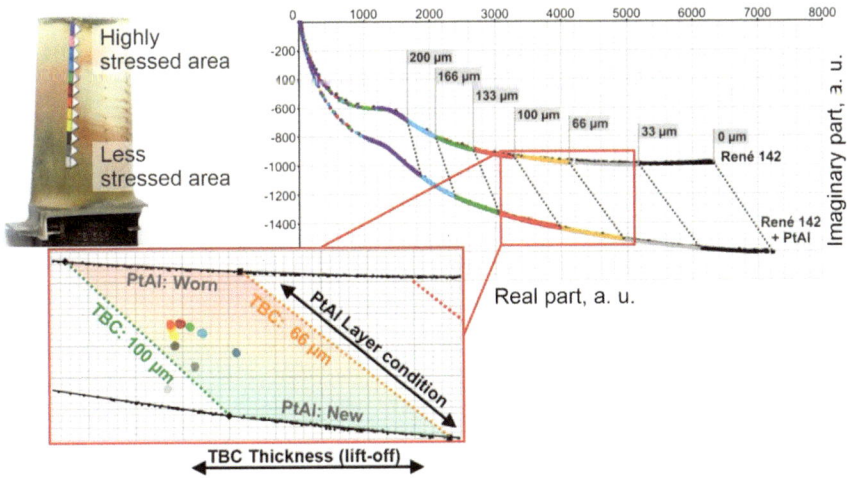

Fig. 13 Assessment of the coating system using HF-EC measurement technique; in a single process, both corrosion protection and thermal barrier coatings (TBC) can be evaluated (Frackowiak et al. 2018)

By using a high-temperature HA-EC sensor, in situ microstructure monitoring could be realised. Using an adapted isothermal LCF test with the test sequence shown in Fig. 14c, the sensor was able to record data in recurring cooling phases. Upon cooling, the amplitudes of the 3rd harmonic were recorded, and thus the respective Curie temperature could be determined cyclically. Several tests with varying mechanical strains were carried out such that time effects and changes due to the mechanical stresses could subsequently be evaluated.

Figure 15 illustrates data obtained from these different tests. In the left diagram, the amplitude of the 3rd harmonic at a sample temperature of 25 °C is plotted versus the absolute number of cycles. It is evident that an increased mechanical stress leads to an increased growth of the amplitude. If the x-axis is now transferred from absolute number of cycles to percentage values for the fatigue progress and the respective values of the Curie temperature are used instead of the 3rd harmonic, an intersection point becomes visible. When a Curie temperature of approximately 90 °C is reached, each of the specimens tested has reached about 80% of its fatigue life.

Apparently, the Curie temperature is a usefull indicator of fatigue damage, as oxidation of the subsurface zone in this case is closely correlated with fatigue crack initiation and early growth. Thus, by continuously monitoring the electromagnetic parameters, useful input for fatigue life prediction can be provided (Barton et al. 2022).

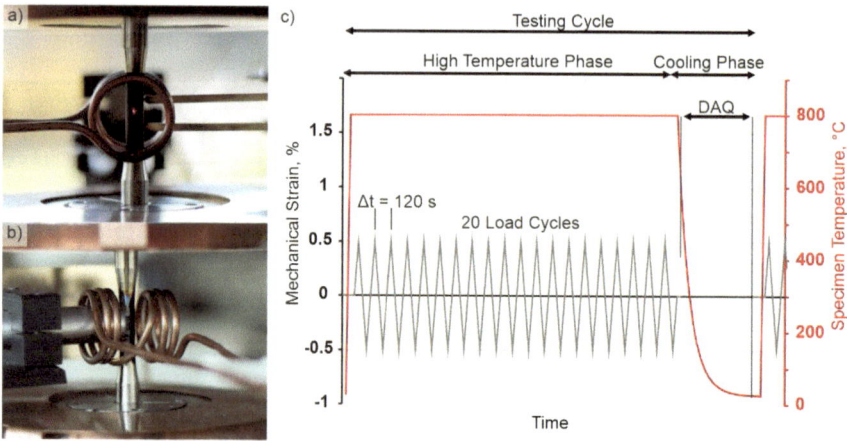

Fig. 14 Fatigue test cycle with combined HA-EC monitoring: in **a** the HA-EC sensor is in the waiting position and the measuring spot of the pyrometric temperature detection is indicated by the laser spot; in **b** the HA-EC sensor is in data acquisition (DAQ) mode during the cooling phase marked on the graph in (**c**)

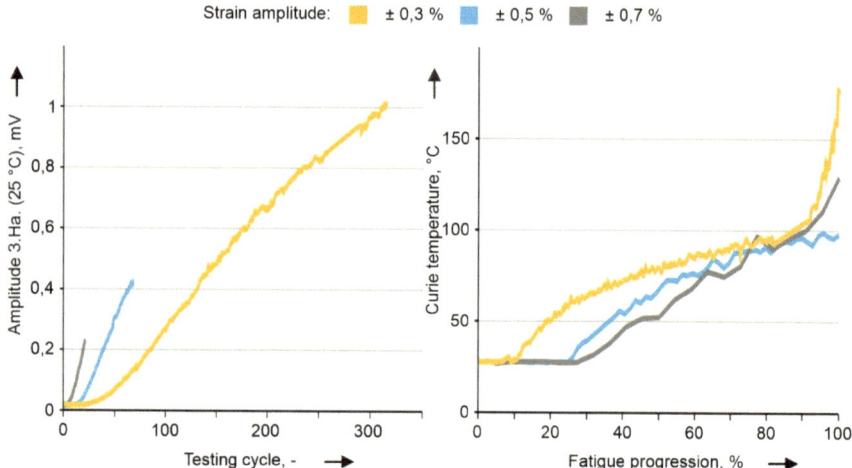

Fig. 15 Left: evolution of the amplitude of the 3rd harmonic with the number of test cycle; right: change in Curie temperature as a function of the normalized fatigue life

5 Conclusions

In work the development of non-destructive testing techniques for the differentiated evaluation of the coating system of a high-pressure turbine blade has been presented. Low-frequency harmonic analysis of eddy current signals is used to evaluate the condition of the base material. This can sensitively detect ferromagnetic phases that have developed in the paramagnetic base material as a result of high-temperature corrosion processes. In addition to the integral determination of such phases, they can also be detected locally. In this way, damage can be evaluated in terms of its position and distribution. This testing technique is also suitable for determining the Curie temperature in components made of nickel-based materials. The maximum change in the microstructure can be derived from this. These testing techniques make it possible to evaluate the condition of a turbine blade, for example, and thus make a net decision as to whether this blade can continue to be used with or without repair, or whether it must be replaced (Barton 2022).

HF induction thermography was developed to detect cracks and damage in the protective layers. High-frequency, pulsed excitation generates an eddy current distribution in areas close to the surface, which is influenced by defects such as cracks or delaminations. This can be detected by means of a thermographic camera. In this way, these defects can be detected and evaluated by imaging. In order to be able to image the defects in high resolution, a thermographic camera with a macro lens was used and component-matched inductors were also developed. The phase and amplitude information of the individual image pixels was used for evaluation.

With the high-frequency eddy current test with test frequencies in the MHz range, the protective coatings can be evaluated. Sensors adapted to the task were developed for this purpose. In the case of the ceramic thermal barrier coating, degradation manifests itself in a reduction in thickness, which can be detected via a lift-off sensor. The formation of oxide on the corrosion protection layer reduces its electrical conductivity, which can also be detected by the high-frequency eddy current test. In this way, the condition of both protective layers can be recorded in parallel in one measurement and the condition can be evaluated from the result. The combination of these testing techniques provides data for the selection of the appropriate regeneration process depending on the component condition.

Furthermore, the findings from the condition assessment of the turbine blade were used to evaluate the quality of repair processes under thermo-mechanical load. For this purpose, non-destructive testing techniques were integrated into a suitable test rig. This made it possible to detect cracks at an early stage and to characterize the change in microstructure during cyclic loading.

Acknowledgements Funded by the Deutsche Forschungsgemeinschaft (DFG, German Research Foundation) SFB 871/3 119193472.

References

Barton, S., Zaremba, D., and Maier, H. J. (2021). Microstructural degradation in the subsurface layer of the nickel base alloy 718 upon high-temperature oxidation, Materi- als at High Temperatures, 38 (3), p. 147-157. doi:https://doi.org/10.1080/09603409.2021.1895541

Barton, S. (2022). Zerstörungsfreie Bewertung des Randzonenzustands und Schädi- gungsgrads in Nickelbasislegierungen infolge von Hochtemperaturkorrosion. Dissertation, Leibniz Universität Hannover, ISBN: 978-3-95900-703-0

Barton, S., Weiss, M. K.-B., and Maier, H. J. (2022). In-Situ Characterization of Micro- struc-tural Changes in Alloyv 718 during High-Temperature Low-Cycle Fatigue, Metals, 12 (11), p. 1871. doi: https://doi.org/10.3390/met12111871

Bräunling, W. J. (2015). Flugzeugtriebwerke: Grundlagen, Aero-Thermodynamik, ideale und reale Kreisprozesse, Thermische Turbomaschinen, Komponenten, Emissionen und Systeme. doi:https://doi.org/10.1007/978-3-642-34539-5

Bruchwald, O., Frackowiak, W., Reimche, W., and Maier, H. J. (2016). Applications of High Frequency Eddy Current Technology for Material Characterization of Thin Coatings. In Proceedings of the THE "A" Coatings Conference, Garbsen

Carter, T. J. (2005). Common failures in gas turbine blades. Engineering Failure Analysis, Volume 12, Issue 2, p. 237–247. ISSN 1350–6307

Cosack, T. (2009). Schutzschichten auf Turbinenschaufeln im Flugtriebwerk, MTU Aero Engines Publikation.

Delaunay, F., Berthier, C., Lenglet, M., and Lameille, J.-M. (2000). SEM-EDS and XPS studies of the high temperature oxidation behaviour of inconel 718. Mikrochim Acta. 132 (2–4), 337–343. doi:https://doi.org/10.1007/s006040050027

Frackowiak, W., Barton, S., Reimche, W., Bruchwald, O., Zaremba, D., Schlobohm, J., Li, Y., Kaestner, M., and Reithmeier, E. (2018): Near-Wing Multi-Sensor Diagnostics of Jet Engine Components. ASME Turbo Expo 2018: Turbomachinery Technical Conference and Exposition. Oslo, Norway. https://doi.org/10.1115/GT2018-76793

Garat, V., Deleume, J., Cloue, J.-M., and Andrieu, E. (2005). High temperature inter- granular oxidation of alloy 718. In: Superalloys 718, 625, 706 and various derivatives. TMS, p. 559–569. https://doi.org/10.7449/2005/Superalloys_2005_559_569

Jia, Q. and Gu, D. (2014). Selective laser melting additive manufactured Inconel 718 superalloy parts. Opt Laser Technol, 62, 161–171. doi:https://doi.org/10.1016/j.optlastec.2014.03.008

Kumar, A., Chernatynskiy, A., Hong, M., Phillpot, S. R., and Sinnott S. B. (2015). An ab initio investigation of the effect of alloying elements on the elastic properties and mag- netic behavior of Ni3Al. Computational Materials Science. 101, p. 39-46. doi:https://doi.org/10.1016/j.com matsci.2015.01.007

Maaß, M. and Nehring, J. (2004). Oberwellenanalyse bei der magnet-induktiven Wir- belstromprü- fung in der Praxis. DGZfP-Jahrestagung, ZfP in Forschung, Entwicklung und Anwendung, p. 17–19.

Melchert, N., Weiss, M. K.-B., Betker, T., Frackowiak, W., Gansel, R., Keunecke, L., Reithmeier, E., Maier, H. J., Kästner, M., and Zaremba, D. (2021). Combination of optical metrology and non-destructive testing technology for the regeneration of aero engine components. tm-Technisches Messen, 88 (4), p. 237–250. https://doi.org/10.1515/teme-2020-0093

Mercier, D., Lesage, J., and Decoopman, X. (2006). Eddy currents and hardness testing for eval-uation of steel decarburizing. NDT E Int. 39 (8), p. 652–660. https://doi.org/10.1016/j.ndteint. 2006.04.005

Netzelmann, U., Walle, G., Lugin, S., Ehlen, A., Bessert, S., and Valeske, B. (2016). Induction ther-mography: principle, applications and first steps towards standardisation, Quantitative InfraRed Thermography Journal, 13 (2), p. 170-181. https://doi.org/10.1080/17686733.2016.1145842

Reed, R. (2006). The Superalloys: Fundamentals and Applications. Cambridge: Cam- bridge University Press. https://doi.org/10.1017/CBO9780511541285

Reimche, W., Bernard, M., Bombosch, S., Scheer, C., and Bach, Fr.-W. (2008). Nachweis von Anrissen in der Randzone von Hochleistungsbauteilen mit Wirbelstromtechnik und induktiv angeregter Thermographie. HTM Journal of Heat Treatment and Materials 63 (5), p. 284–297. https://doi.org/10.1515/htm-2008-0007

Reimche, W., Bruchwald, O., Frackowiak, W., Bach, Fr.-W., and Maier, H. J. (2009). Non-Destructive Determination of Local Damage and Material Condition in High-Performance Components, IITM J. Heat Treatm. Mat 68 (2), p. 59–67, https://doi.org/10.3139/105.110176

Reimche, W., Bruchwald, O., Frackowiak, W., Bach, Fr.-W., and Maier, H. J. (2013). Non-destructive determination of local damage and material condition in high-performance components: High-frequency eddy current and induction thermography techniques. HTM Journal of Heat Treatment and Materials, 68 (2), p. 59–67. https://doi.org/10.3139/105.110176

Rocha, A. D., Peres, R. S., Barata, J., Barbosa, J., and Leitão, P. (2018). Improvement of multistage quality control through the integration of decision modeling and cyber- physical production systems," in 2018 International Conference on Intelligent Systems (IS), p. 479–484, IEEE. https://doi.org/10.1109/IS.2018.8710492

Schlobohm J., Bruchwald O., Frackowiak W., Li Y., Kästner M., and Pösch, A. (2016). Turbine blade wear and damage – An overview of advanced characterization tech- niques, Materials Testing 58 (5), p. 389–394. doi: https://doi.org/10.3139/120.110872.

Schlobohm, J., Bruchwald, O., Frąckowiak, W., Li, Y., Kästner, M., Pösch, A., Reimche, W., Maier, H. J., and Reithmeier, E. (2017). Advanced Characterization Techniques for Turbine Blade Wear and Damage. Procedia CIRP, Volume 59, p. 83-88. doi:https://doi.org/10.1016/j.procir.2016.09.005.

Stegemann, D., Reimche, W., Feiste, K. L., and Heutling, B. (1998). Determination of mechanical properties of steel sheet by electromagnetic techniques. In: Green RE, editor. Nondestructive characterization of materials VIII. Boston, Springer US, 269–275. https://doi.org/10.1007/978-1-4615-4847-8_43

Wang, K. S. (2013) Towards zero-defect manufacturing (ZDM) - a data mining ap- proach. Adv. Manuf. 1, p. 62–74. doi:https://doi.org/10.1007/s40436-013-0010-9

Multiscale Measurement of Blade Geometries with Robot-Supported, Laser-Positioned Multi-sensor-Techniques

Tim Sliti, Markus Kästner, and Eduard Reithmeier

Abstract The regeneration of aircraft engine components requires a thorough assessment of the current condition. Based on this, a suitable repair strategy can be selected. To provide a measurement system, which can be used for the inspection of worn components, various optical measurement methods were combined to create a multi-sensor system. To completely reconstruct complex geometries a 6-axis industrial robot and an additional rotational axis are applied. This robot-assisted multi-sensor system is used to digitise and characterise turbine blades of aircraft engines. The inspection process is non-destructive and different features are measured to acquire a holistic model. The sensors are used to reconstruct the 3-D geometry in different scale ranges and characterise the surface based on reflection properties. Afterwards, the data of each individual sensor are transferred into a uniform coordinate system. To ensure high sensitivity to wear and damage, a model-based system calibration and a data interface for subsequent diagnostics and simulations are essential to provide a reliable assessment of the performance and durability of the inspected components.

Keywords Metrology · Multiscale · Fringe projection profilometry · White light interferometry · Wear assessment

1 Introduction

The complexity of modern machines and their components is growing. While this trend improves the overall performance of higher level systems, it introduces challenges in terms of maintenance. The level of complexity often causes larger part costs. Thus the regeneration of these parts may be more economical and with increasing material scarcity also more ecological.

T. Sliti (✉) · M. Kästner · E. Reithmeier
Institute of Measurement and Automatic Control, Leibniz University Hannover, An der Universitaet 1, 30823 Garbsen, Germany
e-mail: tim.sliti@uni-hannover.de
URL: https://www.imr.uni-hannover.de

© The Author(s) 2025   29
J. R. Seume et al. (eds.), *Regeneration of Complex Capital Goods*,
https://doi.org/10.1007/978-3-031-51395-4_3

This work focuses on the regeneration of turbine blades used in aircraft engines. To assist the regeneration process, a combination of different optical sensors is combined to assess the current state of the component. This information can further be used to detect defects and estimate the reliability and performance. Based on the gathered information, a suitable regeneration strategy is chosen.

2 Motivation

The assessment of worn components is challenging due to a wide range of their possible conditions. Mechanical stress and thermal or chemical wear cause deviations from the original state and can impact the overall performance and reliability (Tabakoff et al. 1998; Kurz and Brun 2000; Laguna-Camacho et al. 2016). Detecting these changes and characterizing their appearance with non destructive optical methods is the goal of this project.

However, deviations can occur in varying shapes, sizes and types. Thus multiple geometric scales have to be taken into consideration, when reconstructing the object.

While macroscopic defects (multiple centimetres) like larger cracks and dents may be detectable with one sensor, a different system is necessary to reconstruct defects which only have a size of a few micrometers.

In order to take into account various types of defects, a combination of multiple different sensors (S_1,..., S_n) is necessary, since a single sensor is not capable of covering all of the required scale ranges. The basic concept of multiscale geometry inspection and fusion to a holistic dataset in a common coordinate system, which was developed in this work, is shown in Fig. 1.

Different scale ranges require different measurement techniques and global data registration necessitates the identification of spatial relationships of the individual sensors to each other to allow the fusion of data gathered from multiple sensors. Based on the available measurement methods, a suitable sensor can be chosen in order to meet the requirements to reconstruct a defect.

3 Multi-sensor Design

To provide a holistic model, a combination of geometric and non-geometric data has to be acquired. The developed sensor system consists of three sensors, which provide geometric data. A fringe projection unit is used to cover the macro- and mesoscale ranging from sub millimetre to multiple centimetres. A low-coherence interferometer is used to reconstruct micro scale features of the measurement object. In addition to these sensors an illumination sensor is used to provide information about non-geometrical surface properties. The structure, function principle and results of each individual sensor will be introduced in the next sections.

Fig. 1 Fusion of different
sensors combined into a
multi sensor system to cover
multiple scale ranges

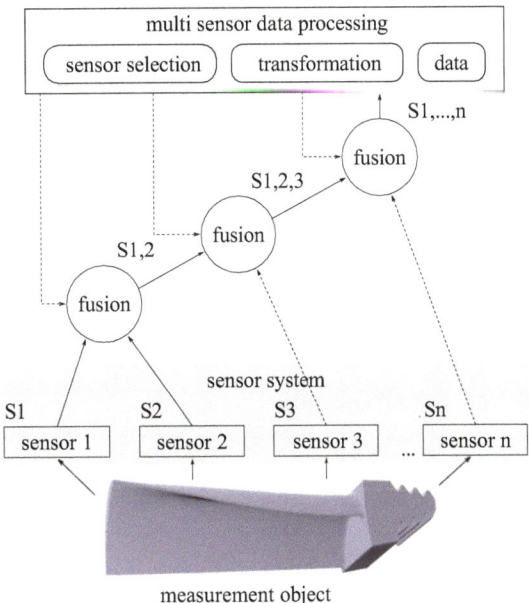

3.1 Illumination Sensor System

This sensor is mainly designed to assess the local reflective properties under different
illumination scenarios. Therefore, the sensor is designed with 42 white light-emitting
diodes (LEDs). Each one is collimated and placed on a hemisphere pointing towards
to the centre (see Fig. 2). To capture the illuminated scene an industrial camera
(AVT Manta G-1236B) is mounted in the centre of the hemisphere surface. Each
light source is individually controllable, which leads to a large number of possible
combinations and thus, lighting scenarios.

Fig. 2 Inside of the half
sphere of the illumination
sensor. A camera is placed in
the middle, 42 LEDs are
arranged in four rings around
the centre

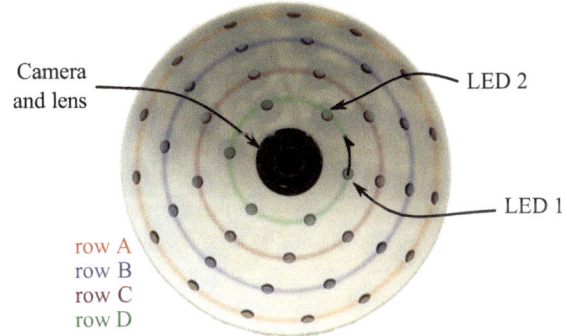

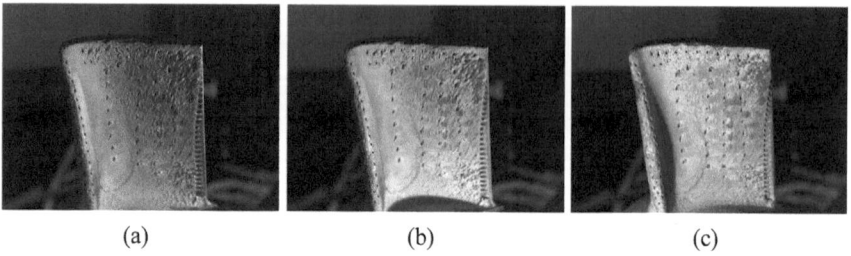

| (a) | (b) | (c) |

Fig. 3 Camera image of a turbine blade illuminated with varying light directions (Intensities have been slightly adjusted to improve visibility) **a** From right side. **b** From the bottom. **c** From the left

3.1.1 Data Acquisition

To perform a measurement a set of pre-defined light configurations is executed and an image is recorded, resulting in an image stack with one image for each configuration. While more complex scenarios are possible, this approach mainly focuses on images when enabling only one LED at a time. Depending on the enabled LED, the amount of reflected light changes the intensity of different regions. This is caused by a combination of light direction and geometry which, based on macro and micro structures on the surface, lead to shadowing or affect the overall reflected light. These changes can be observed in the example images given in Fig. 3.

3.1.2 Measurement Principle

In order to take advantage of the reflective properties of the objects surface, different algorithms are applied to extract information about the measurement object. They are divided into two approaches for a macroscopic and a microscopic characterization. Further it is assumed that the positions of each LED in respect to the camera is sufficiently well-known from the computer-aided design (CAD). Thus the light vectors can be calculated with the assumption of perfectly collimated light and a fixed working distance for the camera when placed in focus. These assumptions may deviate from the actual conditions due to assembly-related deviations, not perfectly collimated LEDs and varying distances to the measurement object, but experiments have indicated, that these simplifications are feasible for this application.

   The evaluation of the data is based on the following surface reflectance model utilising the surface normal $\mathbf{n}$, the aforementioned incident light vectors $\mathbf{l_n}$ with $n \in [1, 42]$ for each LED and a view vector $\mathbf{v}$. Figure 4 shows these vectors for a single surface point. Additionally the angles θ and φ are defined for in and outgoing rays, respectively light and view vector. The parameter vector $\mathbf{k}$ is used to describe the surface properties. Its length is dependant on the applied model. Utilizing these parameters the shape of the measurement object can be approximated by applying a photometric stereo approach. This algorithm requires a set of images with varying light direction to determine the surface normal for each pixel. However, basic approaches

Fig. 4 Simplified surface
reflectance model

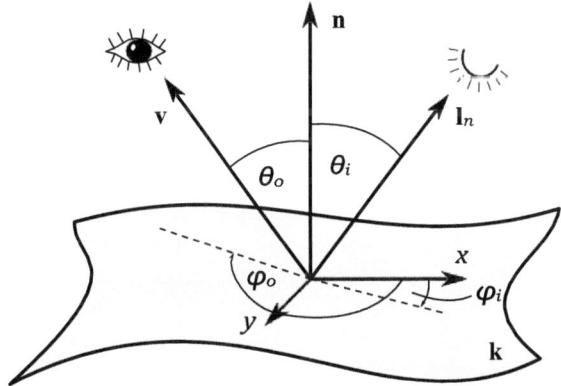

assume that the surface is mainly diffuse with low specular features. Research has been carried out to overcome this limitation (Herbort and Wöhler 2011; Zheng et al. 2020). While highly specular objects are still challenging, the optimization process was adjusted to be more robust.

The objective function is to find the surface normal by utilizing the set of pixel intensities for each light vector. A least squares based approach was published by Woodham (1979), which provides good results for worn turbine blades but is limited at strong specular reflective conditions.

The resulting surface normals can further be used to create an unscaled 3-D model of the measurement object. Since there is no metric information, only qualitative statements can be made. This, however, is already sufficient to detect macro scale deviations of the overall geometry e.g. missing parts of the nominal model. Experiments have shown, this method is not well suited for fine structures in micrometer range since the algorithm tends to smoothen filigree features.

In addition, the estimated normals can further be applied to characterize the objects surface properties. This is achieved by approximating a reflectance function for each pixel. For this a bidirectional reflectance distribution function (BRDF) is utilised. The basic formula in Eq. 1

$$BRDF(\theta_i, \varphi_i, \theta_0, \varphi_0) \tag{1}$$

describes the ratio of reflected to incident light, with respect to the vectors incident and azimuth angles, θ and φ, and the surface normal vector. These functions are mainly used in computer graphics to render photorealistic images of different materials. In general a BRDF model requires the incident light beam $\mathbf{l}$, the surface normal $\mathbf{n}$, the viewing vector $\mathbf{v}$ and some model specific parameters $\mathbf{k}$ (Guarnera et al. 2016), see Fig. 4. With the previously determined surface normals all but the model parameters, which are used to describe the reflectance, are known. There has been extensive research about identifying model parameters from real world data (Ward 1992; Westin et al. 1992; Lafortune et al. 1997). However, in this case the data base is usually collected using gonioreflectometers, which provide a fine sampling along possible incident and viewing angles (Guarnera et al. 2016; Schröder and Sweldens 1995; Ngan et al. 2005; Bieron and Peers 2020).

The data provided by the illumination sensor is rather limited and thus may not be sufficient to find the 'actual' model parameters. But the data basis is sufficient to do an approximation of reflectance models. Figure 5 shows the pixel intensities of one pixel of the image for all 42 LEDs. While the fitted intensities are not perfectly overlapping with the actual data, the trend is reproduced well. These approximated parameters from the model fit can then be divided into multiple classes, which can be used to do a multi-class segmentation of the object surface. For this state of the art cluster algorithms like K-Means (Hartigan and Wong 1979) or DBSCAN (Ester et al. 1996) can be applied to determine cluster borders in parameter space. Based on the used algorithm an expected number of classes has to be given or is derived from the data. An example result of a K-Means clustering with three classes is shown in Fig. 6b. Here the regions are marked with different colours. While the reflectance of rough surfaces, e.g. sand-blasted metal, is mainly diffuse, surfaces with lower roughness, e.g. polished metal, is shiny. Both effects are represented by parameters of the reflectance model, which are used for the classification. Thus, the classes correspond to areas with different roughnesses. This difference is shown in Fig. 7 where the marked spots from Figure 6a have been measured with a confocal laser scanning microscope (Keyence VK-X200) as a sample for the classified regions. The resulting roughness parameters for the measurements are listed in Table 1. These values suggest, that the red region has a higher roughness than the green areas. As stated by Bons (2010) the roughness affects the performance of the gas turbine. Therefore, the resulting surface classification is useful to identify differing areas which can further be examined with punctual roughness measurements. A model-driven estimation of the local roughness parameters according to the identification of the reflection parameters is not reliably possible with the existing data base.

Fig. 5 Actual and fitted pixel intensities for one pixel

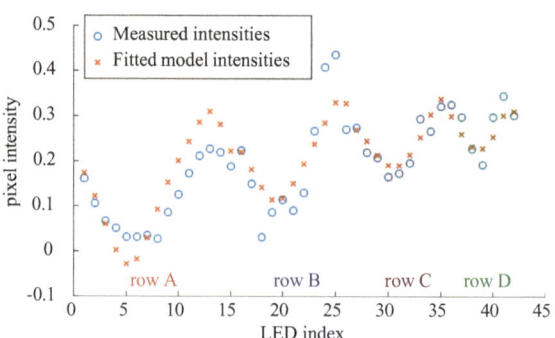

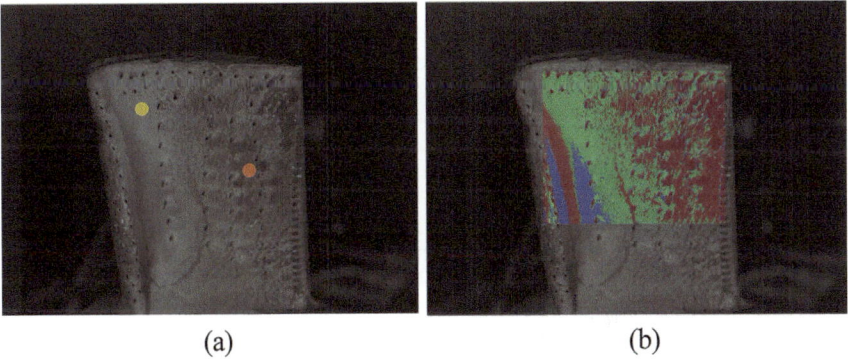

Fig. 6 **a** One image from the sequence. *Orange* and *yellow* marked areas indicate measurement spots. **b** Partially classified section of the turbine blade. Different regions can be distinguished. *Red* sections are in areas with rough structures, while these are missing in *green* regions. *Blue* coloured pixels often occur in transitioning areas from *red* to *green*

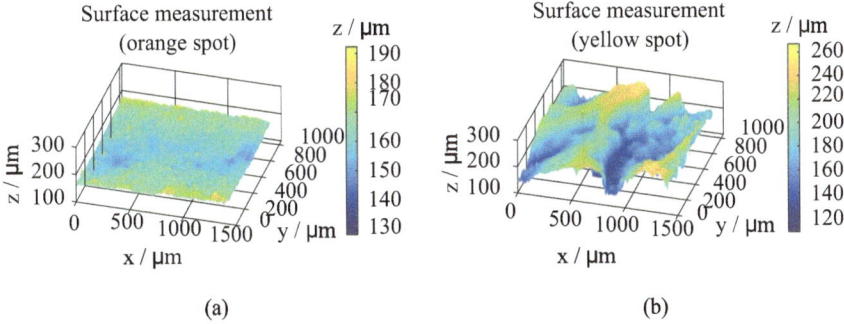

Fig. 7 **a** Surface measurement of the *yellow* area. **b** Measurement data for the *orange* spot. (cf. Fig. 6a)

Table 1 Roughness parameters for the measured regions (cf. Figs. 6a and 7)

Region	Sq (μm)	Sp (μm)	Sv (μm)	Sz (μm)
Orange	10.749	34.8023	33.3354	68.1377
Yellow	3.4026	14.8741	21.0471	35.9212

3.1.3 Conclusion Illumination Sensor

Since the reflectance of a surface is dependent on the surface roughness, the results can be used to make a qualitative distinction between regions with differing roughnesses. These sections can then be measured in future steps to assign quantitative values. Thus, this sensor is mainly used to gather qualitative information about the measurement object. A rough geometric assessment is possible with the addition of distinguishing between different surface regions by examining the reflectance.

3.2 Fringe Projection Sensor

A regular fringe projection system (FPS) consists of a camera and a projector unit to project structured patterns onto the measurement object, see Fig. 8. In this case two industrial monochrome cameras (AVT Manta G-419B) are coupled with a programmable projector module (Wintech PRO4500). The second camera is equipped with a different focal length to expand the scale range of the acquired data. Both systems are calibrated by identifying the optical parameters and the spatial relation between the components by applying state of the art camera, projector and stereo calibrations like Zhang (2000) or methods proposed by Hartley and Zisserman (2003). To encode the projector pixels a 8-step phase shifting sequence following Peng (2007) is applied. Here, 8 phase shifted sine patterns with increasing frequencies are projected onto the measurement object.

3.2.1 Data Acquisition and Registration

Triangulation determines a corresponding surface object point for each camera pixel by 3-D ray intersection, resulting in a high-density three-dimensional, metric point cloud. For the unambiguous spatial determination of the corresponding viewing beams of both camera and projector, the projection patterns are captured and evaluated via a phase unwrapping pipeline.

For a better assessment of the turbine blades state, a 3-D model is necessary. Since the fringe projection sensor only provides surface measurements from a single viewing point, multiple measurements have to be combined. To register these, point

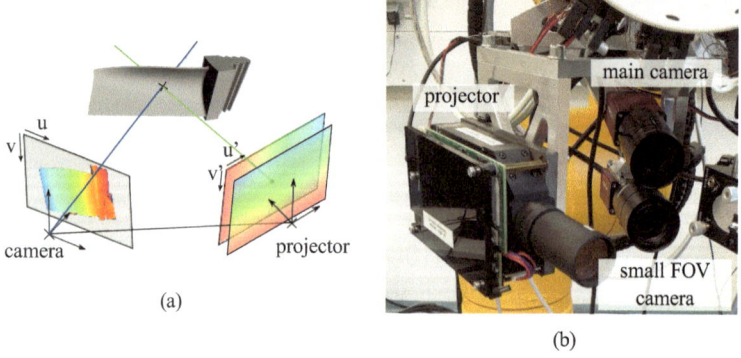

(a)

(b)

Fig. 8 **a** Fringe projection principle. **b** Fringe projection sensor

correspondences have to be estimated. There are various algorithms which can be used to determine these correspondences. Depending on the algorithm itself, different kind of data and requisites may be necessary to apply them. Below, those strategies will be divided into two categories: 2-D based and 3-D based algorithms.

The first uses 2-D organised data, e.g. image data, to detect features and calculate descriptors, which then can be used to estimate correspondences. The data used for the feature detection is the colour information, which can be greyscale or RGB. Some of the most common algorithms used for image feature detection are ORB (Rublee et al. 2011), SURF (Bay et al. 2008) or SIFT (Lowe 2004).

The other category handles unorganized 3-D data, e.g. point clouds. Here each 3-D point and its neighbourhood is taken into consideration to calculate feature descriptors. Usually these require the surface normals of each point, which can easily be approximated (Mitra and Nguyen 2003). Algorithms of this kind are FPFH (Rusu et al. 2009), 'spin images' (Johnson and Hebert 1999) or SHOT (Salti et al. 2014). These feature descriptors can be used for object recognition (Gupta et al. 2019), while recently neural networks are frequently used for this task (Liang and Hu 2015).

However, the state of the measurement object can vary a lot and thus the texture and number of geometrical features present. This means that the registration using only natural occurring features is not robust to changing measurement objects. To overcome this issue artificial textures are applied to the measurement object by projecting random patterns onto it (Betker et al. 2020). Hereby, more features can be detected with 2-D feature detectors. The advantage of this approach is that no manual steps are necessary to apply any markers next to or on the object. However, neither the position with respect to the object nor the optical properties may change during the measurement process.

The setup for the random pattern projection is shown in Fig. 9a. Multiple projectors (Texas Instruments DLPDLCR2000EVM) are placed around the mounting system and aimed at the object. Each projector has a different generated random pattern. To ensure a high density of features different kinds of pattern designs and 2-D feature detectors have been examined. Overall, binary patterns performed better than greyscale. Partly because sharp borders are more robust against effects introduced by defocus, noise and the mixture with the actual object's texture. The best results were obtained with a combination of randomly placed overlapping black rectangles on white background and a SIFT feature detector and descriptor. Since the 3-D reconstruction is calculated in the camera coordinate system, additional colour information can be added by acquiring an image with active random pattern projection. This way each point holds not only the spatial coordinates, but also an arbitrary number of, in this case, greyscale intensities. These greyscale images can be used to determine 2-D correspondences by applying the aforementioned 2-D SIFT feature detector. Since each reconstructed 3-D object point corresponds to a camera pixel, any 2-D assignment can subsequently be transferred to the registration of the point clouds. An example pair of images respectively surface measurements is shown in Fig. 9b. To increase visibility not all found correspondences are shown.

Due to the chosen pattern the algorithm detects a large amount of corresponding points. Although the majority of correspondences is plausible, a certain amount of

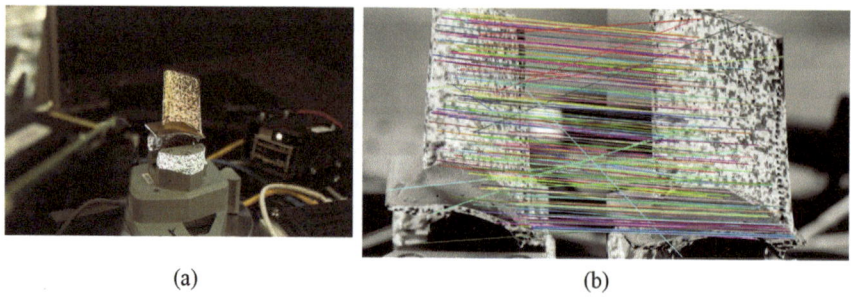

(a)   (b)

Fig. 9 **a** Turbine blade inside the mounting system surrounded by multiple projectors which project random patterns onto the object. **b** Pair of measurements as gathered by the fringe projection camera. Each coloured line represents an estimated correspondence

false connections are present. This problem is addressed by utilizing a random sample consensus (RANSAC) approach to estimate the rigid body transformation between both measurements. The combination of high density features, distributed on the surface, and an outlier-aware transformation estimation results in a robust alignment of multiple 3-D measurements, which is independent of natural features. To further improve the transformation and reduce the remaining alignment error, an iterative closest point (ICP) algorithm is applied. In this work a coloured ICP (Park et al. 2017) is used to further benefit from the projected patterns.

However the pairwise registration of measurements is error-prone and even small alignment errors add up when forming the complete 3-D model. This can result in a loop closing problem, where first and last measurement of a sequence are not aligned as expected. This problem is addressed by applying a multiway registration as proposed by Choi et al. (2015) which performs a graph optimization between all segments. Further neighbouring measurements are combined into fragments to increase the overlap between segments.

Running the presented registration pipeline results in blade measurements as seen in Fig. 10. To reduce the number of points in the merged point cloud, close points are merged with a weighted voxel based filter. The respective weight is chosen by the reconstruction quality of each point, which is derived from the signal quality during measurement.

3.2.2   Data Evaluation

In this section some options to process the data provided by the fringe projection unit will be discussed. The goal of the evaluation step is to detect defects or damaged regions on the turbine blade. Given the nominal geometry, the 3-D measurement can be aligned to it. This allows the estimation of deviations to the nominal structure of the turbine blade. An example deviation map is shown in Fig. 11. Since the actual nominal geometry from the manufacturer is unknown, another worn blade was chosen to represent the nominal geometry. Therefore, deviations were calculated between

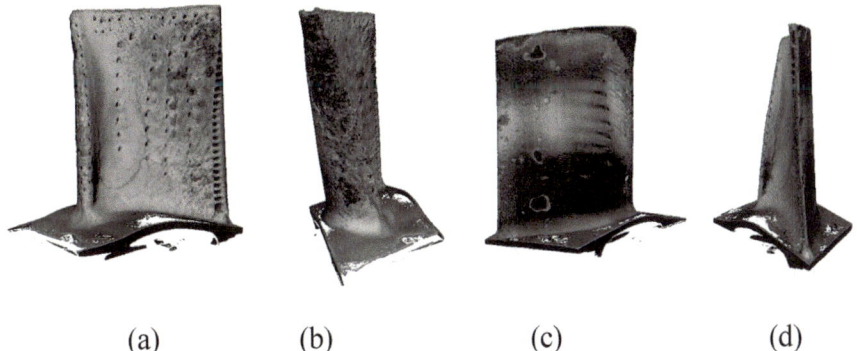

| (a) | (b) | (c) | (d) |

Fig. 10 a–d Different views of multiple measurements combined into one 3-D model

multiple worn blades to illustrate the process. Regions with missing material are coloured in blue, added material is represented with red. Green regions have low deviations to the nominal structure.

While this representation allows a quick assessment of damaged an undamaged regions there are some considerations to be made. First, the nominal geometry has to be given. Furthermore the initial alignment of nominal and actual geometry may be influenced by errors introduced by variations of the structure. To ensure a good alignment reference points which are not influenced by the operational stress would be necessary. Nevertheless this strategy can, depending on the requirements, be sufficient to draw conclusions about the blades condition.

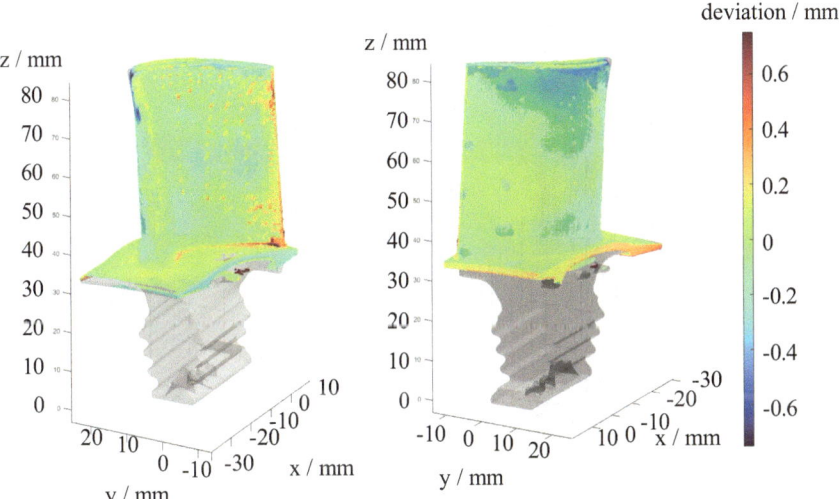

Fig. 11 Deviations of the measured 3-D model to a given nominal geometry

Because of the aforementioned drawbacks another strategy has been researched. With emerging research in the field of artificial intelligence, various algorithms and neural networks have been released to handle a multitude of tasks. To use these methods in the field of defect detection, convolutional neural networks (CNN) were chosen. To be more specific, image segmentation networks. Since single measurements of the fringe projection sensor are in a matrix structure this data can be used as input data for CNNs.

The goal is to detect defects in single measurements using this approach. Firstly a'defect' has to be defined in order to create according labels. Because the worn turbine blades have numerous of defects and a clear line between defect and intact regions is hard to define, the cooling holes of the blade are chosen instead. These are much more distinct and thus easier to label without expert knowledge. In some way the form of the cooling holes resembles the geometry of a defect. They interrupt an otherwise continuous surface by introducing high curvatures in a certain area.

With this definition a dataset of multiple measurements has been labelled. A sample is shown in Fig. 12. The annotations are based on the greyscale image, which can then be extended to the 3-D data. Different combinations of input data have been examined. A promising combination is the colour information with approximated normals. A prediction of the trained model and the difference in prediction and manual labels is shown in Fig. 13. It can be seen that more regions are marked than in the original labels. Subjectively these predictions are plausible and usually represent regions with larger curvatures. However the available training dataset is limited and the overall performance could further be improved with more data. But even with this rather small dataset the network seems to learn the rules for a local deviation, which can also be interpreted as a defect. Therefore, it is expected that it is possible to train this type of neural network to segment more general deviating regions.

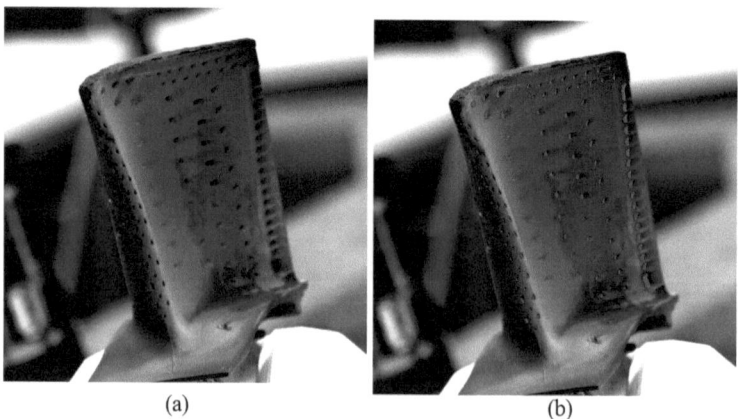

(a)   (b)

Fig. 12 **a** The raw greyscale image as recorded by the camera of the fringe projection unit. **b** The red sections show the labelled cooling holes. Regions without reconstructed 3-D data are omitted

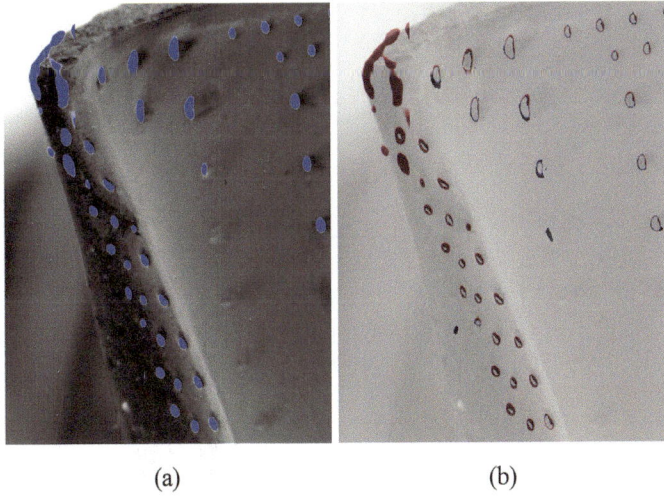

(a)   (b)

Fig. 13 a Prediction given by the trained neural network. **b** Difference between labels and predictions. Red indicates additional predictions, blue indicates missing labels

3.2.3   Conclusion Fringe Projection Sensor

The main task of the fringe projection system is the reconstruction of turbine blades in 3-D metric space. As shown in the previous section, it is possible to use this data to detect defects or deviations. This can be achieved through the use of neural networks or regular deviations estimations by estimated point distances. Thus this sensor is mainly used to gather geometrical data in macro and meso scale ranges.

3.3   Low-Coherence Interferometer

The Low-Coherence-Interferometer (LCI) is an interferometer in a Michelson config-uration. The basic setup is shown in Fig. 14. A regular industrial camera (Basler acA1920-48gm) in combination with a telecentric lens is used to capture the inter-ferences which form on the surface. The objective has a comparatively large working distance as similar systems. This simplifies the positioning of the sensor in relation to the complex geometries and reduces risks of collisions.

The low coherent light source has a wave length of 665 nm. The light beam is collimated and sent into a 50/50 beam splitter to split reference and measurement beam. The reference beam is then send to a deflection mirror which is implemented as a 50/50 beam splitter and gets reflected on the reference mirror which is also realised with a beam splitter, but with a 90% transmission 10% reflection ratio. As a result the reference beam intensity is reduced since the surface of our measurement objects is very rough and does not reflect a lot of light. With these adjustments the

Fig. 14 Low-coherence
interferometer setup

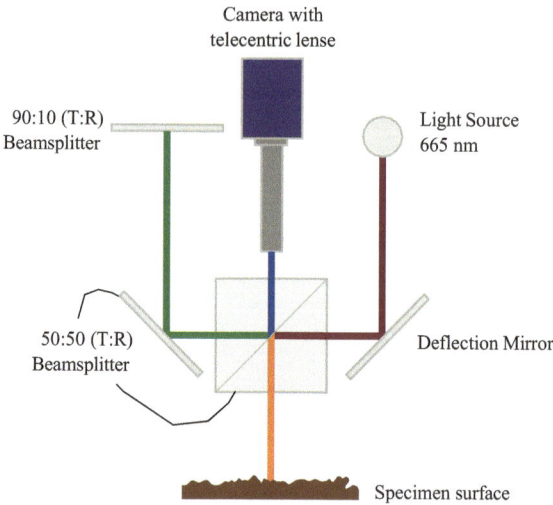

beam intensities of measurement and reference ray are aligned in order to improve
the overall contrast of the occurring interferences.

3.3.1 Data Acquisition and Evaluation

The LCI is mounted onto a high precision linear axis (PI L-509) to perform the
scanning process for each measurement. This way the optical path length of the
measurement beam is changed, which allows sampling the depth of the surface. For
each step of the scanning process an image is recorded and put into an image stack,
which can further be evaluated. Li et al. (2015) presented a GPU-based evaluation
strategy to calculate the corresponding depth maps. Paired with the magnification
properties of the telecentric lens it is possible to estimate 3-D data.

Some example results are given in Fig. 15. The left measurement shows a section
of the pressure side of the turbine blade. The right side shows an area of the leading
edge. Smaller defects and transitions into cooling holes are visible. Depth resolutions
lower than 200 nm are possible but strongly depend on the signal strength which again
is dependent on the surface properties. The lateral resolution is determined by the
pixel size and lens magnification which leads to 1.2 μm.

3.3.2 Conclusion Low-Coherence-Interferometer

With its rather small measurement area of around 3 mm^2 and a long measuring
duration, the LCI is not suitable for digitizing complete geometries or surfaces,
but more for local inspections which require a high depth resolution. Thus the LCI
is used as a complementary sensor to gather data in a micro scale in particularly

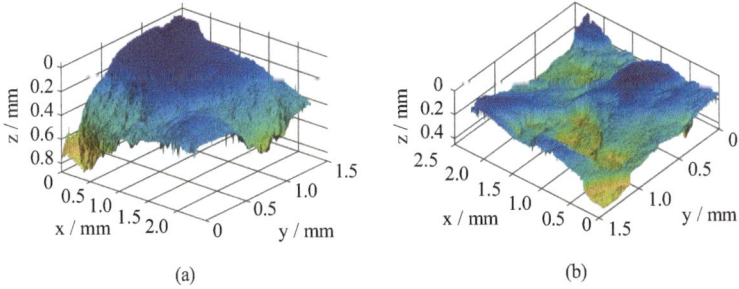

(a)   (b)

Fig. 15 a Measurement of the leading edge. **b** Section of the pressure side

relevant regions of the turbine blade such as damages, cooling holes or areas with fine structures which can influence the air flow.

3.4 Multi-sensor Measuring Head

In the previous sections the different sensors were introduced. To combine these into one multi-sensor system the individual requirements regarding working distance, orientation or field of view have to be taken into account. Given the interface of the robot, a measuring head has been developed. A CAD is shown in Fig. 16b. All sensors are mounted on the end-effector of a 6-axis industrial robot (Stäubli TX90). This allows a flexible positioning of the sensor head with respect to the specimen and guarantees appropriate positioning. The robot kinematic is extended with a rotational axis, which rotates the measurement object and the random pattern projectors used for the data alignment. The combination of robot and rotational axis makes it possible to fully reconstruct each individual specimen. With this design each sensor can be chosen by rotating the end-effector. LCI and fringe projection unit were placed in a way that their fields of view overlap to allow direct interactions, see Fig. 16a. The illumination sensor is separated to prevent collisions when handling other sensors. Thus, with this setup is possible to use the individual sensors within the multi sensor setup without directly affecting the other sensors.

4 Data Fusion

Every sensor operates on its own scale range and provides measurement data. However, the measurements of each sensor are currently in the respective sensor coordinate system (see Fig. 16). This means the measurement data is scattered in space and does not refer to a unified coordinate system. Thus it is not possible to immediately combine the information gathered from different sensors. In order to

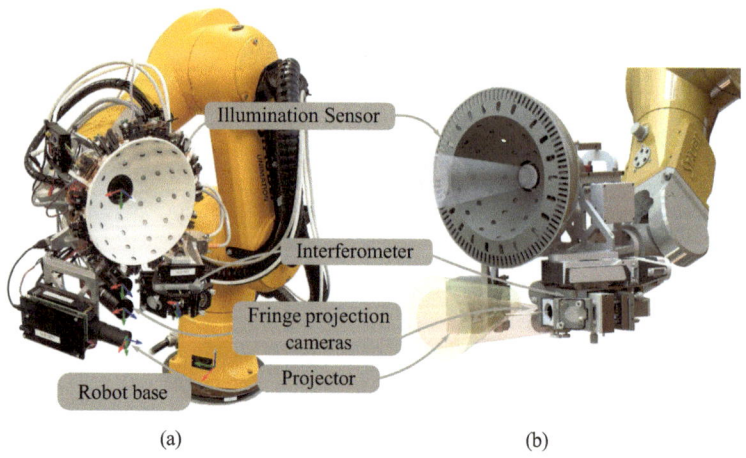

Fig. 16 **a** Measurement system setup with important coordinate systems. **b** Rendering of the measuring head CAD. Cones are used to visualize working distances and field of view

achieve this goal, each sensor coordinate system has to be calibrated to refer to a unified coordinate system. The next sections will outline this calibration process for each sensor.

4.1 Sensor Hand-Eye Calibration

For camera-based sensors state of the hand-eye calibration methods can be applied to calculate the position of the camera coordinate system in respect to a robots base coordinate system (Tsai et al. 1989). For this the robot is moved and for each pose a stationary calibration target is captured with the camera. In this work a 2-D dot pattern with known dot distance is used as calibration target. From the known properties of the target the transformation between it and each camera pose can be estimated. The set of camera to target transformations and forward kinematic based transformations of the robot facilitate the hand-eye calibration of the fringe projection and illumination sensor. Both cameras can be modelled with the pinhole camera model. The camera and lens of the LCI, however, requires a different model. In addition the depth of field of the telecentric lens is a lot smaller and the required number of different robot poses can not be achieved, because the blurriness prevents the transformation estimation. In addition the camera centre of a telecentric lens cannot be explicitly determined, but has to be set manually. Therefore a different calibration procedure has to be used for the LCI.

To calibrate the interferometer in respect to the other sensors, a 3-D calibration strategy is applied. For this a pair of LCI and fringe projection measurements is used. Both perform a measurement of a 3-D calibration target with distinct features, which allows an unambiguous alignment of both measurements. The resulting transformation which is necessary to align LCI into fringe projection data can be used to determine the 3-D calibration of the LCI coordinate system. Thus the fringe projection can be used as reference coordinate system for the LCI, which closes the transformation chain to the robot base.

The calibration of each sensor makes it possible to transform data from one coordinate system to the other. In order to transform fringe projection data (FPS) into the coordinate system of the interferometer (LCI), the homogenous transformation matrix $^{LCI}T_{FPS}$ is applied by multiplying it with the fringe projection data $\mathbf{p}$ (cf. Eq. 2).

$$\mathbf{p}_{FPS, LCI} = {}^{LCI}\mathbf{T}_{FPS} \cdot \mathbf{p}_{FPS} \qquad (2)$$

The subscript of data is used to show the coordinate system. For better understanding the data origin is kept as first subscript for transformed data. Sub- and superscripts applied to transformation matrices determine the source respectively destination co-ordinate system.

4.2 Calibration of the Rotational Axis

While the hand-eye calibrations are sufficient to transform data when only moving the robot, the rotary axis is not yet considered. To include the additional rotational movement, the axis is integrated into the kinematic chain and calibrated. For this the rotational axis in respect to the robot has to be identified. This is achieved by mounting the calibration target onto to the axis and recording multiple images while rotating the axis is small steps. For each axis position the target centre can be estimated utilizing the calibrated camera and known target properties. The 3-D information of each estimated centre is used to calculate a three dimensional circle fit. The normal of this circle is used to describe the location of the rotational axis. The direction of the normal is chosen with respect to the actual rotation of the axis following the right-hand rule. The missing degree of freedom along the axis is set manually.

4.3 Combination of Measurement Data

With the calibrated systems it is now possible to transform all gathered data into a unified coordinate system. Figure 17 shows the registered measurement of LCI and FPS. Larger structures on the blade surface can be observed in both surface reconstructions. Roughness measurements can thus be used to complement existing data with higher depth resolution. The transformation of the data is performed by using Eq. 2:

Fig. 17 LCI (*dark grey*)
measurement registered onto
surface measured (*light grey*)
with FPS

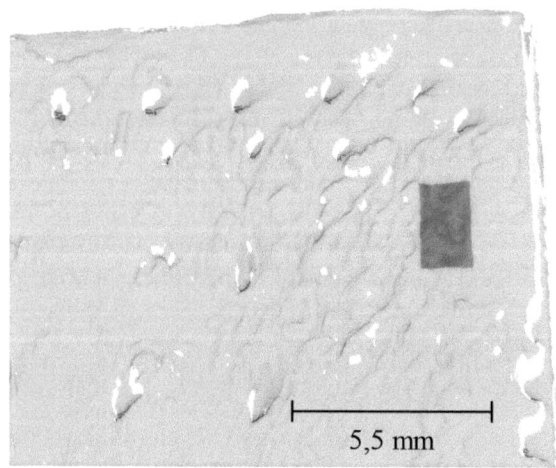

5,5 mm

$$\mathbf{p}_{FPS, RS} = {}^{RS}\mathbf{T}_{FPS} \cdot \mathbf{p}_{FPS} \tag{3}$$

$$\mathbf{p}_{LCI, RS} = {}^{RS}\mathbf{T}_{LCI} \cdot \mathbf{p}_{LCI} \tag{4}$$

In this case both measurements are transformed into the stage coordinate system (RS) to include rotations which may have been performed between both measurements. Since some geometrical features are available this coarse alignment can further be improved by applying e.g. a point-to-plane ICP.

The previous example demonstrates the combination of multiple scales of 3-D data.

The determined transformations additionally allow the merging of 2-D and 3-D measurements as present with the results of the illumination sensor. This, however, requires the addition of the intrinsic camera parameters to perform a proper projection of the data into the respective camera coordinate system. Melchert et al. (2020) demonstrated this projection step to utilize the surface normals from the FPS to improve the classification results and map them onto a 3-D measurement.

Figure 18 shows this process. The classification image which is derived from 2-D data shown on the left can be projected onto the 3-D points. With that it is possible to increase the amount of information of each point of the surface measurement and thus of each point of the 3-D model. Based on this enhanced model and the segmentation results of the CNN (see Fig. 13a), a more comprehensive defect detection can be carried out.

However, due to the fact that multiple calibrations depend on each other, the registration process is prone to error. Since a lot of calibration approaches are based on robot mounted cameras, even small errors of the camera calibration influence the hand-eye calibration, the identification of the rotary stage or the stereo calibration results. In addition the robot has assembly-related deviations in segment lengths and

Fig. 18 2-D classification from illumination sensor with the projection of 2-D labels onto a 3-D point cloud (Melchert et al. 2020)

Increased roughness

Nominal roughness

Shadowed regions

axis alignment. To reduce this effect, the robot is factory-calibrated. Nevertheless these aspects impact the overall registration performance and have to be noted.

5   Conclusions and Outlook

The application of optical measuring instruments of different scales and modalities in a common, global coordinate system opens up a wide range of novel inspection possibilities. In this way, the advantages of different measuring principles can complement each other while taking into account the resolution, measuring field size, accuracy and measuring duration in an optimal way and ensure a fast and meaningful diagnostic process. The detection and characterization of defects can be derived from multiple layers of information. Approaches for a multilayered defect detection and interpretation of combined measurement data are subject of further research.

In addition, individual sensors can profit from the multi sensor setup. A good example are the fringe projection unit and the illumination sensor. While the illumination sensor is capable of estimating the surface normals of the object on its own, the results are strongly dependant on the surface properties and favour low frequency structures. The surface reconstruction of the fringe projection system, on the other hand, can provide normals with a much higher certainty.

The interaction of multiple sensors opens up possibilities for measurement planning. While FPS and illumination sensor can be used to get a good overview of the measurement object, the LCI is introduced for local detail measurements. The position of these surface roughness measurements can be derived from the overview data e.g. by identified damaged regions based on the classification results. This on-demand sensor selection can greatly improve the measurement process, since only necessary measurements are performed.

Acknowledgements Funded by the Deutsche Forschungsgemeinschaft (DFG, German Research Foundation)—SFB 871/3—119193472. In addition the authors are grateful to all laboratory assistants and students who contributed to the realisation of this project.

References

Bay, H., Ess, A., Tuytelaars, T., and Van Gool, L. (2008). Speeded-up robust features (surf). *Computer vision and image understanding*, 110(3):346–359.

Betker, T., Quentin, L., Kästner, M., and Reithmeier, E. (2020). 3d registration of multiple surface measurements using projected random patterns. In *Optics and Photonics for Advanced Dimensional Metrology*, volume 11352, page 113520C. International Society for Optics and Photonics.

Bieron, J. and Peers, P. (2020). An adaptive brdf fitting metric. In *Computer Graphics Forum*, volume 39, pages 59–74. Wiley Online Library.

Bons, J. P. (2010). A Review of Surface Roughness Effects in Gas Turbines. *Journal of Turbomachinery*, 132(2). 021004.

Choi, S., Zhou, Q.-Y., and Koltun, V. (2015). Robust reconstruction of indoor scenes. In *Proceedings of the IEEE Conference on Computer Vision and Pattern Recognition (CVPR)*.

Ester, M., Kriegel, H.-P., Sander, J., Xu, X., et al. (1996). A density-based algorithm for discovering clusters in large spatial databases with noise. In *kdd*, volume 96, pages 226–231.

Guarnera, D., Guarnera, G., Ghosh, A., Denk, C., and Glencross, M. (2016). Brdf representation and acquisition. *Computer Graphics Forum*, 35(2):625–650.

Gupta, S., Kumar, M., and Garg, A. (2019). Improved object recognition results using sift and orb feature detector. *Multimedia Tools and Applications*, 78(23):34157– 34171.

Hartigan, J. A. and Wong, M. A. (1979). Algorithm as 136: A k-means clustering algorithm. *Journal of the royal statistical society. series c (applied statistics)*, 28(1):100–108.

Hartley, R. and Zisserman, A. (2003). *Multiple view geometry in computer vision*. Cambridge university press.

Herbort, S. and Wöhler, C. (2011). An introduction to image-based 3d surface reconstruction and a survey of photometric stereo methods. *3D Research*, 2(3).

Johnson, A. E. and Hebert, M. (1999). Using spin images for efficient object recognition in cluttered 3d scenes. *IEEE Transactions on pattern analysis and machine intelligence*, 21(5):433–449.

Kurz, R. and Brun, K. (2000). Degradation in gas turbine systems. *Journal of Engineering for Gas Turbines and Power*, 123(1):70–77.

Lafortune, E. P., Foo, S.-C., Torrance, K. E., and Greenberg, D. P. (1997). Non-linear approximation of reflectance functions. In *Proceedings of the 24th annual conference on Computer graphics and interactive techniques*, pages 117–126.

Laguna-Camacho, J., Villagrán-Villegas, L., Martínez-García, H., Juárez-Morales, G., Cruz-Orduña, M., Vite-Torres, M., Ríos-Velasco, L., and Hernández-Romero, I. (2016). A study of the wear damage on gas turbine blades. *Engineering Failure Analysis*, 61:88–99.

Li, Y., Kästner, M., and Reithmeier, E. (2015). Development of a compact low coherence interferometer based on gpgpu for fast microscopic surface measurement on turbine blades. In *Optical Measurement Systems for Industrial Inspection IX*, volume 9525, pages 164–170. SPIE.

Liang, M. and Hu, X. (2015). Recurrent convolutional neural network for object recognition. In *Proceedings of the IEEE Conference on Computer Vision and Pattern Recognition (CVPR)*.

Lowe, D. G. (2004). Distinctive image features from scale-invariant keypoints. *International journal of computer vision*, 60(2):91–110.

Melchert, N., Kästner, M., and Reithmeier, E. (2020). Robot-assisted BRDF measurement and surface characterization of inhomogeneous freeform shapes. In de Groot, P. J., Leach, R. K., and Picart, P., editors, *Optics and Photonics for Advanced Dimensional Metrology*, volume 11352, pages 37–42. International Society for Optics and Photonics, SPIE.

Mitra, N. J. and Nguyen, A. (2003). Estimating surface normals in noisy point cloud data. In *Proceedings of the Nineteenth Annual Symposium on Computational Geometry*, SCG '03, page 322–328, New York, NY, USA. Association for Computing Machinery.

Ngan, A., Durand, F., and Matusik, W. (2005). Experimental analysis of brdf models. *Rendering Techniques*, 2005(16th):2.

Park, J., Zhou, Q.-Y., and Koltun, V. (2017). Colored point cloud registration revisited. In *Proceedings of the IEEE International Conference on Computer Vision (ICCV)*.

Peng, T. (2007). *Algorithms and models for 3-D shape measurement using digital fringe projections.* University of Maryland, College Park.

Rublee, E., Rabaud, V., Konolige, K., and Bradski, G. (2011). Orb: An efficient alternative to sift or surf. In *2011 International Conference on Computer Vision*, pages 2564–2571.

Rusu, R. B., Blodow, N., and Beetz, M. (2009). Fast point feature histograms (fpfh) for 3d registration. In *2009 IEEE international conference on robotics and automation*, pages 3212–3217. IEEE.

Salti, S., Tombari, F., and Di Stefano, L. (2014). Shot: Unique signatures of histograms for surface and texture description. *Computer Vision and Image Understanding*, 125:251–264.

Schröder, P. and Sweldens, W. (1995). Spherical wavelets: Efficiently representing functions on the sphere. In *Proceedings of the 22nd annual conference on Computer graphics and interactive techniques*, pages 161–172.

Tabakoff, W., Hamed, A., and Shanov, V. (1998). Blade deterioration in a gas turbine engine. *International Journal of Rotating Machinery*, 4(4):233–241.

Tsai, R. Y., Lenz, R. K., et al. (1989). A new technique for fully autonomous and efficient 3 d robotics hand/eye calibration. *IEEE Transactions on robotics and automation*, 5(3):345–358.

Ward, G. J. (1992). Measuring and modeling anisotropic reflection. In *Proceedings of the 19th annual conference on Computer graphics and interactive techniques*, pages 265–272.

Westin, S. H., Arvo, J. R., and Torrance, K. E. (1992). Predicting reflectance functions from complex surfaces. In *Proceedings of the 19th annual conference on Computer graphics and interactive techniques*, pages 255–264.

Woodham, R. J. (1979). Photometric stereo: A reflectance map technique for determining surface orientation from image intensity. In Nevatia, R., editor, *SPIE Proceedings*. SPIE.

Zhang, Z. (2000). A flexible new technique for camera calibration. *IEEE Transactions on Pattern Analysis and Machine Intelligence*, 22(11):1330–1334.

Zheng, Q., Shi, B., and Pan, G. (2020). Summary study of data-driven photometric stereo methods. *Virtual Reality & Intelligent Hardware*, 2(3):213–221. 3D Visual Processing and Reconstruction Special Issue.

Exhaust Jet Analysis

Konstantinos Armanidis, Sebastian Kurth, Viet Nghiem, and Joerg R. Seume

Abstract The aim of project A3 of the Collaborative Research Centre (CRC) 871 is to develop a novel method for automated condition assessment of aircraft engines. The method is based on a combination of numerical simulations, Background Oriented Schlieren (BOS) measurements, and the use of pattern recognition algorithms. The density distribution in the exhaust gas stream of the engine is first measured using the BOS method and then compared with a damage library which was generated numerically through CFD. The automated classification of damage cases can be realized using pattern recognition algorithms. In the third funding period of the CRC, existing reconstruction algorithms used for the calculation of the density distribution in the exhaust gas were improved to increase the accuracy of the BOS measurements. Subsequently, the method was tested experimentally. First, experiments with a model combustion chamber show that 98.5% of all damage cases can be correctly classified by a suitable choice of integral parameters to describe the density distribution in the exhaust gas jet. In a second experimental test, the method was applied to a research engine in order to take realistic impacts into account.

Keywords Aircraft engines · Condition assessment · Exhaust jet analysis

1 Introduction

Due to the growing connectivity and mobility in society, a global increase in air traffic has been observed. According to a forecast made by the International Air Transport Association (2017) (IATA), passenger traffic is expected to double in the next 20 years, which will lead to an increasing demand for commercial aircraft. This

K. Armanidis · S. Kurth · V. Nghiem · J. R. Seume (✉)
Institute of Turbomachinery and Fluid Dynamics, Leibniz University Hannover, An der Universitaet 1, 30823 Garbsen, Germany
e-mail: seume@tfd.uni-hannover.de

K. Armanidis
e-mail: armanidis@tfd.uni-hannover.de
URL: https://www.tfd.uni-hannover.de/en

© The Author(s) 2025
J. R. Seume et al. (eds.), *Regeneration of Complex Capital Goods*,
https://doi.org/10.1007/978-3-031-51395-4_4

will also lead to higher cost of maintenance. Aircraft engines are among the most maintenance-intensive components of an aircraft, as they are subject to particularly stringent safety criteria and have a decisive influence on aircraft performance. It is estimated that 30% of operating costs are attributable to aircraft engines, of which one third is due to their maintenance (Rupp 2001). This makes it clear that there is a potential for cost savings in the maintenance process, which is the incentive for the development of new methods. For the maintainers, precise knowledge of the engine's condition is necessary to enable them to provide targeted repair and, at the same time, to reduce the risk of unexpected damage occurring during the maintenance process. For this purpose, engines are nowadays equipped with Engine Health Monitoring (EHM) systems, which can monitor the engine's performance on the basis of thermodynamic data within the framework of a Gas Path Analysis (GPA) and are used for detecting deviations. Despite the continuous improvement of the monitoring systems, a full condition assessment cannot yet be carried out, since such systems obtain their data from the engine control system. The condition assessment is therefore limited to the measurement of pressures and temperatures at discrete locations of the engine, which represents a small database for a detailed analysis of the engine condition. According to Volponi (2014), an increase in the number of sensors used is not a solution to this problem, as it entails an increase in engine weight and maintenance effort for the additional sensors. By contrast, a non-contact engine inspection offers the possibility of identifying damaged components before the repair for a precise estimation of the maintenance effort. Nowadays, borescopes are inserted through small openings in the engine to assess the damage condition of the engine. However, since the number of borescope openings in engines is limited, a full damage inspection cannot be carried out here either, which leads to a residual risk of the occurrence of unexpected sources of damage.

For these reasons, there is an unsatisfied demand for a condition assessment of the engine prior to disassembly for reliably planning engine maintenance. Within the scope of the present project, a novel method for the detection of defective components in the hot gas path of an aircraft engine has been developed and tested. By combining numerical flow simulations, optical measurements, and pattern recognition algorithms for machine learning, this approach allows detecting component damage before disassembly. The method proceeds in 3 steps:

1. The influence of characteristic component damage on the density distribution in the exhaust gas jet is predicted using numerical flow simulations.
2. The simulations are then used to build a damage library in order to train pattern recognition algorithms.
3. The engine condition can then be analysed and evaluated using these pattern recognition algorithms. For this purpose, the density field in the exhaust jet of an engine is reconstructed using the BOS method and compared with the library to identify the damage case.

With the aid of numerical simulations, Adamczuk et al. (2013a) investigated the influence of defective components on the density distribution in the exhaust gas jet and were able to show that both, a burner failure and different defects in the high-pressure turbine have an influence on the density distribution downstream of the turbine. Building on this, Adamczuk (2014) and Adamczuk and Seume (2016) were able to show that the influence of selected defects can be observed as far downstream as the exhaust gas jet of the engine, which is a basic requirement for a damage analysis with the BOS method. In the work of Hartmann (2012), Adamczuk et al. (2013b) and Adamczuk (2014), the density distribution in the exhaust jet of a helicopter engine was reconstructed using the BOS method. Density non-homogeneity that occurs in the form of a cold air stream in the exhaust gas jet were considered. The applicability of the BOS method to a civil aircraft was also demonstrated by Adamczuk (2014) and Hartmann et al. (2015) by measuring the exhaust jet using a non-tomographic setup consisting of one camera. This set up identified cases of damage such as oil leakage in the bearing seal air.

2 Objectives

The quality of an automatic failure classification by using the BOS method depends on the reconstruction accuracy of the BOS algorithm. There already exist algorithms such as the filtered back projection (FBP) and the algebraic reconstruction (ART), which can be used for reconstructing the density distribution in the exhaust gas flow. Preliminary investigations have shown, however, that especially in regions with large density gradients artefacts are formed, which impair the quality of the reconstruction. Therefore, one objective of this work is to extend already existing algorithms in order to increase the reconstruction accuracy of complex density structures.

Within the framework of an experimental test on a model combustion chamber, it will be examined whether the combination of achievable reconstruction accuracy and the properties of the pattern recognition algorithm are sufficient for an automated classification of combustion chamber failures. For this purpose, a model combustion chamber is used, which offers the possibility of adjusting the parameters of individual burners in order to simulate a damage case.

As a next step towards an industrialization, the applicability of the presented method to an aircraft engine has to be demonstrated. For this purpose, numerical simulations are first used to test whether it is possible to automate the detection of defects and defect combinations in the engine.

By using synthetic BOS measurements based on the numerical simulations, various parameters of the BOS setup such as the number of cameras and the distance to the exhaust gas jet are investigated in order to find the optimal choice of parameters. The results of the synthetic measurements are then used to determine the probability of success of an automatic classification by the chosen pattern recognition algorithm.

In a second experimental test, the applicability of the presented BOS method in a real research engine will be investigated as part of the scope of two measurement campaigns. For this purpose, a special measuring ring was designed and manufactured for the attachment of the BOS measuring equipment. The measurement data from the engine test will be used to evaluate the accuracy of the numerical simulations and the automated defect detection on the basis of a real database.

3 Tomographic Reconstruction of Density Fields

The background is observed by the cameras through the whole density field. The displacement seen by a single camera only gives the integral of the density gradients from the camera. To reconstruct the density field, the projections from several cameras are needed. To evaluate the number of cameras needed and to test existing tomographic reconstruction methods, an artificial density field was used. The artificial density field was created with RANS simulations of a low- pressure turbine. To simulate cold streaks, an inhomogeneous temperature field was set as the inlet boundary condition. The three main cold streaks with decreasing intensity are marked (1, 2, 3) in Fig. 1.

The advantage of using an artificial density field is that the reconstruction can be directly compared to the original density field. Also, the number, location, distance, and angle of view of the cameras can be chosen freely. For the following comparison of reconstruction algorithms, 16 virtual cameras, placed with equidistant spacing on a 180° arc, are used. Special attention was paid to the reconstruction of large density gradients resulting from the cold streaks.

Two reconstruction algorithms, the filtered backwards projection (FBP) and the algebraic reconstruction technique (ART) are compared. The FBP reconstruction algorithm was applied to BOS by Goldhahn and Seume (2006) based on the work of Radon (1917). The three main steps are:

1. calculation of the Fourier transforms of the measured projection integrals
2. back transformation of the high-pass filtered signal
3. back projection with coordinate transformation.

The implementation and use for BOS measurements can be found in Goldhahn (2009). The ART is based on solving a linear system of equations with an iterative

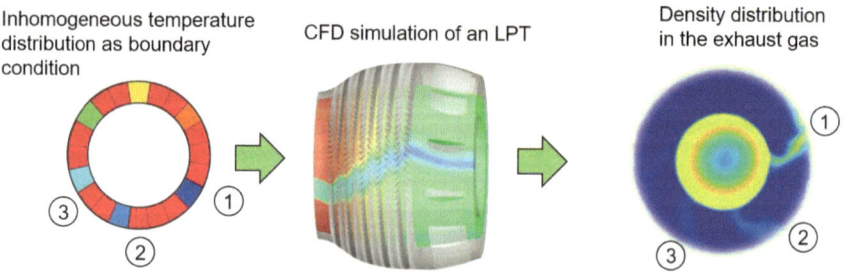

Fig. 1 Artificial density field from RANS simulation of a low pressure turbine (LPT) with inhomogeneous inlet boundary condition

method. Compared to the FBP reconstruction, the ART algorithm is slower, but needs less projections for a good reconstruction (Kak and Slaney 2001; Guan and Gordon 1996). This can be seen in Fig. 2a and b, comparing the FBP and ART with 16 virtual cameras each. The ART shows less reconstruction artefacts and is closer to the artificial density field (Fig. 1). The relative density difference is shown in Fig. 2c and d. Here, the artefacts can be seen as a large relative differences. With the ART, the first and second cold streak can be well detected while the third cold streak cannot.

To improve the quality of the reconstruction, Hartman (2020) developed a method combining the FBP and ART algorithm into an initialised algebraic reconstruction technique (iART). In this method, the gradient field from the FBP reconstruction is used to initialise the ART. In Fig. 3 the results of the reconstruction with 16 virtual cameras are shown. The artefacts, seen in the reconstruction with the FBP and ART method are further reduced and the third cold streak can be detected. Using the root mean square deviation as a measure for the accuracy shows, that the quality of the measurement is improved from $\mathrm{RMS_{FBP}} = 0.1$, over $\mathrm{RMS_{ART}} = 0.0547$ to $\mathrm{RMS_{iART}} = 0.0257$ (Hartman 2020).

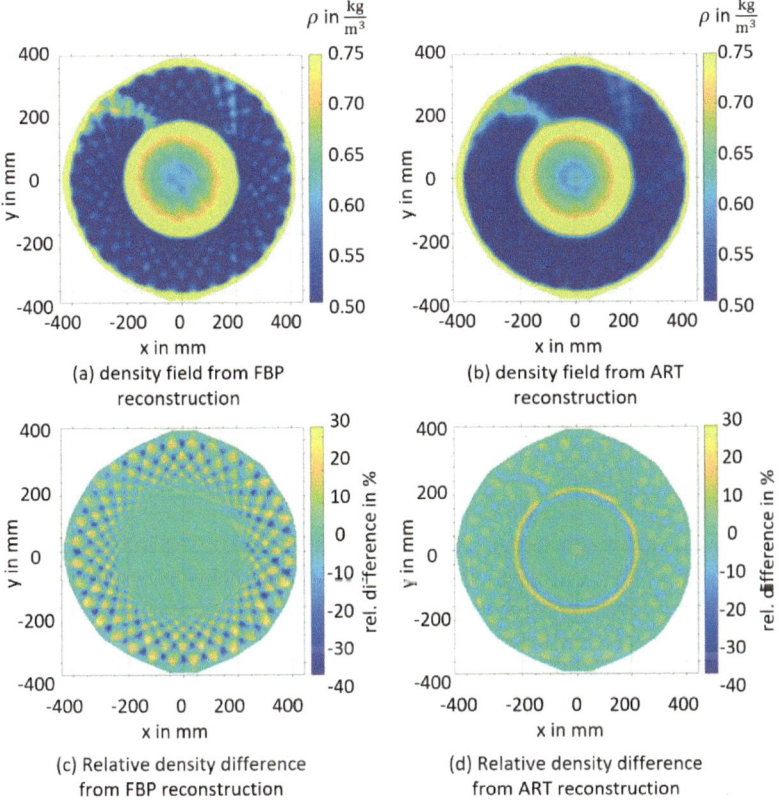

Fig. 2 Comparison of FPB and ART reconstruction of artificial density field (from Hartmann 2020)

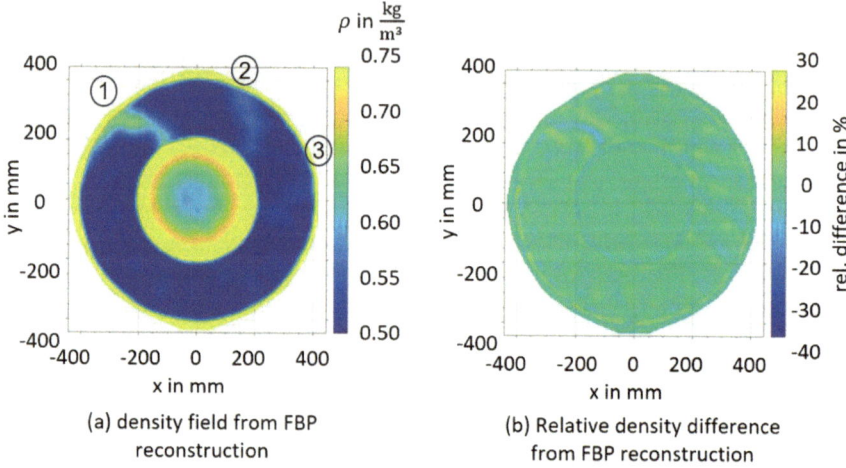

Fig. 3 Reconstruction with the initialised algebraic reconstruction technique (iART) algorithm

4 Application of the BOS Method to a Model Combustion Chamber

In the first experimental test, the combination of the BOS measurement with the support vector machine (SVM) described by Hartmann (2020) is investigated for automated defect detection on a model combustion chamber. First, the BOS measurement is tested for sufficient accuracy to achieve automated classification. Subsequently, integral parameters of the density distribution are introduced to parameterise the defect influence on the density distribution. The parameters are then used to automate the classification using SVM.

Experimental Setup

For the investigations, a model of a combustion chamber was used in the laboratory of the partner institute for combustion, ITV (Fig. 4). This consists of eight swirl-stabilised premix burners distributed over a circle with a circumference of 210 mm. Each burner consists of a 28 mm diameter combustion tube mounted with a swirl generator and a turbulence grid to simulate realistic combustion chamber flow conditions. The experimental set-up allows individual burners to be manipulated in order to simulate defects such as clogging of the fuel nozzles. Furthermore, it is possible to investigate geometric defects by adjusting the position of individual burners. A detailed description of the test rig can be found in Hennecke et al. (2015), von der Haar et al. (2016), Hartmann et al. (2016) and Hartmann et al. (2018).

For the reconstruction of the density distribution, 16 cameras of the type Allied Vision Technologies Manta G201b with 25 mm lenses were used. The cameras are positioned on a semicircle with a radius of 1950 mm around the burner. The angle between two cameras is 11.75°. In front of the cameras a dot pattern with a dot size

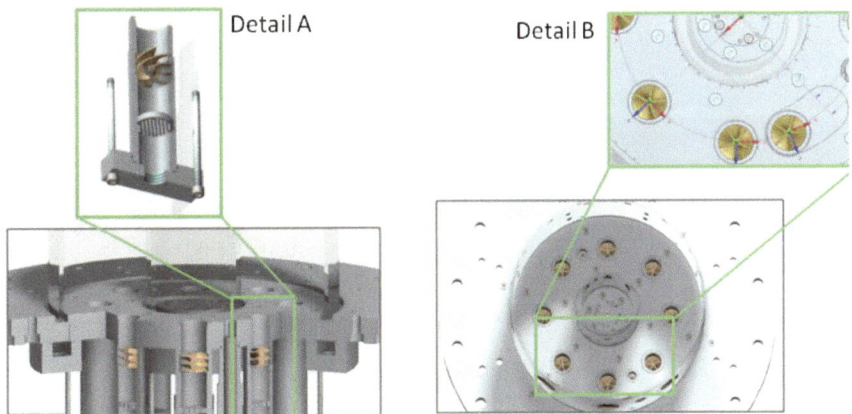

Fig. 4 Model of the combustion chamber. Figure according to von der Haar et al. (2016) and Hennecke et al. (2015)

of 3–4 px is placed. For each operating point, 500 images are generated per camera, at a frame rate of 50 Hz. The exposure time is 150 µs due to the low beam speed.

Results

In the experiments conducted by Hartmann (2020), several parameters of a burner were varied in order to investigate the influence on the density distribution of the exhaust gas jet behind the combustion chamber. A detailed description of the results is documented in Hartmann's dissertation (Hartmann, 2020). At the beginning, the output of a single burner was reduced to simulate a burner failure. FIG shows the exit density together with the marked positions of the burners. The investigations have shown that there is a visible loss of rotational symmetry in the density distribution of the exhaust gas. Due to the local momentum deficit caused by a burner failure, cold air flows into the flue gas jet, which leads to a local increase in density (Fig. 5). A similar behaviour occurs when varying the air ratio of an individual burner. A decrease leads to a reduction of the volume flow at constant burner output and thus to a rich combustion. The reduction of the volume flow causes a momentum deficit which allows fresh air to enter the exhaust jet.

As already shown in Hartmann et al. (2018), geometric defects of burners also have the potential of influencing the density distribution in the outlet of the combustion chamber. The circumferential position of a single burner was varied, resulting in localized regions of higher density, similar to the other two parameter variations. In the case where the burner power was simultaneously reduced, an increase of the region of higher density was observed. It was also possible to show that a burner offset in combination with an increase of the air ratio has no influence on the density distribution. This shows that a mutual compensation of defects can occur, which makes detection difficult.

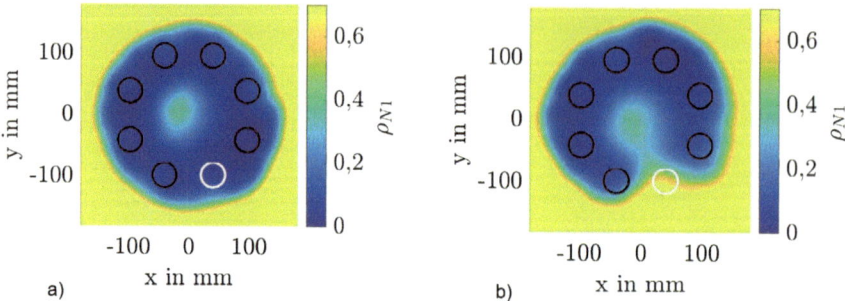

Fig. 5 Influence of the reduction of the burner power on the normalised density distribution in the exhaust gas jet. **a** Reference case **b** Measurement with burner failure (from Hartmann 2020)

Subsequently, Hartmann (2020) investigated the possibility of an automated classification of defects using the combustion chamber model. Integral parameters were introduced to describe the density and the density gradient at the outlet of the combustion chamber. For this purpose, various statistical moments were introduced and tested for their ability to parameterise the influences of defects. Some parameters are shown as examples in Fig. 6. The measurement data from the case studies with different combustion chamber defects were used as training data for a support vector machine (SVM) algorithm to form a separation plane between the different classes as shown in Fig. 6. A total of 400 data sets from the measurements were used. Of these data sets, 100 were used for the reference class and 300 for the defect class. Two thirds of the latter data sets were used for training the SVM algorithm and one third as a test data set.

To find the required number of parameters, the number of parameters used was increased gradually and then the result of the classification was calculated with the test data. Figure 6 shows that a good separation between reference and damage cases can be achieved with only a few parameters. By using two parameters, a classification accuracy of 99% of the defective cases and 87.8% of the reference cases can be achieved. Thus, an overall accuracy of 96.2% can be achieved. The best result can

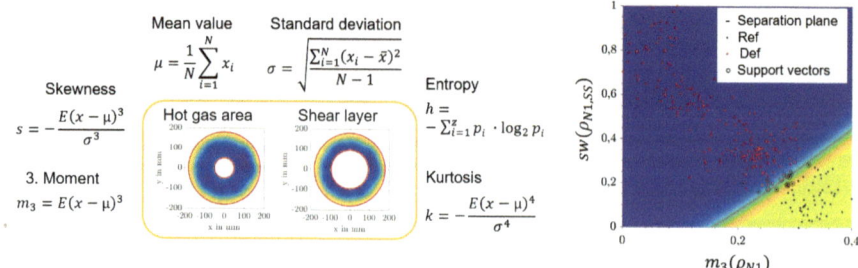

Fig. 6 Selection of statistical moments to describe the normalised density distribution and separation plane between different classes with optimised selection of parameters (Hartmann 2020)

be achieved with five parameters. Here, 98.5% of all cases are correctly assigned, of which 99% are defective cases and 97% are reference cases.

5 Application of the BOS Method on an Aircraft Engine

Next, the BOS method was used for the investigation of an aircraft engine. First, numerical investigations of the entire hot gas path (HGP) are performed, to check whether combustion chamber damage has an influence on the density distribution at the outlet nozzle of an engine.

Numerical Setup

The finite-volume solver TRACE 9.1.538 of the German Aerospace Center (Franke et al. 2005) is used for the numerical simulations of the turbine and the exhaust jet. The numerical setup (see Table 1) of the turbine is resolved with 93 million cells and the exhaust jet with 26 million cells. For all calculations, a second-order accurate Fromm scheme (Darwish 1993) using no limiter is used for spatial discretisation. Turbulence is modelled using the shear stress transport model k-w-SST by Menter (1994) with the correction by Kato and Launder (1993) to correct the overproduction of turbulent kinetic energy at the stagnation region. Rotational effects are considered using the modification by Bardina et al. (1985). Boundary-layer transition on the blades is modelled by a multi-mode model (Kozulovic et al. 2007) and all walls are resolved with the wall-bound cells adhering to $y^+ \leq 1$.

For unsteady calculations, in addition to the previously specified settings, an Euler backward 2nd order time discretisation scheme with 1080 timesteps per period and 30 sub iterations was used. The Courant-Friedrichs-Levy number is set to CFL = 50. These parameters were derived by a timestep study of the mid-span flow path.

Two-dimensional boundary conditions at the high-pressure turbine (HPT) inlet are specified as obtained from a combustion chamber simulation. At the exit guide vane (EGV) outlet, the back- pressure is taken from a one-dimensional performance calculation. For the inter-row coupling between rows of HPT and low-pressure turbine (LPT), as well as between both components, direct interfaces are used for both steady and unsteady simulations, i.e., the steady-state simulation is a frozen-rotor calculation. Unsteady simulations are restarted from the steady results and were performed for nine sectional revolutions after which they achieved convergence. Cooling mass

Table 1 Parameters of the numerical simulation model

Model	Turbine	Exhaust Jet
Total resolution	93 million cells	36 million cells
CFD approach	RANS (Steady)—URANS (Unsteady)	RANS
Row interfaces	Direct	–
Turbulence model	k – w – SST	k – w

flows in the HPT have been neglected to reduce the complexity and the duration of the simulation. From this turbine simulation, the core stream boundary conditions for the exhaust jet simulation are provided. The exhaust jet is simulated in a free stream condition using the data of a performance simulation for the boundary conditions in the free and bypass stream.

Experiments on an aircraft engine

In addition to the numerical simulations, experiments with an aircraft engine are carried out to validate the results of the numerical simulation and to test the BOS method under real conditions. In the second experimental test, the research engine V2500 of the Institute of Jet Propulsion and Turbomachinery (IFAS) at Technische Universität Braunschweig is used for the investigation of the BOS method in the test facility of MTU Maintenance in Langenhagen. Within the scope of an engine test, the fuel supply of a burner is gradually throttled in order to simulate a burner defect, similar to the tests with the model combustion chamber. The resulting inhomogeneous density field at the engine outlet is then reconstructed using the BOS method, Fig. 7. A special measurement ring was designed and manufactured for the test.

The ring has a weight of 509.75 kg and an average diameter of 4.15 m. The entire ring construction consists of two halves assembled together. One half is used as a carrier for 15 DALSA GENIE NANO-M4060 cameras with RICOH FL-BC2518-9 M lenses of 25 mm focal length. The distance between the cameras is 8°, equidistant. A retroreflective foil with a defined dot pattern is applied to the second half. The dot diameter is 2.8 mm and the dot density 35%. The cameras are mounted to the central ring strut of the BOS ring by means of a manually adjustable U-profile support and a holding device. The orientation of each camera can be adjusted so that it coincides with a fixed target position. Each camera position is marked by means of attached calibration crosses on the opposite retroreflective foil. Before the engine test, the alignment of each camera is checked and, if necessary, corrected by means of a calibration in order to avoid measurement errors. Six mounting plates were designed with which the ring can be fixed to the eyelets in the exhaust duct, thus mounting the

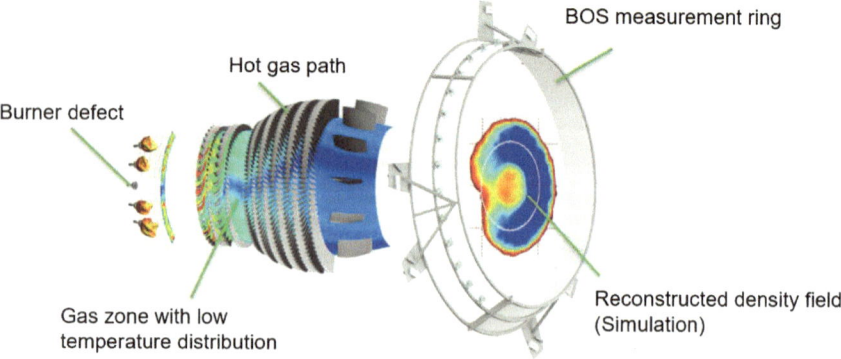

Fig. 7 Investigation of the BOS method on a real aircraft engine

BOS ring in the exhaust duct of the test cell. The experimental setup in the MTU test cell with the instrumented engine is shown in Fig. 8. To prevent axial movement of the ring, retaining clips were used on all 6 plates.

A scheme of the setup is shown in Fig. 9. A 16 Amp three-phase connection was used for the power supply. A power distributor supplies 5 power supply units per phase; each phase is connected to a camera and a ring light. During each measurement, the recorded images were transmitted via a switch to the PC, which is located in the transition room next to the test cell. The cameras can be controlled via the PC by means of a measuring program in order to trigger the flash lights synchronously. Before each measurement, a setup file must be loaded into the program which defines camera settings such as exposure time, the region of interest (ROI) and frame rate of the image recordings. For the present experiment, 1000–1200 images were taken at each operating point with a frame rate of 14 Hz. The exposure time is 110 μs and the ROI 4112 $\times$ 900 px is used.

In addition to the BOS measurement system, a separate monitoring system is used to observe safety-relevant variables on the BOS ring during the test and to determine the pressure and temperature boundary conditions near the ring for the BOS measurement. For this purpose, thermocouples type J were used on the 6 mounting plates to monitor the temperature and vibration sensors from Conplatec (Brüel & Kjaer 4508) to observe the ring vibration caused by the unsteady aerodynamic load from the exhaust jet and the engine acoustics. Prandtl tubes were used to determine the pressure and temperature boundary conditions near the ring.

For the test, the temporal profile of the engine's thrust was determined in order to reach the desired operating points of the engine (Fig. 10). A distinction is made between the four load bands A, B, C, D, whereby load band A with 102.7 kN represents the highest engine thrust and D with 42.75 kN the lowest. The dwell time on all load bands except A is 6 s. On load band A the dwell time was set to 3 s to reduce the risk of damage to the test structure due to the high thermal and mechanical stresses. In order to ensure thermal equilibrium of the engine despite the short dwell time, the engine is first stabilised on load belt B for 3 s before it is moved up to load belt A. Before the test, the engine runs through a system check to ensure that all systems are functioning correctly. Then, the reference thrust curve is first run without burner defect to generate the reference images for the BOS measurement. After the reference measurement is completed, the damage case with 100% closed fuel spray nozzle is first initiated by replacing a single burner. At the same time, a mechanical check of the BOS ring is carried out to assure the integrity of the bolt connections. Finally, the damage case with 50% throttled fuel nozzle is examined.

Results

The main goal of the BOS measurements on the aero engine is to detect the cold streak introduced by the throttled or closed fuel spray nozzle. To investigate the propagation of the cold streak through the engine, first RANS simulations were performed. Results of the simulations are shown in Fig. 11. The normalized density is defined by

Fig. 8 Experimental setup consisting of the V2500 research engine of the IFAS and the BOS ring in the MTU Maintenance test cell. (Picture used with kind permission of MTU Maintenance and IFAS)

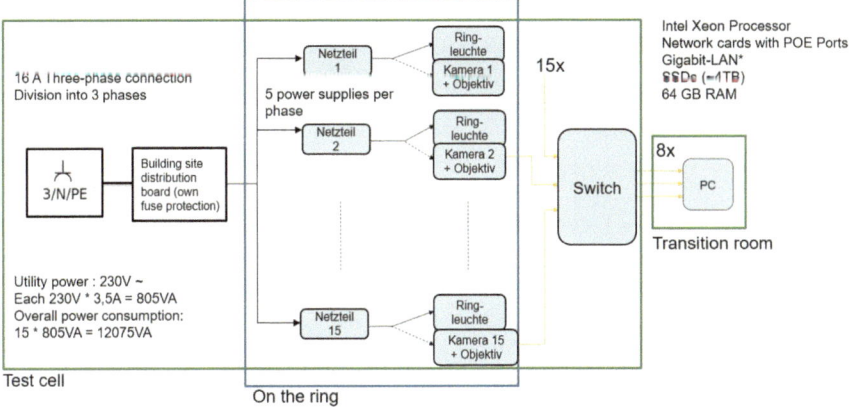

Fig. 9 Schematic of the experimental setup

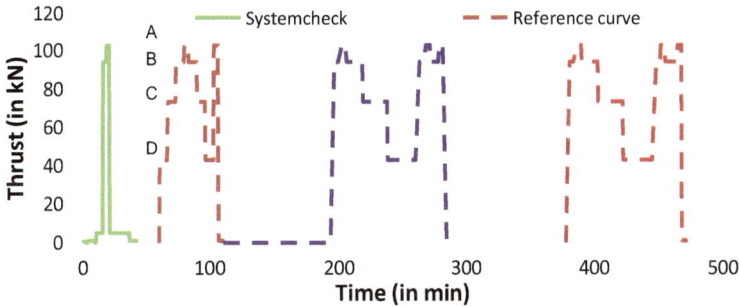

Fig. 10 Thrust curves of the engine during the experimental test (provided by IFAS)

$$\bar{\rho} = \frac{\rho - \rho_{min}}{\rho_{max} - \rho_{min}} \tag{1}$$

High temperatures result in a low normalised density; cold streaks result in an increase of the normalised density. To illustrate the changing shape of the cold streaks, three cross-sections are extracted from the simulation:

A. nozzle inlet
B. nozzle outlet
C. BOS measurement plane.

At the nozzle inlet plane (A), the cold streak can be seen as a round shape in the upper right part of the normalised density field. A swirl is introduced by the exit guide vanes (EGV), deflecting and deforming the cold streak in the nozzle's outlet plane (B). The influence of the EGV can also be seen by the flower shape of this plane. Further downstream at the BOS measurement plane (C), the cold streak is even more

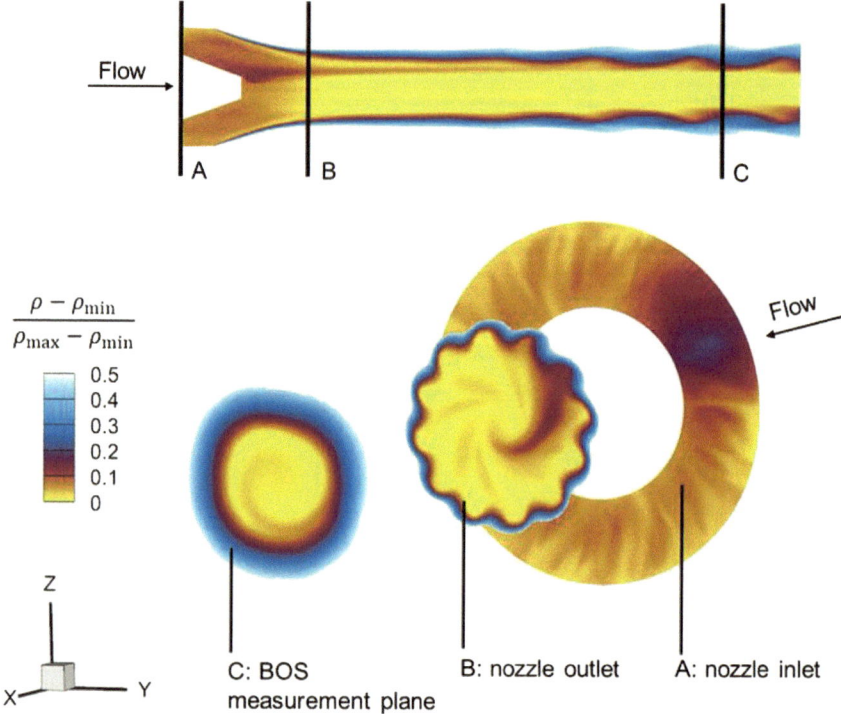

Fig. 11 Results of RANS-simulation of exhaust jet of aero engine

deformed and less distinguishable. The area of the low normalised density plane is also reduced, because the hot gases mix with the bypass and ambient air.

Figure 12 shows the measurement results from the BOS measurements with the aero engine. The normalised density field is shown for the reference case with all fuel spray nozzles opened and with one fuel spray nozzle 50% and 100% closed respectively. All density fields have the same shape. This is expected, as the cold streak is predicted not to alter the shape of the density field. Contrary to the prediction by the RANS simulation, a cold streak is not visible for the measurements with the closed fuel spray nozzle. The measured density field is not as symmetric as the simulation results, and all density fields are shifted upwards. Four main aspects were identified to explain the discrepancies between the simulation and the experiment:

1. Structures in the test facility, like the aero engine's ceiling support, influence the exhaust gas jet and the flow around the engine. These effects are not modelled in the RANS simulation and were underestimated before the experiment. The uneven flow around the aero-engine can explain the upwards shift of the exhaust gas jet and might lead to a deformation of the density field and an underestimation of the mixing behaviour of the exhaust gas jet. A better insight into the aerodynamics of the test facility will be gained by a simulation of the complete

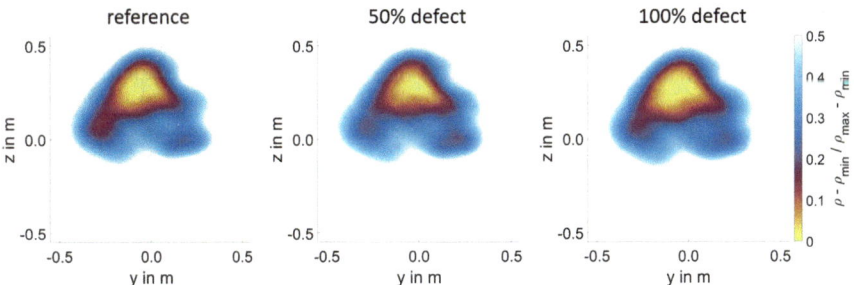

Fig. 12 Reconstruction of normalised density field from aero engine test

test facility currently being carried out by the Institute of Jet Propulsion and Turbomachinery, IFAS at Technische Universität Braunschweig.

2. The upwards shift of the exhaust gas jet is leading to issues concerning the reconstruction of the density field. The ambient temperature and pressure are set as boundary conditions on the boundary of the reconstruction. Because of the upwards shift, the density field is no longer in the middle of the reconstruction area and the reconstruction boundary is intersecting with the density field. This leads to errors at the reconstruction boundary and can deform the reconstructed density field. These issues will be addressed in the upcoming second measurement by a rearrangement of the cameras. In addition, more temperature and pressure measurement points will be added, reaching further into the hot gas stream, to decrease the uncertainties of the boundary conditions.

3. The missing cold streak in the measurement might be explained by an underestimation of the mixing of the exhaust gas in the RANS simulation or by the reconstruction errors. To reduce the influence of the mixing of hot and cold gas and of the upwards shift, it would be preferable to decrease the distance between the engine's outlet and the measurement plane. However, this is not possible in the test facility used for these measurements, due to constructive restrictions.

4. The exhaust gas jet can have an uneven distribution of exhaust gases. The composition of the exhaust gases influences its optical properties. To consider the influence of the inhomogeneous gas, constant values need to be defined locally for upcoming reconstructions.

We are looking forward to addressing as many of the mentioned aspects in the upcoming measurement campaign. Unfortunately, the date for the measurement is after the publication date of this report. We hope, that the interested reader will follow our upcoming publications, where we will present results of this upcoming measurement campaign.

6 Conclusions

During 12 years of research many steps towards the detection of hot gas path defects by background-oriented schlieren (BOS) measurements have been completed. Initially, the algorithms for tomographic reconstruction were improved for better accuracy of BOS measurements. The filtered backwards projection (FBP) and the algebraical reconstruction technique (ART) combined yielded the initialised algebraic reconstruction technique (iART).

iART's mean deviation in the density compared to numerical simulations is 0.0257, which is a significant improvement over FBP and ART, for which the corresponding deviations are 0.1 and 0.0547.

Following the improvement of the reconstruction algorithm, the BOS method was used to detect defects in a model combustion chamber, in which the influence of characteristic burner damage on the density distribution in the exhaust gas jet was investigated. For this purpose, the fuel supply to a single fuel spray nozzle was gradually reduced. This resulted in a momentum deficit, which allows dense ambient air to enter into the exhaust jet of the combustion chamber. The measurement data was then used to train a support vector machine (SVM) algorithm for the automated classification of burner damage. It was shown that by choosing suitable integral parameters to describe the normalised density distribution, a high probability of successfully classifying burner failures can be achieved even with a small number of parameters. The best result was achieved with a choice of 5 parameters. Thus, 99% of all defect cases were correctly classified and 97% of all reference cases.

After the successful application of the BOS method on a model combustion chamber, a real engine was considered. First, numerical investigations were performed using RANS simulations to investigate the influence of combustor damage on the density distribution of the exhaust gas jet. The results showed that a cold streak forms at the combustion chamber outlet as a result of a burner failure. Despite mixing with the surrounding gas in the hot gas path (HGP) of the engine, the cold streak could also be observed at the exit of the engine, which indicates an existing influence on the density distribution in the exhaust gas jet.

Further experiments on the research engine V2500 validate the numerical simulations of the reference configuration. For the application of the BOS method a specially designed measurement ring was used for mounting the cameras. The experimental data also showed that, in contrast to the RANS simulations, there is an asymmetry in the density distribution. Furthermore, the existence of a cold streak in the exhaust gas jet could not be confirmed for the investigation case with 50 and 100% throttled fuel spray nozzle. Finally, four possible causes for the occurrence of this deviation were identified and solutions were described which will be implemented in the upcoming measurement campaign.

Acknowledgements This research is funded by the Deutsche Forschungsgemeinschaft (DFG, German Research Foundation) SFB 871/3 119193472. We gratefully acknowledge the contribution of the DLR Institute of Propulsion and MTU Aero Engines AG for providing TRACE. We would also like to thank MTU Maintenance GmbH for their support with the aero engines tests.

References

Adamczuk, R. R. (2014). Zustandsbeurteilung eines Triebwerks durch die Analyse des Abgasstrahls, Dissertation. PhD. thesis, Leibniz Universität Hannover.

Adamczuk, R. R. and Seume, J. R. (2016). Numerical Evaluation of the Condition of a Jet Engine through Exhaust Jet Analysis. In: Proceedings of the ASME Turbo Expo. ASME, Seoul, South Korea, pp. 1–14.

Adamczuk, R. R., Buske, C., Roehle, I., Hennecke, C., Dinkelacker, F., and Seume, J. R. (2013a). Impact of Defects and Damage in Aircraft Engines on the Exhaust Jet. In: Proceedings of the ASME Turbo Expo. ISBN 978–0–7918–5513–3. https://doi.org/10.1115/GT2013-95079.

Adamczuk, R. R., Hartmann, U., and Seume, J. R. (2013b). Experimental demonstration of Analyzing an Engine's Exhaust Jet with the Background-Oriented Schlieren Method. In: AIAA Ground Testing Conference. American Institute of Aeronautics and Astronautics, Reston, Virginia, pp. 1–10. https://doi.org/10.2514/6.2013-2488.

Bardina, J., Ferziger, J.H., and Rogallo, R. S. (1985). Effect of Rotation on Isotropic Turbulence: Computation and Modelling, Journal of Fluid Mechanics 154, 321–336.

Darwish, M. S. (1993). A new high-resolution scheme based on the normalized variable formulation, Numerical Heat Transfer, Part B: Fundamentals 24(3), 353–371.

Franke, M., Kügeler, E., and Nürnberger, D. (2005). Das DLR-Verfahren TRACE: Moderne Simulationstechniken für Turbomaschinenströmungen, in ‚DGLR-Jahrbuch. Deutscher Luft- und Raumfahrtkongress'.

Goldhahn, E. (2009). Weiterentwicklung der Hintergrundschlierenmethode zu einem quantitativen Verfahren der Vermessung von Dichtefeldern, Leibniz Universität Hannover

Goldhahn E. and Seume, J. R. (2006). The Background Oriented Schlieren Technique: Sensitivity, Accuracy, Resolution and Application to a Three-Dimensional Density Field, Proceedings of 13th International Symposium on Applications of Laser Techniques to Fluid Mechanics, 26–29 June 2006, Lisbon, Portugal (auch veröffentlicht in Experiment in Fluids, July 2007)

Guan, H. and Gordon, R. (1996). Computed tomography using algebraic reconstruction techniques (ARTs) with different projection access schemes: a comparison study under practical situations. In: Physics in medicine and biology, volume 41(9):pp. 1727–1743. ISSN 0031–9155

Hartmann, U. (2012). Experimentelle Anwendung der BOS-Methode am Hubschraubertriebwerk Artouste. Projektarbeit, Leibniz Universität Hannover.

Hartmann, U. (2020) Auflösung und Zuordnung von Defekten im Heißgaspfad von Flugtriebwerken anhand der Dichteverteilung im Abgasstrahl, Hannover: Gottfried Wilhelm Leibniz Universität, Diss., 2020, 116 S., Anh.https://doi.org/10.15488/9979

Hartmann, U., Adamczuk, R. R., and Seume, J. R. (2015). Tomographic Background Oriented Schlieren Applications for Turbomachinery (Invited). In: 53rd AIAA Aerospace Sciences Meeting. American Institute of Aeronautics and Astronautics, Reston, Virginia. ISBN 978–1–62410–343–8. https://doi.org/10.2514/6.2015-1690.

Hartmann, U., Hennecke, C., Dinkelacker, F., and Seume, J. R. (2016). Automatic Detection of Defects in a Swirl Burner Array Through an Exhaust Jet Pattern Analysis. In: Journal of Engineering for Gas Turbines and Power, volume 139(3):p. 031504. ISSN 0742–4795. https://doi.org/10.1115/1.4034449.

Hartmann, U., Von der Haar, H., Dinkelacker, F., and Seume, J. R. (2018). Experimental Defect Detection in a Swirl-Burner Array Through Exhaust Jet Analysis. In: 2018 AIAA Aerospace Sciences Meeting. American Institute of Aeronautics and Astronautics, Reston, Virginia. ISBN 978–1–62410–524–1. https://doi.org/10.2514/6.2018-0303.

Hennecke, C., Hartmann, U., Dinkelacker, F., and Seume, J. R. (2015). Correlation of defects in an annular swirl-burner-array by optical measuring exhaust gases and numerical analysis. In: Deutscher Luft- und Raumfahrtkongress. Rostock, Deutschland.

International Air Transport Association (2017). IATA Annual Review 2017. In: Annual Report.

Kak, Avinash, C., and Slaney, M. (2001). Principles of Computerized Tomographic Imaging. Society for Industrial and Applied Mathematics.

Kato, M. and Launder, B. E. (1993). The Modeling of Turbulent Flow Around Stationary and Vibrating Square Cylinders, in '9th Symposium on Turbulent Shear Flows', pp. 10.4.1–10.4.6.

Kožulović, D., Röber, T. and Nürnberger D. (2007). Application of a Multimode Transition Model to Turbomachinery Flows, in 'Proceedings of the 7th European Conference on Turbomachinery', pp. 5–9.

Menter, F. R. (1994). 'Two-equation eddy-viscosity turbulence models for engineering applications', AIAA Journal of Fluids Engineering 38(8), 269–289.

Radon (1917). Über die Bestimmung von Funktionen durch ihre Integralwerte längs gewisser Mannigfaltigkeiten. In: Akad. Wiss., volume 69.

Rupp, O. (2001). Instandhaltungskosten bei zivilen Strahltriebwerken. In: Maintenance von Flugzeugen und Triebwerken. MTU Maintenance Hannover GmbH, DGLR.

Volponi, A. J (2014). Gas Turbine Engine Health Management: Past, Present, and Future Trends. In: Journal of Engineering for Gas Turbines and Power, volume 136(5):p. 051201. ISSN 0742–4795. https://doi.org/10.1115/1.4026126.

Von der Haar, H., Hartmann, U., Hennecke, C., Dinkelacker, F., and Seume, J. R. (2016). Defect detection in an annular swirl-burner-array by optical measuring exhaust gases. In: Proceedings of the ASME Turbo Expo. Seoul, South Korea.

Adaptable and Component-Protecting Disassembly in the Regeneration Path

Richard Blümel and Annika Raatz

Abstract The disassembly initiates a product's maintenance and regeneration. In order not to cause any additional damage to the components during disassembly, which would lead to higher repair costs or, in the worst case, to destruction and loss of the components, disassembly must be as gentle as possible on the components. Due to ambiguous causes, such as thermal or mechanical product loads during operation, the unknown product condition is a characteristic uncertainty factor in disassembly. This paper presents approaches and methods on how the disassembly of complex capital goods, which is usually carried out manually, can be automated while still being protective on components and adaptable to varying product conditions. Manual disassembly procedures are substituted using micro impacts induced by a piezo actuator. A learning model predicts optimized process parameters based on varying operational usage scenarios.

Keywords Disassembly · Predictive maintenance · Machine learning

1 Introduction

Disassembly is a crucial step toward sustainable life cycle engineering, restoring a product's functionality in subsequent regeneration processes or recovering product components (Seliger 2007). At regular maintenance, repair and overhaul (MRO), components are regenerated, their properties restored and renewed or, if necessary, replaced with new parts. After an initial assessment, the components are disassembled gently, to ensure proper preservation. Contrary to assembly, the product's condition at the end of its life cycle changes to an unknown extent, making it difficult to plan an automated disassembly. Considering aircraft engines, due to environmental conditions, e.g., direct contact to hot exhaust gas mixture and high tensile loads during

R. Blümel (✉) · A. Raatz
Institute of Assembly Technology an Robotics (Match), Leibniz University Hannover, An der Universitaet 2, 30823 Garbsen, Germany
e-mail: bluemel@match.uni-hannover.de
URL: https://match.uni-hannover.de

© The Author(s) 2025

J. R. Seume et al. (eds.), *Regeneration of Complex Capital Goods*,
https://doi.org/10.1007/978-3-031-51395-4_5

operation, its components like high-pressure turbine (HPT) blades are exposed to three major influences: fatigue, corrosion and creep (Bräunling 2015). As a result, the HPT blades inside the turbine disks change their condition. That change is characterized in particular by a solidification[1] of the joint, as it essentially hinders and impedes the separation of the joining partner.

Due to the unknown state of the assembly connections, process planning parameters, like tool dimension, needed forces or disassembly time only become apparent during the ongoing blade dismantling process. Depending on operating parameters such as operating hours, landing and take-off (LTO) cycles, or environmental influences, like flight routes over the sea or desert, the component's joints solidify in an undefined manner and might be further damaged without a suitable disassembly strategy. That would destroy the components before the beginning of any maintenance and regeneration processes.

Especially the turbine blade's special manufacturing for highest stresses makes this component very costly with approximately $ 10,000 each (IAE V2500 engine, according to the turbine's manufacturer).

Besides the uncertain product condition, the shape of the joint differs. Depending on the engine and the engine stage, numerous shapes and dimensions exist. The HPT blade roots, for example, are assembled in fir tree slots into the turbine disks, as shown in Fig. 1 (Benad 2019). The disassembly tasks of any engine thus differ in their unknown condition due to their differing use and operation scenario and their geometric properties. Although the blades rest loosely in the fir tree slot during assembly, the solidification of the joint prevents them from being easily pushed or pulled out during disassembly. Therefore, the disassembly is mainly carried out by manual labor (Schmücker et al. 2021). HPT blades, for instance, are hammered out with hammer strokes. By being carried out manually, the disassembly is adaptable to changing product conditions while being protective to highly valuable parts. Nevertheless, damage to components can still occur during manual disassembly.

In this article, we present our research on developing an automated disassembly by overcoming the uncertain product condition as its obstacle. In the section below, we outline the challenges and the approach to implement an adaptable and component-protecting disassembly.

2   Objective

The CRC 871 focuses on the optimization of maintenance tasks of complex capital goods. By developing a novel process chain characterized by automated processes and integration of a virtual layer into the real layer, an efficient and resource-saving process is achieved (Aschenbruck et al. 2014). Our objective in the sub-project A5

---

[1] In the context of this sub-project, "solidification" refers to the effects of the change in the joint, which makes disassembly significantly impeded, i.e., it becomes detachable with increased difficulty and effort.

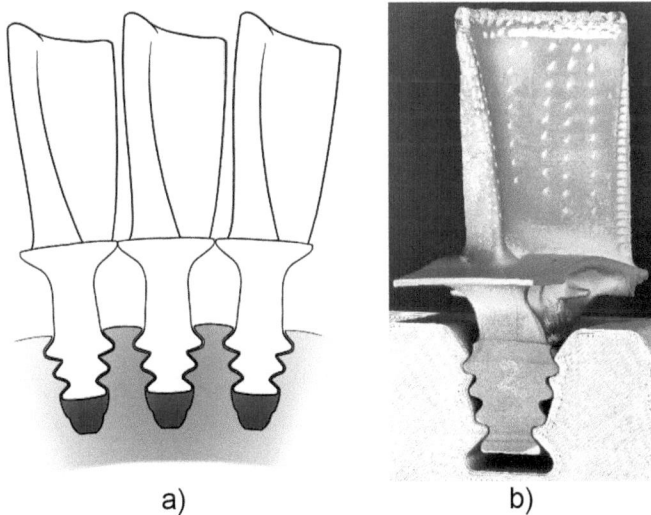

a)   b)

Fig. 1 **a** Sketch of blades inside a turbine disk, **b** Image of a turbine blade with fir tree profile

of the CRC 871 is to establish a technologically plannable disassembly despite characteristic uncertainties. Our approach is to substitute the manual hammer blows with reproducible micro impacts, induced by a piezo stack actuator. The reproducible adaptation of the disassembly process parameters to the operationally varying condition of the joints while adhering to maximum force limits of the blade material allows component-protecting disassembly and its automation. In order to set up the investigation, we developed an experimental environment and a simplified model of a solidified joint. In Sect. 3, the setup of the investigation, including the development of an experimental environment and a simplified model of a solidified joint is illustrated.

For our research on a component-protecting disassembly, we could not use real solidified joints between HPT blades and turbine disks of a used aircraft turbine. Therefore, our first sub-objective is to investigate methods to provide valid and reproducible samples. Thus, a substitute model of the operationally changed and solidified joints provides a basis for further investigation on the disassembly process. Consequently, the second sub-objective focuses on the development of a component-protecting disassembly. For this purpose, we are investigating how the disassembly process causes damage to the components, and how damage to components can be prevented by choosing optimized process parameters. Section 4 outlines the results of the development of an adaptable and component-protecting disassembly.

According to feedback from manufacturers and MRO service providers, no documentation is made of the extent to which the various influences affect the resulting solidification of the joint and disassembly effort. Subsequently, the third sub-objective comprises the summary of the results obtained. In order to adjust the disassembly process parameters adaptively to the joint's condition, we designed and developed a learning model. Based on the varying engine's operational usage scenario, it estimates, for example, required tool dimensions, the disassembly force and process times. With the knowledge of material-specific force limits, components like the HPT blades can be disassembled without damage by adhering to these limits. Hence, by using the learning model we obtain the correlation between the operational data, the resulting solidification and necessary disassembly forces and process parameters. Section 5 presents the development of a learning model to predict disassembly process parameters based on the condition of the solidified joint.

With regard to the automated regeneration process chain, capacity-critical requirements and subsequent processes, such as handling, are also considered. An exemplary partially automated disassembly workstation in a system demonstrator was developed and set up (Sect. 6).

3   Setup

Disassembly is a key step in the remanufacturing process. Regeneration for reuse in regular maintenance intervals or the recovery of valuable resources from end-of-life products becomes possible through disassembly. Airbus' Process for Advanced Management of End of Life of Aircraft (PAMELA) program, for example, showed that up to 85% of aircraft components could be recycled or reused (Airbus 2008). The complex characteristics of disassembly, like the uncertainties and the lack of knowledge about the product's condition, make disassembly planning more challenging than assembly planning (Bentaha et al. 2014) and have been investigated and improved over the last decades. Therefore, disassembly tasks in aircraft MRO are usually executed manually (Schmücker et al. 2021). However, productivity can be improved by overcoming the necessity of manual disassembly through automation.

3.1   Experimental Environment

As introduced, several effects lead into a varying solidification of the joints, resulting in laborious and challenging disassembly. While highly trained and experienced workers perform disassembly, little is documented about the condition of the joints. Intending to carry out disassembly in a reproducible and automated manner, we have developed an experimental environment, as shown in Fig. 2 (Bluemel and Raatz 2021).

It consists of a motor and a ball screw that converts the rotation to a linear motion. The pushing rod exerts the disassembly feed movement on the blade root. The resulting force to disassemble the blade can be measured and monitored with the integrated load cell. Centerpiece of the mechanism is a piezo stack actuator, which can induce and superimpose vibration on the disassembly movement. Researchers showed in simulation and experiments that superimposed ultrasonic vibrations can

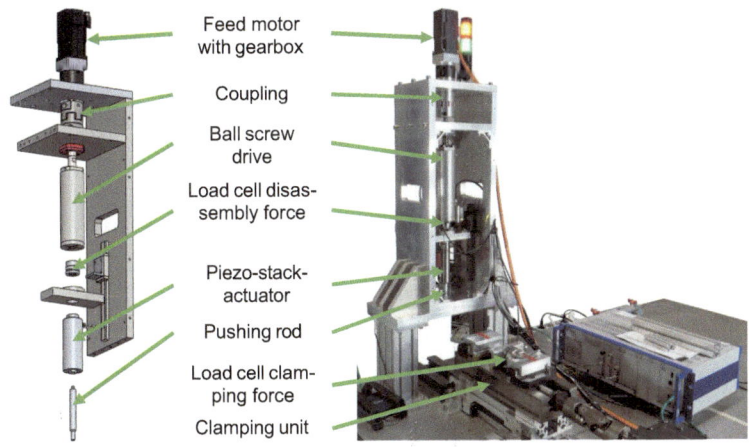

Fig. 2 Experimental environment to perform disassembly tests (Bluemel and Raatz 2021)

significantly reduce the coefficient of friction (Littmann et al. 2001; Popov et al.2009). Saffar and Abdullah (2021), for example, present the application of ultrasonic vibration to reduce process forces during drilling and milling. Experiments in which a bladed turbine disk was placed in an ultrasonic bath first showed a positive result. The blades slid out of the joint after a short time due to their own weight, indicating that the solidification of the joint was significantly reduced. However, further investigations showed a substantial change of the microstructure in the material's surface zone which was severe enough to cause irreversible damage to the components (Lufthansa Technik 2014).

Consequently, vibrations are suitable for lowering the coefficient of friction and thus also reducing necessary disassembly forces. For a component-protecting disassembly process, however, the adjustment parameters must be carefully investigated and adjusted, in order to adhere load limits applicable on the blade root. According to our objectives, we intend to use the piezo stack actuator as a tool to support disassembly. Thereby, we use defined impacts instead of manual hammer strokes by superimposing vibration on the disassembly movement. As we have shown in previous work, it allows the required disassembly forces to be reduced (Wolff et al. 2016). The main characteristics of the piezo stack actuator are the oscillation amplitude, which is varied by adjusting the applied voltage, the period duration from which the frequency can be determined, and the waveform, such as a sinus or triangular wave. Secondary parameters such as the piezo actuator's resonance frequency and the operating temperature determine the maximum set table frequency due to the high heat dissipation. According to the manufacturer, the piezo actuator has a total length of 139 mm and a diameter of 25 mm. The travel range is 120 µm, with a resolution of 3.6 nm. The actuator's resonance frequency is specified at 4.5 kHz. The maximum push force capacity in motion direction is specified at 4,500 N, which represents the maximum force the disassembly experimental environment can apply.

Therefore, the experiments in our study are limited to a maximum force of 4,500 N. In preparation for the experimental investigations carried out in the test environment, we created an analytical model of the solidified joint between the blade root and the turbine disk.

3.2 Simplified Model of a Solidified Joint

Within the research of the CRC 871, our focus is on the disassembly of the HPT blades. The blade root is the part connecting the blade to the turbine disk. Due to its appearance, the profile of the blade foot is referred to as the fir tree profile. Among other blade root profiles, the HPT blade's fir tree profile has the highest bearing capacity appearing during the engine's operation (Li et al. 2019). Varying operation scenarios, as described at the beginning, solidifies the connection, which makes disassembly more complicated.

In order to describe the solidified joint connecting blade root and turbine disk, we use an analytical model in analogy to a friction model (Fig. 3) (Wolff et al. 2015).

We identify the solidification of the joint, caused by wear and tear due to the engine's operation, as a layer formed between the contact surfaces. The solidification of the joint is modelled as a resulting auxiliary material between the blade root and the turbine disk with the resulting surface pressure $ps(z)$ as the central operation-dependent solidifying characteristic and a material and surface dependent coefficient of friction μs. It results in a solidifying force $Fs(z)$ opposing the disassembly movement. Therefore, the disassembly force FD (z) must be higher than the resulting $Fs(z)$, so that the blade root can be pressed out. However, the disassembly force must not exceed a limit, in order to prevent damaging the blade. The contact surface $AC(z)$ is known from CAD data and decreases during the disassembly process along the disassembly path z. It becomes equal to zero when the actual path reaches the total disassembly length $lD(z)$ (Bluemel and Raatz 2021). The solidifying force can be described as follows:

Fig. 3 Simplified model of the solidified joint between blade root and turbine disk

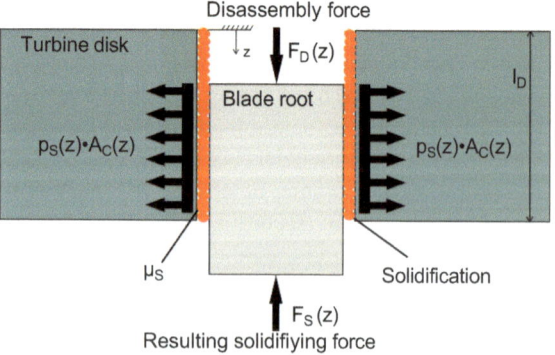

$$F_s(z) = \mu_s \cdot p_s(z) \cdot A_C(z) \qquad (1)$$

In our investigation, the causes and the manner of solidification are not to be characterized or modelled. For the development of the model and our investigation, the degree of solidification, the resulting disassembly force to be applied and disassembly force curves are decisive. In later phases of the investigation, a functional relationship between the degree of solidification, the operational usage scenarios, and the geometry will be developed. However, due to the lack of real engines with individual operational usage scenarios, we used a replacement model for a solidified joint between blade root and turbine disk. In the next section, we present the development of this replacement model.

3.3 Replication of Solidified Joints

In order to investigate the solidified blade-disk joint, we created the simplified solidification model (Sect. 3.2). That model can be used to extract properties that affect the solidification of the assembly joint. Different usage scenarios can be imitated by varying these properties, constituting the resulting solidifying force. Therefore, for the developed replacement model, it is necessary to allow reproducible adjustability of these properties. As in Eq. 1, the solidifying force Fs is influenced by the coefficient of friction μs, the resulting contact pressure as the solidifying characteristic ps and the contact surface AC. For the replacement model, we do not manipulate the contact surface, so AC is omitted, leaving μs and ps as adjusting parameters.

We had samples manufactured with re-designed blade root shapes based on the original HPT blade root to replicate solidified joints, as seen in Fig. 4. The samples consist of an inner part representing the blade root and the outer part representing the turbine disk with the blade's negative contour. The samples were manufactured by wire cutting to achieve high surface accuracy and fitting precision. In contrast to the clearance fit of the real joint, we use a slightly tighter fit for uniform contact.

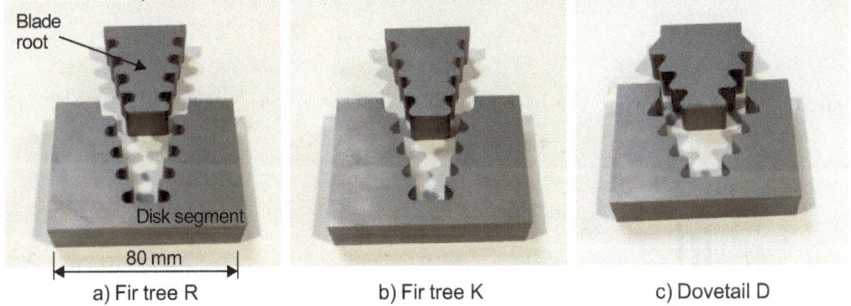

a) Fir tree R   b) Fir tree K   c) Dovetail D

Fig. 4 Samples with the re-designed shape of blade roots: **a** fir tree "R", **b** fir tree "K", **c** dovetail "D"

In order to allow replicated variation of the parameters influencing the solidifica-tion, we considered several methods: We investigated adding auxiliary materials to the gap, artificial ageing, according to a technical standard for accelerated ageing, and exerting an external force, as presented below.

The first method we were investigating was the use of auxiliary materials, i.e., adding glue to the samples gap. We applied glue (Loctite 222) on the contact surface to prepare the samples. Using a clamping device, we applied a force for a defined period of time at a constant ambient temperature of 25 °C, under which the adhesive hardened. With that, we achieved a uniform layer thickness of the adhesive. To carry out the experiments, the inner part representing the turbine blade was pushed out while measuring the required disassembly force. By varying the curing time and clamping force, we intended to substitute the deviating solidification of the joint. For the investigations, we considered two different shapes of the blade root, fir tree "K" and fir tree "R" (Fig. 4).

Figure 5 shows an extract of the results of the subsequent regression analysis to investigate the influence of the input factors on the disassembly force and, eventually, verify the method. The model obtained by the regression is significant, but with a coefficient of determination, representing the goodness of fit, of R^2 equal to 0.738, it has comparatively little meaningfulness. With the calculated standardized regression coefficients, we can compare the input variables despite different units of measure-ment (Siegel and Wagner 2022). As a result, a variation in the shape of the blade root and the curing time have a negligible impact. Since a variation in the shape also means a change in the contact surface, a dependency should also be recognizable in accordance with our friction model.

This method, therefore, does not allow reproducible replication of the joint's solidification. We assume that this is caused by the manual application of the adhesive, the varying surface textures, and environmental conditions.

Along with using auxiliary material to replicate the solidified blade root—turbine disk joint, we investigate the method of artificially ageing. We used a salt mist environment to artificially age the manufactured samples according to a standardized procedure. To ensure rapid corrosion, the samples were made of mild steel. We set the salt mist chamber to 37 °C with a 5% saline solution. We then put samples into the salt mist chamber for 72 h and 168 h, respectively. Similar to the previous

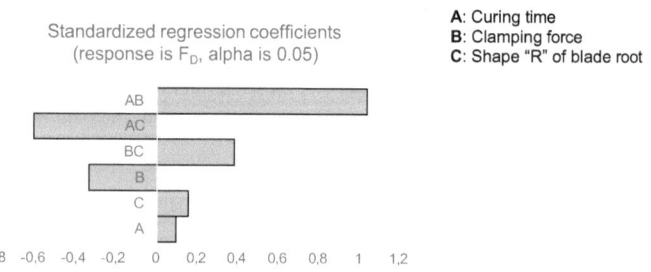

Fig. 5 Pareto chart of parameter's standardized regression coefficients for auxiliary material

experiments, the samples' inner blade root model was pushed out. In addition to the shapes fir tree "K" and "R", we considered the third shape, dovetail "D" of the blade root, for the experiments using artificial ageing (Fig. 4).

Due to the formation of corrosion, an increase in disassembly forces was observed. The measured disassembly force for the samples at 72 h was between 2,500 and 3,500 N. At 168 h, the disassembly force considerably increased over 5,000 N and thus beyond the load limit of the actuator. Figure 6 shows an extract of the results of the subsequent conducted regression analysis. With a coefficient of determination R^2 equals 0.778, the model shows a dissatisfactory fit. However, the standardized regression coefficients for different shapes of the blade root show a greater impact than using adhesives to substitute the joint's solidification.

As a result, the artificial ageing showed restricted applicability to substitute the operational solidified joint. An increase in the required disassembly force was measured, and the force reduction by the piezo stack actuator was observed. However, the limiting factor of the artificial ageing method was the one-time examination of each sample. When disassembled, the sample became scrap and could not be re-prepared. Because of the lack of reproducibility and the costly preparation of the samples, we did not consider this method for further investigation.

Using auxiliary material and the artificial ageing showed unsatisfactory results, lack of reproducibility and reusability. Following the friction model, these methods did show an increase in the disassembly force by changing the coefficient of friction and the contact pressure. However, it is barely possible to adjust a determinable change. Therefore, we developed a clamping unit that exerts an external clamping force *FCl* on the blade root segment, inducing a surface pressure in the gap, as seen in Fig. 7. Instead of synthetically creating the solidified joint by manually adding glue or corrosion, we can repeat the experiments infinite times and can set a certain contact pressure by adjusting the clamping force, which represents the solidification of the connection (Bluemel and Raatz 2021). The clamping unit holds the samples fed into the disassembly experimental environment. It consists of a servo motor with an integrated gearbox connected to a machine vice. Both are placed on a linear guidance to position the sample holder under the disassembly pushing rod manually (see Fig. 2). The sample mount is placed between the vice's jaws and includes a load

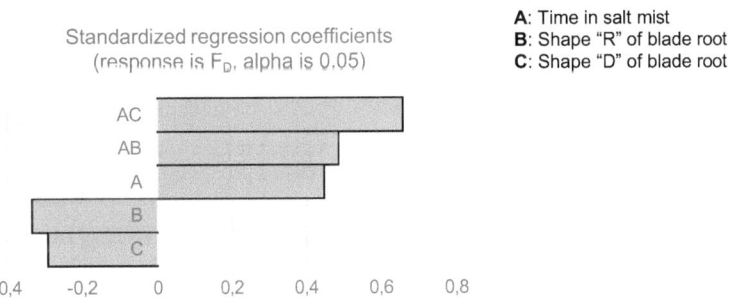

Fig. 6 Pareto chart of parameter's standardized regression coefficients for artificially ageing

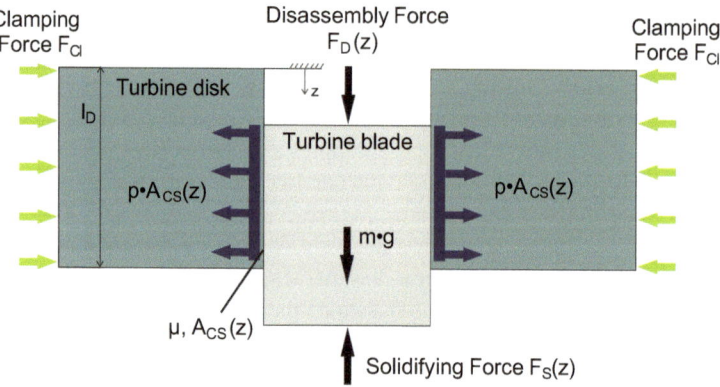

Fig. 7 Adapted model of a solidified joint using a clamping force

cell to measure the normal force exerted on the samples. The samples are placed and clamped with repeatable accuracy using a locking pin and an adjusting screw. By using a load cell to measure the normal force exerted on the samples, we can describe and reproducibly adjust the substituted solidification of the joint, allowing us to reproduce the solidified blade root—turbine disk connection.

We presented in previous work that by varying and adjusting the clamping force, we can mathematically and statistically describe the induced contact pressure representing the solidification of the joint (Bluemel and Raatz 2021). The regression analysis revealed a good model fit with a coefficient of determination R^2 equals 0.981. To obtain results in the later process of our investigations that are more closely related to the IAE V2500 aircraft engine's HPT blade root, used as an application case, we have designed new blade root geometries. Similarly, as in the previous experiments, we designed slightly modified contours to vary the blade root's shape and the contact surface, as seen in Fig. 8 According to the design, we named the original shape "Original", the sample with twice the size "Double", and a contour having the same contact surface but a different contour as the original blade root contour "B-shape" (Bluemel and Raatz 2021). The new samples were made of stainless steel to prevent the occurrence of incidental corrosion. Also, the new samples have a cutting and thus consist of three parts. Similar to Fig. 4, the inner part represents the blade root. On the contrary, the outer part, representing the turbine disk, is separated by the cutting. Below, we present the exemplary results using the original HPT blade root's shape.

The subsequent regression analysis showed the influence of the input parameters and their interactions on the disassembly force. Figure 9 shows an extract of the results. The disassembly force shows the greatest influence. The "B-shape" contour has a negligible impact (0.031) since it has the same surface as the "Original" contour. Whereas the "Double" contour has a slightly larger influence (0.095). In comparison, the clamping force has the major (1.130) influence, as expected.

Consequently, we have defined a reproducible method to replicate the joint's solidification. We can perform disassembly tests with varying joint conditions by

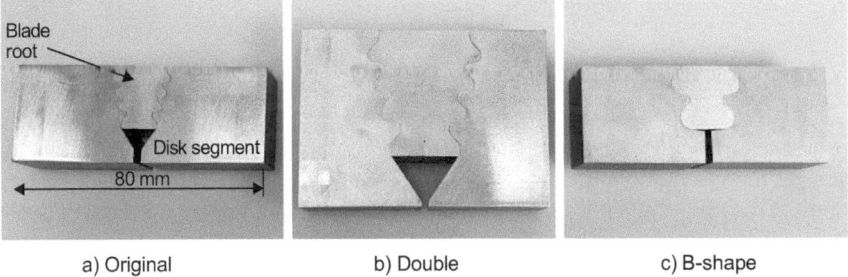

a) Original   b) Double   c) B-shape

Fig. 8 Overview of the different blade root samples similar to Fig. 4 with an additional cutting (Bluemel and Raatz 2021)

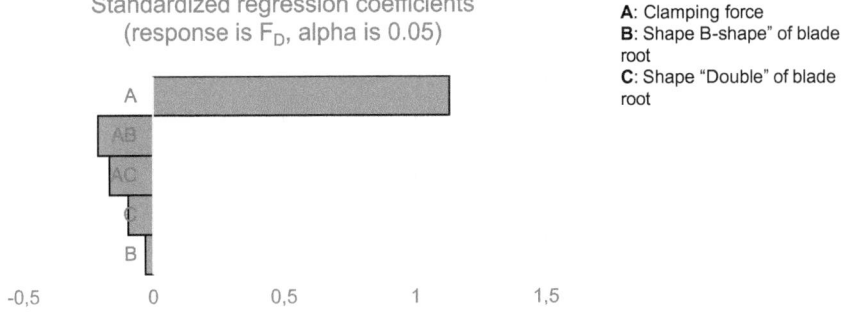

Fig. 9 Pareto chart of parameter's standardized regression coefficients for clamping the samples

replacing the a priori data with the clamping force, substituting the engine's individual operational scenarios. The induced contact pressure generates the force that opposes the disassembly force. With this replacement model, we have achieved the first key objective by implementing the clamping force to validly and reproducibly model and substitute the joint's solidification.

Thus, as in the simplified model in Sect. 3.2, the disassembly force must be above the resulting solidification force to allow the blade root to be disassembled. Likewise, a material-specific force limit must also be adhered in order to prevent damage to the blade root and thus ensure a component protecting disassembly. The following section presents the development and results of a component-friendly disassembly process.

4   Component-Protecting Disassembly

The second key sub-objective focuses on the development of a component-protecting disassembly. As we have shown, the applied disassembly force is an essential factor when detaching the blade root. To separate the joint between blade root and turbine disk, a minimum force must be applied while at the same time adhering to a force limit. That allows the blades to be disassembled without damaging them, ensuring component protection. The following section presents our research on determining the material-specific force limit for HPT blades used in our experiments.

4.1   Determination of Force Limits

The goal of a component-protecting disassembly is to detach the operational solidified joint between blade root and turbine disk without damaging components. Therefore, we determined the force limits that can be exerted on the blade root. The manufacturer's specifications in the engine manuals did not consider the application of a load as we plan to exert on the blade root.

In order to determine the force limit, we prepared already disassembled HPT blades which the manufacturer provided to us. We cleaned, labeled, and photographed the blade root to capture its condition before the investigation. Using a universal testing machine for tensile and compression tests, we performed static load tests. For this, we exerted a defined static force in 10 kN increments on the blade root using a tungsten carbide ram with a diameter of 12 mm and increased it successively until a significant plastic deformation was visible.

Up to a force of 20 kN, traces of the ram were observable by visual inspection. From 30 kN on, more visible imprints were observed until the imprint became clearly visible above an applied force of 60 kN. At higher forces, more and more distinct indentations were formed, which showed a palpable plastic material deformation. Above 70 kN, a significant deformation and thus damage to the blade material resulted.

After the visual inspection, we examined the blade roots using a confocal laser microscope for a more precise inspection. Figure 10 shows excerpts of the results of the tests for a static load of 20 kN and 70 kN: Up to a force of 50 kN, we could not observe any deformation within the measuring range. At forces up to 60 kN, the slightest deformation could be seen, besides indications of leveling of the surface's roughnesses below 10 μm. From a force of 70 kN and above, a definite imprint was measurable.

The investigation into the component protection results can be concluded: For a component-protecting disassembly, the material of the blade root must not be affected in any way. The maximum force can be determined using the presented destructive test method. With that, damage during disassembly can be prevented. In the case of the HPT blades used in our research, the maximum force is less than 20 kN at

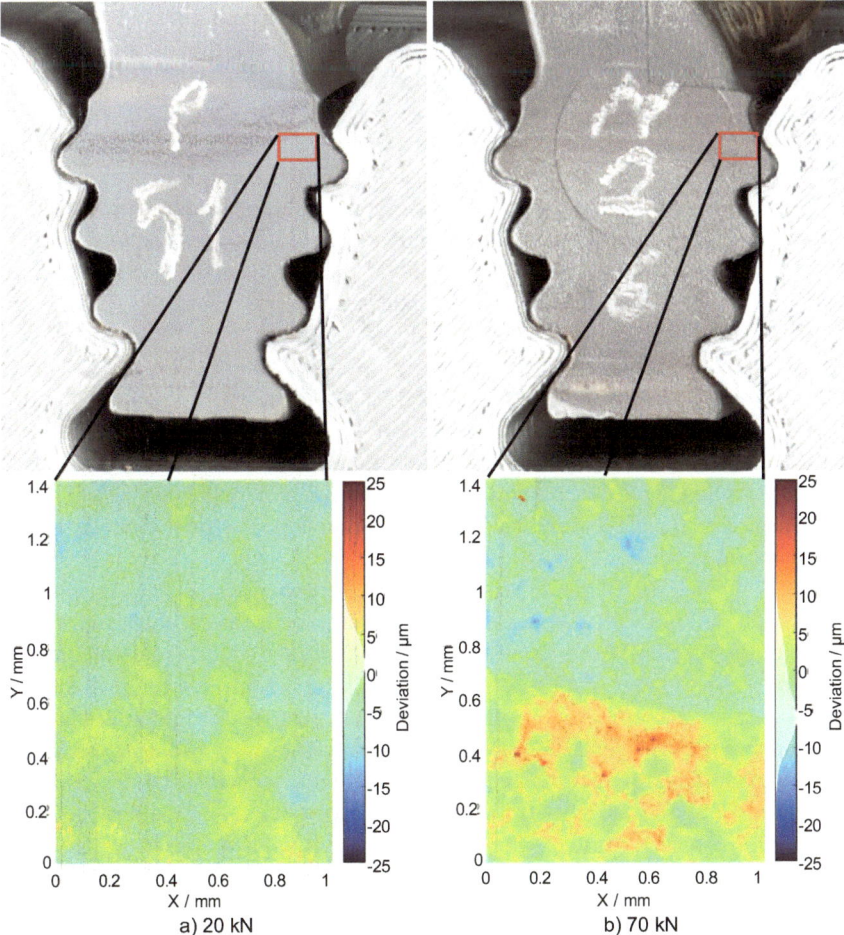

Fig. 10 Examination of the surface using a confocal laser scanning microscope (rotated by 90°) after a force exertion of 20 kN (**a**) and 70 kN (**b**) (Middendorf et al. 2022)

12 mm ram diameter or a static load of 44.2 MPa, since no damage was observed. However, an accurate assessment of the surface and component defects requires the manufacturer's agreement. We have published further in-depth findings in our study (Middendorf et al. 2022).

4.2 Reduction of the Disassembly Force

With the knowledge of the maximum applicable load, the subsequent task to develop the component-protecting disassembly focused on reducing the disassembly force.

The Response Surface Method (RSM) allows identifying and describing the influential disassembly process parameters affecting the force to dismantle the joint between the blade root and turbine disk. The experiments were based on investigations to substitute the operationally solidified joint.

The initial step of the RSM was the specification of the input and output variables (Witek-Krowiak et al. 2014). The disassembly force is the decisive factor for a component-protecting disassembly. Therefore, we determine it as the investigation's output parameter. The selection of the influential parameters is shown in Fig. 11. Besides the disassembly force, we identified the disassembly speed as having a major influence, as it affects the capacity planning. With slow disassembly speed and, therefore, long times, not only the following regeneration tasks are delayed. It also delays following re-assembly tasks, leading to long slack times. We also include the vibration-determining adjustment variables of the piezo stack actuator-amplitude, frequency, waveform. The mentioned a priori data, such as operating hours, and LTO cycles, are replaced by the described clamping force, which induces the contact pressure replicating the joint's solidification state. In order to further investigate the dependence on the blade root's shape and to decouple geometric properties, we take the shape into consideration as well.

The following step of the RSM is the design of an experimental plan. The design of experiments allows with a limited number of experimental runs adequate informative value of the results (Witek-Krowiak et al. 2014). Using a face-centered composite design of experiments (CCF), we executed 72 experiment runs for each shape of the blade root. Exemplarily, we present the results for the "Original" form, which corresponds to the original HPT blade root. The following regression analysis showed a good model fit with a coefficient of determination R^2 equals 0.961.

Figure 12 shows the standardized regression coefficients of the analysis. As in the previous studies, the clamping force as a substitute solidification model has the

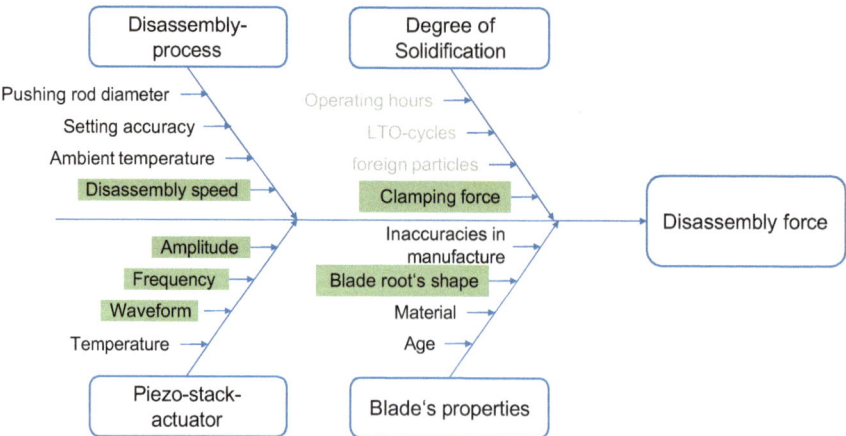

Fig. 11 Ishikawa-diagram for the influences on the disassembly force

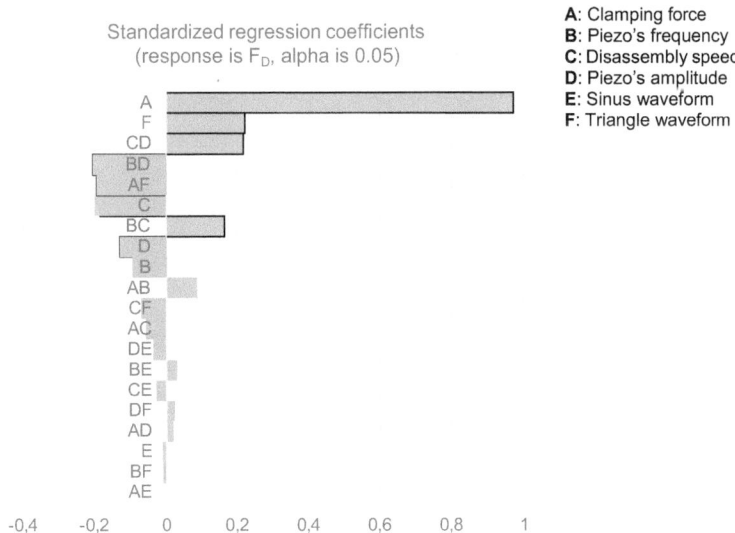

Fig. 12 Pareto chart of parameter's standardized regression coefficients for RSM analysis

weightiest influence. The influence of the piezo's amplitude and disassembly speed is similar in comparison. However, the influence of the frequency is noticeably lower. The comparison of the waveforms reveals the triangle waveform to have a greater influence than the sinus waveform.

In order to evaluate which influencing variables reduce the maximum disassembly force, the signs must be observed. The results showed that increasing the piezo's frequency (-0.093), amplitude (-0.129) and their interaction (-0.203) decreased the disassembly force, as expected. Although an increase of the disassembly speed (-0.183) decreased the disassembly force, its interaction with, for example, the piezo's frequency (0.165) and amplitude (0.218) had a negative effect on the reduction of the disassembly force. The interaction between the disassembly speed and the waveform has a negligible influence (-0.029 for sinus and -0.070 for triangle). When comparing the waveform, the sinus (-0.012) decreases, and the triangle (0.222) increases the maximum disassembly force compared to the sawtooth waveform. However, both the sawtooth and triangular wave-forms have an uncomfortable acoustic effect. More complex relationships existed with the interactions of the clamping force. The experiments and the statistical evaluation confirmed the physical assumption of the clamping force having the most considerable influence, which is supported by the standardized regression coefficient of 0.975.

Since the aim of the experimental investigation is the minimalization of the disassembly force, we calculated its maximum reduction in the next step. According to the results of the RSM, we obtained optimized setting parameters, and we were able to calculate the maximum reduction of the disassembly force that can be realized. The RSM analysis showed that the amplitude and frequency must be maximum

Table 1 Optimized parameters for maximum reduction of the disassembly force (Blümel et al. 2023b)

Fixed parameters	
Clamping force (F_{Cl}) in N	4,000
Amplitude (A_{Pi}) in μm	100
Frequency (f_{Pi}) in Hz	60
Varied parameters	
Disassembly speed (vD) in mm/s	1; 5.5; 10
Waveform (WF_{Pi})	Sinus and triangle

and the disassembly speed minimum to reduce the disassembly force. In order to show the disassembly force's complex dependencies and interactions on the disassembly speed, we want to present the results for the levels chosen in our study. Also, we show the dependency on varying the vibration's waveform using the sinus and triangle vibration. Table 1 shows the parameters we set up for this investigation.

To investigate the maximum reduction of the disassembly force, we randomized and repeated a series of experiments, 45 runs total. Table 2 shows the mean values of the results. The tests were performed at a clamping force of 4,000N. Therefore, the calculated values are only valid for this clamping force. At other clamping forces, the values for the reduction also change and can be more or less.

As seen in Table 2, the maximum reduction of the disassembly force was achieved using a disassembly speed of 1 mm/s and sinus waveform for the superimposed vibration. By increasing the disassembly speed, the achievable reduction of the disassembly force decreases when superimposing triangle vibration. When superimposing sinus vibration, the reduction of the force is more complex. Here, the interactions of the individual inputs became noticeable. Thus, in this validation test, the required maximum force was reduced by 20.4% with a sinus wave and 16.5% with a triangular wave compared to disassembly without using the piezo actuator. As seen in the varying reduction for increased speeds, the interactions of the influencing variables have a non-negligible effect on the reduction.

In addition, the disassembly force curve over disassembly time for vD of 1 mm/s was exemplary, plotted in Fig. 13. The test rig's pushing rod was moved to contact with the sample. After this, the disassembly force stagnates briefly and then decreases continuously over time.

Occasionally when running the tests, some blade root samples tilted slightly in the disk segment just before the end of the disassembly process. That resulted in a slight increase in the disassembly force until the samples fell out of the disk segment.

Table 2 Result of the maximum reduction of the disassembly force (Blümel et al. 2023b)

	$v_D = 1$ mm/s	$v_D = 5.5$ mm/s	$v_D = 10$ mm/s
F_D without vibration	2,157 N	2,139 N	2,144 N
F_D with sinus vibration	1,717 N (-20.4%)	1,950 N (-8.8%)	1,931 N (-9.9%)
F_D with triangle vibration	1,802 N (-16.5%)	1,910 N (-10.7%)	2,022 N (-5.7%)

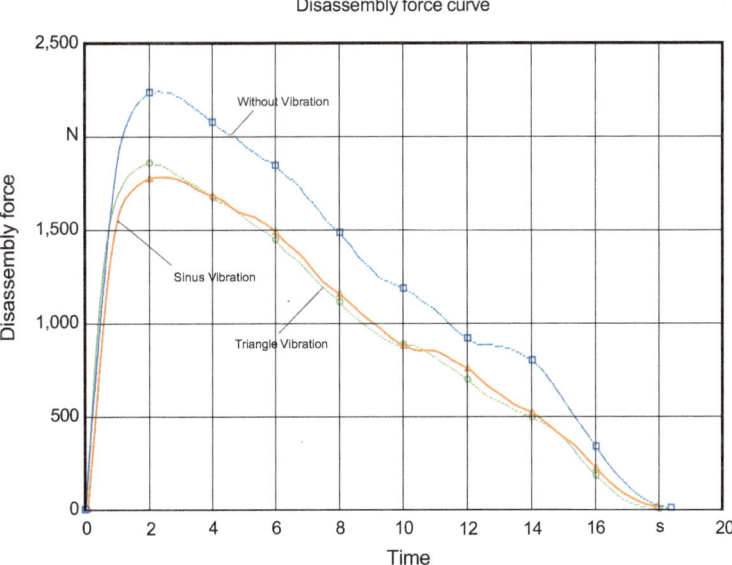

Fig. 13 Maximum reduction of the disassembly force for varying waveforms with vD of 1 mm/s

Overall, the graph also shows that the use of a piezo actuator reduces the disassembly force. According to our second sub-objective, we demonstrated a method for component-protecting disassembly by reducing the maximum required disassembly force. As long as material-specific force limits are adhered, the disassembly speed can be increased to reduce cycle times, despite increasing the disassembly force. Thereby still component protection is achieved while optimizing cycle times.

Based on the results of the RSM on the reduction of the disassembly force, we developed a learning model. Using the condition of the joint, affected and altered by the engine's operation, it estimates disassembly forces and times, as presented in the section below. Thus, the optimized parameters are calculated individually for each occurring joint's condition.

5 Learning Model to Predict Blade Specific Process Parameters

As we have shown, optimized process parameters for reducing the maximum disassembly force and thus component-protecting disassembly can be calculated as a function of the clamping force. The clamping force is a replacement model of the real solidification between the blade root and turbine disk caused by the engine's operation. Therefore, our third sub-objective is the interconnection between the operational characteristics and the resulting optimized process parameters to dismantle the joints. However, since the joint's condition varies depending on the individual engine's operations, an estimation of the respective a priori data is necessary. Therefore, the aim is to use that data to estimate the necessary disassembly force and its resulting reduction by optimized selection of process parameters. In order to achieve the aim, we have developed a learning model, as seen in Fig. 14. The a priori data, substituted by the clamping force as the replacement model for solidified joints, serves as input. Based on the data, the learning model predicts and determines the optimized process parameters, adaptive to the joint's condition. In a later industrial application, the replacement model can be re-replaced with the real operational data (a priori data), which will then be correlated with documented disassembly forces of real aircraft engines' disassembly tasks.

The basis for the development of the learning model are the results of the previously conducted research and RSM analysis, as we presented in (Blümel et al. 2023b). We trained a learning model that, similarly to the experimental investigation described in Sect. 4.2, uses a multiple linear regression algorithm with the results of the RSM analysis. In addition, we performed further disassembly runs with randomized values for the input values to obtain a test subset for the trained model. That gives us an approx. 75 to 25% split between the training and test data set. In order to evaluate the trained learning model, we calculated the coefficient of determination R^2 and the symmetric mean absolute percentage error (sMAPE), being a coefficient to evaluate machine learning and regression studies (Chicco et al. 2021). An R^2 equals 0.9248, and a sMAPE equals 8.594% indicate good predictive performance. Using the same data set, we have also developed a feed forward neuronal network with two hidden layers. With a calculated R^2 equals 0.9251 and a sMAPE

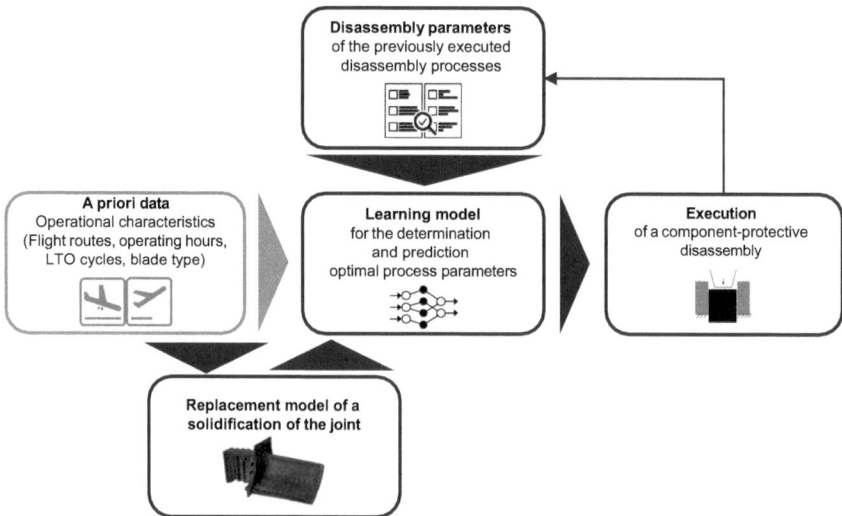

Fig. 14 Disassembly process using a learning model to predict process parameters, based on operational data (a priori data), substituted by a replacement model

equals 8.635%, it indicates a slightly better fit. However, the neural network's root mean square error (RSME) being higher than the regression's indicate a more inaccurate result. We have concentrated on the regression being more descriptive. Yet, a more extensive database might improve the neural network's results. The design and operation of the learning model is shown in Fig. 14. When the maintenance task is planned, the disassembly planning is provided with the operational characteristics of the engine (a priori data), in our case, substituted by the clamping force of the replacement model. Using the IAE V2500 high-pressure turbine's first stage as an example, the material-specific maximum applicable disassembly force is known and also provided to the learning model as a priori data. The disassembly parameters of the previously executed disassembly tasks, which were used to train the learning model's algorithm, represent the database. With the trained algorithm, the learning model calculates the optimized parameters to reduce the disassembly force. The disassembly speed determines the time of each dismantling process and is the key factor affecting the disassembly process and following tasks planning. The learning model attempts to keep the disassembly speed at maximum, while adhering to the force limit to achieve shortest process times. By optimized selection of the piezo stack actuator's parameters, amplitude, frequency, and waveform, the disassembly force is minimized. After each successful disassembly task, the learning model calculates the difference between the measured and material-specific maximum disassembly force. If the difference is in between a specified safety margin, the disassembly speed will be increased while adapting the vibration's parameters to continue minimizing the disassembly force for the following disassembly runs (Blümel et al. 2023b).

To illustrate the functionality of the learning model, we present an exemplary application on the "Double" HPT blade root contour in a disassembly environment. Figure 15 shows ten individual disassembly executions, for example, ten blades that are disassembled one after another. At the beginning, the disassembly is executed at a previous defined low speed, which is gradually increased by the learning model in following disassembly executions, as illustrated with the black squares. It can be seen that the green line as the material-specified maximum disassembly force, is not exceeded by the measured maximum disassembly force, indicated by the green bars. By adjusting the vibration's parameters of the piezo stack actuator, even the disassembly speed can be increased. That allows increasing the speed to a previously defined maximum (6 mm/s in the shown example). All subsequent blades can then be disassembled with these parameters.

In addition, the learning model can recalculate the degree of solidification, in this example, the clamping force after each successful disassembly run. The blue bars show the measured clamping force, whereas the blue stars show the recalculated clamping force, replacing the joint's solidification. The comparison of the measured and recalculated solidification allows traceability and an evaluation of the learning model's performance. The learning model thus has information about the degree of solidification, which the model can calculate. From the variation of the measured clamping force, the learning model's adaptability to varying joint conditions is recognized. The parameters of each disassembly run are stored in the learning model's database and can be used to improve its performance (Wolff et al.2019).

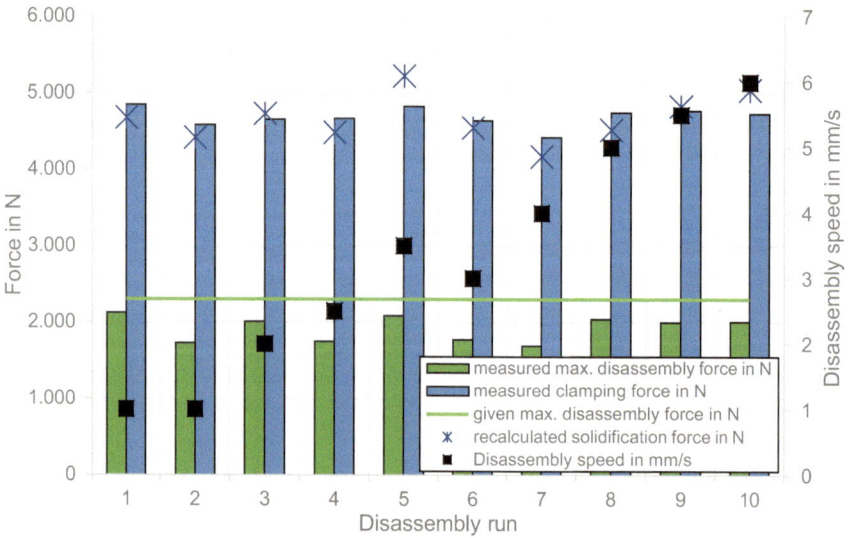

Fig. 15 Diagram of an exemplary disassembly process including ten runs

Based on the operational data of the engine, the learning model can, therefore, estimate the disassembly force and calculate the optimized process parameters. If an engine is to be disassembled with an identical operational usage scenario, the setting parameters learned can be reused. However, for a disassembly task of an engine with a different usage scenario, the learning model will estimate the disassembly force and calculate optimized process parameters using an initially low speed, as described previously. Thus, if the flight is unknown to the learning model, within a short approximation interval of, in our case, ten out of 64 blades, the learning model adapts the disassembly speed as well as the vibration's parameters amplitude, frequency and waveform, to ensure optimal disassembly times.

For the adaptability of the learning model to different engine types and, therefore, blade root contours, we investigated the transferability of geometric dependencies. In extensive FEM simulation studies, we investigated a parameter with geometric information (geometric parameter, Wolff et al. 2018). Figure 16 shows the FEA for two different contours. Using the coefficient of friction and an offset, i.e., an interference, of the contacting surfaces, we substituted the solidification of the joint. The geometric parameter was identified by varying the simulated solidification of the joint and analyzing geometric dependencies. That geometric parameter allows the decoupling of the joint's solidification from geometric properties as well as estimating the disassembly force of a yet unknown contour based on an already disassembled contour. The parameter, independent of the condition of the solidified joint, indicates how much the required disassembly force differs from a contour "A" to a contour "B". To illustrate, Eq. 2 shows an example regression equation for estimating the disassembly force with x_1 and x_2 being arbitrary inputs. The standardized coefficients

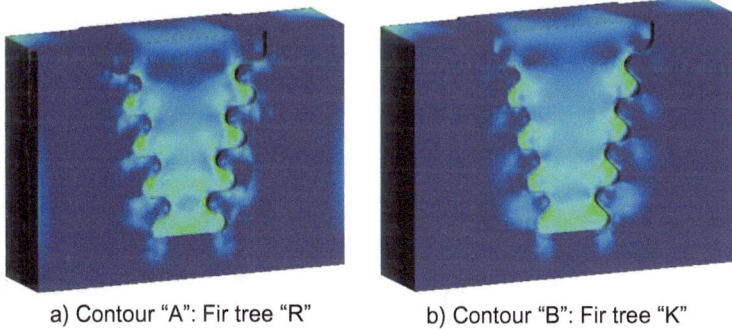

a) Contour "A": Fir tree "R"   b) Contour "B": Fir tree "K"

Fig. 16 Exemplary result of the FEM simulation for two different contours

β_i are the estimators to indicate each input's weight. The blade root is integrated as a binary variable: When disassembling contour "A", d_A will become equal to one, and d_B equal to zero. That leads to omitting $\beta_B \cdot d_B$ and adding β_A to β_0. Contour "B" is vice versa (Bluemel and Raatz 2021).

$$F_D = \beta_0 + \beta_1 \cdot x_1 + \beta_2 \cdot x_2 + \cdots + \beta_A \cdot d_A + \beta_B \cdot d_B \qquad (2)$$

Aided by the FEA, depending on geometric properties, we can estimate the contour-specific regression coefficient β_J of yet not disassembled contours. Thus, the coefficient β_J is decoupled from and independent of operational data. The estimate in advance increases process reliability, as the learning model is not required to calculate the yet not disassembled contours regression coefficient β_J through progressive learning.

For a vivid presentation of the developed adaptable and component-protecting disassembly, we engineered and built a disassembly workstation, as shown in the next section.

6 Development of an Automated Disassembly Workstation

We engineered and built a disassembly workstation integrated into the CRC 871's system demonstrator to summarize the results. In addition to the disassembly environment, the workstation includes a robot-assisted handling system (Fig. 17a).

The disassembly workstation consists of a section of a turbine disk in which a blade can be prepared as an example. The disassembly environment is identical in structure and function to the test environment. However, a robot gripper supports the disassembly process: During the disassembly, the blade is held securely so that it does not drop when it is completely detached. The blade held by the gripper is then transferred to the handover station in the workpiece carrier developed in the CRC 871 and is available for the subsequent steps of the regeneration process.

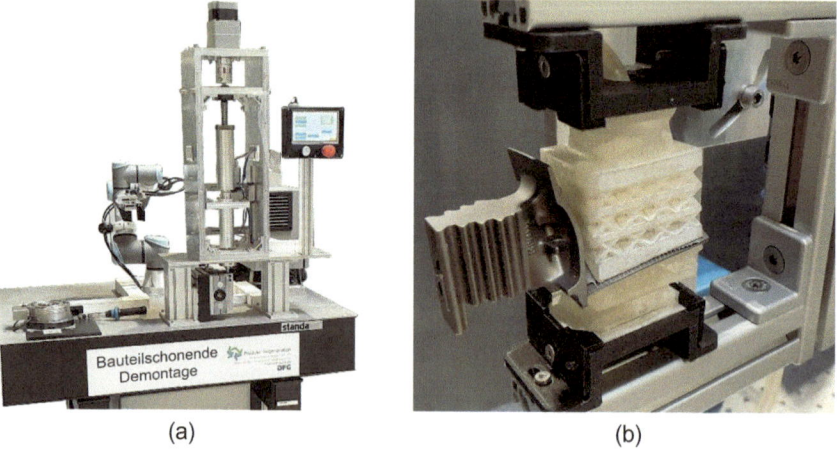

(a)   (b)

Fig. 17 **a** Automated disassembly workstation, **b** Component-protecting gripper based on soft robotics (Blümel et al. 2023a)

In order to ensure component protection also during handling tasks, the robot gripper is developed from soft robotics, as shown in Fig. 17b (Blümel et al. 2023a). The soft surface of the gripper jaws allow a flat and adaptable contact with the complex shape of the blade's surface.

An actuator inspired by an origami structure controls the opening and closing of the gripper. In addition, the gripping force can be adaptively adjusted. Compared to a rigid gripper, this can compensate for slight tilting that can occur when holding the blade during the disassembly process. The flexibility also ensures that the blade is held evenly and can be held more securely overall. The soft gripper can also insert the blade into the workpiece carrier. However, current work and research is still focused on improving the service life by investigating and selecting suitable gripper materials and optimized gripper design.

7 Conclusions

The presented research on component-protecting and adaptable disassembly demonstrates possibilities to integrate the disassembly characterized by uncertainties into the regeneration chain. The varying product histories and usage scenarios lead to significant differences in disassembly processes. Therefore, disassembly is usually carried out by manual labor. Considering aircraft engine high pressure turbines, we established a technologically plannable disassembly despite characteristic uncertainties. Since the engine type and scenario data are known a priori, our objective was to interlink this a priori data with disassembly planning parameters.

Due to the lack of operational altered and solidified joints between blade root and high-pressure turbine disks, we developed a reproducible method to replicate the joint's solidification. A clamping force was used to replace operational solidified joints between blade root and turbine disk. Thus, we were able to perform experiments on component protecting disassembly with varying joint conditions.

For component-protecting disassembly, i.e., to prevent damage on components, it is necessary to keep the required process forces to a minimum. As known from the literature, vibrations, depending on their parameters, reduce the coefficient of friction of objects in contact. We were able to successfully demonstrate transferring that effect to the disassembly of operationally solidified joints. By using a piezo stack actuator, which superimposes vibrations on the disassembly movement, a reduction of the maximum force required to disassemble the joining partners was achieved. Reducing the disassembly force ensures that material-specific force limits are adhered, which prevents damage. Since the joint's condition varies depending on the individual usage scenario, we developed a learning model to adapt process parameters to varying a priori data.

The learning model was trained using experimental data and integrated into an automated disassembly environment. Using a regression model, the learning model estimates the disassembly force based on the joint's condition, known from the geometric data and engine's operational characteristics. With that data, the learning model chooses optimized parameter settings to adhere to material-specific force limitations. That allows adaptive adjustment to the uncertain and varying product conditions. Ultimately, the learning model predicts needed process times for each disassembly execution, enabling process planning to be carried out before initial disassembly steps.

Eventually, we illustrated the application and suitability of the method on a system demonstrator. With this system demonstrator, the transferability of the developed learning model to an exemplary automated disassembly workstation including subsequent handling processes, integrated into the process chain developed in CRC 871, could be successfully demonstrated.

Acknowledgements Funded by the Deutsche Forschungsgemeinschaft (DFG, German Research Foundation) SFB 871/3 119193472.

References

Airbus, SAS (2008). Process for advanced management of end-of-life of aircraft (PAMELA). France: Airbus academy.

Aschenbruck, J., Adamczuk, R., and Seume, J. R. (2014). Recent Progress in Turbine Blade and Compressor Blisk Regeneration Procedia CIRP 22256-262. https://doi.org/10.1016/j.pro cir.2014.07.016

Benad, J. (2019). Numerical methods for the simulation of deformations and stresses in turbine blade fir-tree connections. Facta Universitatis, Series: Mechanical Engineering, 17(1):1.

Bentaha, M. L., Battaïa, O., Dolgui, A., and Hu, S. J. (2014). Dealing with uncertainty in disassembly line design. CIRP Annals, 63(1):21–24

Bluemel, R. and Raatz, A. (2021). Experimental validation of a solidification model for automated disassembly. In Herberger, D.; Hübner, M., editor, Proceedings of the Conference on Production Systems and Logistics: CPSL 2021, pages 339–348. Hannover: Institutionelles Repositorium der Leibniz Universität Hannover.

Blümel, R., Morales, D. S. G., Raatz, A. (2023a). Development of a Gripper for component-friendly Handling of Complex Capital Goods. IEEE International Conference on Soft Robotics (RoboSoft), Singapore

Blümel, R., Zander, N., Blankemeyer, S., and Raatz, A. (2023b). Prediction of Disassembly Parameters for Process Planning Based on Machine Learning, Production at the Leading Edge of Technology: Proceedings of the 12th Congress of the German Academic Association for Production Technology (WGP), Springer International Publishing p. 613–622

Bräunling, W. J. (2015). Flugzeugtriebwerke. Springer Berlin Heidelberg.

Chicco, D., Warrens, M.J., and Jurman, G. (2021) The coefficient of determination R- squared is more informative than SMAPE, MAE, MAPE, MSE and RMSE in regression analysis evaluation. PeerJ. Computer science, 7, e623

Li, J., Chen, F., and He, S. (2019). Computer aided evaluation of fir tree blade root profile based on particle swarm algorithm. Journal of Physics: Conference Series,1237(2):022146.

Littmann, W., Storck, H., and Wallaschek, J. (2001). Sliding friction in the presence of ultrasonic oscillations: superposition of longitudinal oscillations. Archive of Applied Mechanics (Ingenieur Archiv), 71(8):549–554.

Lufthansa Technik. (2014). Own documentation after visit of the disassembly department of Lufthansa Technik and conversation, Hamburg.

Middendorf, P., Blümel, R., Hinz, L., Raatz, A., Kästner, M., and Reithmeier, E. (2022). Condition assessment and damage characterization of turbine blades during inspection cycles and component-protective disassembly processes. Sensors 22(14):5191. MDPI AG.

Popov, V. L., Starcevic, J., and Filippov, A. E. (2009). Influence of ultrasonic in-plane oscillations on static and sliding friction and intrinsic length scale of dry friction processes. Tribology Letters, 39(1):25–30.

Saffar, S. and Abdullah, A. (2021). Experimental investigation on ultrasonic assisted drilling (UAD). Journal of the Brazilian Society of Mechanical Sciences and Engineering, 43(7).

Schmücker, R., Meyer, H., Roedler, R., Raddatz, F., and Rodeck, R. (2021). Digitalization and data management in aircraft maintenance based on the example of the composite repair process. In Deutscher Luft- und Raumfahrtkongress 2021. Deutsche Gesellschaft für Luft- und Raumfahrt - Lilienthal-Oberth e.V.

Seliger, G. (2007). Sustainability in Manufacturing. Springer Berlin Heidelberg.

Siegel, A. F. and Wagner, M. R. (2022).Practical Business Statistics. Elsevier Science Technology, 8 edition.

Witek-Krowiak, A., Chojnacka, K., Podstawczyk, D., Dawiec, A., and Pokomeda,K. (2014). Application of response surface methodology and artificial neural network meth- ods in modelling and optimization of biosorption process.BioresourceTechnology, 160:150–160.

Wolff, J., Borchert, G., and Raatz, A. (2015). Demontage bei verfestigten Verbindungen/disassembly in the case of solidified connections. Werkstatttechnik Online, 105(09):604–609.

Wolff, J., Yan, M., Schultz, M., and Raatz, A. (2016). Reduction of disassembly forces for detaching components with solidified assembly connections. Procedia CIRP, 44:328–333.

Wolff, J., Kolditz, T., Fei, Y., and Raatz, A. (2018). Simulation-based determination of disassembly forces. Procedia CIRP, 76:13–18.

Wolff, J., Kolditz, T., and Raatz, A. (2019). A Learning Method for Automated Disassembly. In: Ratchev, S. (eds) Precision Assembly in the Digital Age. IPAS 2018. IFIP Advances in Information and Communication Technology, vol 530. Springer, Cham. https://doi.org/10.1007/978-3-030-05931-6_6

Impact of Mixing on the Signature of Combustor Defects

Panagiotis Ignatidis, Henrik von der Haar, Christoph Hennecke, and Friedrich Dinkelacker

Abstract Defects in the combustion chamber can negatively influence the performance of an aircraft engine and increase component stress in the turbine. One aim of the Collaborative Research Center 871 is to provide early prediction about the condition of the engine by analysing the signature of the exhaust gas jet. This includes the usage of machine learning techniques and helps to optimise maintenance times and to reduce costs. This topic is linked to the question, how defects in the combustion chamber affect the flow field and how the defect signature is mixed out in the hot gas path. Examples are shown for a simplified ring burning chamber, where several experimental and numerical studies have been done. Additionally, one failure case is described in detail here, where the methodology is applied to a real size gas turbine burning chamber and its subsequent turbine. Furthermore the diffusion theory is generalized to situations with complex geometrical boundary conditions, for instance from the turbine passage channel geometry. This approach is applied on the investigated example case and shows complex thermal diffusion coefficients, being in the order of 10,000–100,000 times larger than the molecular diffusion coefficient. Even these large values allow the determination of burning chamber defects from the exhaust flow pattern.

Keywords Combustion chamber · Combustor defect detection · Machine learning · Complex diffusion

P. Ignatidis · H. von der Haar · C. Hennecke · F. Dinkelacker (✉)
Institute of Technical Combustion, Leibniz University Hannover, An der Universitaet 1, 30823 Garbsen, Germany
e-mail: dinkelacker@itv.uni-hannover.de
URL: https://www.itv.uni-hannover.de/

P. Ignatidis
URL: https://www.itv.uni-hannover.de/

H. von der Haar
URL: https://www.itv.uni-hannover.de/

C. Hennecke
URL: https://www.itv.uni-hannover.de/

© The Author(s) 2025
J. R. Seume et al. (eds.), *Regeneration of Complex Capital Goods*,
https://doi.org/10.1007/978-3-031-51395-4_6

1   Influence of Defects on the Exhaust Gas and Temperature

Defects in the combustion chamber produce specific signatures in the distribution of exhaust gas components and temperatures. In order to investigate how these signatures are created, numerical and experimental investigations were carried out on an atmospheric model combustion chamber. The model combustion chamber consists of eight swirl burners in a ring array, from which one is individually operable, to simulate a burner defect in a defined way (von der Haar et al. 2016; Hennecke et al. 2017; Hennecke 2018). The model combustion chamber is capable to produce up to 25 different defects with variations in the thermal power, fuel–air-ratio and burner position of this variable burner. A model of the combustion chamber and the manipulated burner can be seen in Fig. 1.

For the lab experiments the concentration of certain species was measured using the Fourier Transform Infrared Spectroscopy (FTIR). This is an invasive 0-dimensional measurement technique where a ceramic suction tube is placed in the exhaust jet to collect samples. The exhaust gas composition is then determined at this measurement point. Figure 2 shows the model combustion chamber in operation with the ceramic suction tube for the FTIR measurements. Using a robotic arm, nine different measuring positions are approached to obtain a certain spatial resolution. A schematic view of the model combustion chamber with the burner and measurement position can be seen in Fig. 3. The red burner can be manipulated in the power or it can be tilted. For some experiments instead the blue burner is shifted from its ideal position.

In parallel, three-dimensional numerical CFD simulations were carried out for the model combustion chamber. For this purpose, a tetrahedral mesh was generated for the swirl burners and a hexahedral mesh for the combustion chamber which

Fig. 1 Geometry of the model combustion chamber with eight burners, the manipulated burner is marked in the red circle (von der Haar et al. 2021; von der Haar 2021)

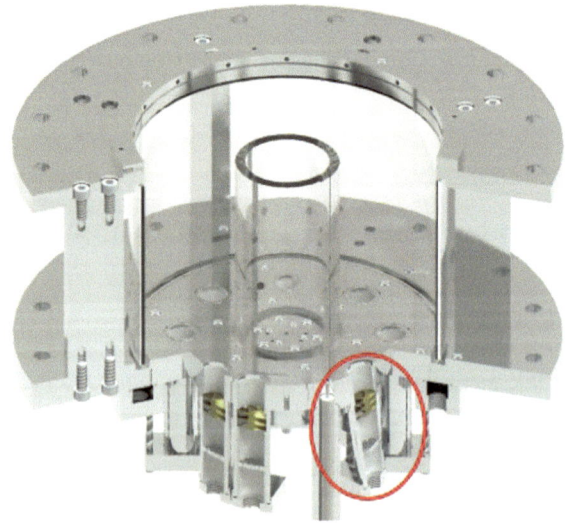

Fig. 2 Model combustion chamber in operation with ceramic suction tube for FTIR experiments

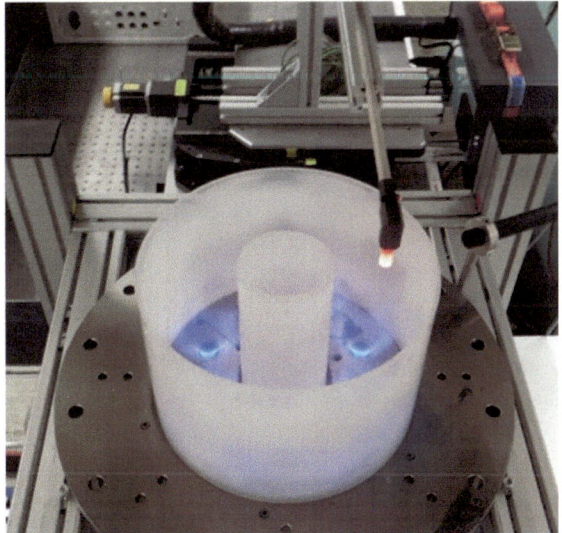

Fig. 3 Schematic top view of the model combustion chamber with burner positions, points of measuments (x), manipulated and tiltable burner (red) and shiftable burner (blue) (von der Haar et al. 2021; von der Haar 2021)

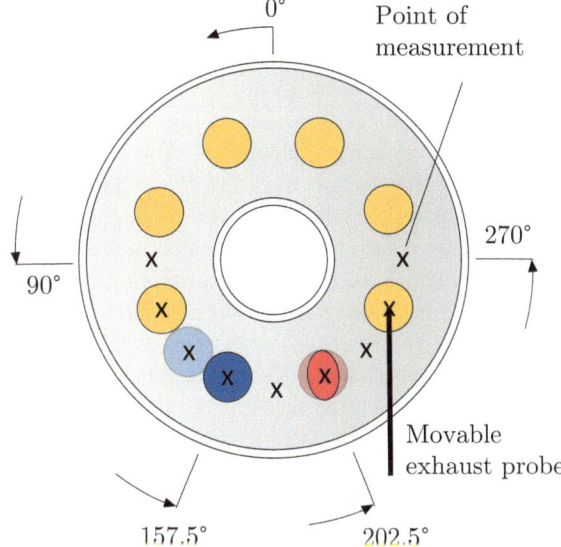

consist of 9 million elements in total. A mesh independence study was performed (von der Haar 2021). The investigated defects in the experiments were passed on as boundary conditions for the simulations. The meshing was carried out in the software ANSYS ICEM while the calculations were performed in ANSYS Fluent 17.2. In the combustion chamber, transient Scale-Adaptive-Simulations (SAS) were carried out with a pressure based solver. The 2-step mechanism 2S-CH4-CM2 (Boudier

2007) was used for the calculation of the main combustion process, while NOx was simulated in a postprocessing step with the model include in ANSYS Fluent.

Figure 4 shows the exhaust gas concentrations from the FTIR measurements and the CFD simulations at the positions according to Fig. 3. The dashed red line indicates the position of the manipulated burner. Some defects demonstrated here are listed in Table 1. A complete list of defects used for damage detection and classification and additional information about this work can be found in (von der Haar et al. 2021; von der Haar 2021). P is the thermal power and λ is the air–fuel ratio for the manipulated burner (Single) and the remaining seven burners (Array). Typically, the "normal" burners were operated with 15 kW thermal power each. The profiles show great similarity, even when the exact concentrations are not obtained by the simulation. The comparison of the experimental and the numerical data shows, that the simulation predicts the processes qualitatively correctly. The position of the disturbance by the manipulated burner agrees well in the simulation and experiment. Furthermore, each defect produces a specific distribution and the different cases can be distinguished from each other on the basis of the obtained results. Both, the experimental and numerical data sets can be used for defect detection and classification. After demonstrating that burner defects have a specific influence on the local exhaust gas composition in the model combustion chamber, this method is transferred to the combustion chamber of the V2500 engine (Hennecke 2018; von der Haar 2021). For this purpose, a numerical model of a 90° segment is build and discretised with a Delaunay mesh with 45 million elements. The numerical domain used for the CFD simulations can be seen in Fig. 5. The segment includes five burners, with the middle one being manipulated. The investigated defects include a partially or completely clogged fuel line (P50, P0), which result in a reduced thermal power and a cold gas streak. Furthermore, defects in the flame stabilisation were investigated. Two swirled air streams are generated in the fuel spray nozzle, which stabilise the flame through recirculation in the primary zone. The outer swirler (see Fig. 6, yellow arrows) is manipulated for the investigation. For this purpose, the swirler blades are removed (P100-Swirl) or the swirl air duct is blocked (P100-AirDuct), which leads to a destabilisation of the flame, creating a hot jet.

Only the middle burner will be manipulated, the remaining burner operate in normal mode. Steady-State RANS and transient SAS simulations were performed for these defects. The resulting outlet temperature and CO profiles from the RANS simulations can be seen in Figs. 7, 8 and 9. The temperature and CO_2 distribution correlates very strongly, so that the CO_2 distribution is not shown. Both can be used for damage classification. It can be seen that a clogged fuel line leads to a significant local temperature decrease. Damages in the swirler blades and the combustion air duct lead create local hot spots and increased CO concentrations. The circumferential position of the defective fuel spray nozzle and the signature on the combustor outlet remains constant at for all defects. The impact position in the combustion chamber outlet remains static at 45°.

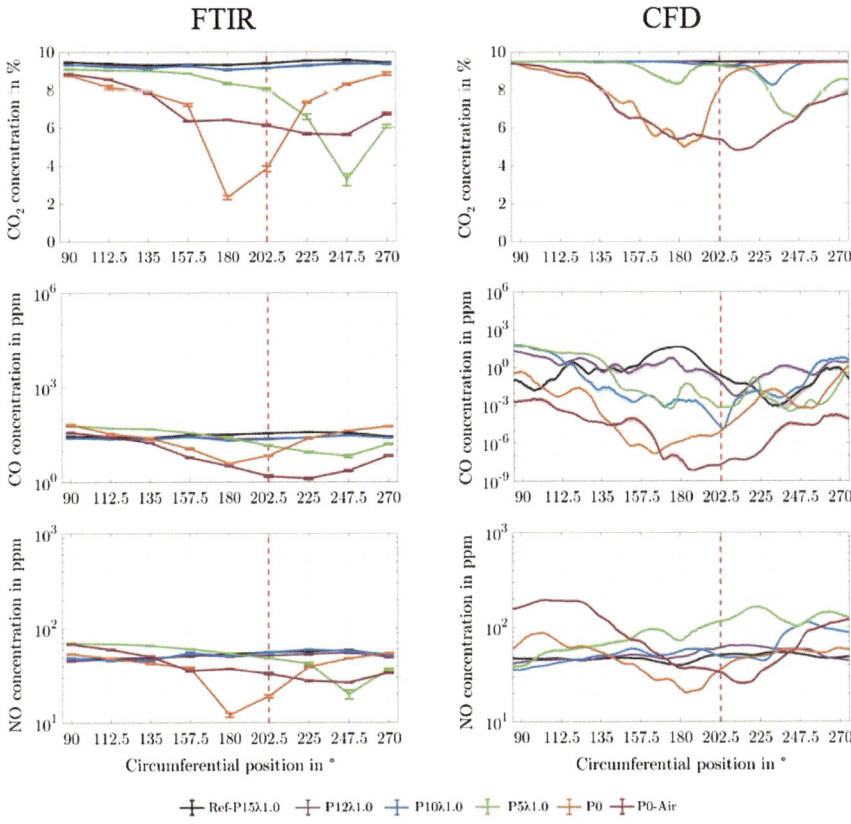

Fig. 4 Circumferential profile of the exhaust gas concentrations from the FTIR measurements (left) and the CFD simulations (right) (von der Haar et al. 2021; von der Haar 2021)

Table 1 Some investigated operating points and defects (selection) (von der Haar et al. 2021; von der Haar 2021)

Name	P_{Single} (kW)	λ_{Single}	P_{Array} (kW)	λ_{Array}
Ref − P15λ1.0	15	1.0	7 × 15	1.0
P12λ1.0	12	1.0	7 × 15	1.0
P10λ1.0	10	1.0	7 × 15	1.0
P0	0	–	7 × 15	1.0
P0 − Air	0	∞	7 × 15	1.0

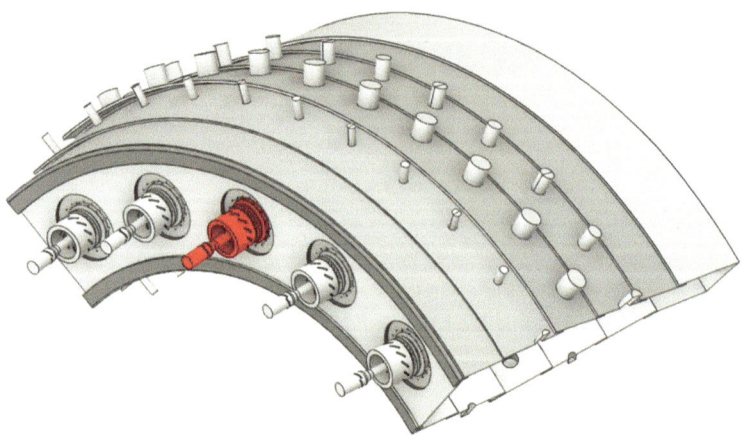

Fig. 5 Numerical domain of the V2500 combustion chamber, manipulated burner marked in red (Ignatidis et al. 2022)

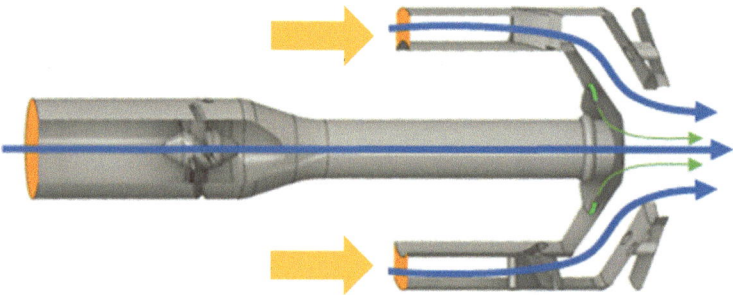

Fig. 6 Cross section view of the fuel spray nozzle. Air flow is marked in blue, fuel flow marked in green

Fig. 7 Temperature contour plot at the combustion chamber outlet for reference case P100 (top), P50 (middle) and P0 (bottom)

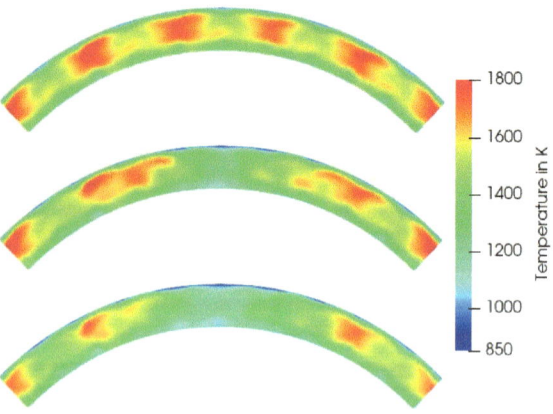

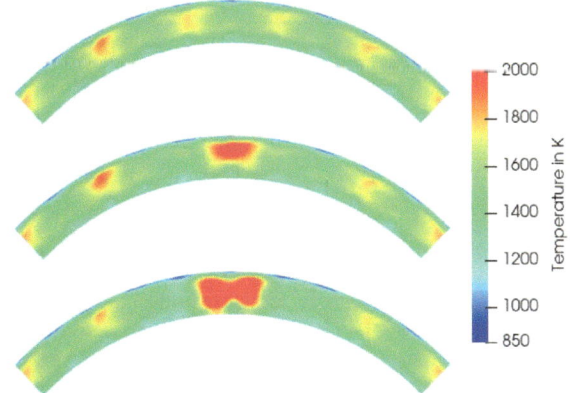

Fig. 8 Temperature contour plot at the combustion chamber outlet for reference case P100 (top), P100-AirDuct (middle), P100-Swirler (bottom)

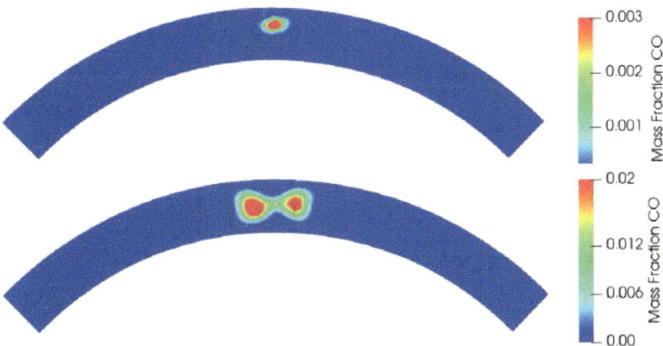

Fig. 9 CO contour plot at the combustion chamber outlet for defect P100-AirDuct (top) and P100-Swirl (bottom)

2 Damage Detection and Classification

The investigations on the model combustion chamber and the V2500 combustor show that defects have a specific influence on the exhaust jet. This allows a clear assignment of the defect from the exhaust gas. A method was developed with which damage detection and classification can be performed using machine learning techniques (von der Haar et al. 2021; von der Haar 2021). A Support-Vector-Machine (SVM) algorithm was applied on the model combustion chamber with the data sets from the FTIR measurements. The detection and classification was carried out with the CO, CO_2 and NO concentrations at three positions (180°, 202.5° and 225°, see Fig. 3).

First, a defect detection was done using a One-Class SVM. For that, all defects are seen as one class. Only the reference data is used for training. A big advantage of this method is that no damage library has to be created and that unknown defects can also be detected. With the help of this method, defects can be detected but not classified. A proper selection of features is necessary for a reliable defect detection and classification. The CO, CO_2 and NO concentrations are available at three evaluation positions, resulting in ten different feature combinations. The possible feature combinations can be found in Table 2. The damage detection and classification is performed with all combinations in order to select the best features.

In 85% of the cases it was possible to correctly detect the defects. The best classification rate is achieved for the combination #1 (3 × CO_2). The One-Class-SVM results for all feature combinations are shown in Figs. 10 and 11 (von der Haar et al. 2021; von der Haar 2021).

In the next step, a defect classification was done using Multi-Class SVM. For this purpose, each defect is seen as a separate class. The SVM is trained on a damage

Table 2 Feature combinations for defect detection and classification (von der Haar et al. 2021; von der Haar 2021)

Position	180°			202.5°			225°		
Combination	CO_2	CO	NO	CO_2	CO	NO	CO_2	CO	NO
#1	x			x			x		
#2		x			x			x	
#3			x			x			x
#4	x	x	x						
#5				x	x	x			
#6							x	x	x
#7	x	x	x	x	x	x			
#8				x	x	x	x	x	x
#9	x	x	x	x	x	x	x	x	x
#10	9 × CO_2 (90°–270°)			9 × CO (90°–270°)			9 × NO (90°–270°)		

Fig. 10 One-Class-SVM FTIR

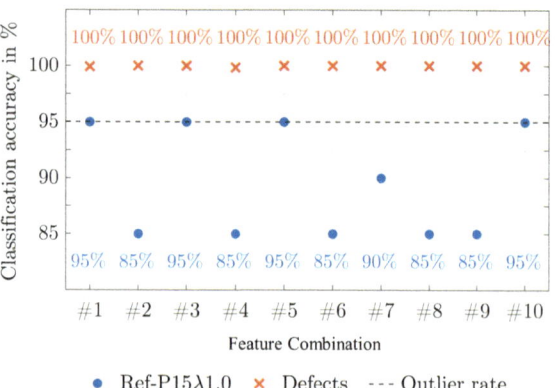

- Ref-P15λ1.0   × Defects   --- Outlier rate

Fig. 11 One-Class-SVM CFD

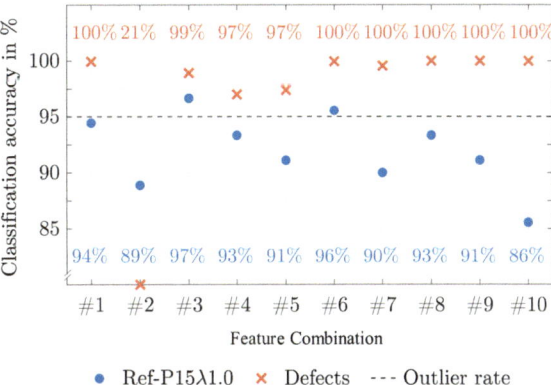

library using the defects (listed above). Compared to the one-class SVM, the damages are not only recognised but also classified. The results of the Multi-Class SVM are shown in Figs. 12 and 13 (von der Haar et al. 2021; von der Haar 2021). In over 85% of the cases, the defects were correctly classified with the exception of feature combination #2. With feature combination #9, all test data are correctly classified.

The method was transferred to the defects in the V2500 combustion chamber. The damage detection and classification here is also based on the CO, CO_2 and NO concentrations at three evaluation points. Figure 14 shows the CO_2 distribution at the combustor outlet and the evaluation points. The defect detection with the One-Class-SVM and damage classification using Multi-Class-SVM performed well. Apart from the defect 'P100-AirDuct', all defects were classified correctly. The Multi-Class-SVM result matrix can be seen in Fig. 15.

Defect detection with One-Class-SVM and classification with Multi-Class-SVM worked very well with both the model combustor and the V2500 combustor. The method is suitable to reliably detect and classify defects based on exhaust concentrations (von der Haar 2021).

Fig. 12 Multi-Class-SVM FTIR

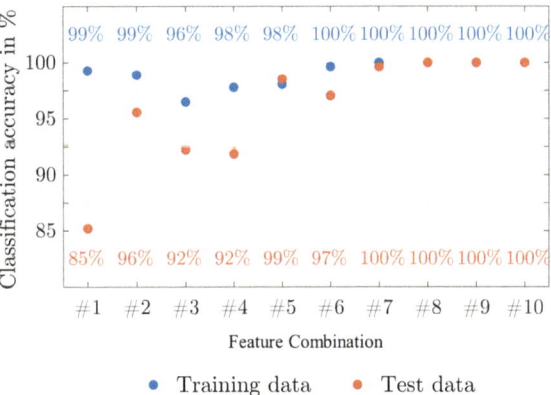

Fig. 13 Multi-Class-SVM CFD

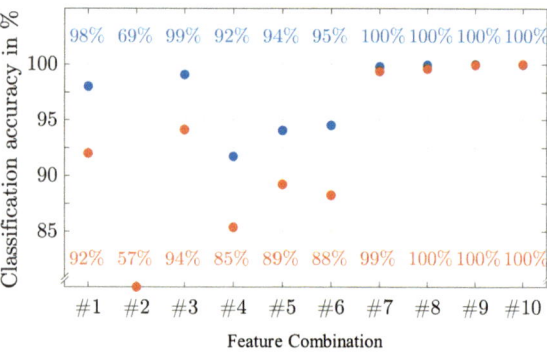

Fig. 14 CO$_2$ distribution at combustor outlet for defect 'P0' with evaluation points for classification (von der Haar 2021)

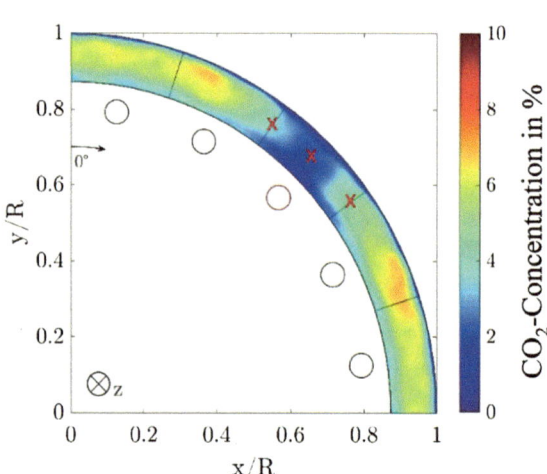

Fig. 15 Result matrix for Multi-Class-SVM classification of defects (von der Haar 2021)

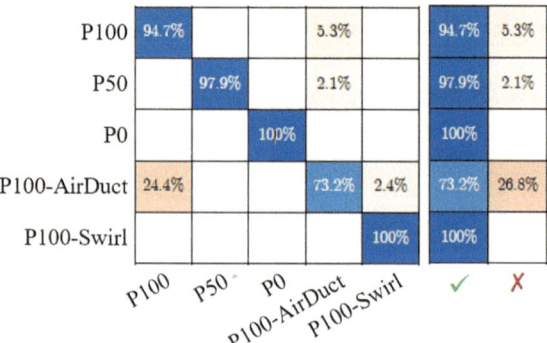

3 Defect Signature Propagation in the Turbine

Now that it is known how defects in the combustion chamber affect the flow field and how the defects can be detected and classified, it remains to be clarified how these defect signatures are propagated through the turbine to the outlet. For this purpose, an approach was developed to quantify the defect signatures in order to understand their progression through the turbine. In cooperation with the Institute of Turbomachinery and Fluid Dynamics (TFD), simulations of the downstream turbine were carried out and the approach applied (Ignatidis et al. 2022, Ignatidis 2024). The defect of the clogged fuel line was investigated, which produces a cold gas streak. The resulting flow field at the combustion chamber outlet is used as an inlet boundary condition for the turbine simulation. The temperature field was evaluated in each row in the turbine. For every angle, spatial averaging was done in radial direction. The cold gas streak produces a temperature profile that decreases locally and can be fitted with a Gaussian distribution curve according to Eq. 1.

$$\theta(\xi) = T_0 + \frac{\theta_0}{\sqrt{2\pi\sigma^2}} \exp\left(-\frac{(\xi - \varphi)^2}{2\sigma^2}\right). \tag{1}$$

Here, ξ is the circumferential position, φ represents the streak position, T_0 is the mean temperature of the surrounding gas and θ_0 a normalization factor (negative for cold gas streaks). The standard deviation σ is a measure of the gas streak width and can be used to calculate the full-width at half maximum (minimum) FWHM using $2\sqrt{2\ln 2}\sigma$. The progression of the streak position and the FWHM now provide information about the mixing of the signature in the turbine. Figure 16 shows the circumferential temperature at the combustor outlet with the corresponding fitted Gaussian curve. The defective burner is at $\xi = 45°$.

The propagation and the mixing of the cold gas streak through the turbine is influenced by the blade curvature and the blade rotation. In order to examine these influences in a differentiated manner, two turbine simulations were carried out. A steady simulation with frozen-rotor was performed to show the geometric influences. An unsteady simulation shows how the mixing and streak shift is affected by the blade rotation. For both simulations, the width of the streak (its FWHM) and the streak position φ were determined at all rows. The progression is shown in Figs. 17

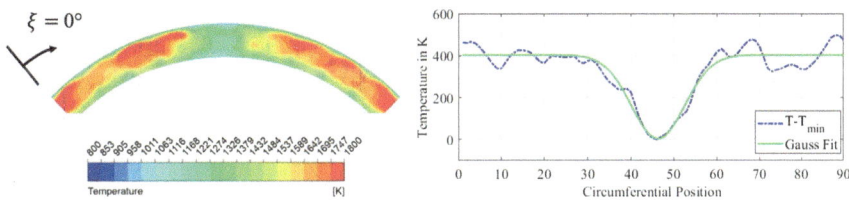

Fig. 16 Temperature distribution at combustion chamber outlet (left), circumferential temperature with Gaussian curve (right) (Ignatidis et al. 2022)

and 18. The rows R1–R4 represent the high-pressure turbine (HPT), R5–R15 the low-pressure turbine (LPT) and 'HPTLPT' the interface between HPT and LPT. The evaluation positions are located upstream of the corresponding row. Both blade curvature and speed are higher in the HPT than in the LPT which will affect the propagation of the gas streak.

The FWHM plot (Fig. 17) shows that the disturbance width increases in the HPT, while it remains almost constant in the LPT for the steady simulation case and only slightly increases for the unsteady simulation case. Thus, the mixing increases with the blade curvature and rotational speed. The displacement of the streak position φ (Fig. 18) is minimal in the steady simulations, showing little influence from the blade curvature. However, the influence of the blade rotation becomes clear for the unsteady simulation. The streak shifts differently in the HPT and LPT, which is due to the different rotational speeds. The total mixing and streak displacement in the turbine is determined to be FWHM(R15) – FWHM(R1) = 9.7° and φ(R15) – φ(R1)

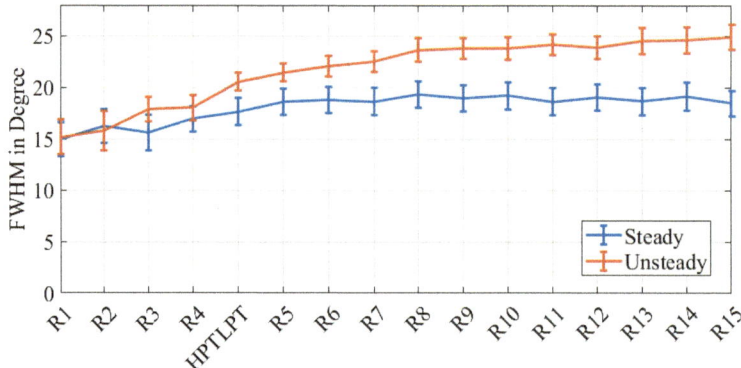

Fig. 17 Disturbance FWHM for steady and unsteady simulation (Ignatidis et al. 2022)

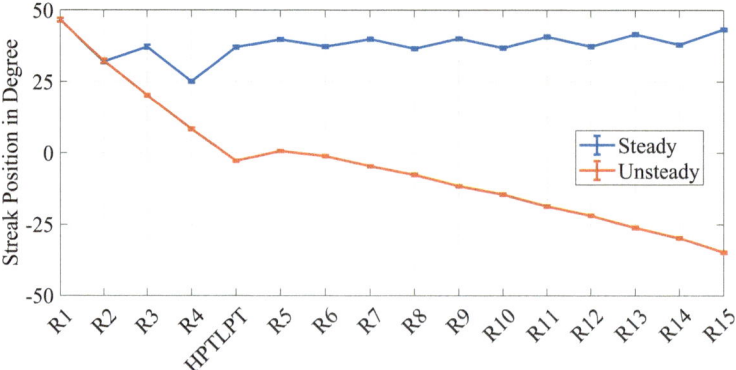

Fig. 18 Streak position for steady and unsteady simulation (Ignatidis et al. 2022)

Table 3 FWHM and streak position φ in HPT and LPT (Ignatidis et al. 2022)

			HPT (R1–R4)	LPT (R5–R15)
FWHM/stage	Steady		1.2°	0°
	Unsteady		2.7°	0.7°
φ/stage	Steady		−4.8°	0.7°
	Unsteady		−24.7°	−7.1°

$= 81.4°$. In addition, the mixing and streak displacement are considered per stage and can be found in Table 3.

Now that the mixing and streak displacement per stage are known for that rotational speed, it is possible to determine the total streak displacement and mixing for this operating point using a generalized approach with the number of stages n. To do this, Eqs. 2 and 3 are first used to calculate the total displacement φ_{Total} and total mixing $FWHM_{Total}$.

$$WHM_{total} = n_{HPT}(FWHM/Stage)_{HPT}\, n_{LPT}(FHWM/Stage)_{LPT} \qquad (2)$$

$$\varphi_{total} = n_{HPT}(\varphi/Stage)_{LPT}\, n_{LPT}(\varphi/Stage)_{LPT} \qquad (3)$$

$$\varphi_{Defect} = \varphi_{EGV} - \varphi_{Total} \qquad (4)$$

If a cold gas streak is detected in the exhaust gas jet and its position is determined, Eq. 4 can be used to calculate the position of the defective burner backwards. The method will be demonstrated using the temperature distribution downstream of the Exhaust Guide Vane (EGV, Fig. 19). Here, a cold gas streak is determined at the position $\varphi_{EGV} = 35°$. Equation 4 results in the defect burner position φ_{Defect} of 50°. The actual defective burner position is at 45° (see Fig. 16). Despite the error of 5°, the defective burner can be determined from the temperature distribution downstream the EGV, since the burner segments have a width of 18°.

It should be mentioned that the values for the mixing and streak displacement per stage in Table 3 are only valid for the examined rotational speeds at this engine operating point. The HPT has a higher rotational speed than the LPT. For a more general statement, investigation with additional operating points will be needed to be carried out.

The analysis of the temperature distribution in the turbine shows how the blade curvature and the rotational speed influence the mixing and the position of the gas streak. This new approach allows the localization of a defective fuel spray nozzle from the temperature distribution in the exhaust gas.

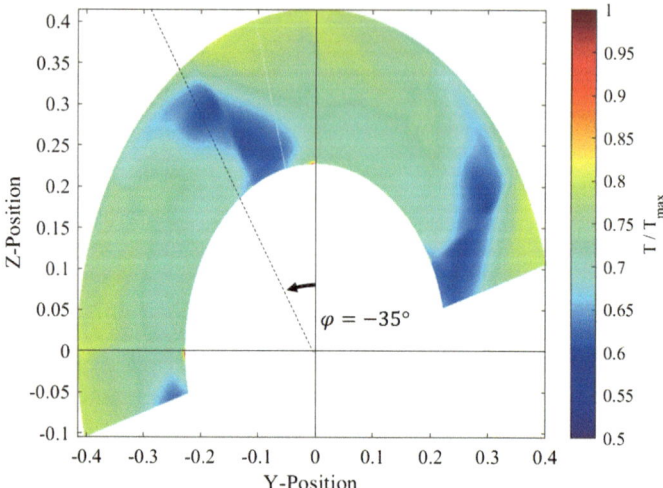

Fig. 19 Normalized temperature distribution at EGV outlet (Ignatidis et al. 2022)

4   Complex Diffusion Model for Defect Propagation

The above example shows that the width of the defect signature remains relatively limited (Fig. 17), while a first order assumption could be that for instance the cold gas streak mixes rather soon away, so that it would not be visible anymore at the exit of the turbine. In order to be able to generalize the mixing behaviour of such disturbances, an approach is developed, which is based on a generalized diffusion theory (Dinkelacker 2022).

In flows without any turbulence, a thermal streak (and similarly a streak of another type of chemical species) would mix out based on molecular diffusion processes, which are induced from thermal diffusion (or species diffusion), and which are described by Fourier's law of thermal diffusion (or Fick's law of species diffusion). The molecular diffusion process is typically very small, for gases at normal temperature and pressure (300 K, 1 bar) the molecular diffusion coefficient is in the order of 20×10^{-6} m^2/s. For the typical passage time through the turbine (in the order of 5 ms) this diffusional process would not be observable at all due to its very small value.

In flows with turbulence it has become common, to model the effects of turbulent mixing on such diffusional mixing processes with the help of a turbulent viscosity concept. For instance for Reynolds averaged numerical simulations a turbulent viscosity parameter ν_t is modeled as a function of two turbulence parameters k and ϵ with $\nu_t = c_\mu \cdot k^2/\epsilon$, with the constant $c_\mu = 0.09$. Such approaches are leading to estimations of turbulent mixing, which work as an approximation, if the turbulent flow is not too far away from straight flows with sufficiently developed turbulence conditions (Ferziger and Perić 2002).

Fig. 20 Concept of complex diffusion for the turbulent mixing of the defect flow within the complex boundary conditions of a turbine. They lead to a complex mixing behaviour, being described as a generalized complex diffusion process with an effective complex diffusion constant D*

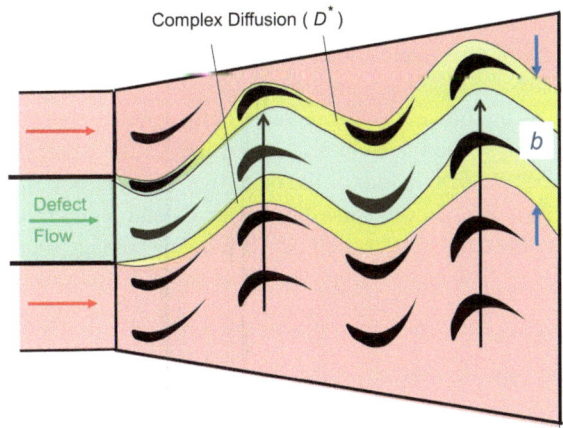

For flows within burning chambers of jet engines or within turbines, the boundary conditions are leading to very complex flow pattern with recirculation, dilution jets, and complex wall boundary layers between the turbine channels, such that an approximation of the resulting turbulent mixing process seems to be nearly impossible. Within our project we proposed to generalize the procedure of turbulent mixing to such complex situations with the aim to reach a reduced modeling approach. We label this approach as "complex diffusion model", see Fig. 20.

As it is clear that here the details of the boundary situation will influence the turbulent mixing process significantly, such an approach can assist for the description of first order approximations only. We developed this approach empirically from the results of numerical simulation of the defect propagation process, either within the burning chamber or within the subsequent turbine segments.

The complex diffusion approach is based on the diffusion theory, where the diffusion equation is given for the number density n of molecules of one type with (Bird et al. 2007).

$$\frac{\partial n}{\partial t} = D \cdot \frac{\partial^2 n}{\partial x^2} \tag{5}$$

Here the species diffusion equation form, 2. Fick's law, is described, but for thermal diffusion the same is valid for the temperature field, if the thermal diffusion constant is used instead of the species diffusion constant. The mathematical solution depends on the initial and boundary conditions. For the example of an initial delta function of the defect pattern, the solution (obtained within the Fourier space, see mathematics textbooks) leads to a stable Gaussian distribution—similar to Eq. 1, where the denominator within the exponent contains the product $4 \cdot D \cdot t$, with the diffusion constant D and time t. For a defect pattern within the ring-shaped geometry of a typical aviation engine (both within the combustor and the turbine passage section) a circumferential defect pattern can be described with the width B or the corresponding circumferential angle β (see Fig. 21). The mixing process within the complex flow and geometry conditions in the combustor and the turbine section leads to an increase of the disturbance width, being described with the generalized effective diffusion constant D^*. It can be shown, that the increase of the width from B_1 to B_2 within a certain section can be described analytically as solution of the generalised diffusion law (Dinkelacker 2022). If the width of the disturbance (described by its FWHM) is expressed in the form of the corresponding angles β_1 and β_2 and the mean radius r_1 of the channel at the first cross section, and if the flow passage time between the two cross sections $t_2 - t_1$ is known from the mean velocity and distance, than the effective diffusion coefficient can be determined with the following equation

$$D^* = 14.24 \cdot \frac{r_1^2 \cdot \left[\left(\frac{\beta_2}{360°} \right)^2 - \left(\frac{\beta_1}{360°} \right)^2 \right]}{t_2 - t_1} \tag{6}$$

With this approach it is possible to evaluate the resulting effective diffusion coefficient D^* even under the complex boundary conditions occurring within the burning chamber or the turbine of the airjet engine (Dinkelacker 2022).

The approach has been applied for the following situations for the determination of the effective thermal diffusion coefficient.

- Flow within the burning chamber
- High pressure turbine section (HPT)
- Low pressure turbine section (LPT)

Within the burning chamber the third part of the combustor is analysed, where the dilution air is added to the exhaust gas to cool it down before the turbine inlet.

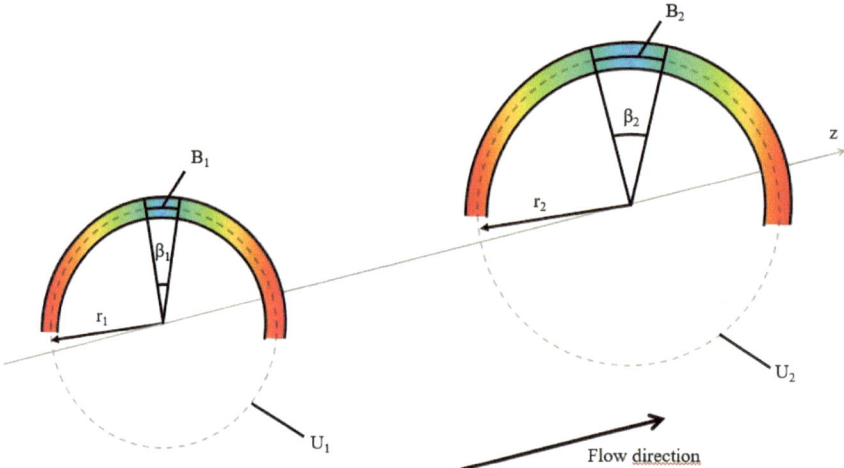

Fig. 21 Determination of the effective diffusion coefficient from measurement of the width B_1 and B_2 (FWHM) of a defect disturbance at two different downstream positions 1 and 2 in a ring channel

Table 4 Determination of the effective thermal diffusion coefficient for the third part of the burning chamber (mixing of dilution air), high pressure turbine (HPT) and low pressure turbine (LPT). Also the ratio to the molecular thermal diffusion coefficient is given

	Plane	FWHM	Radius r_1	t (ms)	D^* (m²/s)	D_m (m²/s)	D^*/D_m
Combustor	X7	13°	0.28 m	1.6	0.24	18×10^{-6}	13,000
	X12	14.7°	–				
HPT	R1	15.2°	0.3 m	1.1	1.82	19×10^{-6}	95,000
	HPTLPT	20.5°	–				
LPT	R5	21.5°	0.31 m	2.7	0.64	21×10^{-6}	30,500
	R15	25°	–				

For the high pressure and low pressure turbine, the transient simulations shown above are analysed. In Table 4 the width of the disturbance is given for the analysed planes (X7 and X12 are two relevant planes within the burning chamber, not in detail shown in this report) together with the necessary geometry data and the approximated transit time of the flow. It is found that the resulting complex thermal diffusion coefficient reaches values between 0.24 and 1.82 m²/s for the burning chamber section and the turbine section. These values are significantly larger than the molecular thermal diffusion coefficients, which are also given in the table as a function of temperature and pressure for the relevant conditions. The ratio between complex diffusion coefficient and molecular diffusion coefficient reaches values of 13,000 for the burning chamber, 95,000 for the HPT and 30,500 for the LPT section. These large values are on one hand coming from the turbulent mixing in the flow, being commonly described with the turbulent viscosity, which is typically known to reach values of 100–1000 times that of the molecular viscosity. On the other hand here the influence of the complex additional mixing processes is included, coming from the irregular mixing flow within the combustor and from the multichannel flow within the moving turbine blade sections.

Experimental confirmations are not yet given. Real size gasturbines experiments have been planned in subproject A3, but have not been possible so far.

The advantage of this concept of complex diffusion is, that it would allow already from low-dimensional estimations the predetermination of the complex diffusion coefficient. With that it would be easy to predict the possible detectability of defect pattern also in other geometries.

5 Conclusions

In subproject 'A6: Mixing and Combustor Signature' it was investigated, how damage signatures are created in the combustion chamber and how they can be detected and classified from the exhaust gas jet. The initially asked questions were answered.

How does the damage in the combustion chamber affect the exhaust gas composition and temperature distribution? The experimental and numerical investigations on the model combustion chamber and the V2500 combustor show that defects create a characteristic signature in the temperature and species distribution, which allows a clear allocation of damages. How is a combustion chamber damage detected and classified? Applying Support-Vector-Machine methods allow the automatic detection and classification of defects based on the distribution of the exhaust gas components. How does the disturbance signature propagate through the turbine section and how is it mixed out? The numerical investigations on the V2500 turbine show how defect signatures propagate in the hot gas path and how the mixing and displacement are affected by the blade curvature and rotating speeds. In addition, an approach was developed with which a defective fuel spray nozzle can be localised from the temperature distribution in the exhaust gas jet at a known rotational speed.

Furthermore the diffusion theory is generalized to situations as in the burning chamber or turbine, where the turbulent flow mixing process is not only determined from molecular diffusion and turbulent mixing but also from the complex geometrical boundary conditions, for instance from the turbine passage channel geometry. This approach is applied on the investigated example case. It is found that the complex thermal diffusion coefficient of the investigated cold gas defect is in the order of 10,000–100,000 larger than the molecular diffusion coefficient, being especially large in the high pressure turbine section. Even these large values still allow the determination of burning chamber defects from the exhaust flow pattern.

Acknowledgements Funded by the Deutsche Forschungsgemeinschaft (DFG, German Research Foundation)—SFB 871/3—119193472.

References

Bird, R. B., Stewart, W. E., and Lightfoot, E. N. (2007). *Transport phenomena.* Wiley, New York.

Boudier, G. (2007). *Methane/Air Flame with 2-Step Chemistry: 2S-CH4-CM2.* Toulouse, CERFACS.

Dinkelacker, F. (2022). The diffusion equation and the determination of diffusion constants for application in complex situations. Technical report, Leibniz University Hannover.

Ferziger, J. H. and Perić, M. (2002). *Computational methods for fluid dynamics.* Springer, Berlin, Heidelberg, 3rd, rev. ed edition.

Hennecke, C. (2018). *Methodik einer Zustandsbeurteilung von Triebwerksbrennkammern.* Dissertation, Leibniz University Hannover.

Hennecke, C., von der Haar, H., and Dinkelacker, F. (2017). Failure Detection in an Annular Combustion Chamber with Experimental and Numerical Methods. *Journal of Aeronautics & Aerospace Engineering,* 6(2). 1000193.

Ignatidis, P. (2024). *Modellierung der Mischvorgänge im Heißgaspfad von Flugtriebwerken zur Detektion von Brennkammerschäden im Abgasstrahl.* Dissertation, Leibniz University Hannover.

Ignatidis, P., Nghiem, V. D., Goeing, J., Oettinger, M., Friedrichs, J., Seume, J. R., and Dinkelacker, F. (2022). Mixing Behaviour and Propagation of Combustion Chamber Defects in an Aircraft Engine. In *Proc. of Global Power and Propulsion Society, GPPS TC 2022 0103*, Chania, Greece.

von der Haar, H. (2021). *Untersuchung einer speziesbasierten Abgasstrahlanalyse zur automatisierten Detektion von Brennkammerschäden in Flugtriebwerken*. Dissertation, Leibniz University Hannover.

von der Haar, H., Hartmann, U., Hennecke, C., Dinkelacker, F., and Seume, J. R. (2016). Defect Detection in an Annular Swirl-Burner-Array by Optical Measuring Exhaust Gases. In *Proc. ASME Turbo Expo 2016, GT2016–57847*, Seoul, South Korea.

von der Haar, H., Ignatidis, P., and Dinkelacker, F. (2021). Experimental and Numerical Based Defect Detection in a Model Combustion Chamber through Machine Learning. *International Journal of Gas Turbine, Propulsion and Power Systems*, 12(4):1–9.

Interaction Between Production Processes and the Product's Functional Characteristics (Project Area B)

Near Net Shape Turbine Blade Repair Using a Joining and Coating Hybrid Process

Martin Nicolaus, Kai Möhwald, and Hans Jürgen Maier

Abstract Brazing is an established repair technology for turbine blades (air foils and vanes) in the industry and includes multiple process steps that may also require a high degree of manual work. In this study, the development of a near net shape joining and coating hybrid technology for the repair of turbine blades is presented. With this technology, it is possible to shorten the current state-of-the-art process chain for repairing turbine blades. The worn turbine blade receives a repair coating applied by thermal spraying consisting of a nickel based filler metal, a hot gas corrosion protective layer, a bond coat and finally a thermal barrier coating. Subsequently the coated turbine blade is heat treated and a simultaneous brazing and aluminizing process is carried out. The technology presented here brings about technical and economic advantages and allows to shorten the state-of-the-art process chain for repairing turbine blades.

Keywords Thermal spraying · Brazing · Aluminizing · Hybrid technology

1 Introduction

The thermal spray technology is a high performance coating process. According to the standard DIN EN ISO 14917 (2017), thermal spraying generally refers to processes, in which spray additives are heated to a plastic or molten state inside or

M. Nicolaus (✉) · K. Möhwald · H. J. Maier
Institut für Werkstoffkunde (Materials Science), Leibniz University Hannover, An der Universitaet 2, 30823 Garbsen, Germany
e-mail: nicolaus@iw.uni-hannover.de
URL: https://www.iw.uni-hannover.de

K. Möhwald
e-mail: moehwald@iw.uni-hannover.de
URL: https://www.iw.uni-hannover.de

H. J. Maier
e-mail: maier@iw.uni-hannover.de
URL: https://www.iw.uni-hannover.de

J. R. Seume et al. (eds.), *Regeneration of Complex Capital Goods*,
https://doi.org/10.1007/978-3-031-51395-4_7

outside the spray gun or torch and then projected onto a prepared surface. Thermally sprayed coatings are used to reduce wear, enhance corrosion protection and apply thermal barrier coatings. Especially the latter are used in aviation and power plant industries. Turbine blades of aero engines are operated in harsh environments such as high temperature and mechanical loads for long-term. As a result, various defects like erosion, distortion, wear, cracks and impact dents occur. In order to increase the service life of such components, maintenance, repair and overhaul is becoming increasingly important in view of the growing competition in the industry. Turbine blades made of nickel-based alloys are mainly used in high-pressure turbines in the aviation industry and in power plant design for stationary gas turbines. Established repair processes are welding and brazing (Henderson et al. 2004; Huang and Miglietti 2011). The repair brazing of turbine blades was the focus of this study. Figure 1 shows the essential steps for repair brazing of turbine blades (Stolle 2004).

The process steps are as follows: the coating of the worn turbine blade is stripped down to the base material. The filler metal is applied manually in form of pastes, tapes or melt spun foils, which are also nickel-based alloys. After the brazing process in a high-vacuum furnace, excess filler metal is removed by machining. Subsequently a hot gas corrosion protective coating (e.g. NiCoCrAlY) is applied by thermal spraying, followed by an aluminizing process, to enhance the resistance against hot gas corrosion by the formation of the β-phase (NiAl) (Miracle 1993; Zhan et al. 2009). The aluminizing process is carried out by a chemical vapor deposition process (CVD) by pack cementation, consisting of aluminum, alumina (Al_2O_3) and an aluminum halide (Pytel et al. 2012). Often, special aluminizing furnaces are also used. The last layer is the thermal barrier coating (TBC, e.g. $ZrO_2 \cdot 8Y_2O_3$). After the coating of the turbine blades, the cooling holes are repositioned, for example by laser drilling. This repair

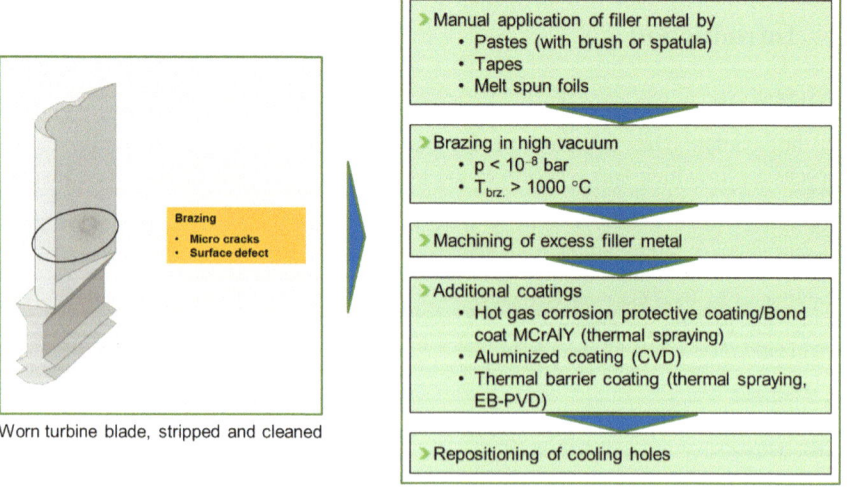

Worn turbine blade, stripped and cleaned

Fig. 1 Steps of turbine blade repair brazing process (Stolle 2004)

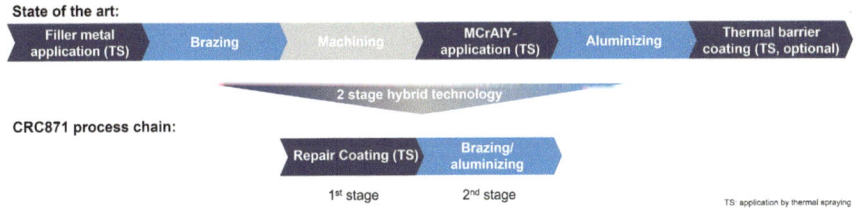

Fig. 2 Process chain for repair brazing turbine blades

process is expensive and includes several process steps. The aim of this study was to develop a hybrid joining and coating technology, consisting of two stages, which allows to shorten the process chain for repair brazing of turbine blades. Reducing the process chain is achieved by applying the materials required for the repair by thermal spraying. The repair coating taken into account consisted of the nickel-based filler metal, the hot gas corrosion protective layer (NiCoCrAlY) and optional additional layers, like aluminum and the TBC. The coated turbine blade was subjected to a heat treatment and a simultaneous brazing and aluminizing process was carried out. The feasibility of the developed hybrid technology could be demonstrated in previous studies (Nicolaus et al. 2017a, b, 2018). Figure 2 shows the principle of shortening the process chain for repair brazing of turbine blades.

Inconel 718 flat specimens were used as the substrate. This nickel-based alloy was chosen as its properties are well investigated (Brooks and Bridges 1988; Schirra 1991; Rao et al. 2001; Česnik et al. 2008). The filler metal used was Ni19Cr10Si (also known as Ni650 or B-Ni5). The advantage of this filler metal is that it consists only of three alloying elements, where silicon is the melting point depressant. For the hot gas corrosion protective layer, a MCrAlY (M = Ni and/or Co) alloy was used, which is state-of-the-art. In prior studies, two approaches for the simultaneous brazing and aluminizing process were considered. In the first approach the brazing/aluminizing process was carried out in a pack cementation. The feasibility of this procedure could be shown, but to get a microstructure nearly free of pores, a process duration of about 36 h is necessary (Nicolaus et al. 2017a). The combined brazing/aluminizing process showed that the flow and wetting ability of the thermally sprayed filler metal was maintained, providing good crack infiltration capability. The second approach was to apply the aluminum layer by thermal spraying as well, so that the pack cementation could be avoided. In this case, the process duration could be decreased to less than 30 min (Nicolaus et al. 2017b). While the brazing and aluminizing process is carried out, the NiAl-phase (β-phase) is formed so that the turbine blade is better protected against hot gas corrosion. A more detailed analysis of the coatings' microstructures also suggest the additional formation of various intermetallic phases, like Ni_3Al (γ'-phase), Al_9Co_2 and Al_3Co (Nicolaus et al. 2017b). The formation of these phases is due to a more pronounced aluminum diffusion into the MCrAlY layer with increasing temperature. Figure 3 illustrates the cross sections of the microstructure at different temperatures.

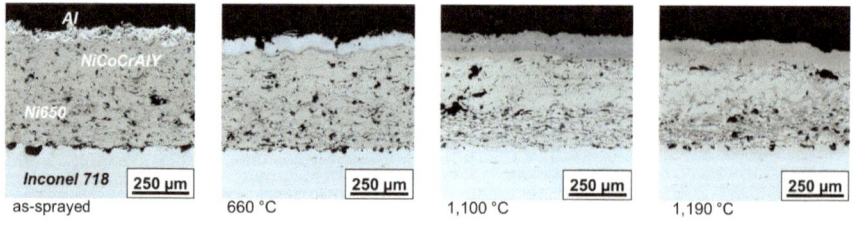

Fig. 3 Micrographs of coated samples at different temperatures

The feasibility of carrying out a simultaneous brazing and aluminizing with thermally sprayed aluminum could be shown. The process duration could be reduced compared to the process carried out in a pack cementation. Still, some pores are formed in the micro-structure of the brazed/aluminized coating due to diffusion- and segregation processes as well as the Kirkendall effect (Nakajima 1997).

The microstructure of a thermally sprayed coating is influenced by the spraying parameters. However, in this hybrid technology, the heat treatment parameters have an influence on the microstructure as well. In the present study, the influence or sensitivity of the coating parameters typical for thermal spraying and the impact of the heat treatment parameters of the microstructure of the coating were analyzed. Due to the fact that the combined brazing and aluminizing process could also be carried out successfully by thermally sprayed aluminum, it seems pertinent to integrate the TBC, consisting of yttria stabilized zirconia (YSZ, $ZrO_2 \cdot 8Y_2O_3$), in this hybrid technology. The results of a mutual heat treatment of the repair coating system Ni650/MCrAlY/Al/TBC carried out in a vacuum furnace and a shielding gas furnace using an argon/silane mixture as an inert gas are presented below. The interactions of the process parameters on the properties of the repair coating were investigated using electron microscopy analysis as well as adhesive tensile testing.

2 Materials and Methods

2.1 Materials

Inconel 718 flat specimens (30 mm × 30 mm × 2 mm) were used for the substrate. To activate the surface of the material, the samples were corundum blasted using EKF54 with a grain size from 250 to 355 μm. To remove the grid, the blasted specimens were cleaned with isopropanol in an ultrasonic bath, dried with pressurized air and afterwards coated by thermal spraying. The brazing material was the filler metal Ni650 (also known as B-Ni5), which consists of the alloying elements nickel, chromium and silicon, where the latter one is the melting point depressant. The hot

Table 1 Chemical composition (in wt%) of the materials used

Material	Inconel 718	Ni650 (B-Ni5)	MCrAlY
Ni	56.5	71.0	47.5
Co			23.0
Cr	19.0	19.0	17.0
Si		10.0	
Al	0.5		12.0
Y			0.5
Fe	18.0		
Nb + Ta + Ti	6.0		

Table 2 Coating parameters

Coating parameters	Values for B-Ni5, MCrAlY, Al	Values for TBC
Current (A)	390	420
Argon (L min^{-1})	40	40
Hydrogen (L min^{-1})	10	15
Powder feed rate (g min^{-1})	25–50	50
Traverse velocity (m s^{-1})	0.9–1.8	1.2
Nozzle distance (mm)	130	130

gas corrosive protection layer was a MCrAlY layer and the thermal barrier coating (TBC) an yttria stabilized zirconia ($ZrO_2 \cdot 8Y_2O_3$). The latter layers are state-of-the-art for turbine blades. The chemical compositions are listed in Table 1.

2.2 Coating Equipment

The coating of Inconel 718 specimen with the nickel-based filler metal Ni650 (thickness 250 μm), the hot gas corrosion protective material (MCrAlY, thickness 300 μm), aluminum (thickness 80 μm) and TBC (thickness 100 μm) was realized by atmospheric plasma spraying (APS, Delta torch, GTV-Verschleißschutz, Luckenbach, Germany). Table 2 shows the coating parameters employed.

2.3 Heat Treatment

The coated specimens underwent a heat treatment in a vacuum furnace (PVA Tepla, Wettenberg, Germany) with a pressure of 10^{-3} Pa within the furnace chamber. The heating rate was 20 K/min, the brazing/aluminizing temperatures were 1,090 and

1,190 °C and the dwells were 5 and 15 min. This was followed by free cooling under vacuum. For the heat treatment under an inert gas atmosphere, a shielding gas furnace (Kohnle HTE 1200-200/80–1500, Birkenfeld, Germany) was used.

2.4 Design of Experiments

To optimize the coating and heating parameters in the vacuum furnace with a focus on the microstructure of the filler metal, a design of experiments (DoE) approach was used. Parameters which influence the microstructure are: powder particle size (P) of the filler metal, powder feed rate (F), traverse velocity (V) of the torch, brazing temperature (T) and brazing time (t). Using two settings of each parameter with a high (+) and low (–) value plus the specimens in the state as-sprayed, the result is a $2^5 + 8 = 40$ full factorial DoE. The process parameters for the DoE are shown in Table 3.

Table 4 shows the principle approach of the DoE.

Table 3 Process parameters DoE

Parameter		Unit	+ Value	– Value
Powder particle size	P	μm	63–106	<63
Powder feed rate	F	g min^{-1}	50	25
Traverse velocity	V	m s^{-1}	0.9	1.8
Brazing temperature	T	°C	1,190	1,090
Brazing time	t	min	5	15

Table 4 DoE approach employed

Sample No.	P	F	V	T	t	
1	+	+	+			As-sprayed
2	+	+	+	+	+	
3	+	+	+	+	–	
4	+	+	+	–	+	
5	+	+	+	–	–	
6	+	+	–			As-sprayed
...						
39	–	–	–	–	+	
40	–	–	–	–	–	

2.5 Characterization of the Coatings

To characterize the coated and brazed/aluminized samples, materialographic cross sections were prepared for microstructural analysis. Both light microscope (Axioplan 2 microscope, Zeiss, Oberkochen, Germany) and electron microscope (FE-REM SUPRA 40 VP, Zeiss, Oberkochen, Germany) images were taken. This provided data about the bonding of the coating to the base material as well as the porosity. The porosity was determined using the software ImageJ (2021). The concentration profile and the distribution of the alloying elements were obtained by EDX analyses. The tensile adhesive strength of the thermally sprayed coatings were determined according to DIN EN ISO14916 (2017). Figure 4 depicts the specimen geometry used for the tensile tests.

Two loading blocks made of mild steel with a diameter of 40 mm were used for the adhesive test sample. A coated Inconel 718 disc was fixed between these loading blocks with an adhesive bond (HTK Ultra Bond 100, HTK Hamburg GmbH, Germany). The prepared test specimens were clamped into a universal testing machine (Walter and Bai AG, Switzerland) and the adhesive strengths of the repair coating were determined.

Fig. 4 Specimen for adhesive tensile test (DIN EN ISO 14916)

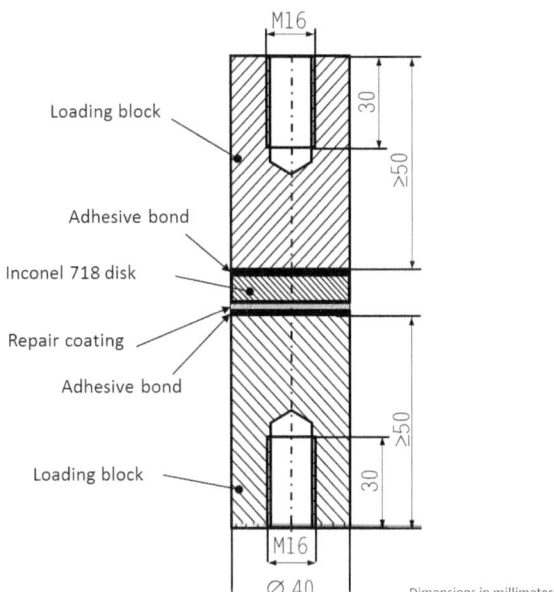

3 Results and Discussion

3.1 Formation of the Microstructure in the Coating and Heat Treating Processes

Figure 5 exemplarily depicts the cross sections of a coated Inconel 718 flat specimens in the as-sprayed state.

Starting from the Inconel 718, the nickel-based filler metal is followed by the MCrAlY hot gas corrosion protective layer and finally the aluminum. In the as-sprayed state, the boundary between the filler metal and the MCrAlY layer is difficult to recognize. As mentioned in the previous section, a DoE was carried out to reduce the pores in the filler metal. Figure 6 shows the measured porosity in the filler metal of the samples.

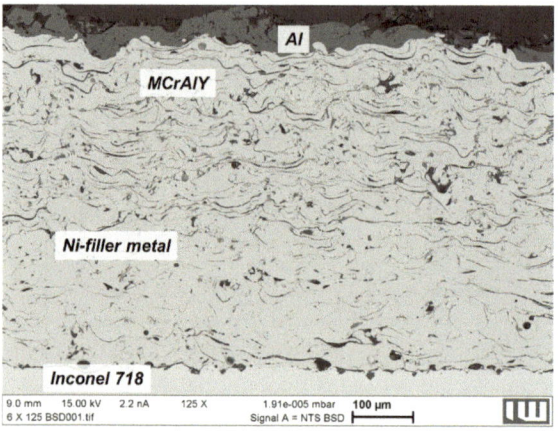

Fig. 5 Cross section of a coated Inconel 718 as-sprayed specimen (SEM picture)

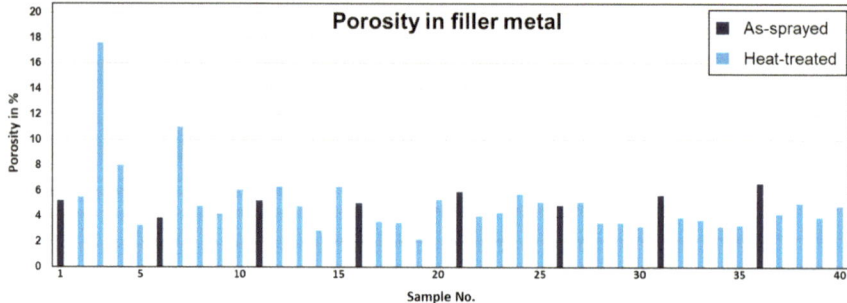

Fig. 6 Porosity of coated specimens, as-sprayed and heat-treated

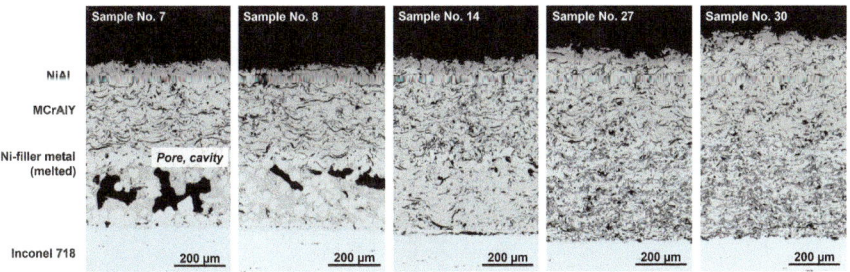

Fig. 7 Heat treated specimens with different coating and heat treatment parameters

The average porosity of the samples in the as-sprayed state was 5.3% with little scatter, while the porosity of the heat treated samples demonstrated strong variations, especially of the samples 2–20. These samples were coated with the coarse filler metal powder and the porosity varied from 2.2 to 17.6%. The porosity of the samples 21–40, which were coated with the fine filler metal powder, was nearly constant having a value of 4.2%. Figure 7 shows the cross sections of selected specimens.

After the heat treatment, the boundary between the filler metal and the MCrAlY became clearly visible, because the filler metal had melted and changed the microstructure. Moreover, a grey area with a thickness of approximately 50 μm can be observed on the surface, due to the formation of the NiAl-phase (β-phase). Samples No. 7 and 8 were heat treated at 1,190 °C and pores and cavities were developed. Sample No. 14 underwent a heat treatment at 1,090 °C and only few pores or cavities were observed. The pores or cavities occur due to diffusion processes caused by local differences in concentration of the alloying elements. Furthermore, segregation effects and volume contractions while the filler metal solidifies are responsible for the formation of pores. Sample No. 27 and 30 are representative for the specimens coated with fine filler metal powder. There are no changes in the microstructure. On the one hand, there are less pores in the filler metal layer. On the other hand, the filler metal layer is interspersed with grey seams. These are due to the oxidation of the fine powder during the coating process. Oxides counteract the flow ability of the filler metal, so that the coating and heat treating parameters have no pronounced influence on the microstructure. When using coarse powder, dark and bright areas can be observed in the filler metal layer and the compositions of these areas are different. Exemplarily, Fig. 8 illustrates a brazed seam (sample no. 12) along with the compositions of these areas. The microstructure is similar to sample no. 7 (cf. Fig. 7). Theses samples were heat treated at high temperature (1,190 °C) and long dwell times (15 min).

In the ternary phase diagram Ni–Cr–Si different solid solutions and phases can be formed (Gupta 2006; Schuster and Du 2000). Because of the heat treatment, for both compositions within the brazed seam, alloying elements from the Inconel 718 dissolve in the filler metal and are distributed over the filler metal layer. Diffusion processes cause an isothermal solidification that leads to the areas with composition

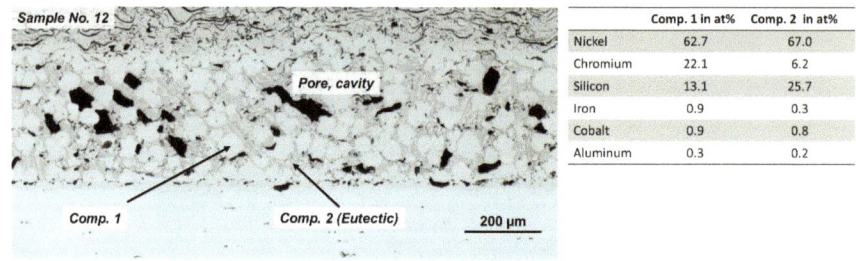

Fig. 8 Brazed seam with chemical composition of the areas marked

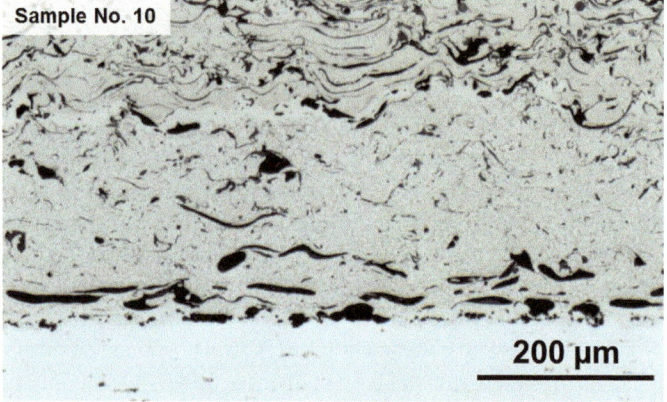

Fig. 9 Brazed seam at low heat treatment temperature and short dwell (1,090 °C, 5 min)

1. These are formed primarily at the boundaries between Inconel 718/B-Ni5 and B-Ni5/MCrAlY. Due to these diffusion processes, some remaining areas in the liquid state are enriched with silicon. These areas then have the composition 2, which is formed more inside the brazed seam. This composition can be assigned to the eutectic composition in the binary system Ni–Si (Nash and Nash 1987).

When using heat treatment parameters with low values (1,090 °C, 5 min), the isothermal solidification at the boundary of the filler metal to the Inconel 718 and the MCrAlY layer is not pronounced and the areas with compositions 1 and 2 are distributed equally all over the brazed seam. Due to the low temperature and short dwell, the diffusion process is slowed down, which leads to the microstructure shown in Fig. 9.

Due to the Kirkendall effect (Nakajima 1997), some bonding defects of the filler metal to the Inconel 718 can be observed. However, the formation of pores in the brazed seam is substantially reduced, which can also be attributed to a higher viscosity of the filler metal at lower temperature. A detailed analysis of the DoE and the impact of the formed micro-structure is given in (Nicolaus et al. 2021). The DoE offers a set of parameters, which leads to an optimized, low porosity microstructure within

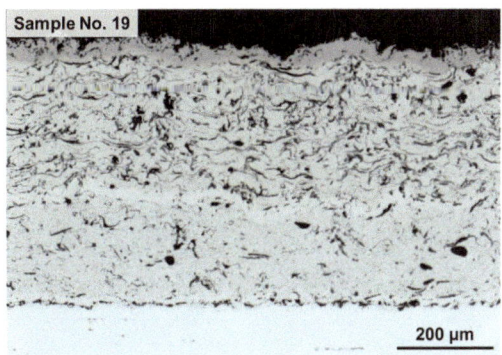

Coating parameters	Values for B-Ni5
Current in A	390
Argon in L·min⁻¹	40
Hydrogen in L·min⁻¹	10
Powder feed rate in g·min⁻¹	25
Traverse velocity in m·s⁻¹	0.9
Nozzle distance in mm	130
Powder particle size in μm	> 63
Brazing temp. in °C	1,090
Dwell in min.	15

Fig. 10 Microstructure of a specimen heat treated with optimized parameters

the brazed seam. The microstructure and the parameters are shown in Fig. 10. The porosity was determined to be 2.2% (cf. Fig. 6).

To get the lowest porosity, the brazing temperature must be set to a low value and an extended dwell time should be employed. The powder feed rate and the traverse velocity must be set to low values. The reason for the latter is still unclear. It should be noted that the DoE must be adapted to the material system used to account for the individual differences in segregation and diffusion processes.

3.2 Including the Thermal Barrier Coating into the Hybrid Process

It could be demonstrated that the aluminizing can be carried out using a thermally sprayed aluminum layer and that the usage of a pack cementation or a special aluminizing furnace is no longer needed. There are some additional aspects that should be considered if the thermal barrier coating should be integrated into this hybrid process as well: (i) The ternary phase diagram of aluminum-zirconium-oxygen indicates the formation of intermetallic Al-Zr-phases (Harmelin 1993; Zhao and Sun 2001), which leads to the assumption that these phases can improve the bonding of the ceramic thermal barrier coating to the MCrAlY layer. (ii) It is also described in the literature that the aluminizing of a MCrAlY layer by pack cementation increases the lifetime of a thermal barrier coating (Lih et al. 1992).

Based on these data, the developed repair process so far was extended, so that the thermal barrier coating could be integrated into the hybrid technology. Starting from the previous repair coating, the extended layer system filler metal/MCrAlY/Al/ TBC, with M = Ni/Co and TBC = $ZrO_2 \cdot 8Y_2O_3$ resulted. Figure 11 shows a coated Inconel specimen with a SEM picture of the cross section.

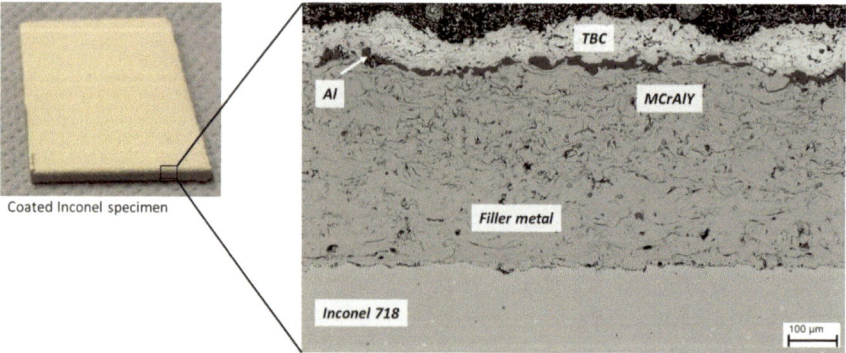

Fig. 11 Coated Inconel 718 specimen with SEM image demonstrating the individual layers

3.2.1 Heat Treatment in a Vacuum Furnace

The coated specimen underwent a heat treatment with the parameters obtained from the DoE mentioned in Sect. 3.1. Figure 12 shows the result of the heat treatment carried out in a vacuum furnace.

After the heat treatment of the coated specimen in vacuum, delamination of the ceramic thermal barrier coating was evident. An analysis of the delaminated area revealed that the delamination progresses between the Al/MCrAlY and the TBC layer. The more detailed analysis of the delaminated coating in Fig. 13 shows the element distribution of aluminum and oxygen at 550 and 1,090 °C.

At 550 °C, which is below the melting point of aluminum, no delamination of the TBC occurs. At 1,090 °C, the concentration of the aluminum at the top of the MCrAlY layer is higher than at 550 °C. Aluminum diffuses into the MCrAlY layer due to a gradient in the chemical potential between these layers. At 550 °C, oxygen

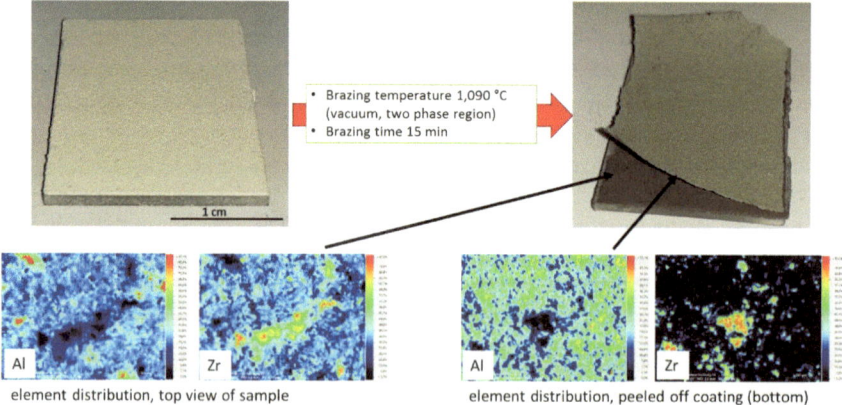

Fig. 12 Element distribution maps of heat treated sample including the TBC

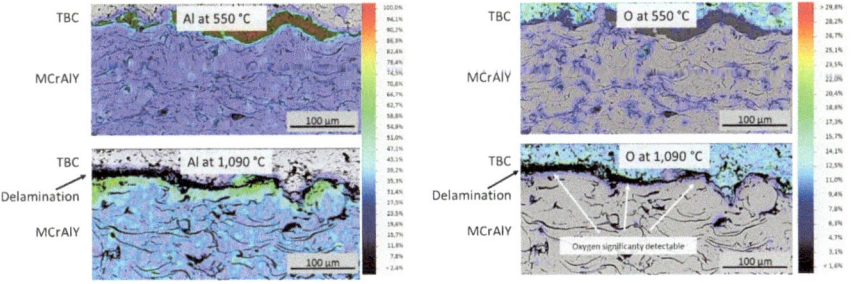

Fig. 13 Element distribution of Al and O in heat treated sample at different temperatures

was detected in the MCrAlY layer. At 1,090 °C, oxygen is significantly present at the top of the MCrAlY, but hardly detected within the MCrAlY layer. At higher temperatures, oxygen dissolves and is distributed homogeneously in the MCrAlY layer, thus leading to a dilution of oxygen and the concentration finally is below the detection limit. The detectable oxygen at the top of the MCrAlY layer can be explained as follows: The coated Inconel 718 sample was heat treated in vacuum at 1,090 °C. The pressure within the furnace was 10^{-8} bar and residual oxygen is present. Yttria stabilized zirconia is an oxygen ion conductor (Yoon et al. 2013) and oxygen diffuses via the TBC towards the aluminum. This leads to the formation of alumina (Al_2O_3) and causes a delamination of the TBC. The detectable oxygen at the top of the MCrAlY layer at 1,090 °C is not "free" oxygen, but the result of the alumina formation. Regarding the other alloying elements like Ni, Co and Cr, the formation of different chemical compounds and phases is possible. Figure 14 shows the corresponding element distribution at 1,090 °C.

Beside the formation of NiAl (β-phase) and Al_2O_3, the element distribution suggests that different oxides like NiO, $NiAl_2O_4$ and other mixed oxides can be formed, taking into account all alloying elements. This is in agreement with information from literature (Lv et al. 2022; Saltykov et al. 2004). According to the phase diagram Ni-Al (Nash and Nash 1987), the β-phase is a stoichiometric, thermodynamically stable composition, so that the aluminum can be substituted with a NiAl

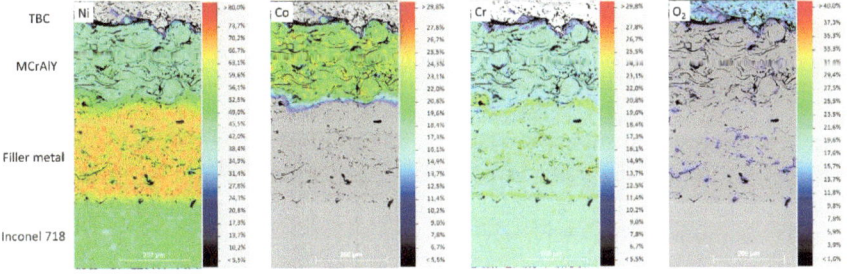

Fig. 14 Distribution of additional alloying elements of the sample heat treated at 1,090 °C

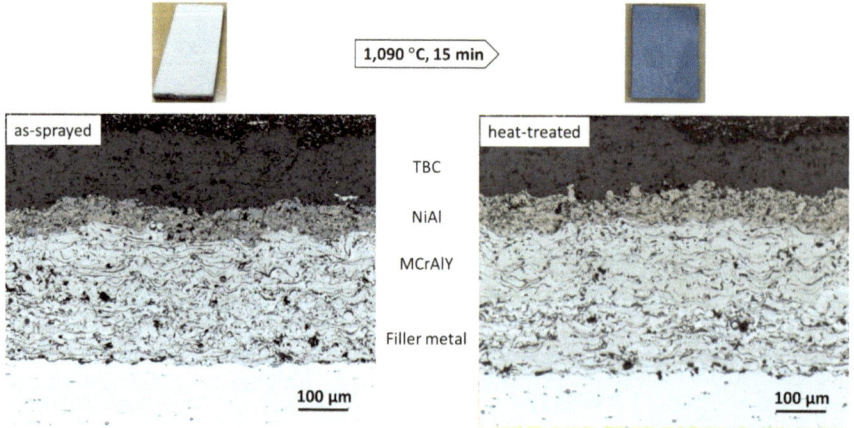

Fig. 15 Heat-treated sample, using a NiAl layer

layer. This led to the following repair coating system: filler metal/MCrAlY/NiAl/ TBC. The coated Inconel specimen underwent a heat treatment (1,090 °C, 15 min) and no delamination occurred. The result and the corresponding cross sections are shown in Fig. 15.

The reason for the formation of this microstructure is the kinetic of the oxidation of the bond coat used (Al or NiAl). The standard Gibbs free energies of the formation of NiAl, NiO and Al_2O_3 are −131 kJ/mol, −234 kJ/mol and −1300 kJ/mol, respectively (Róg et al. 2003; Holmes et al. 1986; Yang et al. 2014). From a thermodynamic point of view, the equilibrium is on the side of these components. Clearly, thermodynamic data does not directly provide the kinetics of chemical reactions. Using aluminum as a bond coat, NiAl is formed at the boundary to the MCrAlY layer, while aluminum oxidizes at the boundary to the TBC without an intermediate step. Using NiAl as a bond coat, NiO and Al_2O_3 are formed at the boundary to the TBC. However, several stages are needed to oxidize the NiAl layer (Unocic et al. 2017), which lower the rate of oxidization, and thus no delamination occurs. Further investigations must be carried out to better understand what happens at the boundary NiAl/TBC in detail (e.g. formation of additional intermetallic phases, like ZrAl, ZrNi etc.).

3.2.2 Heat Treatment in a Shielding Gas Furnace

The results presented so far are based on a heat treatment carried out in a vacuum furnace. The knowledge gained was exploited to transfer the heat treatment process into a shielding gas furnace. Silane (SiH_4) doped argon was used as the shielding gas. The addition of SiH_4 in the single-digit ppm range is sufficient to quantitatively remove oxygen and water impurities in conventional process gases. In this way, extremely high vacuum adequate process conditions can be established (Bach et al. 2010–2011). The brazing temperature was 1,120 °C and the dwell was 25 min, which

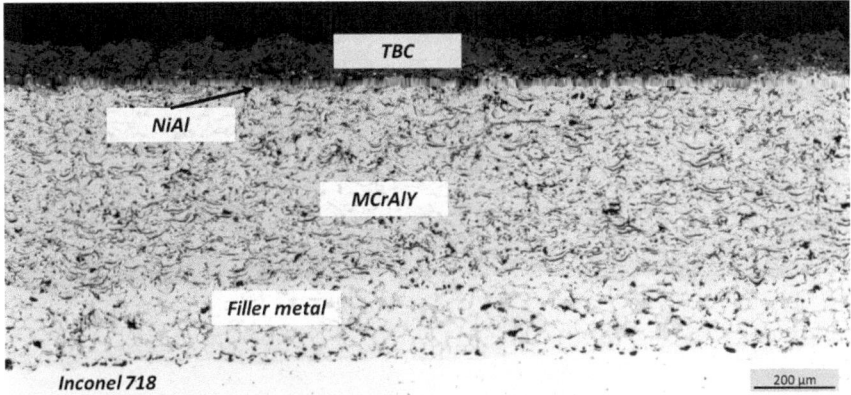

Fig. 16 Microstructure of a sample heat-treated in a shielding gas furnace

represents a standard brazing process under these conditions. Figure 16 shows the cross section of a sample, heat treated in a shielding gas furnace.

The microstructure of the sample is nearly the same as can be seen in Fig. 8 but with less pores and can be explained by the lower brazing temperature used.

3.3 Tensile Adhesive Strength

The previous section described the impact of the heat treatment on the microstructure and the element distribution in the coating. For further assessment, tensile adhesive strength tests of the applied thermal barrier coating in the as-sprayed state and the heat-treated repair coating in a vacuum- and a shielding gas furnace were carried out according to DIN EN ISO 14916 (2017). The results are summarized in Fig. 17.

The tensile adhesive strength of the thermal barrier coating in the as sprayed state was determined to 16 ± 1 MPa. This is a typical value for ceramic coatings and is in the range of data reported in the literature (Limarga et al. 2005; Lima and Guilemany 2007; Pugacheva et al. 2019). Figure 17 also shows a typical failure pattern of the tested coating due to adhesive and cohesive failure. The sample, which was heat treated in the vacuum furnace, shows an improved strength of 22 MPa at first glance but has a scatter of ± 10 MPa. This is caused by cracks formed in the TBC, which led to premature failure of the coating. Obviously, the process parameters must be adjusted, but the trend towards higher tensile adhesive strength is already apparent. In case of heat treated samples in a shielding gas furnace, the tensile adhesive strength is even higher and reached 55 MPa with a scatter of ± 4 MPa. The true value of the tensile adhesive strength could not be determined because the failure of glue applied to the coated test specimens. Research is underway to transfer the repair process into actual industrial application.

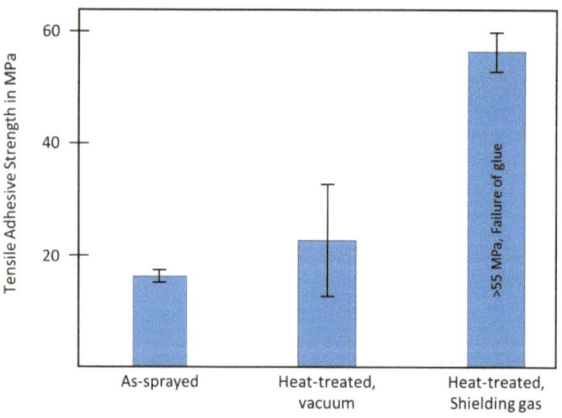

Fig. 17 Adhesive tensile strength of differently heat-treated samples

4   Conclusions

This study demonstrated the development of a hybrid technology for repairing turbine blades of the high pressure turbine. This hybrid technology employs a simultaneous brazing and aluminizing process. Specifically a thermally sprayed repair coating, consisting of the filler metal, the hot gas corrosion protective coating, a nickel aluminum layer and finally the thermal barrier coating is applied. The main results can be summarized as follows:

1. The feasibility of the developed hybrid technology could be shown.
2. With this technology, it is possible to shorten the state-of-the-art repair brazing process of turbine blades.
3. The aluminizing process can be integrated in the hybrid process. This eliminates the need for pack cementation or use of a special aluminizing furnace.
4. The diffusion processes and phase formations in the repair coating of a simultaneous brazing and aluminizing process can be tailored to achieve a well adhering coating with low porosity.

Acknowledgment Funded by the Deutsche Forschungsgemeinschaft (DFG, German Research Foundation) SFB 871/3 119193472.

References

Bach, F. W., Möhwald, K., and Holländer, U. (2010). Physico-chemical aspects of surface activation during fluxless brazing in shielding-gas furnaces. Key Engineering Materials 438:73–80

Bach, F. W., Möhwald, K., Holländer, U., Schaup, J., Roxlau, C., and Langohr A. (2011). Boron and phosphorous free nickel based filler metals for brazing stainless steel in shielding gas furnaces. International Journal of Materials Research 08:964–971

Brooks, J. W. and Bridges, P. J. (1988). Metallurgical Stability of Inconel Alloy 718. In: Superalloys 1988. The Metallurgical Society, TMS, Pittsburgh, PA, pp 33–42

Česnik, D., Bratuš, V., Kosec, B., Rozman, J., and Bizjak, M. (2008). Heat treatment and fine-blankin Inconel 718. RMZ Mater. Geoenviron. 55:163–172

DIN EN ISO 14917:2017 (2017). Thermal spraying - Terminology, classification. Beuth, Berlin, Germany

DIN EN ISO 14916:2017 (2017). Thermal spraying – Determination of tensile adhesive strength. Beuth, Berlin, Germany

Gupta, K. P. (2006). The Cr-Ni-Si (Chromium-Nickel-Silicon) System. J. Phase Equilibria Diffus. 27(5):523–528

Harmelin, M. (1993). Al-O-Zr Ternary Phase Diagram Evaluation. In: Effenberg G (ed) MSI Eureka, MSI, Materials Science International Services GmbH, Stuttgart

Henderson, M. B., Arrell, D., Larsson, R., Heobel, M., and Marchant, G. (2004). Nickel-based superalloy welding practices for industrial gas turbine applications. Science and Technology of Welding and Joining 9(1):13–21

Holmes, R. D., O'Neill, H. St. C., and Arculus, R. (1986). Standard Gibbs free energy of formation for Cu_2O, NiO, CoO, and Fe,O: High resolution electrochemical measurements using zirconia solid electrolytes from 900-1400 K. Geochimica et Cosmochimica Acta 50:2439–2452

Huang, X. and Miglietti, W. (2011). Wide Gap Braze Repair of Gas Turbine Blades and Vanes - A Review. J. Eng. Gas Turbines Power, 134(1), 010801–010801-17

ImageJ (2021) Image Processing and Analysis in Java. https://imagej.nih.gov/ij/. Accessed 20 Jan 2021

Lih, W., Chang, E., Wu, B. C., and Chao, C. H. (1992). The Effect of Pack-Aluminization on the Microstructure of MCrAlY and the Performance of Thermal Barrier Coatings. Surface & Coatings Technology 50:277–288

Lima, C. R. C. and Guilemany, J. M. (2007). Adhesion improvements of Thermal Barrier Coatings with HVOF thermally sprayed bond coats. Surface & Coatings Technology 201:4694–4701

Limarga, A. M., Widjaja, S., and Yip, T. H. (2005). Mechanical properties and oxidation resistance of plasma-sprayed multilayered Al_2O_3/ZrO_2 thermal barrier coatings. Surface & Coatings Technology 197:93–102

Lv, Y.L., Ren, Y. J., Zhou, M. N., Feng, K. K., Chen, J., and Niu, Y. (2022). The Oxidation of Four Co-20Ni-xCr-yAl (x = 8,15 wt.%; y = 3,5 wt.%) Alloys Under 1 atm O_2 at 800 °C and 900 °C. Oxid Met (2022). https://doi.org/10.1007/s11085-022-10106-6

Miracle, D. B. (1993). The Physical and Mechanical Properties of NiAl, Overview No. 104. Acta metall. mater. 41:649–684

Nakajima, H. (1997). The discovery and acceptance of the Kirkendall Effect: The result of a short research career. JOM 49:15–19

Nash, P. and Nash, A. (1987). The Ni−Si (Nickel-Silicon) system. Bulletin of Alloy Phase Diagrams 8:6–14

Nicolaus, M., Möhwald, K., and Maier, H. J. (2017a). Regeneration of high pressure turbine blades. Development of a hybrid brazing and aluminizing process by means of thermal spraying. Procedia CIRP 58:72–76

Nicolaus, M., Möhwald, K., and Maier, H. J. (2017b). A Combined Brazing and Aluminizing Process for Repairing Turbine Blades by Thermal Spraying Using the Coating System NiCrSi/NiCoCrAlY/Al, Journal of Thermal Spray Technology 26(7):1659–1668, https://doi.org/10.1007/s11666-017-0612-z

Nicolaus, M., Rottwinkel, B., Alfred, I., Möhwald, K., Nölke, C., Kaierle, S., Maier, H. J., and Wesling, V. (2018). Future regeneration processes for high-pressure turbine blades, CEAS Aeronaut J 9(1):85–92, https://doi.org/10.1007/s13272-017-0277-9

Nicolaus, M., Möhwald, K., and Maier, H. J. (2021). Thermally Sprayed Nickel-Based Repair Coatings for High-Pressure Turbine Blades: Controlling Void Formation during a Combined Brazing and Aluminizing Process. Coatings 11(725). https://doi.org/ https://doi.org/10.3390/ 11060725

Pugacheva, N. B., Guzanov, B. N., Obabkov, N. V., Bykova, T. M., and Michurov, N. S. (2019). Studying the Structure and Adhesion Strength of Thermal Barrier Coating. In: AIP Conference Proceedings 2176, 030013. https://doi.org/10.1063/1.5135137

Pytel, M., Góral, M., Nowotnik, A., Sieniawski, J., Drajewicz, M., and Ziaja, M. (2012). Heat Treatment and CVD Aluminizing of Ni-Base René 80 Superalloy. J. Achiev. Mater. Manuf. Eng. 51(1):30–38

Rao, G. A., Kumar, M., Srinivas, M., and Sarma, D. S. (2001). Effect of Solution Treatment Temperature on the Microstructure and Tensile Properties of P/M (HIP) Processed Superalloy Inconel 718. In: Superalloys 718, 625, 706 and Various Derivatives. The Minerals, Metals & Materials Society, TMS, Pittsburgh, PA, pp 605–616

Róg, G., Borchardt, G., Wellen, M., and Löser, W. (2003). Determination of the activities in the (Ni + Al) alloys in the temperature range 870K to 920K by a solid-state galvanic cell using a CaF_2 electrolyte. J. Chem. Thermodynamics 35:261–268

Saltykov, P., Fabrichnaya, O., Golczewski, J., and Aldinger, F. (2004). Thermodynamic modeling of oxidation of Al–Cr–Ni alloys. Journal of Alloys and Compounds 381:99–113

Schirra, J. J. (1991). Effect of Heat Treatment Variations on the Hardness and Mechanical Properties of Wrought Inconel 718. In: Superalloys 718, 625, 706 and Various Derivatives. The Minerals, Metals & Materials Society, TMS, Pittsburgh, PA, pp 431–438

Schuster, J. C. and Du, Y. (2000). Experimental Investigation and Thermodynamic Modling of the Cr-Ni-Si System. Metall Mater Trans A 31A:1795–1803

Stolle, R. (2004). Conventional and advanced coatings for turbine airfoils, MTU Aero Engines Munich. http://www.academia.edu/7789702/Conventional_and_advanced_coatings_ for_turbine_airfoils. Accessed 19 Apr 2021

Unocic, K. A., Shin, D., Unocic, R. R., and Allard, L. W. (2017). NiAl Oxidation Reaction Processes Studied In Situ Using MEMS-Based Closed-Cell Gas Reaction Transmission Electron Microscopy. Oxid Met 88:495–508

Yang, W., Dong, S,, Luo, P., Yangli, A., Liu, Q., and Xiw, Z. (2014). Effect of Ni addition on the preparation of Al_2O_3–TiB_2 composites using high-energy ball milling. Journal of Asian Ceramic Societies 2:399–402

Yoon, S., Noh, T., Kim, W., Choi, J., and Lee, H. (2013). Structural parameters and oxygen ion conductivity of Y_2O_3–ZrO_2 and MgO–ZrO_2 at high temperature. Ceramics International 50(8):9247–9251

Zhan, Z., He, Y., Li, L., Liu, H., and Dai, Y. (2009). Low-temperature formation and oxidation resistance of ultrafine aluminide coatings on Ni-base superalloy. Surface & Coatings Technology 203:2337–2342

Zhao, Y. T. and Sun, G. X. (2001). In Situ Synthesis of Novel Composites in the System Al-Zr-O. J. Mater. Sci. Lett. 20:1859–1861

Dexterous Regeneration Cell

Berend Denkena, Tim Schumacher, and Markus Hein

Abstract Complex capital goods such as components from aircraft engines or gas and steam turbines usually have free-form geometry features, individual material deposits, and poor machining accessibility. The recontouring of these components requires increased machine and process flexibility. In sub-project B2, "Dexterous Regeneration Cell", a dexterous milling repair cell was researched, representing a central component of the real regeneration path in the Collaborative Research Centre (CRC) 871. "Dexterousness" is understood as the ability to carry out a self-optimizing, best possible repair machining. The combination of advanced methods for process design, novel machine tool technologies and adaptive machining functionalities allows the reliable 5-axis recontouring of individual damage cases despite repair-specific variances, influences from upstream processes, and flexibility of the workpiece or tool. In this paper, a force model is created using artificial intelligence to determine variances. Further process knowledge is obtained to minimize shape deviations. Furthermore, the displacement in the recontouring process of turbine blades is carried out with the help of a magnetically guided spindle.

Keywords Deflection compensation · Electromagnetic guide · Process design · Process force models

B. Denkena (✉) · T. Schumacher · M. Hein
Institute of Production Engineering and Machine Tools, Leibniz University Hannover, An der Universitaet 2, 30823 Garbsen, Germany
e-mail: Denkena@uni-hannover.de
URL: https://www.ifw.uni-hannover.de/de/

T. Schumacher
URL: https://www.ifw.uni-hannover.de/de/

M. Hein
URL: https://www.ifw.uni-hannover.de/de/

1   Introduction

The repair and maintenance of aircraft engines are safety-related but also of economic interest to the airlines and engine manufacturers. High-value products like aircraft engines require effective maintenance, repair, and overhaul services to increase economic sustainability and efficiency (Uhlmann et al. 2013). Compressor and turbine blades made from Ti–6Al–4V or Inconel 718 are an example of products for which a repair is highly motivated. Because of the complex geometry and their integral design, these repair processes are a challenging task and differ from new part production (Denkena et al. 2015). A typical repair process chain for blades consists of four phases, namely pretreatment, material deposit, recontouring and post-treatment (Eberlein 2007; Kelbassa et al. 2002).

By recontouring, excess material that was deposited on the part by welding or soldering is removed. This process is challenging due to freely formed geometric features and individual material deposits. Therefore, it is often executed by skilled workers with handheld tools. In addition to the need for component-specific process planning (SP C1), form deviations due to cutting force are the key challenge for automated recontouring on milling machines (Denkena et al. 2021). Therefore, the aim of this sub-project is the process-safe 5-axis recontouring of individual damage cases despite repair-specific variances, influences from preceding processes, and compliances of the workpiece and tool.

Every milling process is associated with forces due to the material separation at the cutting edges. These process forces act in the contact zone between the milling cutter and the workpiece and deform all components in the flow of forces. The tool and the workpiece are deformed, including the machine tool structure, which closes the force flow between the tool and the workpiece (Fig. 1). The deflection of the tool and the workpiece leads to a shift in the Tool Center Point (TCP) and results in a deviation of the finished workpiece shape from the nominal shape.

The shape deviations caused by displacement are minimized by adequately adapting the process parameters. In addition, adaptive machining approaches can be used to reduce errors. The state of the art for these two areas is briefly outlined in the sections below.

1.1   Modelling of Milling Processes, Process Force Models and Process Design

For predictive error avoidance and, therefore, for process planning, a high level of knowledge about the milling process is required. Following the material deposit, the excess material must be removed. The removal is done via milling or grinding. For the repair of compressor or turbine blades, 5-axis ball end milling is often used to restore the complex shape (Eberlein 2007; Gao et al. 2008). Furthermore, in addition to high geometric accuracy and low surface roughness, compressive residual stresses

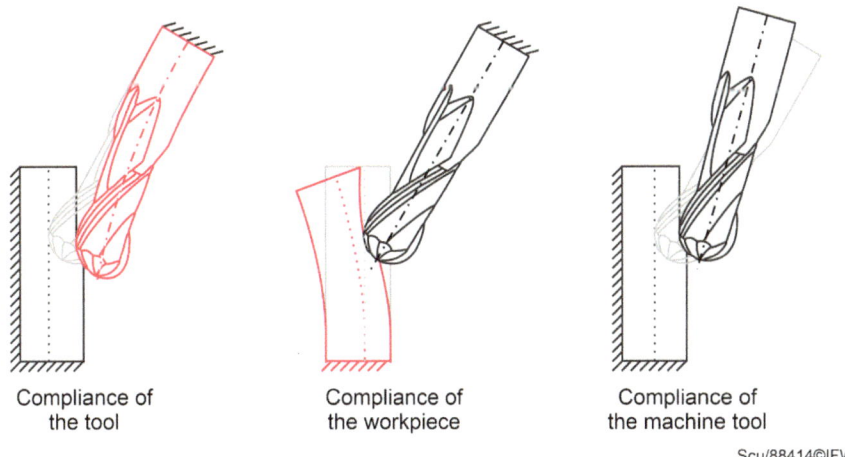

Compliance of
the tool

Compliance of
the workpiece

Compliance of
the machine tool

Scu/88414©IFW

Fig. 1 Tool, workpiece, and machine tool deflection in milling processes

in the component surface can be achieved with ball end milling tools (Denkena et al. 2014b). Subsequently, post treatment processes such as shot peening can be used to improve surface integrity. As already mentioned, the machining of blades poses some challenges, in particular, shape deviations, which are primarily caused by the deflection of the workpiece and tool during machining. The low Young's modulus of titanium as well as Inconel and wall thicknesses less than 2 mm may result in vibrations. Adjusted process parameters and special clamping systems are therefore often necessary for a successful recontouring. In principle, it is very important to avoid vibrations in the recontouring process of compressor and turbine blades due to the long projection length of the blades. In addition, the recontouring of these components also involves different material properties between the base material and the deposited material, which can lead to local variances in the milling process. This also poses a special challenge to the recontouring process. Using 5-axis processes, selecting an appropriate tool orientation during process planning can help avoid vibrations and increase surface quality (Ozturk et al. 2009). There are a few studies on the influence of tool orientation on process dynamics in 5-axis milling. For example, Ozturk and Budak (2007) developed a model to predict cutting forces and investigated the effect of tool inclination on chatter stability during 5-axis ball end milling. Tuysuz et al. (2013) extended the model by considering the tool indentation effect. Layegh et al. (2015) analyzed the effect of tool orientation during the 5-axis ball end milling of flexible parts. They proposed mechanics-based tool posture maps, which reveal part deflection and cutting torque values depending on the inclination angle. All these studies have their focus on machining new parts and do not specifically address the recontouring process.

Thus, boundary conditions like different material properties between the deposited material and the base material are not considered. In this respect, it has already been shown along with the CRC 871 that the angle of attack during the recontouring of turbine blades has a significant influence on the process stability (Denkena et al. 2016).

In conjunction with 5-axis ball end milling in the recontouring process, there are many influencing factors on the resulting component quality. In addition to the process parameters, the tool geometry, the machine rigidity, and the previous material deposit process also influence the surface quality and the existing process forces (Denkena et al. 2020; Engin and Altintas 2011). Variances in the recontouring process can be determined by deviations in the process forces. Therefore, the expected process forces must be continuously compared with the real process forces. The difference between the predicted and the measured process forces can then be used to determine local variances in the workpiece properties. For process force prediction, Machine Learning (ML) can be applied.

The use of ML in machining is already widespread. Many researchers have successfully applied ML techniques in studying various manufacturing problems. In Cho et al. (2005), the authors used Support Vector Machines (SVMs) to provide a detection system which is able to define process anomalies during a milling process based on power consumption and process forces. In their investigations they show that, compared to a multilinear regression model, the Support Vector Regression (SVR) model performs with higher accuracy. They concluded that SVR models can lead to a reduction in machine downtime and thus to a reduction in production costs. Hashemitaheri et al. (2020) compared the SVR and the Gaussian Process Regression (GPR) for the prediction of cutting forces and maximum tool temperature. They observed that both SVR and GPR models predict these parameters accurately although the former is marginally better than the latter. Further, Jurkovic et al. (2018) compared three different ML models to predict cutting parameters in a high-speed turning process. They used SVR, polynomial (quadratic) regression and Artificial Neural Networks (ANNs) based on input parameters of cutting forces, surface roughness, and tool life. In their study, the SVR and polynomial regression models show a higher accuracy in the prediction than the ANN model. In addition, Charalampous (2020) used four ML models to predict the cutting forces in milling AISI 4140 QT. He studied the accuracy of SVR, ANN, Polynomial Regressor and Random Forest. His results show the best performance was recorded from the SVR, followed by the ANN, the Polynomial Regressor, and lastly the Random Forest model. He reached an accuracy of about 96% with the SVR model. Krizek et al. (2007) applied the ML models response surface methodology, evolutionary algorithms, SVR and ANN to their data sets. They aimed to find the best approach to determine the cutting force based on the input variables spindle speed, feed rate, and depth of cut. The best percent deviation was reached by the SVR model. In summary, ML with different models is already being used to predict the process forces in cutting processes. However, it is not yet possible to identify the most suitable model for all processes. In addition, there is still no knowledge about the use of ML for process force prediction in the ball end milling of Inconel and titanium.

1.2   Adaptive Machining for Reduced Shape Deviations

Since repair cases are individual and underlie variances, a priori information from offline process simulations also shows a certain degree of uncertainty. This does not question their suitability for general process design but makes them insufficient for the adaptation of, e.g., a specific toolpath. On the one hand adaptive machining approaches react to the process forces in realtime (online), on the other hand they obtain the latest data from the running process. Therefore, they can make short-term adaptions and deal with unpredictable effects like tool wear or variations in the material properties.

Adaptive machining approaches for reduced shape deviations can be classified according to their working principle into process force restriction and deflection compensation. The process forces are usually restricted by reducing the feed rate in an automatic control loop. This way, productivity is sacrificed for accuracy (Dittrich et al. 2019; Denkena and Boujnah 2018). Furthermore, the displacement cannot be fully avoided since chip formation is always accompanied by forces and the structural stiffness of the machining setup is always limited. The compensation of deflection is a different approach. It accepts the deformation of the machine tool components and continuously counteracts the deflection by corrective movements. Thus, the shape error can be reduced without any loss of productivity (Boujnah 2019). Reactive deflection compensation has already been investigated for various areas of application. Usually, the displacement between the tool and the workpiece is determined indirectly. The process forces are measured and the deflections of tool or workpiece are calculated based on a compliance model. The necessary compensation movement is then carried out either by the NC axes of the machine tool or by additional actuators.

However, Brecher et al. (2019) and Boujnah (2019) have shown that control delays limit the effect of reactive deflection compensation when abrupt changes in the process forces occur. Since signal delays cannot be completely avoided, attempts were made to overcome this issue with an online force prediction. Hähn and Weigold (2020) combined a reactive compensation with an offline process simulation to reduce errors due to predictable force changes in Robot machining. Altintas (2011) has described an adaptive generalized predictive control of the process forces in milling. To avoid high peak forces, Stemmler et al. (2016) proposed a model predictive control (MPC) approach for the feed rate. For deflection compensation, a comparable approach is not yet known. Although a short-term force prediction for online compensation was considered (Hähn and Weigold 2020), it has not been implemented and researched yet. Predictive and anticipatory approaches for deflection compensation were also investigated within the sub-project B2 of CRC 871. These approaches have already been successfully implemented in flank milling processes (Mücke 2020; Denkena et al. 2021).

1.3   Approach

In the sub-project B2, "Dexterous Regeneration Cell" a dexterous repair cell is being researched, representing a central component of the physical regeneration path. In this project, new types of machines and cutting technologies were developed. The aim is the process-reliable 5-axis recontouring of individual damage cases despite repair-specific variances, influences from preceding processes and deflection of the workpiece and tool. The approach to reach this aim is a 2-step solution with a predictive error avoidance and reactive error compensation as shown in Fig. 2. Predictive error avoidance is carried out as part of the process planning. The results of process simulations are used to design processes that result in the lowest possible shape deviations. This includes planning the individual tool path, the process parameters, the process kinematics (tool orientation) and the force-stiffness alignment.

Since the process forces and thus the deflection cannot be completely avoided, the remaining deflection is compensated during machining in a second step. For this reactive error compensation, the process forces are recorded, converted into compensation values and superimposed on the position of the tool by an electromagnetic guide, which serves as an additional actuator. This technology was developed and investigated throughout the CRC 871 regarding its characteristics, its behavior in milling processes and its additional functionalities (Denkena et al. 2014a, b, 2016; Flöter 2017). The predictive error avoidance is explained in Sects. 2 (Modelling of process forces using Machine Learning) and 3 (Process design to minimize shape deviation). Further, Sect. 4 (Electromagnetic guide system for adaptive machining) introduces the regeneration cell with the electromagnetic guide, and the following Sects. 5 (Deflection compensation in flank milling) and 6 (Deflection compensation in 5-axis recontouring processes) discuss the reactive error compensation.

2   Modelling of Process Forces Using Machine Learning

The tool deflection can be determined with the help of the process forces. For this reason, knowledge about the existing process forces at different process parameters during the recontouring process is necessary. Therefore, four ML models were used

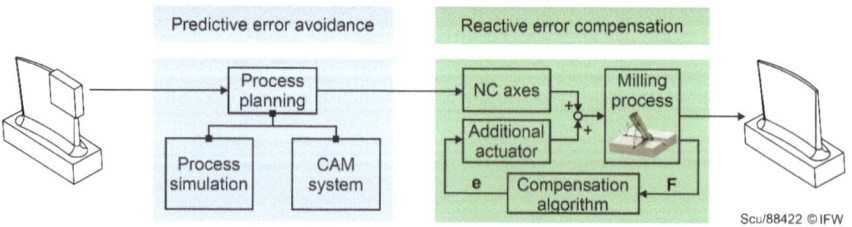

Fig. 2 Two-step solution for the reduction of shape deviations

for predicting process forces during 5-axis ball end milling under stable cutting conditions. The main objective is to identify a suitable ML model that accurately predicts the process forces, thereby enabling variance detection. For this purpose, the methods SVR, ANN, and the Automatic Machine Learning models Auto_ML and Tpot were used. The required training data was generated by varying the process parameters during the 5-axis ball end milling of Inconel. All milling investigations were carried out on a DMG Mori Milltap 700 5-axis CNC milling center. The ball end mills used are provided by SECO and have a diameter of 6 mm and six teeth. The cutting tools have a helix angle of 38°, a cutting edge rounding $\bar{S} = 15\ \mu\text{m}$ and a TiAlN coating. The DMG Mori Lasertec 65 3D DED was used to manufacture the workpieces. Rectangles were welded onto a base plate made of Inconel 718. Powder PWA 1480 with a powder diameter of 60–105 μm was used for the laser build-up welding process. A power of $P = 400$ W and a feed of $f_1 = 800$ mm/min as well as a delivery rate of $f_p = 2$ g/min were selected. The manufactured rectangles have a length of 40 mm, a width of 30 mm, and a height of 3 mm. After the welding process, the rectangles were face-milled to ensure equal infeed. The corresponding test setup for the investigations is shown in Fig. 3.

A Kistler-type 9257A three-axis piezoelectric dynamometer was used to measure the process forces during the milling process. The measurement device cooperates with one Kistler-type 5011 signal amplifier per axis. Additionally, the dynamometer was connected with a LabVIEW data acquisition system in which the force signals were acquired and transformed into a numerical format. The cutting procedures

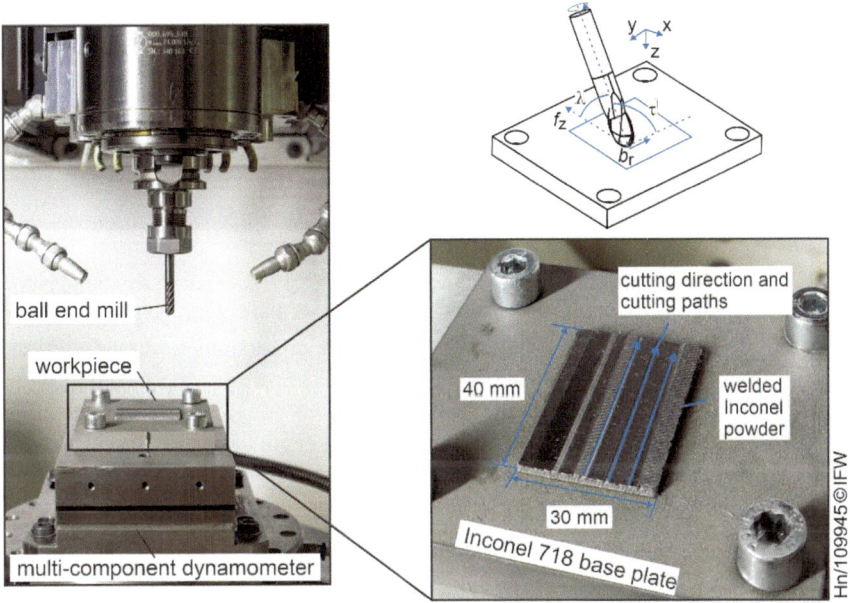

Fig. 3 Experimental setup for data collection of the ML models

were performed in dry milling without any applied cooling liquid. The reason for choosing one of the dry machining processes is a trend towards environmentally friendly manufacturing technology, as this avoids pollution from cutting fluids and reduces the overall cost.

During the cutting tests, the process parameters step over b_r, depth of cut a_p, feed per tooth f_z, lead angle λ and tilt angle τ were varied. The cutting speed v_c was held constant at 30 m/min during all investigations. A total of 258 process parameter combinations were carried out with two repetition tests each. One test consisted of machining a complete path with a length of 40 mm on the welded material. To keep the influence of tool wear low, the number of paths a tool can machine at constant process parameters without significant changes in the process forces was determined in advance. After eleven paths, a significant increase in the process forces was observed. For this reason, each tool was used for nine paths. For the process parameters investigated, one cutting edge of the tool was always engaged in the workpiece. This keeps the measured forces at a constant level. Therefore, the mean value of the process forces F_x, F_y and F_z is used to evaluate the process forces over the engagement range along the path. For the analysis of the prediction accuracy of the respective ML models, 10 milling tests of the 258 performed tests were extracted. These tests were excluded from the training step and utilized for the validation phase of the developed ML models. Further two criteria, Mean Absolute Error (MAE) and Mean Relative Error (MRE) were used to evaluate the forecasting efficiency. These criteria are calculated using the following Eqs. (1) and (2):

$$MAE = \frac{1}{N} \sum_{k=1}^{N} \left| y_k - \hat{y}_k \right| \tag{1}$$

$$MRE = \frac{1}{N} \sum_{k=1}^{N} \frac{\left| y_k - \hat{y}_k \right|}{y_k} \tag{2}$$

where N is the total number of data points, y_k and $\hat{y}_k$ represent the ith observation and models prediction, respectively.

A total of four methods were applied: Support Vector Regression, Artificial Neural Networks and the Automatic Machine Learning models Auto_ML and Tpot. It is observed that all four models generally provide predictions accurately. Further, Fig. 4 demonstrates that the Automatic Machine Learning models Auto_ML and Tpot achieve higher accuracy in predicting the forces with a small data size available in this investigation. With an average MRE of 2.2%, Tpot has an MRE that is at least 70% lower than that of the ML methods SVR and ANN. The developed methods allow the expected process forces to be continuously compared with the real process forces. This means that the difference between the predicted and measured process forces can be used to identify local variances in the workpiece properties during machining.

With the help of the ML model the generated variances in the milling process are detected. Variances can occur during the welding process. To investigate the

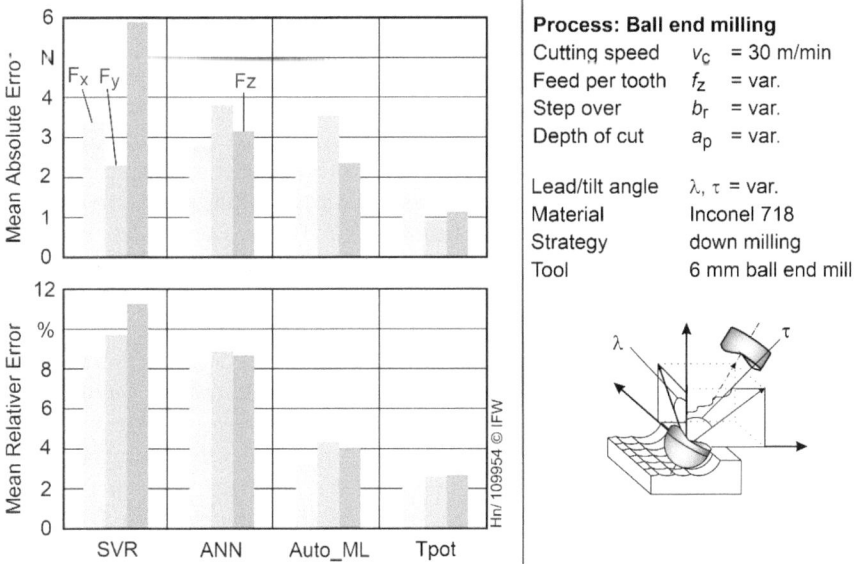

Fig. 4 Comparison of the investigated ML methods

effects, the process parameters power ($P = 400/800/1,200$ W) and feed ($f_1 = 400/800/1,200$ mm/min) were each varied in three steps. The hardnesses of these samples were then measured with the Struers Duramin measuring device according to DIN EN ISO 6507-1. Table 1 shows the resulting hardness values for the respective process parameter combinations of the welding process. The hardness values are averaged from six randomly distributed indentations per surface. It is noticeable that the hardness of the workpieces decreases with an increase in power P of the welding process. Furthermore, it can be concluded that the feed rate also has a significant influence on the workpiece hardness. Increasing the feed rate in the build-up welding process also reduces the hardness of the workpiece.

In the next step, the influence of the hardness on the process forces was analyzed. For this purpose, milling tests were carried out with six different hardnesses for one process parameter combination. The hardness 667 HV1 was generated with welding

Table 1 Differences in hardness after the welding process

Power P (W)	Feed f_1 (mm/min)	Delivery rate f_p (g/min)	Hardness HV1
400	800	2	667
800	800	2	635
1,200	800	2	622
400	400	2	713
400	1,200	2	583

process parameters used previously in this section ($P = 400, f_1 = 800$ mm/min, f_p $= 2$ g/min).

In Fig. 5, it is apparent that the forces generally increase with an increase in hardness. However, it can also be seen that not all force components increase equally. The force F_x does not change significantly with values between 583 and 635 HV1, as can be seen in Fig. 5. Only for hardnesses 667 HV1 and higher is there an increase in this force component as well. On average, the force in the milling process increases by 42.5% from hardness of 583 HV1 to a hardness of 713 HV1. The smallest increase in force is 4.4% for the force component F_x. An increase in hardness of 20% leads to an average increase in force of 42.5% for the process parameters used here. A deviation in hardness in the build-up welding process consequently leads to deviations in the process forces. These can be determined with the ML model. If a deviation of 10% of the predicted process forces compared to the real process forces occurs, the machining process is stopped. Since the greatest deviation of these process parameters is seen in the process force F_y at the different hardnesses, a tolerance of 10% is set along with this force component. In this way, slight process force fluctuations are not detected, but larger deviations due to the hardness differences are recognized (Fig. 5). At lower hardnesses, the process forces are outside the tolerance and can thus be recorded. This allows local variances in the process to be detected and possible errors in the recontouring process to be avoided.

Another possible cause for deviations in the process forces during milling is tool wear. The cutting tools can reach an advanced state of wear during their use. To determine the influence of wear on the resulting process forces, a flank wear width of 100 μm was provoked on the tools. These were also used for three different process parameters of the cutting process (Fig. 6). For all process parameters investigated, the process force increases significantly when the worn tools are used. Parameter sets 1 and 3 exhibit a similarly large increase in the force of 29.1% and 28.6%, respectively.

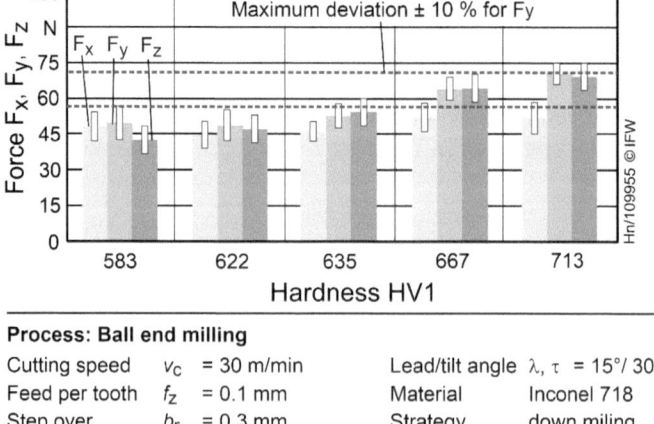

Process: Ball end milling

Cutting speed	v_c	= 30 m/min	Lead/tilt angle	λ, τ	= 15°/ 30°
Feed per tooth	f_z	= 0.1 mm	Material		Inconel 718
Step over	b_r	= 0.3 mm	Strategy		down miling

Fig. 5 Influence of the hardness on the process force

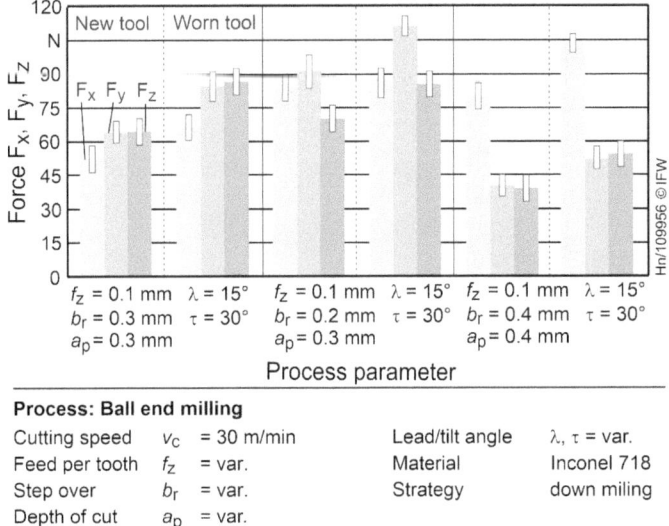

Fig. 6 Forces when using worn tools

Parameter set 2 exhibits a smaller average increase in the force of 16.7%. The force components also increase to different extents here. For example, the force component F_x only shows an increase of 4.8% for the second parameter set. To determine and recognize the deviation, all three force components of the cutting process must be considered. This means that a deviation from the predicted process forces in the real process can be attributed to a worn tool or deviations in the welding process. Thus, local variances in the recontouring process can be detected early and major errors in the recontouring chain can be avoided.

3 Process Design to Minimize Shape Deviations

A significant factor for shape deviations during ball end milling is, in addition to the chipping and the occurring vibration of the tool-workpiece system, especially the process kinematics. In recontouring, changing the process kinematics offers the potential for reducing shape deviation. The causes of the shape deviations were investigated during the entire duration of CRC 871 (Denkena and Flöter 2012; Nespor 2015). With the knowledge gained about the causes of shape deviations, the process design can be carried out to minimize the shape deviations, as shown in Fig. 7. The aim was to determine a combination of process parameters that result in low shape deviations and increased productivity at the same time. For a small shape deviation in the transition area between the weld and the base material, a small step over, a small tool radius, a small cutting edge rounding and small contact angles should therefore

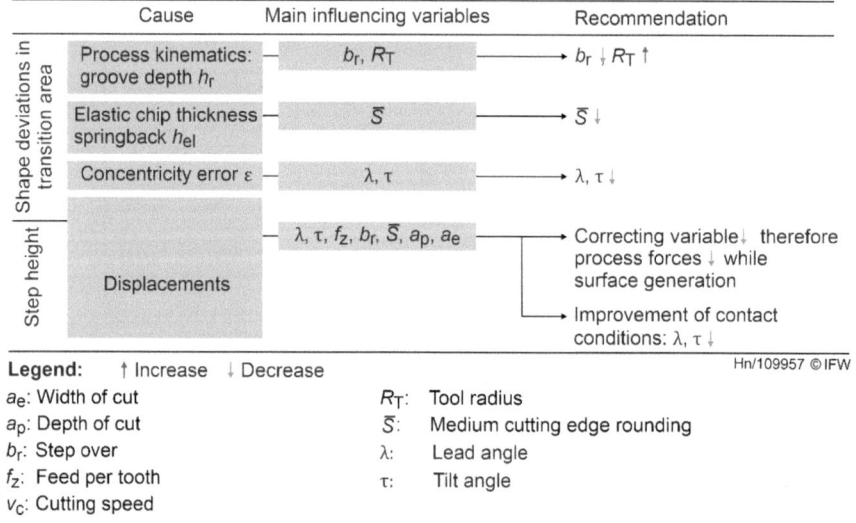

Fig. 7 Causes of the shape deviations and recommendations for minimization (Mücke 2020)

be selected. To avoid a high step height due to displacement, all process parameters except for the cutting speed should be minimized as far as possible.

The measured topography of the reference process considered is shown in Fig. 8. A sharp tool was used to reduce the process force-induced displacements due to the long toolholder. For these investigations, a weld seam with a width of 10 mm and height of 1 mm was applied to a base plate, which was subsequently machined into individual paths. The target roughness in this example was set to $f_z = b_r = 0.5$ mm. This results in an average step height of $\Delta H = 32.2$ μm and a maximum step height of $\Delta H_{max} = 63.2$ μm with a roughness of $Sz = 25.5$ μm. In addition, concentricity errors, process kinematics and the elastic chip thickness springback were not compensated for, resulting in an error at the transition area of ≈ 10 μm. The material removal rate is used to evaluate productivity. The material removal rate at maximum cutting depth $a_p = 0.5$ mm for the reference process is $Q_w = 318.3$ mm^3/min. To maintain the kinematic roughness of the process, i.e. $f_z = b_r = 0.5$ mm = const., the recontouring process was optimized with the help of process simulation. To consider the workpiece surface generation effects in process planning, a time discrete material removal simulation MRS, which uses a multi-dexel model to discretize the workpiece, was applied (Denkena et al. 2019). For this purpose, the lead and tilt angles were first reduced in accordance with the recommendations shown in Fig. 8. In addition, a variation of the cutting speed was carried out to identify a range with the highest possible productivity with given process stability and minimal shape deviations. Furthermore, the strategy for up milling was adapted. In addition, the runout error, the process kinematics, and the elastic chip thickness springback were considered for a minimum error at the transition area. The result of the simulation to

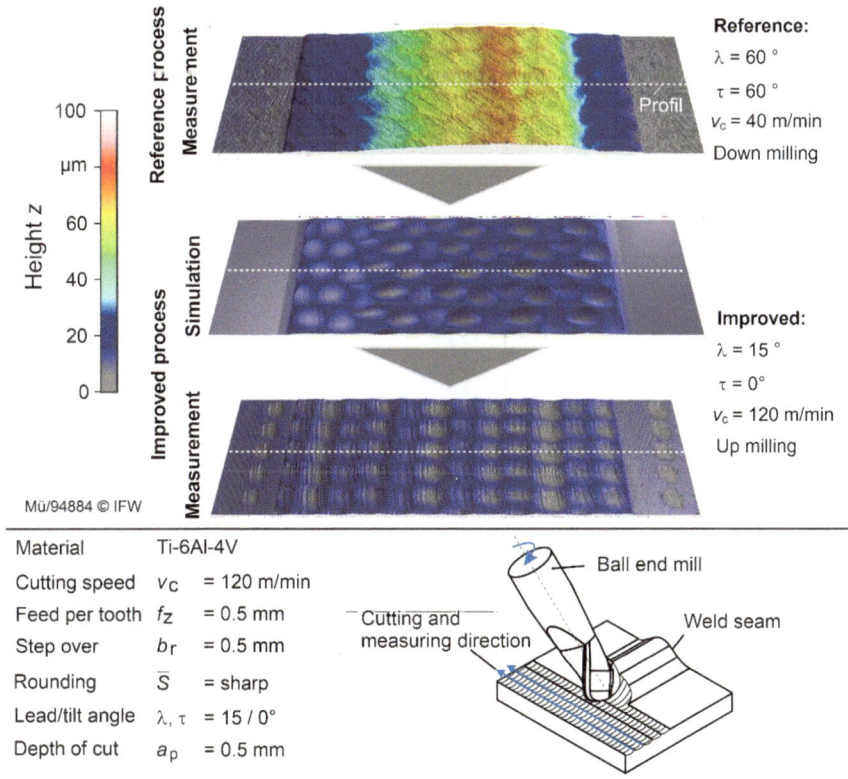

Material		Ti-6Al-4V
Cutting speed	v_C	= 120 m/min
Feed per tooth	f_z	= 0.5 mm
Step over	b_r	= 0.5 mm
Rounding	$\overline{S}$	= sharp
Lead/tilt angle	λ, τ	= 15 / 0°
Depth of cut	a_p	= 0.5 mm

Fig. 8 Comparison of the topographies of the reference and the improved process (Mücke 2020)

improve the process is shown in the middle of Fig. 8. The corresponding experimental result is shown below.

By comparing the topographies of the improved process with the topography of the reference process and the corresponding profile lines (Fig. 9), the benefit of the virtual improvement of the process becomes clear. As can be seen from the profile lines, the error at the transition area is prevented by considering the runout error, the process kinematics and the elastic chip thickness recovery. This allows a smooth transition between the base material and the recontoured area. The maximum step height of the reference process was reduced by 86% from $\Delta H_{max} = 60.7\,\mu$m to $\Delta H_{max} = 8.5\,\mu$m. Likewise, the average step height was reduced from originally $\Delta H = 31.0\,\mu$m to $\Delta H = 0.3\,\mu$m, and thus, by 99%. The predicted parameters from the simulation are in close accordance with the parameters of the actual process. In addition, the shape deviations described by the roughness parameters Sz and Sa were also significantly reduced. Thus, the maximum height of the roughness Sz was reduced by 31%, while the mean arithmetic height Sa of $\approx 2\,\mu$m remained unchanged. These parameters also match the measurement results of the experimental investigation. Because minimal shape deviations were achieved in this example at a higher cutting speed $v_c = 120$ m/

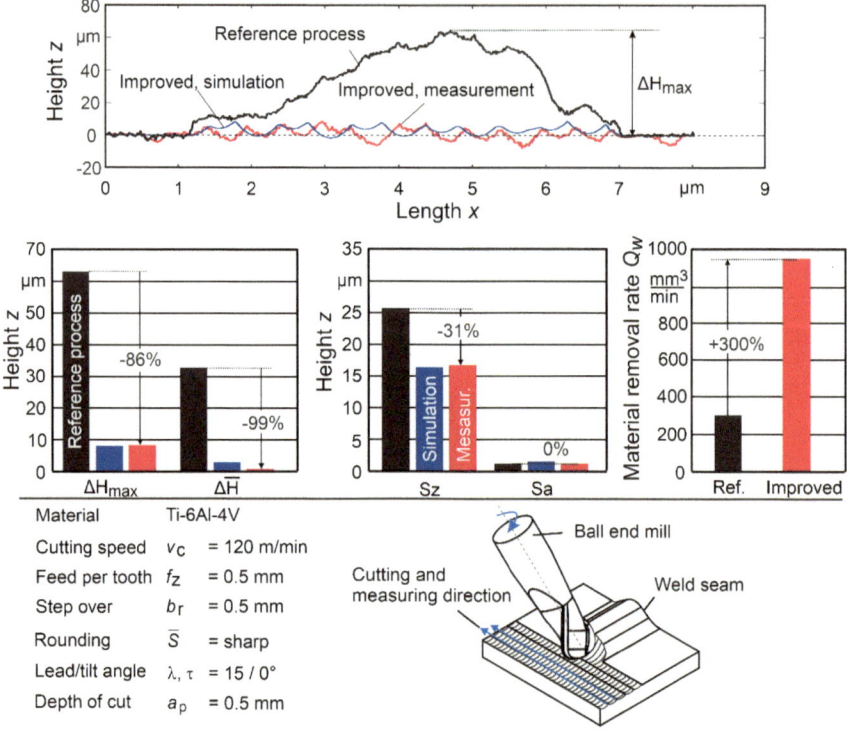

Fig. 9 Evaluation of the minimization of the shape deviations (Mücke 2020)

min, productivity was also increased by 300%. The material removal rate is $Q_w =$ 954.75 mm³/min.

The results show that a significant reduction in shape deviations can be achieved by applying a novel simulation-based process design. The simulation allows for the first time to take into account the knowledge about chip formation and the dynamics of the process. This allows a process-safe recontouring of individual damage cases despite repair-specific variances. However, the process design for complex components is restricted by further constraints like accessibility. Therefore, the deflection of the workpiece and tool should be additionally compensated during machining, as shown in the next sections.

4 Electromagnetic Guide System for Adaptive Machining

The "dexterous regeneration cell" is based on the 5-axis milling machine prototype shown in Fig. 10. The workpiece is clamped on a rotary swivel table with A and C axes, which are mounted onto the Y-axis. The X and Z axes are arranged on the

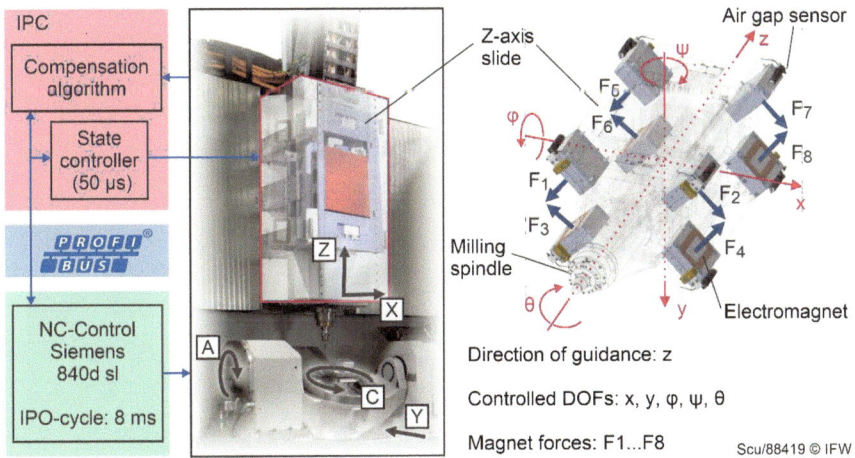

Fig. 10 Dexterous regeneration cell with electromagnetic guide (Denkena et al. 2021)

tool side. The special feature of this machine tool is the electromagnetic guide of the Z-axis. Eight electromagnets with a maximum force capacity of 15 kN each are arranged on the Z-axis slide. By appropriate control of the magnetic forces $F_1, ..., F_8$, the slide hovers completely contactless in its housing. In addition to the translatory movement of the Z-Axis, it can also be precisely positioned in 5 degrees of freedom $(x, y, \varphi, \psi, \theta)$ by the electromagnetic guide. While the five NC axes (X, Y, Z, A, C) of the machine tool are controlled by a standard NC control, the electromagnetic guidance is controlled by a separate industrial PC (IPC). The correction values for the deflection compensation are also calculated on the IPC.

The advantage of the magnetic guide compared to the NC axes for deflection compensation is the high positioning dynamics and the short reaction time. The process forces are measured using a Kistler 9257B dynamometer on the machine table and are then fed to the compensation algorithm via a charge amplifier (Kistler 5015a) and an AD-converter (BECKHOFF EL3702) at a sample rate of 20 kHz. The calculation of the compensation set point takes place on the IPC within the PLC-cycle of 50 μs. For compensation using the NC axis, the actual set-point is transferred to the NC control via PROFIBUS, as depicted in Fig. 10. There, it is superimposed to the axis position within the interpolator. Alternatively, the compensation set point is directly used as a reference input for the state-space controller of the electromagnetic guide.

Figure 11 shows how fast a calculated compensation value can be executed. The reference signal for the position set-point is a step of 50 μm. The step response of the Y-axis stays within the 10% error band around the end value (25 μm) after 74.2 ms. The long delay is due to the signal transmission via the fieldbus and the signal processing within the NC control at an interpolation cycle of 8 ms. In contrast, the electromagnetic guide finishes the 50 μm step in about 11.9 ms, which corresponds to a positioning time reduction of 84%. This allows a faster compensation of variances.

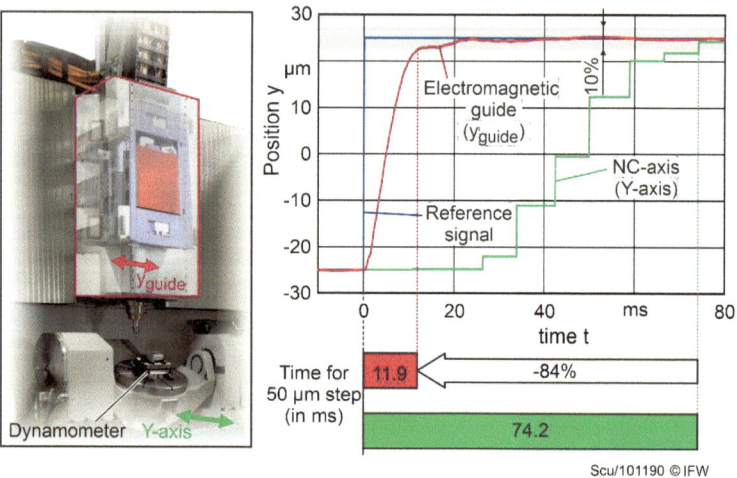

Fig. 11 Reaction time for position set-point step of 50 μm (Denkena et al. 2021)

As a result, better surface quality can be achieved on the workpieces in the entire recontouring chain.

5 Deflection Compensation in Flank Milling

The effect of the dynamic advantage of the magnetic guide over the NC axes was investigated for deflection compensation on the experimental setup presented in Fig. 12. A steel plate was clamped onto the dynamometer. The front flank was prepared with a step of 0.4 mm height and then face-milled with a 10 mm diameter end mill. The resulting flank profile was then measured with a skidless surface finish gauge (Mahr LD130 perthometer). The direction of movement of the milling cutter was the negative x-direction. The contour data depicted in the diagram was therefore generated from right to left. The geometric reference ($y = 0$ μm) for the measurement was derived from a reference surface of the prepared workpiece. In the experiment, the surface in front of the step was passed by the milling cutter without radial infeed. Under ideal conditions, no material would be removed from this surface since there would be no radial immersion. In reality, however, a slight material removal takes place, and the contour shows an undersize of about 5 μm.

At the prepared step in the workpiece surface, the width of the cut a_e and the process forces increase abruptly, and the milling cutter is deflected. Without any compensation, a step of 38 μm remains in the measured contour after machining (blue contour plot). The other measurements show the contours that were created with deflection compensation activated. For the green contour plot, the compensation movement was carried out by the NC axes of the machine tool, and for the red one by

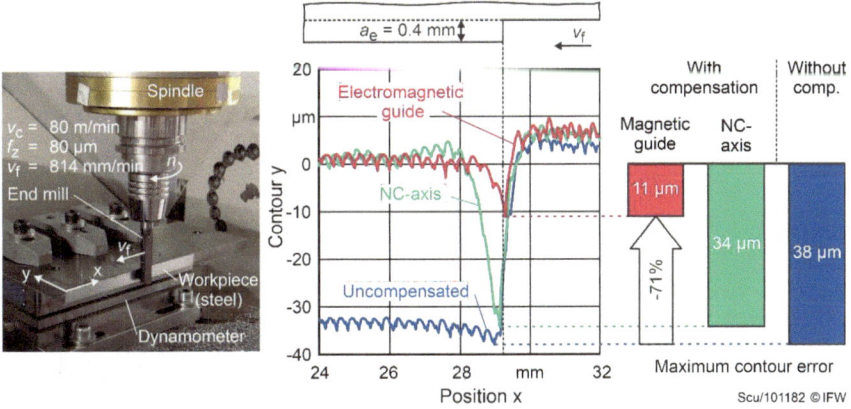

Fig. 12 Flank milling experiment with abrupt change in the width of cut a_e

the electromagnetic guide. In both cases, the compensation reacted with a delay to the increase in force, so that a significant contour deviation remained directly at the step transition. While the compensation by the NC axes reduced the error at the transition only insignificantly, the magnetic guide reacted much faster, so that the error could be reduced by 71%. These results show that reactive deflection compensation can reduce contour errors caused by static force components without any loss of productivity. This allows a reduction of the shape deviation in the entire recontouring process chain. However, since it is a causal process with a finite reaction time, corrective movement can only take place after a change in force has been detected. This time delay leads to significant errors in case of abrupt changes in the cutter engagement. The electromagnetic guide offers higher performance in comparison to the NC axis due to its shorter reaction times. This simple experiment was discussed in (Denkena et al. 2021) in more detail. Furthermore, an anticipatory approach for deflection compensation was proposed and investigated. This method reduces the error in the transition even further (Denkena et al. 2021).

6 Deflection Compensation in 5-Axis Recontouring Processes

In contrast to the previous, simple experiment, determining the compensation values for recontouring processes on real turbine blades is much more challenging. With rotary and tilting movements of the machine table during 5-axis simultaneous machining, the orientation of the workpiece coordinate system (WCS) is constantly changing concerning the machine coordinate system (MCS) (Fig. 13). While the forces are measured in the WCS, the positioning movements of the magnetic guide are specified in the MCS. Furthermore, the static weight force of the components

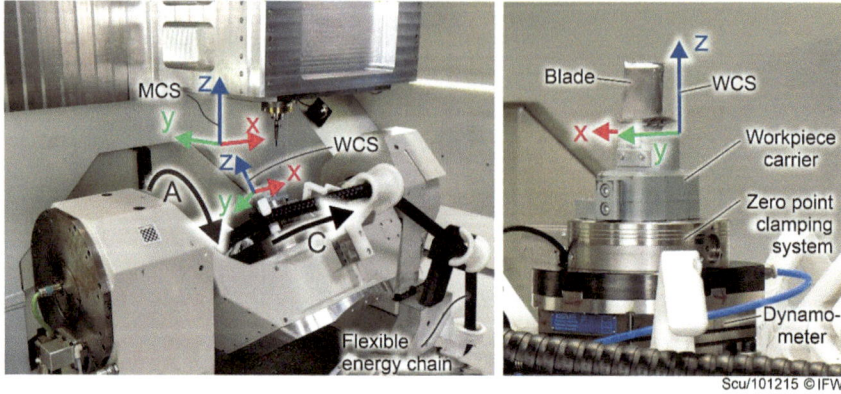

Fig. 13 Setup for turbine blade recontouring

on the dynamometer becomes a variable disturbance concerning the moving WCS. The current orientation must therefore be considered for the compensation. The following subparagraphs therefore discuss the information flow within the controller, the position-dependent compliance model of the workpiece and the results of the compensation in a recontouring process on a real turbine blade from a Rolls-Royce V2500 turbo fan engine.

6.1 Information Flow

The entire information flow for deflection compensation is shown in Fig. 14. The NC control provides the set-points for the 5 NC axes following the planned tool-path. The raw signal $\vec{F}_{raw,WCS}$ from the dynamometer was cleaned from the weight force component $m\vec{g}_{WCS}$ on the IPC and then used to calculate the deflections. The displacement of the workpiece $\vec{\Delta}_{WP,WCS}$ was calculated in the WCS and then transformed into the MCS. The displacement of the tool $\vec{\Delta}_{T,MCS}$ was calculated after transformation into the MCS. The tool and workpiece deflections were then superimposed and filtered in the MCS. Finally, they were applied to the magnetic guide as a correction value.

The transformation between the coordinate systems was done by rotation matrices that depend on the current axis angles of the A and C axes. The tool-side compliance at the ball end mill can be determined experimentally and was assumed to be approximately constant. The compliance of the workpiece depends strongly on the point of force application onto the blade. Therefore, a position-dependent compliance matrix is required to determine the deflection of the workpiece. This matrix is discussed in the next section.

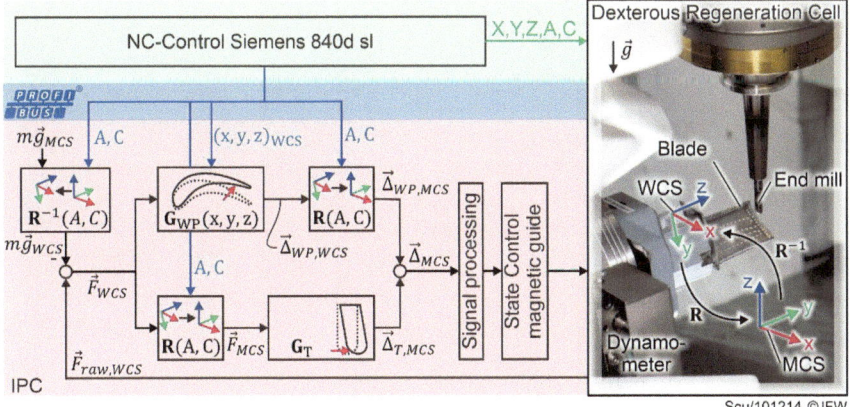

Fig. 14 Information flow

6.2 Workpiece Compliance Model

The position-dependent compliances of the workpiece were determined in a finite element simulation. The reference coordinate system was located at the corner of the blade holder (Fig. 15). Using APDL scripting language, each node in the tip surface was successively defined as a load node and three load cases were simulated. A static load of $F = 1$ N was applied first in the x-, then in the y-, and finally in the z-direction. In the end, a result table contains the displacements (u, v, w) of the load nodes in all three spatial directions for each load direction (x, y, z). A node displacement in the x-direction due to a load in the y-direction is denoted as u_y, for example.

A 3-by-3-compliance matrix with six independent entries can then be constructed for every node.

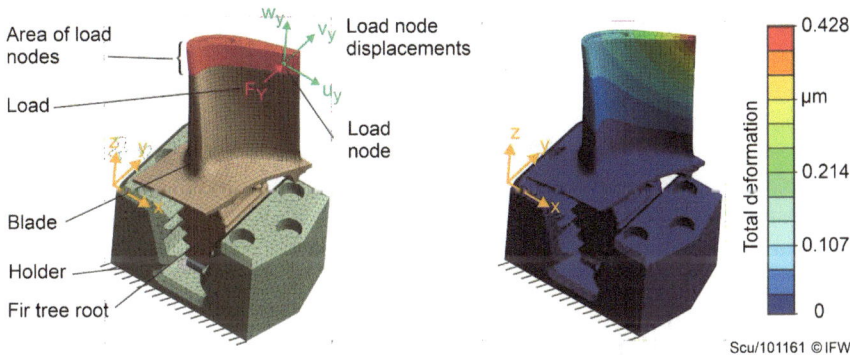

Fig. 15 Determination of the position-dependent workpiece compliance

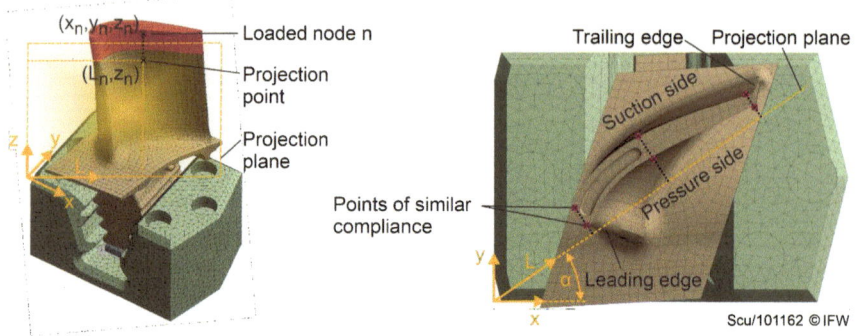

Fig. 16 Inter- and extrapolation of the workpiece compliance

$$
\begin{bmatrix} \Delta x \\ \Delta y \\ \Delta z \end{bmatrix} = \frac{1}{1N} \begin{bmatrix} u_x & u_y & u_z \\ v_x & v_y & v_z \\ w_x & w_y & w_z \end{bmatrix} \cdot \begin{bmatrix} F_x \\ F_y \\ F_z \end{bmatrix} = \underbrace{\begin{bmatrix} g_{xx} & g_{xy} & g_{xz} \\ g_{yx} & g_{yy} & g_{yz} \\ g_{zx} & g_{zy} & g_{zz} \end{bmatrix}}_{G} \cdot \begin{bmatrix} F_x \\ F_y \\ F_z \end{bmatrix} \tag{3}
$$

$$
g_{xy} = g_{yx}, g_{yz} = g_{zy}, g_{zx} \tag{4}
$$

To use the compliance matrices G of the ideal blade for real-time compensation, interpolation must be performed between the nodes. Furthermore, the toolpath may also run through the deposited material. In this case, the compliance must also be extrapolated to force application points that do not lie exactly in the ideal tip surface.

To achieve a distinct mapping between tool position and compliance, the three-dimensional relation was reduced to a two-dimensional surrogate model. Therefore, the tip points of the blade were projected into a plane. This projection plane was constructed as a parallel to a compensation plane through all tip points (Fig. 16). All points of force application can then be referenced by their coordinate along the z and L-axis, which span the projection plane, as shown in Fig. 16.

The top view on the right of Fig. 16 shows that opposite points on the suction and pressure side with the same L and z coordinates are very close to each other (red mark). Therefore, the same compliance can be assumed for them approximately.

It was also shown that the entries of the compliance matrices over L and Z can be approximated very well with fourth degree polynomial functions. This reduces the compliance model of the blade to 6 polynomials, which must be evaluated in real-time for the current cutter position.

$$
G(x, y, z) = \begin{bmatrix} g_{xx} & g_{xy} & g_{xz} \\ g_{yx} & g_{yy} & g_{yz} \\ g_{zx} & g_{zy} & g_{zz} \end{bmatrix}_{x,y,z} \approx \begin{bmatrix} P_{11}(L, z) & P_{12}(L, z) & P_{13}(L, z) \\ P_{12}(L, z) & P_{22}(L, z) & P_{23}(L, z) \\ P_{13}(L, z) & P_{23}(L, z) & P_{33}(L, z) \end{bmatrix} \tag{5}
$$

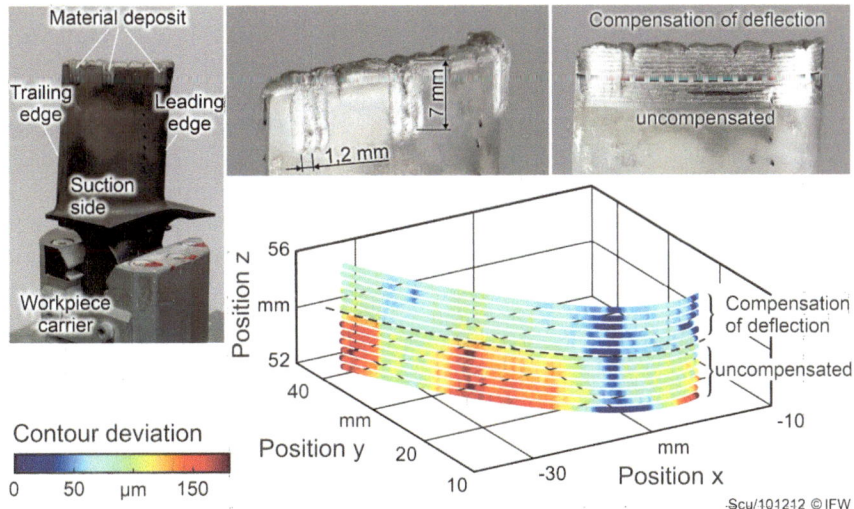

Fig. 17 Result of the compensation

6.3 Results

The results of the compensation on the turbine blade are shown in Fig. 17. Three areas were welded for material deposition on the suction side of the blade. The entire tip-surface was then milled with an infeed of 200 μm. Compensation was switched on in the upper area of the machined surface. It was deactivated in the area below. Finally, the entire blade tip was scanned with a coordinate measuring machine (Leitz Reference Xi 1076). To evaluate the result, the diagram shows a comparison of the measurement data with the nominal shape or the nominal toolpath of the machine tool.

Due to the relatively large infeed, especially at the weld seams, there were also large shape deviations of up to 200 μm. Although displacement compensation reduces those deviations, a large shape error remains. One reason for this is that the compliance model only considers a highly simplified blade geometry without the inner cooling channels and cavities. Further compliance, such as that of the workpiece carrier or the contact between the components, was neglected. Thus, the workpiece-side compliance was underestimated by the model. The second cause is that the determined displacements cannot be fully compensated for by the magnetic guide, as this only allows movement in the x–y plane. Thus, only the projection of the total displacement on the x–y plane can be compensated. Nevertheless, the error could be reduced significantly by up to 50% through the deflection compensation, although the blade is more flexible in the upper area. This increases the accuracy of the entire recontouring process chain.

7 Conclusions

The deflection of the tool and workpiece is the main source of error in recontouring. This paper discusses two complementary approaches for improved process reliability of 5-axis recontouring of individual damage cases with repair-specific variances, influences from preceding processes and deflection of the workpiece and tool. At first it was shown that the choice of process parameters has a significant influence on shape accuracy during recontouring. With the help of a process simulation, the selection of individual process parameters can be supported to minimize shape deviations. In the recontouring of a welding seam, for example, the maximum shape deviation could be reduced by 86%. However, displacement effects in recontouring are unavoidable and unforeseeable to a great extent. Reacting to unpredictable influences is essential because of the variances from upstream repair processes. Based on those challenges, reactive compensation has many advantages compared to other adaptive machining approaches. The unavoidable and unpredictable deflections can be compensated reactively to further reduce shape deviations. Due to its high actuating dynamics, the electromagnetic guide offers a great advantage especially in case of abrupt changes in the process forces.

The investigations on a turbine blade showed that the developed methods led to a robust reduction of the shape deviations despite model uncertainties and unknown model parameters. The error due to deflection could be reduced by up to 50% through the deflection compensation.

Acknowledgements Funded by the Deutsche Forschungsgemeinschaft (DFG, German Research Foundation) CRC 871/3 119193472.

References

Altintas, Y. (2011). Manufacturing Automation. Cambridge University Press, Cambridge

Boujnah, H. (2019). Kraftsensitiver Spindelschlitten zur online Detektion und Kompensation der Werkzeugabdrängung in der Fräsbearbeitung. Ph.D. Thesis, Leibniz Universität Hannover, Hannover

Brecher, C., Wetzel, A., Berners, T., and Epple, A. (2019). Increasing Productivity of Cutting Processes by real-time Compensation of Tool Deflection due to Process Forces. Journal of Machine Engineering. 19, 16–27

Charalampous, P. (2020). Prediction of Cutting Forces in Milling Using Machine Learning Algorithms and Finite Element Analysis. Journal of Materials Engineering and Performance, 30(3), 1082–1088

Cho, S., Asfour, S., Onar, A., and Kaundinya, N. (2005). Tool breakage detection using support vector machine learning in a milling process. International Journal of Machine Tools and Manufacture, 45, 241–249

Denkena, B. and Boujnah, H. (2018). Feeling machines for online detection and compensation of tool deflection in milling. CIRP Annals, 67, 423–426

Denkena, B. and Flöter, F. (2012). Adaptive Cutting Force Control on a Milling Machine with Hybrid Axis Configuration. Procedia CIRP. 4, 109–114

Denkena, B., Gümmer, O., and Flöter, F. (2014a). Evaluation of electromagnetic guides in machine tools. CIRP Annals 2014. 63, 357–360

Denkena, B., Nespor, D., Böß, V., and Köhler., J. (2014b). Residual stresses formation after recontouring of welded Ti-6Al-4V parts by means of 5-axis ball nose end milling. CIRP J Manuf Sci Technol, 7, 347–360

Denkena, B., Böß, V., Nespor, D., Flöter, F., and Rust, F. (2015). Engine blade regeneration: a literature review on common technologies in terms of machining. International Journal of Advanced Manufacturing Technology, 81, 917–924

Denkena, B., Nespor, D., Grove, T., and Pape, O. (2016). Surface topography after recontouring of welded Ti-6Al-4V parts by means of 5-axis ball nose end milling. The International Journal of Advanced Manufacturing Technology. 85, 1585–1602

Denkena, B., Grove, T., and Pape, O. (2019). Optimization of complex cutting tools using a multi-dexel based material removal simulation, Procedia CIRP. 82, 379–382

Denkena, B., Pape, O., Krödel, A., Böß, V., Ellersiek, L., and Mücke, A. (2020). Process design for 5-axis ball end milling using a real-time capable dynamic material removal simulation. Production Engineering, 15, 89–95

Denkena, B., Bergmann, B., and Schumacher, T. (2021). Anticipatory Online Compensation of Tool Deflection Using a Priori Information from Process Planning. J. Manuf. Mater. Process. 5(3), 90

Dittrich, M.-A., Denkena, B., Boujnah, H., and Uhlich, F. (2019). Autonomous Machining – Recent Advances in Process Planning and Control. Journal of Machine Engineering. 19, 28–37

Eberlein, A. (2007). Phases of high-tech repair implementation. 18th International Symposium on Airbreathing Engines (ISABE)

Engin, S, and Altintas, Y. (2011). Mechanics and dynamics of general milling cutters. International Journal of Machine Tools and Manufacture, 41, 2195–2212

Flöter, F. (2017). Potentiale einer elektromagnetischen Führung in Fräsmaschinen und ihr Nutzen für die Reparaturbearbeitung. Ph.D. Thesis, Leibniz Universität Hannover, Hannover

Gao, J., Chen, X., Yilmaz, O., and Gindy, N. (2008). An Integrated Adaptive Repair Solution for Complex Aerospace Components Through Geometry Reconstruction. International Journal of Advanced Manufacturing Technology, 36(11–12), 1170–1179

Hähn, F. and Weigold, M. (2020). Hybrid compliance compensation for path accuracy enhancement in robot machining. Prod. Eng. Res. Devel. 14, 425–433

Hashemitaheri, M., Mekarthy, S., and Cherukuri, H. (2020). Predition of specific cutting forces and maximum tool temperatures in orthogonal machining by Support Vector and Gaussian Process Regression Methods. Procedia Manufacturing, 48, 1000–1008

Jurkovic, Z., Cukor, G., Brezocnik, M., and Brajkovic, T. (2018). A comparison of machine learning methods for cutting parameters prediction in high speed turning process. Journal of Intelligent Manufacturing, 29, 1683–1693

Kelbassa, I., Gasser, A., Backes, G., Keutgen, S., Kreutz, E.-W., and Pirch, N. (2002). Repair and (Re)conditioning of Compressor and Turbine Blades by CO2 and Nd:YAG Laser Radiation. Schriften des Forschungszentrums Jülich. Energy Technology, 21, 751–758

Krizek, Z., Jurkovic, Z., and Brezocnik, M. (2007). Analytical study of different approaches to determine optimal cutting force model. Archives of Materials Science, 28, 69–74

Layegh, K., Yigit, I., and Lazoglu, I. (2015). Analysis of Tool Orientation for 5-Axis Ball-End Milling of Flexible Parts. CIRP Annals, 64(1), 97–100

Mücke, A. (2020). Gestaltabweichungen nach der Rekonturierung reparaturgeschweißter Bauteile. Ph.D. Thesis, Leibniz Universität Hannover, Hannover

Nespor, D. (2015). Randzonenbeeinflussung durch die Rekonturierung komplexer Investitionsgüter aus Ti-6Al-4V. Ph.D. Thesis, Leibniz Universität Hannover, Hannover

Ozturk, E. and Budak, E. (2007). Modeling of 5-Axis Milling Processes. Machining Science and Technology, 11(3), 287–311

Ozturk, E., Tunc, L. T., and Budak, E. (2009). Investigation of Lead and Tilt Angle Effects in 5-Axis Ball-End Milling Processes. International Journal of Machine Tools and Manufacture, 49(14), 1053–1062

Stemmler, S., Abel, D., Adams, O., and Klocke, F. (2016). Model Predictive Feed Rate Control for a Milling Machine. IFAC-PapersOnLine. 49, 11–16

Tuysuz, O., Altintas, Y., Feng, H. Y. (2013). Prediction of Cutting Forces in Three and Five-Axis Ball-End Milling with Tool Indentation Effect. International Journal of Machine Tools and Manufacture, 66, 66–81

Uhlmann, E., Bilz, M., and Baumgarten, J. (2013). MRO – Challenge and Chance for Sustainable Enterprises. Procedia CIRP, 11, 239–344

Influence of Complex Surface Structures on the Aerodynamic Loss Behaviour of Blades

Hendrik Seehausen and Joerg R. Seume

Abstract Operating airfoils under mechanical stress in combination with oxidation and corrosion, abrasion wear, and subsequent regeneration results in complex surface structures that influence the performance of aircraft engines. For predicting this impact on performance the authors propose a reduced order model capable of assessing the effect of surface roughness as a basis for making decisions before or during the regeneration process. The accuracy of this model is increased by using improved roughness sensitive transition and turbulence models created for Reynolds-Averaged Navier–Stokes simulations. Experimental studies are carried out to determine the local and integral effect of complex surface structures on blades and on a flat plate. Direct numerical simulations are also performed to study the effect of complex surface structures on a turbulent boundary layer and contribute to improving the accuracy of prediction, which is achieved by using re-calibrating models and taking into account the effect of skewness and anisotropy of complex surface structures on turbine blade losses.

Keywords Complex surface structures · Transition and turbulence model · Aerodynamic loss behaviour

1 Introduction

During operation, an aircraft engine experiences many different operating conditions that can lead to deterioration. Among them is an increase of surface roughness, consisting of highly complex surface structures. Depending on the location in

H. Seehausen (✉) · J. R. Seume
Institute of Turbomachinery and Fluid Dynamics, Leibniz University Hannover, An der Universitaet 1, 30823 Garbsen, Germany
e-mail: seehausen@tfd.uni-hannover.de
URL: https://www.tfd.uni-hannover.de/en/

J. R. Seume
e-mail: seume@tfd.uni-hannover.de
URL: https://www.tfd.uni-hannover.de/en/

© The Author(s) 2025
J. R. Seume et al. (eds.), *Regeneration of Complex Capital Goods*,
https://doi.org/10.1007/978-3-031-51395-4_9

the engine and the regeneration process, the surfaces show isotropic (stochastically irregular) and anisotropic (oriented) structures. With a confocal laser microscope, the surface of a blade can be optically measured and converted into a height map. For the Computational Fluid Dynamics (CFD) simulation, the calculation of the equivalent sand-grain roughness k_s of the optically measured height map has been considered appropriate. The equivalent sand-grain roughness k_s was first introduced by Nikuradse (1933) and Schlichting (1936), who studied the influence of surface roughness on the flow at different Reynolds numbers and established a relation with the aerodynamic drag. Each technical roughness can be assigned an equivalent sand-grain roughness based on the aerodynamic drag. In an overview, Bons (2010) summarises the most common correlations for the conversion of a technical roughness into the equivalent sand-grain roughness, e.g. the *Shape and Density* parameter Λ_s of Sigal and Danberg (1990), which specifies the shape and density of roughness elements on the surface. Significant research on this topic has also been carried out within this project of the Collaborative Research Centre 871 (CRC 871). For instance, Hohenstein and Seume (2013) and Gilge et al. (2019a) propose the use of the *Shape and Density* parameter Λ_s in combination with the approach of Bons (2005):

$$\log\left(\frac{k_s}{k}\right) = -0.43\log(\Lambda_s) + 0.82. \tag{1}$$

Bons (2005) and Hohenstein and Seume (2013) have shown that the absolute roughness height R_z as a scaling factor k in Eq. 1 is suitable in order to be able to map the influence of complex surfaces in CFD. The use of this approach is also in accordance with the results for polished surfaces (Goodhand et al. 2016).

Initially, the local and integral effect of complex surface structures on turbine blades was studied by Hohenstein et al. (2013b) who measured the surfaces of 9 turbine blades of the second rotor stage of a worn high-pressure turbine (HPT) in the dismantled state. The flight region, in which the turbine blades had completed around 24,000 h and 9,000 cycles, is specified with moderate environmental conditions of Europe and North America. Primarily anisotropic structures were found on the first rotor stage. Some of the roughness structures are applied on an aerodynamically scaled turbine blade and investigated in the cascade wind tunnel of the Institute of Turbomachinery and Fluid Dynamics (TFD). By using total pressure probes, particle image velocimetry (PIV), and an innovative laser ablation process, the effect of locally applied real complex surface structures and their combinations on the wake was measured. Gilge and Mulleners (2016) showed that the local pressure gradient has a significant effect on the roughness-induced loss mechanisms. Additionally, Neuhaus et al. (2016) found a shift of slightly anisotropic turbulence for the wake of a smooth blade to isotropic turbulence in the wake of rough blades. In conjunction with an increase in turbulent kinetic energy, this correlates with higher profile losses. Hohenstein et al. (2013b) also compared the experimental results to Reynolds-averaged Navier–Stokes (RANS) simulations, which predicted significant different total pressure losses. Subsequently, a compressor cascade was designed for the cascade wind tunnel and investigated with regard to complex surface structures within this project of the CRC 871. Gilge et al. (2018) show that a roughness on the suction side affects the entire wake and leads to friction losses up to 7.1%. A roughness on the pressure side affects mainly the shear region on the pressure side leading to friction losses up to 2.3%. In general, the roughness's effect depends strongly on the location and is not completely superimposable, which is consistent with the study of the turbine cascade. To analyse the local flow effects, a boundary-layer water tunnel was installed during the project. In an experimental study, Kurth et al. (2018) show that the anisotropy of surface structures has a significant effect on the boundary-layer flow. This effect is not included in the k_s-correlation of Bons (2005) (see Eq. 1). Similar results to Kurth et al. (2018) can be found in Lorenz et al. (2013) and Ahrens et al. (2021).

Based on the results of this project within the CRC 871, a reduced order model for the quantitative prediction of the effect of surface roughness during operation and after regeneration for the regeneration process is developed. Here, the reduced order model is based on RANS simulations. This type of CFD simulation is mainly used for the aerodynamic design of turbomachinery with lower time requirements compared to the regeneration process. Due to the time averaging, RANS simulations require models that approximate the turbulent viscosity μ_t. As the reduced order model is based on RANS simulations, the accuracy of the RANS models must be high and able to capture the effects of surface roughness. Thus, the objective is to provide a RANS-based turbulence and transition model combination that improves the prediction of the effect of complex surface structures. Two approaches must be pursued: standardizing the roughness's characterization and improving the modelling of the influence of complex surface structures. From these approaches, the question arises: Can an improved prediction of the roughness effect be achieved by a change of the modelling, i.e. RANS model, or by changing the k_s correlation? Before answering this question, the next section presents the significance of surface roughness in turbomachinery.

2 Roughness Effect in Multistage Modules

2.1 High-Pressure Compressor

Gilge et al. (2019a, b) used a confocal laser microscope to measure optically the surfaces of worn rotor blades of a ten-stage high-pressure compressor (HPC) in the dismantled state. The blades measured were rotor blades of a front, middle, and rear stage, which had completed 20,000 cycles in geographic regions with mainly low environmental impact (Wensky et al. 2010). On the pressure side, more homogeneous isotropic structures are found and the roughness height decreases along the blades and across the stages. The isotropic structures are caused by impacts and erosion or deposition. On the suction side, the roughness height also decreases across the stages, but the surface structures vary. At the leading edge, anisotropic structures occur, which become more isotropic along the blade (see Fig. 1). The authors explain the anisotropic structures with oil leakages sticking flow particles to the surface. Seehausen et al. (2020) built a model of a typically rough HPC blading. Figure 1 shows the roughness height k_s across the stages divided into the suction and pressure side. The surfaces of Gilge et al. (2019a), measured optically, serve as supporting points.

In a CFD study, the effect of the real complex surface structures measured and interpolated for the HPC blading on the module performance was investigated. The mesh of the ten-stage HPC of the V2500 engine consists of approx. 26 million cells and was provided by Reitz et al. (2018). The numerical model of the compressor contains the cavities of the front stages, as well as the bleed air extractions. In addition, the snubber in the first blade row is taken into account as it has a significant influence on the flow and the blockage of the compressor. The radial tip-gaps of the rotor rows are set to 1% of the respective duct height. The CFD simulations were carried out with the non-commercial flow solver TRACE 9.1 (Franke et al. 2005; Nürnberger 2004; Kugeler et al. 2008) provided by the German Aerospace Center (DLR). Based on the finite volume method, the Favre-averaged Navier Stokes equations are solved on block-structured meshes. The discretization of the convective flows is carried out with the second-order backward difference scheme according to Roe (1981). The diffusive flows are solved with a second-order central difference method. The HPC is calculated fully turbulent and takes into account the equivalent sand-grain roughness k_s via the wall function. This means that the viscous sublayer of the boundary layer

Fig. 1 Roughness distribution across the stages of the HPC after 20,000 cycles in service and examples of measured surfaces Gilge et al. (2019a, b). The figure is adopted from Goeing et al. (2020b)

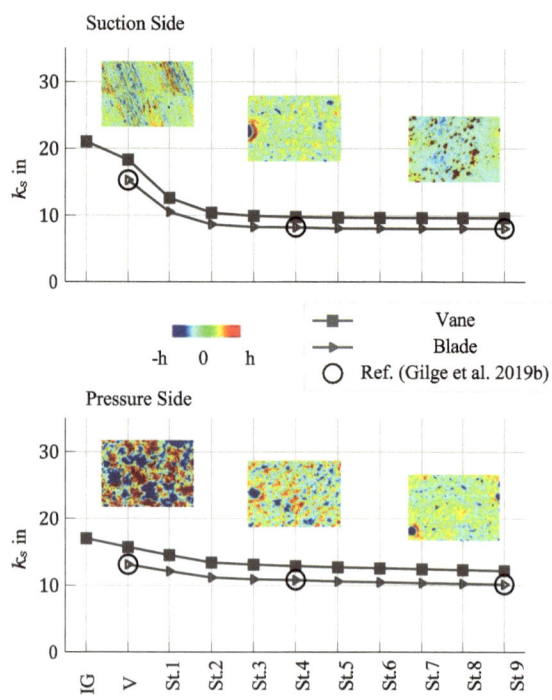

is not directly solved. The wall function is based on the wall shear stress velocity u_τ, which describes a turbulent velocity scale at the wall. If a roughness is assigned to the wall, the logarithmic boundary-layer profile is calculated as a function of the dimensionless equivalent sand-grain roughness $k_s^+ \left(= \frac{u_\tau k_s}{\nu} \right)$ according to:

$$\frac{U}{u_\tau} = u^+ = \frac{1}{\kappa} \ln(y^+) + \frac{1}{\kappa} \ln\left(\frac{1}{k_s^+} \right) + 8.4 \tag{2}$$

Seehausen et al. (2020) created a compressor map, which allows to asses the effect of surface roughness on the module performance (see Fig. 2). The generation of a compressor map requires several CFD simulations with gradually increased or decreased static pressure at the HPC outlet. Here, throttle lines for the operating points 'cruise' and 'take-off' were created with about 15 CFD simulations each. The black squares represent smooth operating points, while green dots show the fully rough map. The fully rough map is shifted towards lower mass flows with a reduction of 5.73% in mass flow $\dot{m}_{corr}$ at 'take-off' and 4% in mass flow $\dot{m}_{corr}$ at 'cruise'. Additionally, the map narrows between the reduced surge and choke margins, leading to a reduction of 4% (12.3%) in total pressure ratio π_{tt}, and a reduction by 2.36% (2.68%) in polytropic efficiency η_{poly} at 'take-off' ('cruise'). Consequently, increased surface roughness leads to an increase in the specific fuel consumption. Reitz et al. (2018) investigated the effect of a macro-scale geometry change through doubled tip-gaps on the performance of the HPC. To reduce computational resources, Reitz et al. (2018) used a method to scale performance maps based on scaling factors Li et al. (2011). The operating points scaled for doubled tip-gaps are represented by the magenta coloured crosses. The effect of surface roughness on the performance reduction is about 1% greater than the effect of doubled tip-gaps. The effect of doubled tip-gaps on the pressure ratio is quite small due to the boundary conditions set in the numerical setup, i.e. total pressure at the inlet and static pressure at the outlet. Goeing et al. (2020a) extended the investigation of Seehausen et al. (2020) by four different sets of rough HPC blades and vanes of a fully rough HPC. Subsequently, a map of scaling factors is obtained to represent the roughness effect of different states on compressor maps. Using this map reduces the computational effort and it is sufficient to import the smooth compressor map into the performance analysis of project D6 "Interaction of combined module variances and influence on the overall system behaviour" within this CRC 871. The results of this study are also used in Goeing et al. (2020b) to asses the effect of combined module variances.

Seehausen et al. (2020) further analyse the loss-sources in a multistage HPC with respect to surface roughness. A stage roughness variation with surface roughness uniformly distributed over the blade's suction side of single stages is performed. The simulations show that the first stage has the biggest effect on the compressor performance at both operation conditions, 'take-off' and 'cruise'. Additionally, it is shown that the stage losses are not superimposable. Seehausen et al. (2020) conclude that surface roughness changes the stage-by-stage matching resulting in bigger overall performance in front stages than in rear stages. These results also confirm that

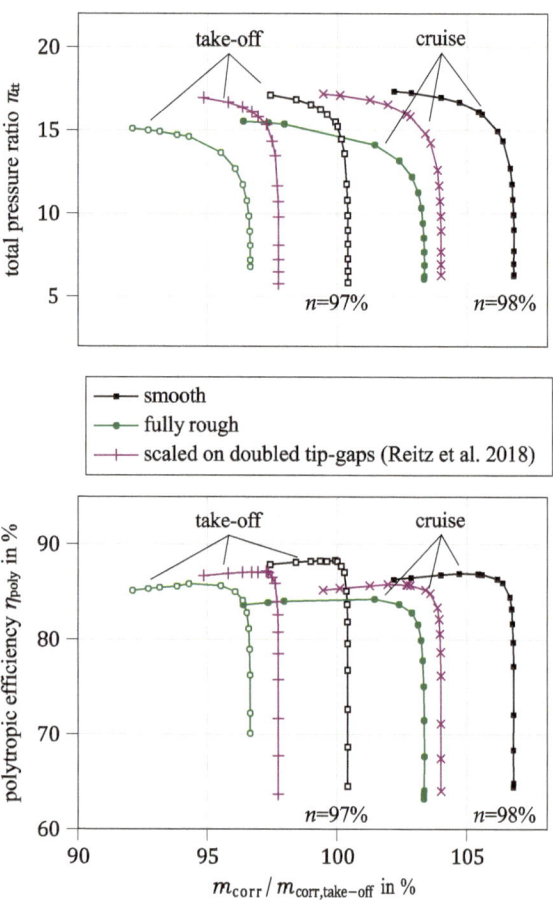

Fig. 2 Compressor maps with different deterioration modes. Inspired by Seehausen et al. (2020)

predicting the local roughness effect accurately is necessary when evaluating the module performance.

2.2 High-Pressure Turbine

In this project of the CRC 871, the authors also studied the effect of surface roughness on the performance of a high-pressure turbine (HPT). In order to demonstrate the process of blade regeneration, a system demonstrator was set up in project S "System Demonstrator" within this CRC 871. A part of the regeneration process is the performance evaluation in the virtual layer of the system demonstrator. To include the effect of surface roughness on the module performance, a reduced order model was required. The description of the reduced order model for the virtual regeneration process of a V2500 HPT blade can be found in Goeing et al. (2022). Based on scaling factors obtained by RANS simulations of the V2500 HPT with different modes of deterioration, a neural network was trained. Subsequently, project D6 "Interaction of combined module variances and influence on the overall system behaviour" within this CRC 871 used this neural network to predict the module performance for the performance simulation of the engine. Figure 3 shows the HPT setup for the RANS simulations consisting of 8M cells, which required about 84 CPUh for a converged solution. It was found that the roughness effect is not as dominant in an HPT as for the HPC with significantly more stages.

Nevertheless, modern blade profiles aim for large laminar boundary layers that lead to low heat transfers and low heat loads. This is accompanied by low aerodynamic losses for laminar boundary layers. In general, surface roughness leads to an opposite effect by introducing disturbances that destabilize the boundary layer. Hohenstein (2014) investigated the effect of locally applied surface structures on the boundary layer and the profile losses. The results show that roughness structures can lead to a shift of the location of transition, a change of the transition mode, and a reduction of the suction peak. These results are in accordance with the results of Stripf (2007). Stripf (2007) studied the effect of surface roughness on the transition with respect to the heat transfer. Although a different concept is used to evaluate

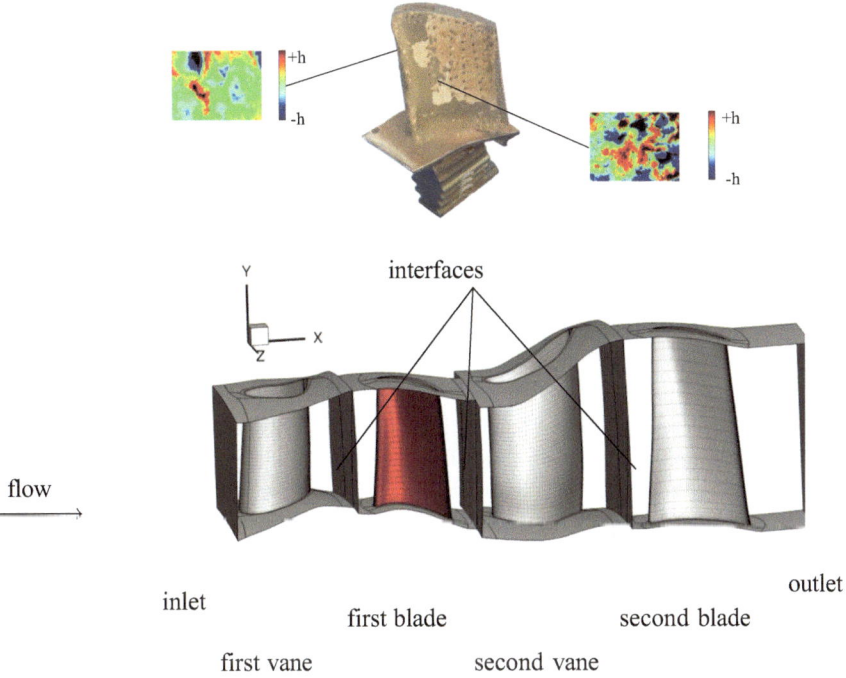

Fig. 3 HPT geometry with blade under consideration highlighted in red and blade geometry with exemplary measured roughness structures. The figure is adopted from Goeing et al. (2022)

surface structures (in this case a discrete element method), the importance of accurately predicting the roughness's effect on the laminar-transitional boundary layer is shown.

3 Modelling the Roughness Effect

3.1 Fundamentals

Solving the RANS equations requires models which approximate the turbulent eddy viscosity μ_t. The turbulence model attempts to predict the production and the decay of turbulence, e.g. turbulent kinetic energy k. In this work, the focus is on the k-ω turbulence model by Wilcox (1988) and the SST turbulence model by Menter (1994). Both models are based on two transport equations: one for the turbulent kinetic energy k and one for the specific dissipation rate ω. Outside the boundary layer, a cross-diffusion term is added to the SST model formulation resulting in the k-ϵ formulation. This eliminates the sensitivity to the free stream turbulence of the conventional k-ω formulation. Besides this difference, the SST model considers the Bradshaw assumption that, when calculating the eddy viscosity μ_t, the shear stress τ_{ij} is proportional to the turbulent kinetic energy k. This limits the turbulence production in regions where the conventional formulation overestimates the eddy viscosity. In case of laminar shear-layers, transition models are used for the correct prediction of transition and the modes of transition. To account for the different modes of transition, different approaches for modelling the transition evolved in the past decades (Suzen et al. 2002; Menter et al. 2006, 2015; Kožulović 2007; Stripf et al. 2008; Langtry and Menter 2009). The transition model controls the production of turbulent kinetic energy of the turbulence model and, thus, most of the transition models are designed to trigger the turbulence model instead of predicting the physical effect in detail (Langtry and Menter 2009). In contrast, a few concepts of laminar kinetic energy (LKE) models exist, which present a more physically-based approach for transition modelling (Walters and Leylek 2004; Walters and Cokljat 2008). However, the use of LKE models is still limited to academic test cases. In this work, the γ-Re_θ transition model of Menter et al. (2006) is used. This transition model is based on local variables of each cell without the need for any boundary layer integration and no requirement on the geometry or meshing. The model solves two additional transport equations, i.e. one for the intermittency γ and one for the momentum-thickness Reynolds-number Re_θ. Based on these two transport variables, experimental correlations are used to predict the correct point of transition. In its original formulation, the model is not able to consider a rough-wall boundary condition. Due to the difficulty to combine different transition modes, Menter et al. (2006), Langtry and Menter (2009) concentrated on first order effects, i.e. natural, bypass transition, and separation-induced transition. Extensions for secondary flow effects, such as injection-induced transition or crossflow-induced transition, are proposed by Herbst et al. (2014) and

Müller and Herbst (2014). Boyle and Senyitko (2003), Hohenstein et al. (2013b), Gilge et al. (2017) show that extensions for other second order effects, such as surface roughness, are needed. Dassler (2013) added a third transport equation to the transition model formulation of Langtry and Menter (2009) for modelling the effect of surface roughness. Wilcox (1988) adjusts the dissipation rate ω at the wall depending on the non-dimensional sand-grain roughness k_s^+ to incorporate second-order flow effects in turbulent boundary layers. This method, however, is not suitable for surface structures such as riblets (exemplary for a strongly anisotropic surface structure) due to their drag-decreasing effects. To realize this effect, Koepplin et al. (2017) applied a damping function to the destruction term of the dissipation rate ω. The sections below describe the process of assessing the original modelling of the roughness effect for turbulent and laminar boundary layers. They also answer the above question: Can an improved prediction of the roughness effect be achieved by a change of the modelling, i.e. RANS model, or by changing the k_s correlation? Below, we will first concentrate on a turbulent boundary layer with DNS and then on a laminar boundary layer with experiments. Changes to the modelling approach are made in the non-commercial flow solver TRACE 9.1 of the DLR (Franke et al. 2005; Nürnberger 2004; Kugeler et al. 2008).

3.2   Turbulent Boundary Layer

From a modelling point of view, a decrease of the specific dissipation ω results in an increase of the eddy viscosity μ_T and turbulent kinetic energy k. Thus, higher turbulent fluctuations and viscous losses occur. Wilcox (1988) shows that the modification of the specific dissipation ω at the wall can accurately predict the velocity profile affected by a rough surface. These results could be verified for complex isotropic surface structures by using the immersed boundary method (IBM) for direct numerical simulations (DNS) (Kurth et al. 2021, 2022). In comparison with body-fitted approaches, the IBM allows regular meshes that are independent of the immersed body and can be easily transferred to other surfaces without need for major human interaction. The DNS-IBM approach provided in *foam-extend-4.0* is used to asses the prediction accuracy of the roughness effect in turbulent boundary layers. The IB-method used can be found in Senturk et al. (2019). The DNS is based on a cell-centred finite volume discretization with a PISO algorithm for the solution of the incompressible flow equations. For the second-order accurate time and space discretization, Gauss linear scheme and backward temporal discretization are chosen from the incompressible solver *icoFoam*. To analyse the local flow effect, a channel setup is used. At channel's top, which corresponds to the boundary layer thickness δ, symmetric conditions are imposed. A periodicity is imposed at the lateral boundaries to limit the flow domain and to allow spatial averaging for better flow statistics. The roughness is immersed in the bottom wall with no-slip condition. A linear blending function to blend 10% of the roughness patch on the borders is used to ensure periodicity of the roughness. Specifying the friction velocity Reynolds number

$$\mathrm{Re}_\tau = \frac{u_\tau \cdot \delta}{\nu}, \tag{3}$$

by a constant momentum body force in the momentum equations allows comparing different surface structures at same aerodynamic conditions. Additional settings can be found in Kurth et al. (2022). Next to the DNS-IBM channel setup, a double infinite channel sized two times the boundary-layer thickness in lateral direction and three times in flow direction is built for the RANS computation with approx. 4M cells. In contrast to the DNS-IBM channel setup, the roughness is not physically visible, but rather represented by a wall boundary condition. For different roughness structures, the RANS simulations with a rough-wall boundary condition are performed enforcing the same friction velocity Reynolds number as in the DNS-IBM setup. Subsequently, the velocity profiles are extracted. As depicted in Fig. 4, it is possible to find a k_s value resulting in the velocity profile of the DNS-IBM approach. In a sensitivity study, Kurth et al. (2022) shows that the skewness of a surface has a significant impact on the local effect of roughness. The skewness Ssk of the roughness is a parameter used to quantify, whether a roughness consists primarily of hills ($Ssk > 0$) or valleys ($Ssk < 0$) (see DIN EN ISO 25178). Based on the effect of the skewness on the velocity deficit in the DNS-IBM approach, the k_s correlation (see Eq. 1) can be extended by:

$$\frac{k_{s,\mathrm{SD},Ssk}}{k_{s,\mathrm{SD}}} = 0.0266 \cdot Ssk^2 + 0.1426 \cdot Ssk + 0.7977. \tag{4}$$

In addition to complex isotropic surface structures, anisotropic surface structures occur during operation or regeneration. Within this project, Ahrens et al. (2021) investigated the effect of synthetic anisotropic structures superimposed with real complex surface structures on the flow losses. A sinusoidal function with a period s^+ = 252.5 and an amplitude of $z^+ = 6$ is used for anisotropic surface structures. The

Fig. 4 Comparison of velocity profile between DNS and RANS, with roughness height in DNS represented by a dashed vertical line

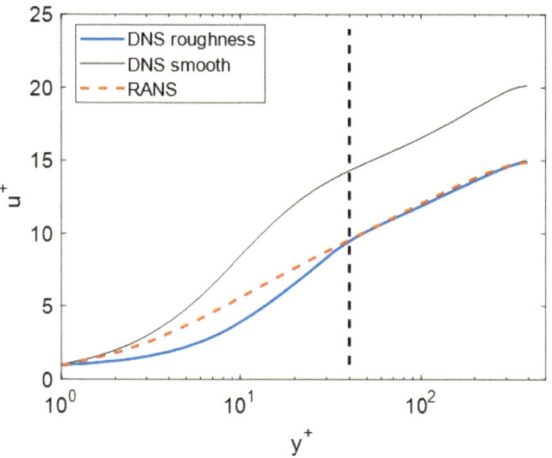

dimensions for these anisotropic surface structures are found on worn compressor blades (Gilge et al. 2019a). It was found that the flow losses increase with higher relative flow angles, but have a non-linear relationship with the orientation of the anisotropic structures. Additionally, the anisotropic structures show a dominant effect over the isotropic roughness effect above a certain relative flow angle. For building new models, the database has to be extended, as only one parameter combination has been investigated so far. Nevertheless, this database is useful for validating new models.

3.3 Laminar Boundary Layer

Alldieck et al. (2020) studied the effect of rough-wall boundary conditions on the RANS-based transition prediction. As could be derived from the full set of equations of the γ-Re_θ transition model, it is not only the transition model responding to a rough-wall boundary condition in laminar boundary layers, but also the turbulence model. A modification of the specific dissipation rate ω by the turbulence model for rough walls must have an effect via the increased eddy viscosity ratio on the roughness-induced transition prediction due to an increased production of γ. This means that the turbulence model can have an accelerating effect on the transition for rough-wall boundary condition. Alldieck et al. (2020) shows that the robustness of the eddy viscosity ratio against surface roughness in the SST model gives low sensitivity to the transition onset prediction of the γ-Re_θ model. This behaviour is advantageous compared to the k-ω model, where the modelled roughness effect strongly depends on the turbulent length scale (Dassler 2013; Bode et al. 2014). Unfortunately, the SST model in its original formulation is not able to account for rough-wall boundary conditions, since the eddy-viscosity limiter ensuring the Bradshaw assumption disables the rough-wall boundary condition of the k-ω formulation (Hellsten and Laine 1997). By introducing an additional limiter which deactivates the Bradshaw assumption in specific regions, such as in the sublayer and roughness layer, Hellsten and Laine (1997) solved this problem. Here, the limiter is implemented in the source code of TRACE 9.1. The model is activated with the command *all immediate ChangeTurbulenceSettings—model MenterSST2003 –HellstenRoughness ON*. Dassler (2013) developed an extension for the γ-Re_θ transition model to consider the rough-wall boundary condition in the transition model. The command to activate the roughness model is given in the TRACE User Guide (2019).

   The roughness modification of the γ-Re_θ transition model by Dassler (2013) is based on the experiments of Feindt (1956) in an annular wind tunnel. Dassler (2013) introduced a third transport equation for the non-dimensional value A_r

$$\frac{D(\rho A_r)}{Dt} = \frac{\partial}{\partial x_j}\left[\sigma_{A_r}(\mu + \mu_T)\frac{\partial A_r}{\partial x_j}\right] \tag{5}$$

allowing the roughness effect to convect and diffuse into the flow field. Similar to the incorporation of a rough-wall boundary condition in the ω-equation of the turbulence model, the value of A_r is correlated with the local non-dimensional sand-grain roughness k^+

$$A_{r,W} = 8 \cdot k_s^+ \tag{6}$$

at the wall. The transported value A_r adds sensitivity to the momentum thickness Reynolds number $\widetilde{Re}_{\theta t}$ via the variable Arg_r. A reduction of the transport value $\widetilde{Re}_{\theta t}$ results in a shift of the transition to an upstream position. Since the model formulation is initially calibrated in conjunction with the k-ω model, a re-calibration is required before applying it with the SST model. The re-calibration process shows that amplifying the function of Arg_r over A_r by a factor of 5 is required, because the SST model damps the roughness effect compared to the k-ω model. These results are presented in Fig. 5. Here, the transition onset Reynolds number Re_{xt} is plotted as a function of the surface roughness height Re_{ks} for different pressure gradients. The transition onset Re_{xt} is defined at the minimum of the friction coefficient c_f. This point agrees with the definition of Feindt (1956). The experimentally measured points of Feindt (1956) are represented by the black circles. The re-calibrated model is tested for different roughness heights and gives a good approximation of the experimentally determined data.

Fig. 5 Transition onset location Re_{xt} depending on the critical roughness Reynolds-number Re_{ks} for different pressure gradients

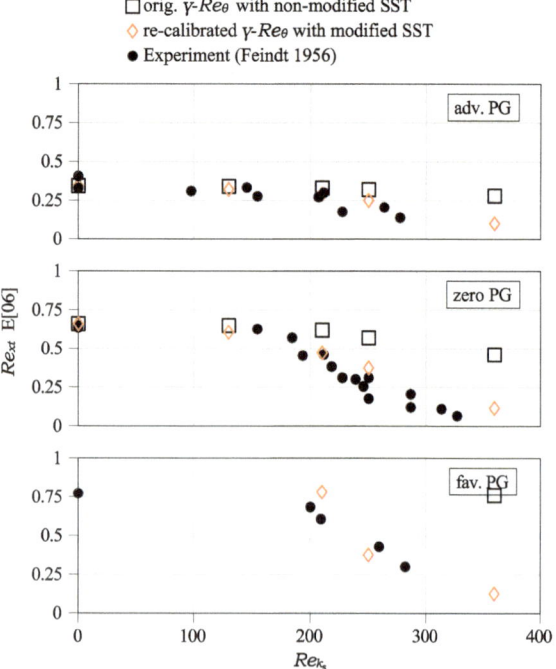

4 Experimental Results and Validation

4.1 Boundary-Layer Water Tunnel

The boundary-layer water tunnel of TFD is used to study the effect of isotropic and anisotropic surface structures on laminar-transitional boundary layers. In this work, the data serve as validation data for complex isotropic surface structures, but also for the development of new models with respect to anisotropic surface structures. Publications by Kurth et al. (2018, 2022) include a description of the water tunnel and present recent work on the effect of surface structures on the local flow. In the present investigation, a flat plate is placed in the measurement section of the water tunnel, and a laminar boundary layer develops on this plate. In accordance with the experiments of Feindt (1956), optically scanned sandpapers with different grain sizes are selected to study the effect on local flow of surface roughness. Each sandpaper is measured at 10 different positions. Subsequently, the roughness correlation of Kurth et al. (2022) is applied to obtain the equivalent sand-grain roughness for each of the samples. The *Shape and Density* parameter is calculated using the roughness parameter Sz. The equivalent sand-grain roughness of the optically measured sandpapers is in a good agreement with the roughness used by Feindt (1956) (see Table 1). In contrast, the samples for the sandpaper P150 show a sand-grain roughness that is too high.

Due to the lower kinematic viscosity of water compared with air, the structures must be scaled. The aerodynamic scaling is obtained by using the RANS simulations of the experiment by Feindt (1956) (Alldieck et al. 2020). A 3D polyjet printer is used for the additive manufacturing of scaled surface structures that have a size of 180 mm × 80 mm (L × W) and approx. 5 mm in height. The roughness structures investigated by Feindt (1956) show high positive skewness values due to the use of sandpapers. According to DIN EN ISO 25178, the roughness height is negative when the point is facing the material from the reference surface. The reference surface of sandpaper is defined by the lowest point of the surface, while the reference surface of complex surface structures of blade profiles can conveniently be assumed to be the mean profile of the measured roughness. Hohenstein (2014) shows that most of the surfaces of worn blades have a skewness of around 0, negative skewness values being slightly more common. The 3D printed surfaces show a skewness of approx.

Table 1 Comparison of equivalent sand-grain roughness k_s of scanned sandpaper with 95% confidence interval

Notation	k_s in mm (Feindt 1956)	k_s in mm (Kurth et al. 2022)
P60	0.480	0.591 ± 0.23
P80	0.280	0.372 ± 0.07
P150	0.135	0.455 ± 0.04
P220	0.100	–
P240	–	0.216 ± 0.01

0.5–1. However, we consider the skewness with the correlation of Kurth et al. (2022) and, thus, are able to predict the roughness effect on the boundary layer.

The isotropic roughness structures P60 and P240 are superimposed with a sinusoidal function with different periods s and amplitudes t to study the effect of anisotropy. A synthetically created surface is chosen for a better isolation of the individual effects on the flow. The ranges of the anisotropic roughness parameters are shown in Table 2. A new parameter Sq_r is introduced to specify the ratio of the anisotropy with respect to the isotropic surface. The parameter is defined as follows:

$$Sq_r = \frac{Sq_{\text{filt}}}{Sq} \tag{7}$$

based on the root mean square roughness height of the filtered and non-filtered surface. The 3D topography is first subjected to a Fourier transform. Then, the spatial frequency with the highest amplitude is filtered. An anisotropy is said to exist when the filtered surface has a significantly lower root-mean-square roughness height than the non-filtered surface. In case of an isotropic surface the ratio of Sq is approx. 1.

The 3D-printed isotropic and anisotropic surface structures are placed in the flat plate. The flow over the rough surfaces is measured using the stereo PIV. A set of 3000 image pairs is recorded with a frequency of 10 Hz. A shift of approx. 10 px/ image and an interrogation window size of 32×32 pixel results in an uncertainty of 0.196 pixel for the evaluation process representing a 95% confidence interval (Westerweel 1997; Raffel et al. 2018). Figure 6 shows an extract of the results. The lateral flow velocity normalized by the free stream velocity is plotted over a period s for different relative flow angles α. The flow is deflected by the wind-wetted flanks of the anisotropic structures. This flow deflection is convected within the boundary layer. A relative flow angle of 60% results in a flow deflection of up to 20%. Kurth et al. (2018) obtain similar results and shows that anisotropy of surface structures can result in a significant flow deviation depending on the parameters of the anisotropy. Based on the experimental setup, a turbulent boundary layer can be assumed. With that, the results are in accordance with the measurements of the 3D-printed surface structures in a laminar boundary layer.

Hohenstein and Seume (2013) show that the equivalent sand-grain roughness k_s can capture the induced drag due to anisotropy. For incorporating the effect of anisotropic surface structures in the RANS models, the roughness correlation of

Table 2 Parameter ranges for the anisotropic surface structures investigated in the boundary-layer water tunnel

Parameter	Lower bound	Upper bound
$Re_{k\,s}$	180	880
Re_s	1842	8289
Re_t	86	442
α	0	90
Sq_r	0.52	1

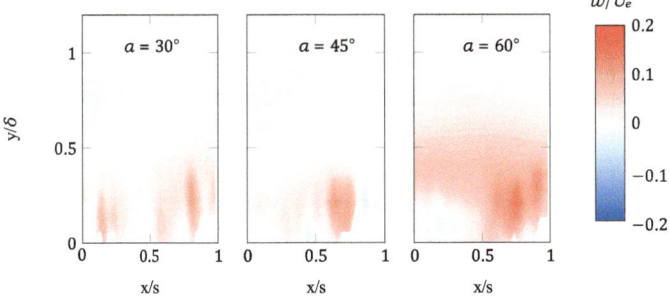

Fig. 6 Contour plot of the lateral velocity induced by the anisotropy of the surface structure normalized with the free stream velocity for $Sq_r = 0.52$

Eq. 8 is extended. By comparing the velocity profiles at specific positions, an equivalent sand-grain roughness is found, which captures the drag increasing effect of the anisotropy. In a first approach, the following correlation is derived for the range of laminar boundary layers:

$$\frac{k_{s,\text{SD},Ssk,A}}{k_{s,\text{SD},Ssk}} = -0.06 \cdot \left(Sq_r - 1\right) \cdot \left(\tanh\left(0.8987 \cdot \pi \cdot \frac{\alpha}{45} - \pi\right) + 6.276\right) + 1. \quad (8)$$

It will be of interest for future studies, to see, whether the correlation also applies to turbulent boundary layers. Moreover, if there are additional parameters that are important for the correlation.

4.2 Cascade Wind Tunnel

The cascade wind tunnel of the TFD was used extensively within this project as it is easily accessible and versatile. In this section, we validate the integral prediction using results of the influence of complex surface structures on the aerodynamics of turbine blades. Please refer to Gilge et al. (2018) for results on the compressor blade investigations. The turbine blade geometry used is designed within an optimization algorithm based on the aerodynamics of a blade from an HPT's second stage (Hohenstein et al. 2013a). The optically measured surface roughness of worn and regenerated blades serves as a data base. Initially, Hohenstein (2014) selected some of the optically measured surfaces to be attached on the turbine blade. The surface structures are aerodynamically scaled and applied on metal strip with 4 mm length in flow direction. A laser ablation process was used for the application. The metal strip is placed in a groove of the blade at different relative chord length positions, i.e. 0, 20, 50, and 85%. The blade of interest is divided in two parts within the cascade. This eliminates external effects by referring to the smooth neighbour blade. A double wedge probe is used to measure the flow quantities in the wake. The integral total pressure loss is calculated based on the wake measurements. Figure 7 shows the relative total pressure loss with respect to a smooth blade for different roughness configurations. The bar plot compares the experiment with the RANS prediction by the old and the new model. Table 3 shows the specifications of the different configurations.

The RANS prediction of the integral effect of complex surface structures shows more accurate results for the new model in the laminar boundary layer than the old model (M1b1 and M2b1). Even though the equivalent sand-grain roughness decreases by using the correlation of Eq. 4, the re-calibrated γ-Re_θ roughness model is able to satisfy the mean experimental results. The improved model reduces the deviation from 0.3 to 0.016% of the measured value for M1b1 and from 0.183 to 0.11% for M2b1. In the turbulent boundary layer, the mean of the experimental results is not that consistent (M4b1 and M4b2). An increase in total pressure loss is expected with increasing surface roughness, as shown by the RANS prediction. This leads to a slight increase in the deviation from 1.08 to 1.27% of the measured value for M4b1, but a decrease in the deviation from 0.828 to 0.737% of the measured value for M4b2. In the experiments, roughness applied at 50% chord showed a suppression of the laminar separation bubble (Hohenstein 2014). The re-calibrated transition model is not able to predict this behaviour (M3b1 and M3b2). One reason for this phenomenon could be the locally changed blade geometry due to the roughness patch glued in

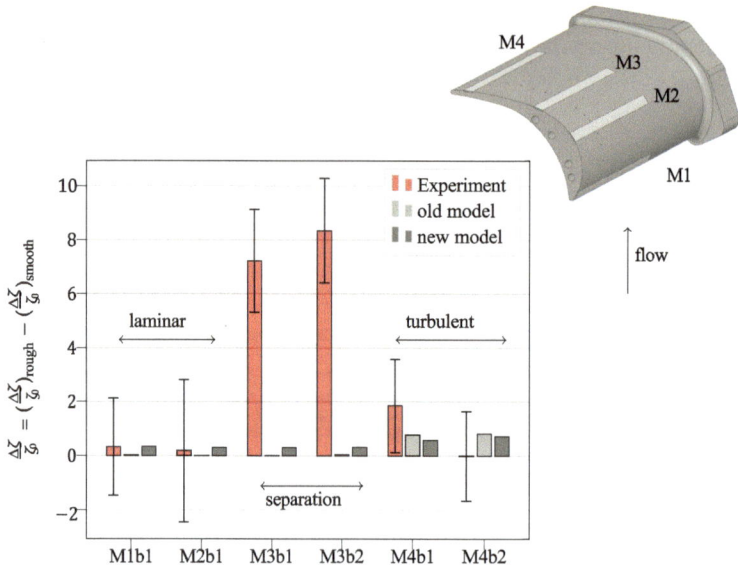

Fig. 7 Relative total pressure loss coefficient with 95% confidence interval for the experiment of different complex surface structures applied on the turbine blade. Old model: SST without extension of Hellsten and Laine (1997), original γ-Re_θ, and k_s correlation according to Eq. 1. New model: SST with extension of Hellsten and Laine (1997), re-calibrated γ-Re_θ, and k_s correlation according to Eq. 8

Table 3 Specifications of the roughness configurations applied on the turbine blade

Notation	Position x/c in %	$k_{s,SD}$ in μm	$k_{s,SDSsk}$ in μm (Kurth et al. 2022)
M1b1	0	200	57
M2b1	0.2	76	37
M3b1	0.5	74	25
M3b2	0.5	200	50
M4b1	0.85	126	30
M4b2	0.85	150	45

the blade groove. However, the new approach significantly increases the prediction accuracy of the mean losses. Nevertheless, the interaction of surface roughness and the suppression of a laminar separation bubble needs further investigations.

5 Conclusions

Aircraft engine performance is influenced by complex surface structures that occur during operation. The authors propose a reduced order model capable of assessing the effect of surface roughness on the performance as a basis for the decision making in the regeneration process. The reduced order model is based on an improved RANS approach for which detailed numerical and experimental studies have been carried out on different test benches.

The modelling deficit for complex surface structures is reduced by considering the skewness of a surface structure when calculating the equivalent sand-grain roughness. The transition model is re-calibrated for a better prediction of a roughness-induced transition. This results in an improved prediction of the effect of complex surface structures on the integral flow parameters. It is found that the conventional rough-wall boundary condition in most of the CFD solvers is applicable when modelling the effects of isotropic complex surface structures.

However, experimental studies show that anisotropic surface structures can lead to a significant flow deflection. The roughness correlation can predict the drag increasing effect in the flow direction due to anisotropy, but not the flow deflection effect. Thus, the authors further strive to improve the model by imposing volume forces to model the flow deflection.

Acknowledgements The present work has been carried out in the subproject B3 in the Collaborative Research Center 871 "Regeneration of Complex Capital Goods", which is funded by the Deutsche Forschungsgemeinschaft (DFG, German Research Foundation)—SFB 871/3—119193472. Moreover, the authors would like to acknowledge the substantial contribution of the DLR Institute of Propulsion Technology and MTU Aero Engines AG for providing TRACE. The results presented here are partially carried out on the cluster system at the Leibniz University IT Service (LUIS). Thus, the authors acknowledge the support of the cluster system team in the production of this work. The authors would also like to thank the North-German Supercomputing Alliance

(HLRN) for the HPC resources that have contributed to the development of the research results presented here.

References

Ahrens, J. D., Kurth, S., Cengiz, K., Wein, L., and Seume, J. R. (2021). Immersed boundary method for the investigation of real surfaces with isotropic and anisotropic roughness components. In *Proceedings of the Global Power and Propulsion Society, Virtual Conference, Xi'an, China.*

Alldieck, R., Seehausen, H., Herbst, F., and Seume, J. R. (2020). The effect of roughwall boundary condition on rans-based transition prediction. In *Proceedings of GPPS*, number GPPS-CH-2020-053.

Bode, C., Aufderheide, T., Kozulović, D., and Friedrichs, J. (2014). The effects of turbulence length scale on turbulence and transition prediction in turbomachinery flows. In *Turbo Expo: Power for Land, Sea, and Air*, volume 2B: Turbomachinery.

Bons, J. P. (2005). A Critical Assessment of Reynolds Analogy for Turbine Flows. *Journal of Heat Transfer*, 127(5):472–485.

Bons, J. P. (2010). A review of surface roughness effects in gas turbines. *Journal of Turbomachinery*, 132(2).

Boyle, R. J. and Senyitko, R. G. (2003). Measurements and predictions of surface roughness effects on the turbine vane aerodynamics. volume 6: Turbo Expo 2003, Parts A and B of *Turbo Expo: Power for Land, Sea, and Air*, pages 291–303.

Dassler, P. (2013). *Modellierung der Grenzschichttransition auf rauen Oberflächen in Turbinen.* PhD thesis, Technische Universität Braunschweig.

Feindt, E. G. (1956). *Untersuchungen über die Abhängigkeit des Umschlages laminar-turbulent von der Oberflächenrauhigkeit und der Druckverteilung.* PhD thesis, Dt. Forschungsanstalt f. Luftu. Raumfahrt e.V.

Franke, M., Kuegeler, E., and Nürnberger, D. (2005). Das DLR-Verfahren TRACE: Moderne Simulationstechniken für Turbomaschinenströmungen. In *DGLR Congress*.

Gilge, P. and Mulleners, K. (2016). Resulting aerodynamic losses of combinations of localized roughness patches on turbine blades. *AIAA Journal*, 54(8):2552–2555.

Gilge, P., Hohenstein, S., and Seume, J. (2017). Experimental investigation of the aerodynamic effect of local surface roughness on a turbine blade. *International Journal of Gas Turbine*, 9:12.

Gilge, P., Seume, J. R., and Mulleners, K. (2018). Analysis of local roughness combinations on the aerodynamic properties of a compressor blade. In *AIAA Aerospace Sciences Meeting*, Kissimmee, Florida, 2018.

Gilge, P., Kellersmann, A., Friedrichs, J., and Seume, J. R. (2019a). Surface roughness of real operationally used compressor blade and blisk. *Proceedings of the Institution of Mechanical Engineers, Part G: Journal of Aerospace Engineering*.

Gilge, P., Kellersmann, A., Kurth, S., Herbst, F., Friedrichs, J., and Seume, J. R. (2019b). Dataset: Surface roughness of real operationally used compressor blade and blisk. *Research Data Repository of the Leibniz University of Hannover.* Available at: https://doi.org/10.25835/008 4372.

Goeing, J., Seehausen, H., Bode, C., Herbst, F., Seume, J. R., and Friedrichs, J. (2020a). Performance simulation of roughness induced module variations of a jet propulsion by using pseudo bond graph theory. In *ASME Turbo Expo: Power for Land, Sea, and Air*, volume 1, Virtual.

Goeing, J., Seehausen, H., Pak, V., Lueck, S., Seume, J. R., and Friedrichs, J. (2020b). Influence of combined compressor and turbine deterioration on the overall performance of a jet engine using rans simulation and pseudo bond graph approach. *JGPP*, 4:296–308.

Goeing, J., Seehausen, H., Stania, L., Nuebel, N., Salomon, J., Ignatidis, P., Dinkelacker, F., Beer, M., Denkena, B., Seume, J. R., and Friedrich, J. (2022). Virtual process for evaluating the influence of real combined module variations on the overall performance of an aircraft engine. *To be published in: Proceedings of the GPPS, Chania, Greece. 18th–20th September 2022.*

Goodhand, M. N., Walton, K., Blunt, L., Lung, H. W., Miller, R. J., and Marsden, R. (2016). The Limitations of Using "Ra" to Describe Surface Roughness. *Journal of Turbomachinery*, 138(10).

Hellsten, A. and Laine, S. (1997). Extension of the k-omega-sst turbulence model for flows over rough surfaces. *22nd Atmospheric Flight Mechanics Conference.*

Herbst, F., Fiala, A., and Seume, J. R. (2014). Modeling Vortex Generating Jet-Induced Transition in Low-Pressure Turbines. *Journal of Turbomachinery*, 136(7).

Hohenstein, S. (2014). *Einfluss komplexer Oberflächenstrukturen auf das aerodynamische Verlustverhalten von Turbinenbeschaufelungen.* PhD thesis, Berichte aus dem Institut für Turbomaschinen und Fluid-Dynamik, Leibniz Universität Hannover.

Hohenstein, S. and Seume, J. R. (2013). Numerical investigation on the influence of anisotropic surface structures on the skin friction. Proceedings of 10th European Conference on Turbomachinery Fluid dynamics & Theromdynamics, Lappeenranta, Finland.

Hohenstein, S., Aschenbruck, J., and Ì, J. R. (2013a). Aerodynamic effects of non-uniform surface roughness on a turbine blade. volume 6A: Turbomachinery of *ASME Turbo Expo: Power for Land, Sea, and Air.*

Hohenstein, S., Gilge, P., Raulf, C., and Seume, J. R. (2013b). Einfluss lokaler Rauheiten auf die aerodynamischen Verluste von Turbinenschaufeln. Beitrag auf dem Deutschen Luft- und Raumhfahrtkongress.

Koepplin, V., Herbst, F., and Seume, J. R. (2017). Correlation-Based Riblet Model for Turbomachinery Applications. *Journal of Turbomachinery*, 139(7).

Kožulović, D. (2007). *Modellierung des Grenzschichtumschlags bei Turbomaschinenstroemungen unter Beruecksichtigung mehrerer Umschlagsarten.* PhD thesis, Ruhr-Universität Bochum.

Kugeler, E., Nurnberger, D., Weber, A., and Engel, K. (2008). Influence of blade fillets on the performance of a 15 stage gas turbine compressor. In *ASME Turbo Expo 2008: Power for Land, Sea, and Air*, pages 415–424.

Kurth, S., Hamann, C., Seume, J. R., and Mulleners, K. (2018). Experimental investigation of the influence of anisotropic surface structures on the boundary layer flow. In *AIAA Aerospace Sciences Meeting, 2018*, Kissimmee, Florida.

Kurth, S., Cengiz, K., Wein, L., and Seume, J. R. (2021). From measurement to simulation - a review of aerodynamic investigations of real rough surfaces by dns. In *Proceedings of the Global Power and Propulsion Society, Virtual Conference, Xi'an, China.*

Kurth, S., Cengiz, K., Moeller, D., Wein, L., and Seume, J. (2022). Modelling the influence of roughness skewness on wall bounded flows by systematic roughness variation. *To be published in: Proceedings of the GPPS, Chania, Greece. 18th–20th September 2022.*

Langtry, R. B. and Menter, F. R. (2009). Correlation-based transition modeling for unstructured parallelized computational fluid dynamics codes. *AIAA Journal*, 47(12):2894–2906.

Li, Y. G., Ghafir, M. F. A., Wang, L., Singh, R., Huang, K., and Feng, X. (2011). Nonlinear multiple points gas turbine off-design performance adaptation using a genetic algorithm. *ASME. J. Eng. Gas Turbines Power*, 133(11):1299–1310.

Lorenz, M., Schulz, A., and Bauer, H.-J. (2013). Predicting Rough Wall Heat Transfer and Skin Friction in Transitional Boundary Layers—A New Correlation for Bypass Transition Onset. *Journal of Turbomachinery*, 135(4).

Menter, F., Smirnov, P., Liu, T., and Avancha, R. (2015). A one-equation local correlation-based transition model. *Flow, Turbulence and Combustion*, 95:1–37.

Menter, F. R. (1994). Two-equation eddy-viscosity turbulence models for engineering applications. *AIAA Journal*, 32(8):1598–1605.

Menter, F. R., Langtry, R. B., Likki, S. R., Suzen, Y. B., Huang, P. G., and Völker, S. (2006). A correlation-based transition model using local variables—part i: Model formulation. *Journal of Turbomachinery*, 128(3):413.

Müller, C. and Herbst, F. (2014). Modelling of crossflow-induced transition based on local variables. In Proc. ECCOMAS, Paper No. 2252, Barcelona (Spain).

Neuhaus, L., Gilge, P., Seume, J. R., and Mulleners, K. (2016). Influence of surface roughness on the turbulent properties in the wake of a turbine blade. In *Proceedings of 18th International Symposium on Applications of Laser and Imaging Techniques to Fluid Mechanics, Lisbon, Portugal.*

Nikuradse, J. (1933). Law of flows in rough pipes. *NACA Technical Memorandum 1292.*

Nürnberger, D. (2004). *Implizite Zeitintegration für die Simulation von Turbomaschinenströmungen.* PhD thesis, Ruhr-Universität Bochum.

Raffel, M., Willert, C. E., Werely, S., and Hutchins, N. (2018). *Particle Image Velocimetry - A Practical Guide.* Springer, Germany.

Reitz, G., Kellersmann, A., and Friedrichs, J. (2018). Full high pressure compressor investigations to determine aerodynamic changes due to deterioration. In *ASME Turbo Expo 2018: Turbomachinery Technical Conference and Exposition*, volume 2A: Turbomachinery.

Roe, P. L. (1981). Approximate riemann solvers, parameter vectors, and difference schemes. *Journal of Computational Physics*, 43(2):357–372.

Schlichting, H. (1936). Experimentelle Untersuchungen zum Rauhigkeitsproblem. *Ingenieur-Archiv*, 7(1):1–34.

Seehausen, H., Gilge, P., Kellersmann, A., Friedrichs, J., and Herbst, F. (2020). Numerical study of stage roughness variations in a high pressure compressor. *International Journal of Gas Turbine*, 11(3).

Senturk, U., Brunner, D., Jasak, H., Herzog, N., Rowley, C. W., and Smits, A. J. (2019). Benchmark simulations of flow past rigid bodies using an open-source, sharp interface immersed boundary method. *Progress in Computational Fluid Dynamics, an International Journal*, 19(4):205–219.

Sigal, A. and Danberg, J. E. (1990). New correlation of roughness density effect on the turbulent boundary layer. *AIAA journal*, 28(3):554–556.

Stripf, M. (2007). *Einfluss der Oberflächenrauigkeit auf die transitionale Grenzschicht an Gasturbinenschaufeln.* PhD thesis, Forschungsberichte aus dem Institut für Thermische Strömungsmaschinen Band 38/2007, Universität Karslruhe.

Stripf, M., Schulz, A., and Bauer, H. (2008). Modeling of rough-wall boundary layer transition and heat transfer on turbine airfoils. *ASME J. Turbomach.*, 130(2):021003.

Suzen, Y. B., Xiong, G., and Huang, P. G. (2002). Predictions of transitional flows in low-pressure turbines using intermittency transport equation. *AIAA Journal*, 40(2):254–266.

TRACE (2019). User Guide accessed October 24, 2019, http://www.traceportal.de/userguide/trace/index.html.

Walters, D. and Cokljat, D. (2008). A three-equation eddy-viscosity model for Reynolds-averaged Navier–Stokes simulations of transitional flow. *Journal of Fluids Engineering*, 130(12).

Walters, D. K. and Leylek, J. H. (2004). A New Model for Boundary Layer Transition Using a Single-Point RANS Approach . *Journal of Turbomachinery*, 126(1):193–202.

Wensky, T., Winkler, L., and Friedrichs, J. (2010). Environmental influences on engine performance degradation. *ASME Turbo Expo 2010: Power for Land, Sea, and Air*, pages 249–254.

Westerweel, J. (1997). Fundamentals of digital particle image velocimetry. *Measurement Science and Technology*, 8(12):1379–1392.

Wilcox, D. C. (1988). Reassessment of the scale-determining equation for advanced turbulence models. *AIAA journal*, 26(11):1299–1310.

Dynamical Behaviour and Strength of Structural Elements with Regeneration Induced Imperfections and Residual Stresses

Ricarda Berger and Raimund Rolfes

Abstract Repair techniques allow to refurbish blades of jet engines and extend the service life of engine components. However, the mechanical properties of the repaired blades commonly differ from those of the nominal blades. The changes in vibration behaviour and stress distribution resulting from the repair processes could significantly impair the structural integrity of blades. This work addresses this evaluation of repair-specific influences on the structural integrity of blades of blade-integrated disks (blisks). Numerical methods are employed to model and analyse the structural changes of different repair designs. Using Finite Element (FE) simulations, the influences of different repair techniques and designs are systematically evaluated and compared. To further improve the maintenance procedures of blisks, a computational scheme is developed, which features the optimization of repair designs according to the inspected damage of the blade. Since the corresponding optimization tasks involve multiple conflicting objectives, multi-objective optimization is carried out to identify Pareto-optimal repair designs.

Keywords Blisk · Patch · Blend · Multi-objective optimization

1 Introduction

The repair or replacement of damaged blades is essential to maintain the structural integrity of jet engines and ensure a safe operation during flight. With the increasing use of integrated part designs, such as blade-integrated disks (blisks), on one hand the replacement of single blades becomes unfeasible. On the other hand, the spare

R. Berger · R. Rolfes (✉)
Institute of Structural Analysis, Leibniz University Hannover, Appelstr. 9A, 30167 Hannover, Germany
e-mail: t.griessmann@isd.uni-hannover.de
URL: https://www.uni-hannover.de/; https://www.isd.uni-hannover.de/

R. Berger
URL: https://www.uni-hannover.de/; https://www.isd.uni-hannover.de/

© The Author(s) 2025
J. R. Seume et al. (eds.), *Regeneration of Complex Capital Goods*,
https://doi.org/10.1007/978-3-031-51395-4_10

Fig. 1 Three blade sectors
of a blisk with patch repair
(left), undamaged blade
(middle) and defect at the
leading edge (right)

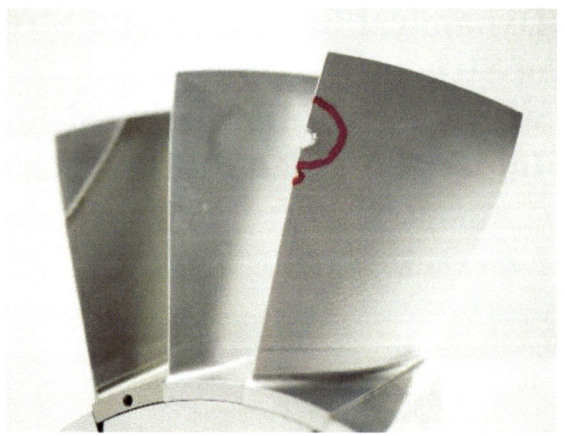

part costs of these complex parts increase rapidly. Therefore, there is growing interest in technologies to repair the damaged blades of the blisk rather than replacing the whole blisk.

Repair technologies developed for blisk blades differ in manufacturing complexity and their application depends on the size of the particular defect (Bussmann and Bayer 2009). A damage pattern typical for blisk blades is depicted in Fig. 1. The blade shown in the front is damaged at the leading edge. Damage patterns like this are most likely caused by particles passing through the engine and are hence known as foreign object damages (FOD).

Depending on the size of the defect, different repair techniques are applicable. Small dents or notches induce local stress concentrations, which cause rapid crack initiation. To prevent crack formation, sharp edges are removed by a milling procedure. This relatively simple repair, named blend repair, leads to remaining geometric changes in the repaired blade and is thus limited to small defects. For more severe damage, the geometric modification due to blending would be too large to maintain the functionality of blades. Therefore, patch repairs are developed, as depicted in the background of Fig. 1. Patching involves the removal of the damaged portion, a welding process to mount the patch material, and a final re-contouring process to restore the aerodynamic contour of the repaired blade. In contrast to blending, a patch repair hence rebuilds the original geometry of the blade.

Consequently, both repair techniques lead to changes in the blade compared to the nominal condition, which affect the remaining life of the repaired components. In the case of blend repairs, the change is mainly characterised by a local geometric modification. The influences of geometric deviations to the nominal state were studied in numerous works. In recent literature, mostly the correlations with natural blade frequencies, mode shapes, forced-response, and vibration amplitude magnification due to mistuning effects were investigated. For example, Brown et al. (2003) found that the natural frequencies of fan blades deviated by 1% when manufacturing imperfections are considered. Other studies on different blade geometries (Heinze et al.

2010) showed comparable results. However, the majority of scientific contributions was limited to nonintended variances caused by manufacturing or wear during operation. A work, which specifically focused on geometric changes in terms of blending, was presented by Beck et al. (2017). They addressed the variations in frequencies of blisk blades with two different sizes of blend shapes. An amplification of vibration amplitudes caused by blade-to-blade variations was determined. A more practical approach towards blend repair shapes was followed by Day et al. (2012). The authors investigated blend repairs carried out on the compressor blades of a stationary gas turbine. A parametric FE model was developed and different blend shapes are compared according to the resulting blade frequency and high-cycle fatigue (HCF) properties utilizing the Goodman relation. The developed computational workflow facilitated the subsequent evaluation of nonstandard blends. A contribution, which goes beyond this parametric assessment of different blend sizes, was later presented by Karger and Bestle (2015). Karger and Bestle firstly combined the structural simulation of blended blades with numerical optimization methods. The shape of the blend, which was parameterized by five geometric design variables, was optimized using a multi-objective optimization approach. The two objective functions of the multi-objective optimization task referred to the minimization of the blend in terms of blade mass and the HCF strength computed in terms of amplitude frequency strengths. The evaluation process was automated using a commercial software and an optimization algorithm implemented in the Matlab Global Optimization toolbox was utilized to solve the problem. Pareto-optimal solutions were determined and hence provided new insights into the influences of blend shapes. They demonstrated that a wide range of different blending shapes (small up to large) may be beneficial and the final design decision should be selected out of the Pareto-optimal set accounting for aerodynamic criteria like efficiency and surge margin.

In contrast to blend repairs, patch repairs received less attention in recent research. Due to the narrow field of application only a few publications were concerned with this hightech repair technique. General considerations on the process chain of blisk repairs by patching were presented by Eberlein (2007). The authors exemplarily showed thermal simulations of the welding process that joints the patch to the blade material and conducted residual stress measurements on patched blades. They further concluded that the geometry of the patch and the position of the welding seam between blade and patch should be selected according to the distribution of vibratory stresses in the blade. A more analytical work was published by Schoenenborn and Reilc (2005), who performed FE simulations to assess the residual stresses of a patched blade. The authors performed uncoupled thermomechanical FE simulation to analyse the effect of welding and heat treatment in the leading edge region of a blisk blade. They further concluded that numerically predicted stresses could complement experimental investigation and thus decrease the risk of blade failure due to HCF. However, a generalized approach on the numerical assessment of patch repair, which includes automatic evaluation schemes as well as optimization methods, has not been followed yet.

Although, existing multi-objective optimization approaches towards improved turbomachinery design showed great potential (Dornberger et al. 2000; Karger and

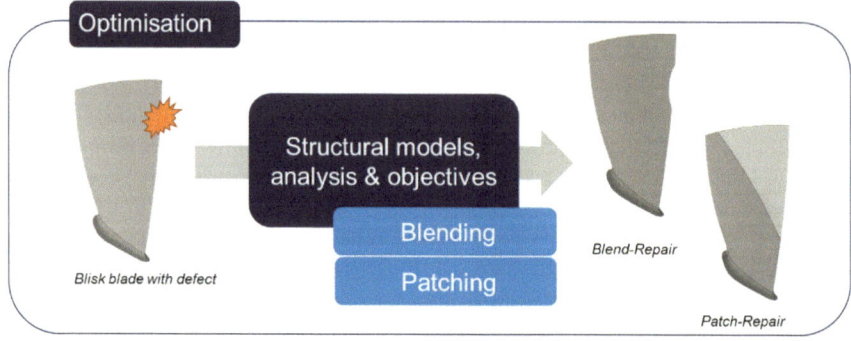

Fig. 2 Concept of numerical optimization of blend and patch repairs of blisk blades

Bestle 2015; Adjei et al. 2020) and the key idea of multi-objective formulations is present in almost all engineering disciplines (Parejo et al. 2012), to the best of the authors' knowledge, there is no universal concept to evaluate different repair designs for blisk blades utilizing state-of-the-art optimization algorithms. In this work a new computational scheme is developed, which exactly addresses this decision making process during the maintenance of blisk blades. The key aspects of this scheme for optimized blisk repairs are illustrated in Fig. 2. The optimization process starts by identifying the damaged portion of the blisk. The defect of the blade is described using the damage model described in Sect. 3.1. Based on the nominal geometry of blades, structural models are developed to model the repair-specific influences. In particular, the geometric modifications of blending and the residual stresses introduced during patching are addressed. Multiple FE analyses are used to determine vibration properties like frequencies or stresses due to welding. The implemented models and modelling assumptions are introduced in Sects. 3.2 and 3.3, respectively.

Moreover, the simulation results provide the basis for the optimization of repair designs. In Sect. 4, the design criteria and objectives are formulated and an optimization task is derived from the structural requirements. Finally, optimal repair designs are identified integrating the structural models, analysis procedures and objectives into the optimization framework EngiO. The framework is a customized optimization software, which is specially designed to handle engineering optimization task like blisk repairs with little effort. The design of the framework is illustrated in Sect. 2. Optimization results for blend and patch repairs utilizing EngiO are presented in Sect. 5. Findings, limitations and aspects for future work are summarized in Sect. 6 of this paper.

2 Optimization Framework EngiO

The initial perquisite for the optimization of blade repairs is a framework where the optimization could be organized with little effort. The framework developed is specifically targeted to optimization tasks like repair optimization. However, it is designed to manage various optimization problems, which may arise in engineering research. The framework named EngiO (Engineering Optimization framework) is a Matlab-based code that is assessable under GNU General Public License via its github repository (https://github.com/isd-luh/EngiO). In the next sections, first the basic design principles are introduced and secondly several features are demonstrated using a multi-objective test function.

2.1 Architecture

The software architecture of EngiO follows an object-oriented programming pattern. The basic framework including main functionalities is illustrated in Fig. 3 using an Unified Modelling Language (UML) diagram. The UML diagram shows dependencies between different implemented classes.

The architecture of EngiO reflects the idea that each optimization process can typically be viewed from two perspectives (Bleuer et al. 2003). The first perspective belongs to the development of new optimization algorithms, their implementation, and their comparison with other state-of-the-art algorithms. The second perspective is related to practical engineering optimization. From a practical point of view, the implementation of custom engineering optimization problems in terms of numerical simulations and user-defined objective functions is of high relevance.

These two perspectives are both considered in the architecture of EngiO and the UML diagram in Fig. 3 reflects this separation between algorithm and problem-specific objective functions as well. On the left side of the diagram, exemplarily three

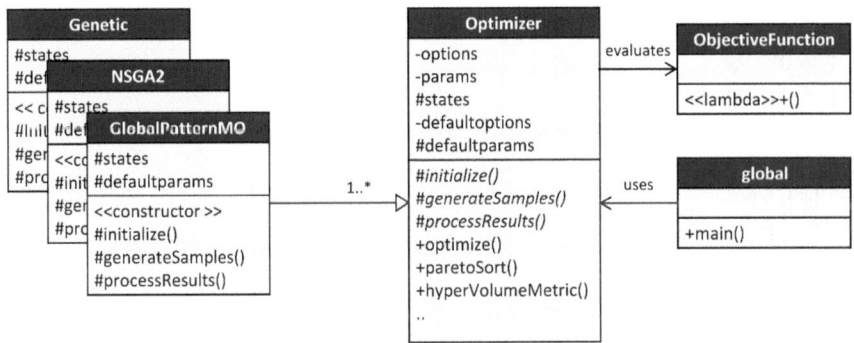

Fig. 3 Software architecture of EngiO

different classes ("Genetic", "NSGA2", and "GlobalPatternMO") of algorithms are shown. Each of these classes includes concrete methods, which are characteristic for the specific algorithm's logic. On the right side, the "ObjectiveFunction" class contains the optimization problem formulation corresponding to the custom optimization task. These objective functions may include analytic expressions or calls of any numerical simulation software like Abaqus or ANSYS.

The central part of EngiO is the "Optimizer" class, which can be seen as the glue code to link the algorithm and the optimization problem part. This "Optimizer" class is the parent class for all optimization algorithms and provides the basic structure for all further implementations. The properties of the class include general optimization settings as well as algorithm specific parameters and states. Further, the class definition includes several abstract and concrete methods. Abstract methods, which are labelled with #, have to be implemented in derived algorithm classes. They are defined according to the observation that all derivative-free optimization processes could be described within three distinct steps. In the first step, the initial samples are evaluated. Subsequently, in each optimization loop, new samples are generated and evaluation results are processed according to the algorithm logic. The algorithm classes inherit these from the parent class for structured implementation of algorithms. Concrete methods, on the contrary, indicated by + centralized routines in the parent class. The most important method implemented in the "Optimizer" class is the method "optimize", because it includes the main iteration loop of the optimization. Further, concrete methods allow fast sorting of solutions for the computation of Pareto-optimal sets and the evaluation of performance indicators like the hyper-volume metric.

The optimization process is started using a main script on a global level. In this main script, optimization settings, such as boundaries or maximum number of function evaluations are specified.

It should be noted that in Fig. 3 only a part of the implemented routines is shown. Additionally, constrained optimization is used complementing the basic framework by a penalty handling approach. Moreover, functions for visualisation of optimization results and well-known optimization test problems are part of the optimization framework. For more detailed information on the structure of EngiO and its application to different engineering tasks, it is referred to Berger et al. (2021a, b).

2.2 Features and Benefits

According to the software architecture, EngiO has several advantages to solve engineering optimization problems. The architecture follows the concept of object-oriented programming and is completely compatible with Matlab and GNU Octave. Thus, the source code has a modular, comprehensible and user-friendly structure. The "Optimizer" class of EngiO provides a unified interface between optimization algorithms and optimization problems. The strict separation of algorithm and problem formulation prevents programming bugs and the basic functionality given by the parent class facilitates the usage of the framework. As stated previously the abstract class definition of the parent class enforce the user to split the algorithm's logic in three distinct parts. This approach has the advantage that all algorithms are structured in a common way. This improves the readability of code and reduces code complexity.

A further benefit of the framework is that the software comes with various implementations of optimization algorithms. The algorithms included are adapted to the framework's architecture and are selected according to the state of the art. Optimization algorithms for single- and multi-objective formulations are implemented. The main focus of EngiO is on optimization tasks involving complex engineering simulations. The related optimization problems most often are not convex and derivatives of the objective function are not available. Thus, only local and global derivative-free algorithms are considered in the interface of the framework. Optimization subjected to constraints is facilitated by a constraint interface complementing the basic framework shown in Fig. 3. The constraints are imposed by adding a penalty on the objective function value, while the specific handling technique is derived from the constraint class.

Another key feature of the optimization framework is the inclusion of several analytic test functions. This allows the performance of algorithms to be tested and compared with each other. The benchmarking of algorithms and their parameter settings is conducted and supports the user decision on a suitable algorithm. Further, the user benefits from the common interface and is able to compare the performance of his own user-defined algorithms with established methods.

Moreover, the interface is further designed such that single- as well as multi-objective optimization problems could be solved, allowing for a broad range of engineering applications. The number of objectives is signalled to the optimizer class calling the objective function with an empty argument. Simple scripts for both cases support the beginner to start with his own optimization setups. Since engineering optimization problems often rely on time-consuming numerical computations, the framework allows to store the current state variables of each iteration. This allows to restart and continue the optimization based on previous optimization results. EngiO also features parallelization aspects. The evaluation of objectives can be performed in parallel to speed up optimization of numerically costly optimization problems. The parallel computing is organized in the "Optimizer" class.

Overall, the design of the optimization framework is based on the requirements of an engineer. The strength of the framework therefore lies in its adaptability to individual needs. Since the framework is regarded as a pure research and teaching code, some basic graphic representations of results are provided.

To illustrate the capabilities of EngiO the results of a multi-objective optimization run utilizing a well-known test function are presented below. The test function was introduced by Poloni (1997) and is an unconstrained two-dimensional two-objective formulation. The optimization problem is stated as follows:

Fig. 4 Pareto frontier of
Poloni test function
optimized using NSGA-II

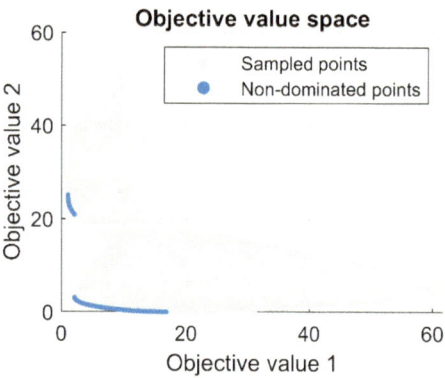

$$\underset{minimize}{} \quad \begin{aligned} F_1 &= 1 + (A_1 - B_1)^2 + (A_2 - B_2)^2 \\ F_2 &= (x_1 + 3)^2 + (x_2 + 1)^2 \end{aligned}$$

$$\text{with} \quad \begin{aligned} A_1 &= 0.5\sin(1) - 2\cos(1) + \sin(2) - 1.5\cos(2) \\ A_2 &= 1.5\sin(1) - \cos(1) + 2\sin(2) - 0.5\cos(2) \\ B_1 &= 0.5\sin(x_1) - 2\cos(x_1) + \sin(x_2) - 1.5\cos(x_2) \\ B_2 &= 1.5\sin(x_1) - \cos(x_1) + 2\sin(x_2) - 0.5\cos(x_2) \end{aligned}$$

$$\text{subject to} \quad \begin{aligned} -\pi &\le x_1 \le \pi \\ -\pi &\le x_2 \le \pi \end{aligned} \tag{1}$$

The optimization problem is solved using the Non-dominated Sorting Genetic Algorithm-II (NSGA-II), which is a state-of-the-art approach in the context of multi-objective optimization (Deb et al. 2000). The optimization process is stopped after 10,000 objective function evaluations are performed. The optimization results of one optimization run are illustrated in Fig. 4.

In Fig. 4, all samples generated during the optimization are plotted in the two-dimensional objective value space. The solutions on the Pareto front, which means that they are non-dominated by other solutions, are further highlighted in blue. In this optimization example, the Pareto front is divided into two parts. This discontinuity in the Pareto front is characteristic for the selected test function. The solutions on the front are well distributed and the comparison with optimization results published by other researchers reveals, that the algorithm identified the Pareto-optimal set correctly. It should be noted that the slope of the Pareto front is individual for each optimization problem formulation (Angus and Woodward 2009).

The selected optimization example can be easily recalculated with EngiO, because the algorithm as well as the test function are part of the framework. The main script used to start the optimization comprises about less than 30 lines of code and has a clear structure. This is mainly allowed by the object-oriented schema and the choice of a high level programming language. Examples for main scripts are also part of the software, which facilitate the application of EngiO in an early stage and allows an easy incorporation into the framework, even for inexperienced users.

3 Structural Models

One way to predict the structural integrity of repaired blades is to model changes in the blade properties, re-analyse stresses and vibration frequencies and predict the remaining lifetime of the blade.

To allow for an automated simulation process several aspects have to be described in a standardized way. The first aspect refers to the description of the damaged portion of the blade. This damage model is introduced in Sect. 3.1 and is utilized for blend as well as patch repairs. In Sects. 3.2 and 3.3 the repair-specific models for blend and patch repairs are introduced.

3.1 Damage Model

Defects of blade may have various forms and shapes. To account for this individual patterns, the defect portion of the blade is described using a point cloud. From optical measurements defect areas can be detected and specified using a set of points in a global coordinate system. This universal specification allows for a very flexible description of blade regions to refurbish. However, in most cases, the damage portion is located at the leading edge of the blade, since particles entering the engine lead to so-called foreign object damage. This kind of damage is illustrated by a point cloud in Fig. 5.

The resulting damage model is utilized to describe the damage portion and facilitates the subsequent search for optimal repair design. Since blend repair as well as patch repair aim to remove the defect completely in each iteration it should be checked whether the repair design fulfils this requirement. In the case of the blend repair, the removed volume should include all damaged points. When a patch repair is conducted the geometry of the patch has to cover the damaged region as well.

Fig. 5 Blade with damaged region at the leading edge. The size and location of the defect is specified using a set of points in global coordinate system (X, Y, Z)

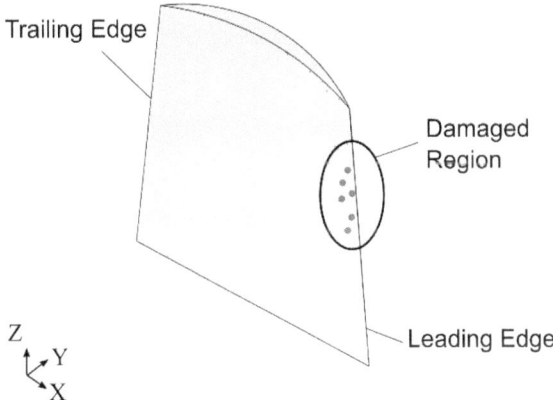

3.2 Blending Model

The model developed for blending of blisk blades is based on the assumption that the blending shape has an elliptical contour. To model this contour a parameterized ellipsoidal volume is utilized. The schematic illustration of a blade and the volume is depicted in Fig. 6.

In accordance with typical damaged regions, as shown in Fig. 5, the ellipsoid is located at the leading edge of the blade and the blending shape corresponds to the cut-out enforced by the ellipsoidal volume. The geometric modification of the blade is thus the result of a Boolean operation between ellipsoidal volume and nominal blade volume.

The position and size of the ellipsoidal tool is defined by three scalar design variables. The first variable e specifies the position of the ellipsoid. The variable in particular determines the distance between the tip of the blade and the centre of the elliptical cut-out. Since only blends at the leading edge are considered, e runs along the leading edge of the blade and is normalized according to the total length of the edge, such that $0 \le e \le 1$. The other two variables define the size of the ellipsoid and cut-out, respectively. The depth of the cut-out resulting from the relative position between ellipsoidal tool and blade is stated by d. The length of the vertical principal axis of the ellipsoid is specified by r_z. The other radii of the ellipsoid are set to fixed values prior to the sampling.

Modelling this geometric change in terms of FE analysis, the mesh of the blade has to be adapted to the new blended blade contour. One way is to utilize an automated meshing procedure and mesh the new blade geometry for each computing example. This approach, however, has the drawback that the simulations for each sample are performed with possibly very different meshes, which could introduce numerical errors. Moreover, evaluating results of simulations performed with similar meshes is much more straight-forward, because displacements can be compared directly. To maintain the majority of the FE mesh related to the nominal geometry a local

Fig. 6 Blade with blendig shape parameterized by three design variables d, e and r_z and the related ellipsoidal tool

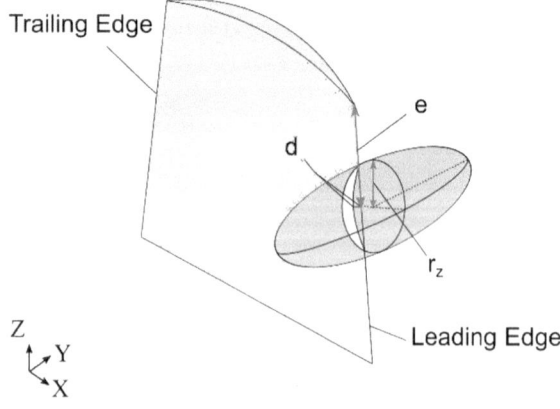

re-meshing procedure was developed. Only the mesh in the proximity of the blend is adjusted to capture the new geometry, while the rest of the mesh remains unchanged.

3.3 Patching Model

From a geometric point of view, the parametric description of patch repairs is not as complex as for blend repairs. Since the blade region, which is replaced by the patch material mainly has a triangular shape the repair is specified by two dimensions only. In Fig. 7 a blade with a patch is illustrated.

As in the previous section, we concentrate on repairs at the leading edge of the blade. Therefore, one variable is again defined in relation to this edge. The variable a determines the height of the patched region and equals the distance from tip to interface between patch and base material. The second variable b corresponds to the width of the patched part and is measured along the chord of the blade. For the purpose of a more intuitive understanding of design variables, a and b are normalised according to the length of the span and chord of the analysed blade.

In Fig. 7, the interface between patch and blade (coloured in red) is idealized as a plane and therefore is of infinitesimal width. In reality, both parts are joined via a welding procedure and, hence, there is the welding bead and the corresponding heat affected zone. The volume of this region depends on the welding process and is strongly influenced by the welding parameters used. To assess these thermal and mechanical effects within this region thermomechanical FE simulations are employed. Firstly, the temperatures caused by the moving heat source are computed via a transient thermal simulation. The heat input of the moving heat source is implemented in a user-defined subroutine. The current position of the heat source is defined using a welding trajectory along the centreline of the weld and a heat flux distribution according to a conical heat source volume. Residual stresses in the blade are computed in subsequent mechanical simulation.

Fig. 7 Blade with patch parameterized by two design variables a and b resulting in a triangular patch geometry

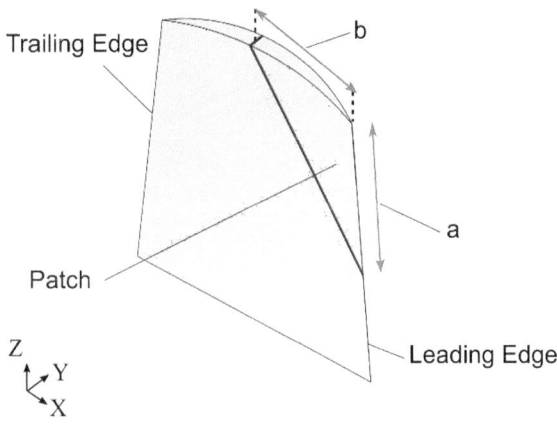

4   Analyses and Objectives

The optimization objectives of the optimization problem formulation result from the design goals of repair procedures. These design goals could further be classified by two groups. The first category refers to structural aspects. In particular, fatigue strength and vibration properties are part of this first category. The second group relates to practical aspects such as manufacturing effort. In the next sections the optimization objectives are introduced and discussed with regard to blending and patching procedures.

4.1   Natural Frequencies and Mode Shapes

One of the most important aspects in the design of rotating machines refers to resonance conditions. During operation of jet engines, compressor blades are subjected to multiple excitation frequencies, which could cause resonances, if the excitation frequency meets one of the natural frequencies of the blisk. This condition is commonly analysed using established concepts like the Campbell diagram or Interference diagram. In the initial design phase the frequencies of the blisk are, hence, tuned such that they do not intersect with the excitation during nominal operation. These vibration properties of blades are predicted utilizing a modal analysis of the geometry. Exemplary results of a modal analysis at standstill are shown in Fig. 8.

To reduce computation time, the modal analysis is not performed using the whole blisk structure. Instead one periodic sector is considered and cyclic symmetry constraints are imposed on the cyclic faces of the sector, neglecting mistuning effects. As depicted in Fig. 8a the sector geometry is meshed using two different element types. The blade and the disk region are meshed using hexahedral elements, while

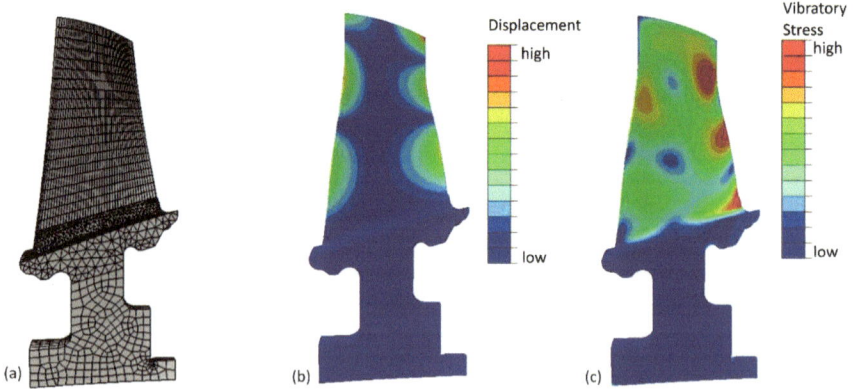

Fig. 8   FE mesh (**a**), mode shape of the 10th eigenmode (**b**), and related modal vibratory von Mises stresses (**c**) of blade sector

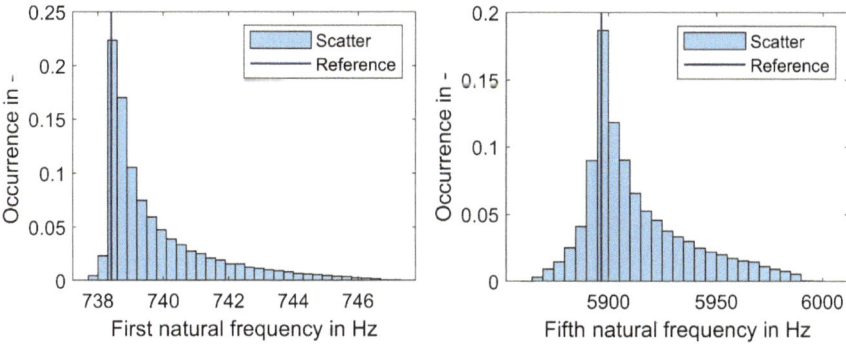

Fig. 9 Scattering in first (left) and fifth (right) natural frequencies due to performed blend repairs at the leading edge

the fillet region requires the use of tetrahedral elements due to the more complex geometry. In Fig. 8b the mode shape of the 10th vibration mode, which refers to a frequency of about 9500 Hz, is shown. Due to the high mode, several local vibration maxima are computed in the blade region, which results in the related vibratory stress distribution Fig. 8c.

In the case of blending, however, the geometry differs from the nominal condition and natural frequencies change. To gain first insights into the effect of these local modifications on natural frequencies, different blend repairs are analysed using the blending model introduced in Sect. 3.2. 6859 samples following a grid sampling scheme are generated and modal analysis is performed for each sample. To capture the stiffening effects due to rotation the modal analysis is performed based on a static analysis imposing rotational forces according to nominal operation. The results of the sampling for two natural frequencies are shown in Fig. 9. The reference frequency denoting the simulation result of the nominal blade is marked by the dark blue line.

The histograms indicate that blend repairs affect the tuning of the first as well as the fifth frequency. Concerning the first frequency mainly a slight increase of the frequency is determined, because the removal of the material mainly reduces the vibrating mass of the blade. In contrast, the distribution of the fifth frequency differs significantly from the distribution calculated for the first mode. While almost any blend repair leads to a higher first frequency, the distribution corresponding to the fifth mode also results in more reduced natural frequencies. This effect is caused by the different mode shapes. The lowest natural frequency corresponds to a pure bending mode. A blend therefore increases the frequency, as the mass reduction predominates over the loss of local stiffness in the blending area.

To improve the blend geometries for damaged blisk blades, two design goals are derived from the statistical study on natural frequency tuning. The first objective results from the comparison between nominal and repaired modal properties. To reduce the risk of mistuning, the repaired blade should behave similarly to the nominal one. Therefore, the difference to the nominal (design) frequencies is considered. The maximal deviations from nominal frequencies are calculated as

$$F^{nom} = \max_i \left(\frac{|f_i(x) - f_i^{nom}(x)|}{\alpha_i} \right) \qquad (2)$$

where α_i is a weighting factor, f_i is the ith natural frequency of the blended blade, f_i^{nom} is the corresponding ith natural frequency of the nominal geometry, and x is the design variable vector with repair-specific parameters. The formulation in (2), hence computes the maximum deviation of multiple natural frequencies with respect to the nominal case. Since the variation of some frequencies is more critical than others the weighting factor could be adjusted. Below, the weighting is performed using the nominal frequencies.

A further design goal refers to the distance between natural frequencies and excitation frequencies, which quantifies the risk of meeting resonance conditions. The minimal distance to the closest excitation frequency during nominal operation is stated as

$$F^{exc} = \min_i \left(\beta_i \cdot |f_i(x) - f_i^{exc}(x)| \right) \qquad (3)$$

where β_i is a weighting factor, f_i is the ith natural frequency of the blended blade, f_i^{exc} is the corresponding closest excitation frequency, and x is again the design variable vector with repair-specific parameters. With the weighting factor, different excitations can be penalized differently.

4.2 High-Cycle Fatigue Strength

Although blades are designed such that resonances are preferably avoided, during operation blades are still subjected to vibrations. These vibrations lead to alternating stresses in the blade material and could finally result in fatigue failure. This failure mode, denoted as high-cycle fatigue, therefore, must be considered in the initial blade design as well as with repaired blisks.

The well-established evaluation concept for HCF, which is also used in this work, is illustrated in Fig. 10.

The diagram shows the dependency between HCF properties and the alternating stress and the mean stress level. The line connecting the endurance limit σ_e and the yield strength σ_y subdivides the diagram in a safe and a failure region. To ensure safe operation, all designs (nominal as well as repaired) have to correspond to the region below this line. Referring to rotating blades, the mean stresses in the initial design are mainly caused by centrifugal forces and the mean stress level depends on the rotational speed of the engine. The alternating stresses originate from blade vibration. Depending on the excited mode, this results in different spatial stress distributions as shown for the 10th mode in Fig. 8.

As indicated by the red arrow in Fig. 10, the stress state in the modified design (dark blue marker) commonly deviates from the nominal state (light blue marker). In

Fig. 10 Constant life diagram with an initial design point and a modified design point corresponding for example to a repaired blade

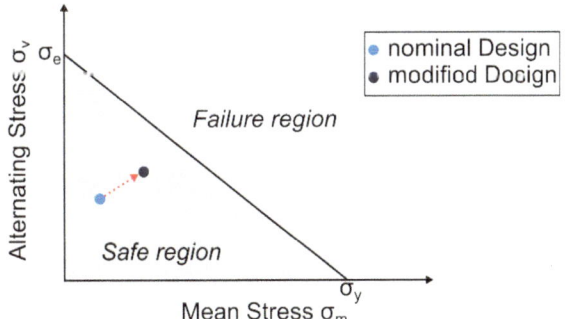

the worst case, both stresses are increased and the modified state is located close or even above the constant life line and failure occurs. One reason for higher alternating stress is the amplification of vibration amplitudes and hence stress amplitudes. For the computation of changes in the vibration amplitudes mistuning as well as aerodynamic conditions have to be considered. Since mistuning (see also subprojects C3 and C6) and aerodynamic are not part of this work, it is focused on the second aspect— the mean stress level. In particular, after a patch repair, the medium stresses in the blade are strongly influenced by the repair process. In the nominal case the blade is assumed to be free from residual stresses, but the welding process involved in patching induces residual stresses in the blade material. These residual stresses are determined in thermomechanical simulations and the welding stresses like they are shown in Fig. 11 are computed.

Fig. 11 Residual stresses (von Mises) remaining in the proximity of the weld after patch repair

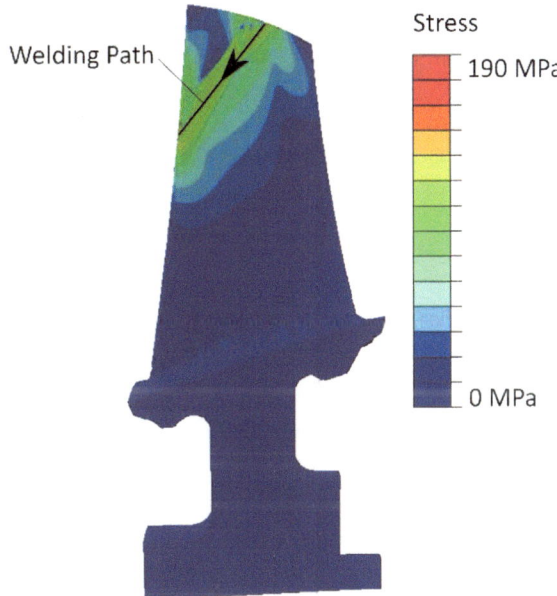

Despite the post-weld heat treatment, the depicted simulation results show that the welding process leads to a significant increase in stresses in the proximity of the welding path. These stress components suppose on the mean stresses due to centrifugal forces and in most case increase the mean stress level in parts of the blade. One design goal of patch repairs is therefore to place the weld in a blade region where the combination of increased mean stresses and vibratory stresses does not lead to failure. The minimal strength of the joint between patch and weld is determined by

$$F^{HCF} = \min_i(af_i) = \min_i \left(\frac{\sigma_e}{\sigma_{v,i}} \left(1 - \frac{\sigma_{m,i}}{\sigma_y} \right) \right) \tag{4}$$

where af_i is the amplitude frequency strength of all FE nodes i, and $\sigma_{v,i}$ and $\sigma_{m,i}$ are the related alternating and mean stresses. The amplitude frequency strength reflects the diagram shown in Fig. 10 and all stresses in (4) are evaluated using von Mises equivalent stresses.

Moreover, the endurance limit of the welded region is significantly lower than for the base material and therefore it is assumed that failure occurs in joint region only.

4.3 Manufacturing Aspects

In practice, a lot of manufacturing aspects drive the optimal repair design of blisk blades. Since the blades of a blisk are closely arranged, there is relatively small space for welding, drilling or any other manufacturing tool. The geometric boundaries of repair design, should therefore be selected in accordance with the particular manufacturing limitations. In addition, different repair designs may vary in manufacturing complexity as well. Considering patch repairs the difficulties involved in welding process also depend on the particular welding path, e.g. the curvature of the blade along the path. A practical design goal for the welding process of patching is hence the length of the weld

$$F^{length} = \ell^{weld\,path}(x) \tag{5}$$

because it corresponds to the manufacturing effort.

5 Optimization and Results

The develop concept for the optimization of blisk repair procedures is exemplarily applied to a real blisk geometry. The blisk geometry was designed at the Institute of Turbomachinery and Fluid Dynamics and operates in an aerodynamic testing facility. The geometry of the blisk and some basic properties are illustrated in Fig. 12.

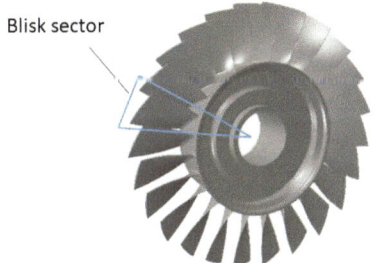

Blisk sector

Property	Value	Unit
Young's Modulus	116.52	GPa
Density	4430	kg/m³
Nom. rotational speed	17100	rpm
Number of blades	24	-
Outer diameter	320	mm

Fig. 12 Blisk geometry (left) used for all numerical examples and corresponding properties of the blisk and the axial compressor (right)

As marked in Fig. 12, the simulations are carried out on one out of 24 sectors of the compressor blisk using the FE solver of Abaqus. The analyses for the prediction of structural properties are performed according to the considerations presented in Sect. 4. Simulation results are processed in custom routines and related optimization tasks are implemented as objective functions in the optimization framework EngiO. Since multiple contradicting objectives are of similar importance the optimization problems are formulated using a two-objective formulation. Boundaries are specified in the main script according to the repair-specific parameters of the mode (Sect. 3). The constraint interface of EngiO is utilized to impose constraints and ensure valid repair designs only.

The multi-objective optimization problems reflecting the structural requirements for blend and patch repairs are described in Sects. 5.1 and 5.2. The optimization results determined utilizing derivative-free optimization algorithms are visualised in the objective value space and Pareto-optimal repair designs are discussed.

5.1 Optimization Results for Blending

The first optimization refers to the optimization tasks corresponding to blend repair optimization. According to the optimization objectives introduced in Sect. 4.1 the tuning of natural frequencies of the repaired blade are optimized. The two-objective optimization problem is stated as

$$
\begin{aligned}
\text{minimize} \quad & \begin{aligned} F_1 &= F^{nom}(x) \\ F_2 &= -F^{exc}(x) \end{aligned} \\
\text{with} \quad & x = [e, d, r_z]^T \\
\text{subject to} \quad & \begin{aligned} g(x) &\leq 0 \\ 0.05 &\leq x_1 \leq 0.5 \\ 0.5 &\leq x_2 \leq 5 \\ 0.5 &\leq x_3 \leq 25 \end{aligned}
\end{aligned}
\tag{6}
$$

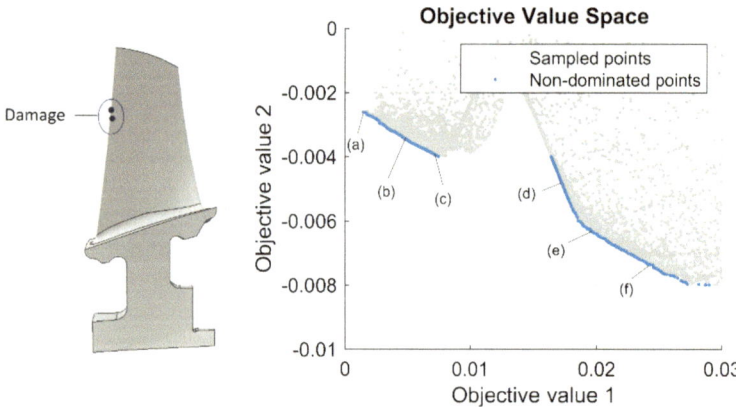

Fig. 13 Damage pattern on the leading edge of a bladed blisk sector (left) and the Pareto front determined for the blend repair optimization in one optimization run (right)

where $g(x) \leq 0$ is an inequality constraint, which is violated when the damage is not completely removed by the blend design. The optimization problem in (6) is formulated as a minimization problem, and consequently for the second objective the expression stated in (3) is multiplied by -1. The design variables and upper and lower boundaries are selected according to the blending model introduced in Sect. 3.2.

Figure 13 shows the damage pattern of the computational example and the results of the optimization problem specified in (6) in the objective value space. The optimization is performed using the state-of-the-art algorithm NSGA-II and allowing for 10,000 objective function evaluations with a population size of 100 individuals. The constraint handling scheme implemented in EngiO is used to penalize infeasible design using a linear penalty approach.

All solutions sampled during the optimization process are highlighted in grey while the non-dominated solutions, which form the Pareto front are coloured in light blue. Similar to the Poloni test problem optimized in Sect. 2, the Pareto front consists of two parts. Consequently, there is no feasible solution in between, which at least to an improvement with respect to one of the two objectives. The practical meaning of the optimization results is illustrated in Fig. 14. Six different Pareto-optimal blend designs, which correspond to the non-dominated points labelled with (a) up to (f) in Fig. 13, are visualised. The first three solutions (a–c) represent the upper left part of the Pareto front. The designs in this part of the front are characterised by a relatively small cut-out and the centre of the blend is located next to the damage. Moreover, the three designs indicate, that the blade frequencies with respect to the excitations (objective value 2) improve with larger blends, while the increased deviation from the nominal shape leads to non-favourable deviations from nominal frequencies (objective value 1). Further, the second part of the Pareto front is associated with blend shapes, which are located closer to the tip of the blade (small e). To still remove the damage, the blending depth and width are hence larger than for most designs of the first part of

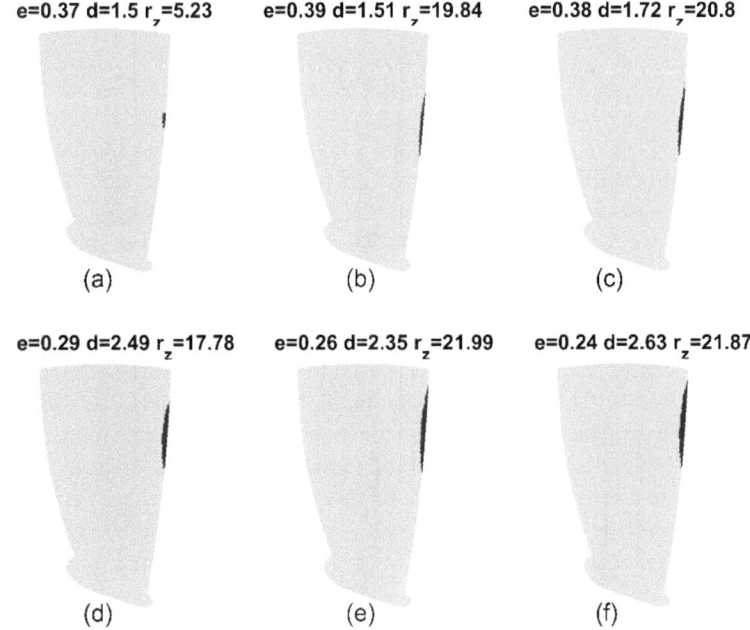

e=0.37 d=1.5 r$_z$=5.23 · e=0.39 d=1.51 r$_z$=19.84 · e=0.38 d=1.72 r$_z$=20.8

(a) (b) (c)

e=0.29 d=2.49 r$_z$=17.78 · e=0.26 d=2.35 r$_z$=21.99 · e=0.24 d=2.63 r$_z$=21.87

(d) (e) (f)

Fig. 14 Six exemplary blend designs (**a**)–(**f**) out of the Pareto-optimal set

the Pareto front. For example, blend design (f) has the largest depth of 2.63 mm. With regard to aerodynamic performance of blended blades, especially the Pareto-optimal solutions, which refer to large cut-outs, have to be re-evaluated by means of CFD simulations before choosing this type of blend design. In the presented example, the solutions on the first part of the front should be preferred to avoid poor aerodynamic properties.

5.2 Optimization Results for Patching

The second optimization considers the design of patch repairs accounting the design goals discussed in Sects. 4.2 and 4.3. Based on the patching model a two-objective optimization task is formulated

$$\text{minimize} \quad \begin{aligned} F_1 &= F^{length}(x) \\ F_2 &= -F^{HCF}(x) \end{aligned}$$

$$\text{with} \quad x = [a, b]^T$$

$$g(x) \leq 0$$
$$subject\ to \quad 0.0 \leq x_1 \leq 1 \tag{7}$$
$$0.0 \leq x_2 \leq 1$$

where $g(x) \leq 0$ is an inequality constraint, which is violated when the damage is not completely removed by the welded patch. Since large HCF strength is favourable and the problem is formulated as a minimization the fatigue strength value is again multiplied by -1. Via the boundaries the two normalized design variables (Sect. 3.3) describing the patch geometry are limited to values between zero and one.

In the left part of Fig. 15 the damage pattern of the computational example is visualised. The same optimization algorithm with the same algorithmic settings as well as the penalty handling approach is used to solve the constrained optimization problem. Moreover, the optimization problem defined in (7) is computed for two different variants of the second objective function. In the first optimization, the fatigue strength corresponding to the first vibration mode is optimized, whereas the second optimization focuses on the fifth vibration mode. In particular, these two modes are studied, because they show an increased risk of high vibrations according to the Campbell diagram. The Pareto front of the first variant is shown in Fig. 15. The non-dominated solutions computed within 10,000 samples show a continuous course. Additionally, it can be seen that no non-dominated point refers solutions with a first objective value greater than 0.4. This indicates, that all patches with welding seams longer than 4 cm are not optimal in any case. This is also reflected by the associated patch geometries, which are shown in Fig. 16.

As indicated by the blue area, all Pareto-optimal patches and therefore all welding beads are located in the upper part of the blades. According to their geometry these patches are also classified as short patches. Additionally, one Pareto-optimal design leading to a solution in the middle part of the Pareto front is shown in Fig. 16. The optimization results are reasonable, since the first mode is a bending mode. The

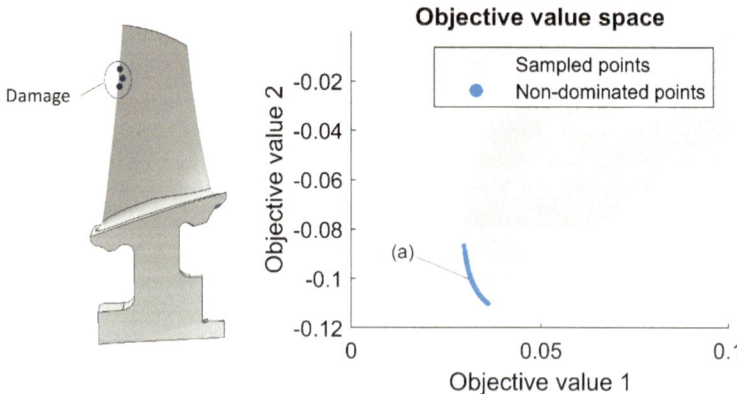

Fig. 15 Damage pattern on the leading edge of a bladed blisk sector (left) and the Pareto front determined for the patch repair optimization accounting for HCF due to first vibration mode (right)

Fig. 16 Patch designs corresponding to the optimization results shown in Fig. 15

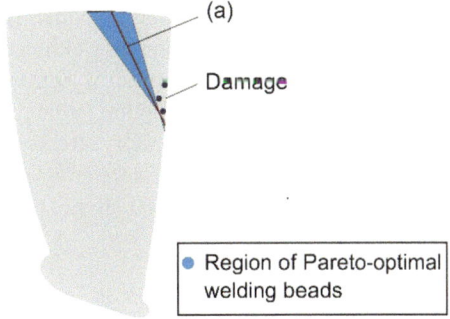

bending results in vibratory stresses next to the blade root and thus beads close to the root are not optimal. The results of the second optimization for the fifth mode are depicted in Fig. 17. The Pareto front has again no significant discontinuities like they are determined for the blend repair. The analysis of Pareto-optimal designs, which are illustrated in Fig. 18, however, reveals, that the designs are different to the designs of the previous optimization.

Fig. 17 Pareto front determined for the patch repair optimization accounting for HCF due to fifth vibration mode

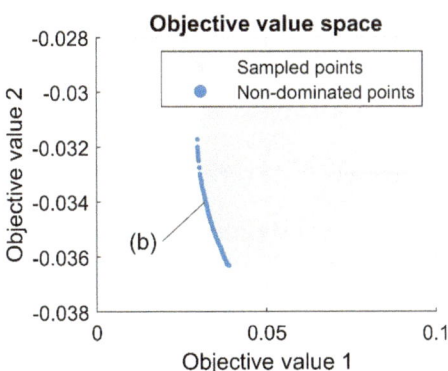

Fig. 18 Patch designs corresponding to the optimization results shown in Fig. 17

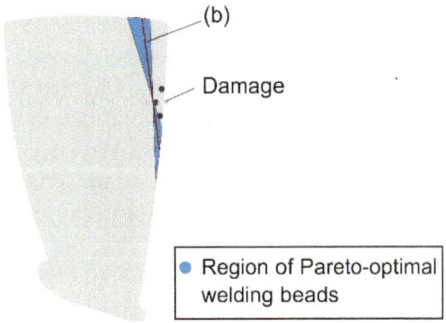

The change in the second objective function leads to a Pareto-optimal set that includes of solutions with longer welding beads. For further illustration an exemplary design (b) is visualized in the objective values as well as in the design variable space. These difference in optimization results between both tasks relate to the mode shape and the resulting vibratory stresses. Higher modes often lead to multiple local blade regions with high vibratory stresses. Consequently, the Pareto-optimal designs correspond to weld beads that are not placed close to these regions.

6 Conclusions and Outlook

In this work, a concept for the systematic evaluation and optimization of blade repairs of blisks by blending and patching was presented. The numerical concept addresses several aspects. First, the optimization framework EngiO, which specifically targets at engineering optimization tasks, was introduced. The object-oriented architecture was described and the benefits of the design were demonstrated using the two-objective Poloni test function. Finally, the capabilities of EngiO provided the basis for all further developments in the field of blisk repairs. Systematically and automated simulation of the two repair processes was performed by parametrization and custom FE simulation models. A damage model was used for unified description of arbitrary damage patterns. With respect to repair-specific changes design goals and requirements are discussed and established evaluation metrics were used to assess the structural properties of repaired blades. Combining multi-objective optimization methods, repair-specific modelling and design objectives Pareto-optimal repair designs were found automatically based on predefined damage patterns. Depending on the particular optimization tasks the Pareto fronts showed to have different appearances. Continuous as well as discontinuous Pareto fronts are determined. Moreover, the Pareto-optimal set and the corresponding repair design allow new insight in the repair design process. In this context, the choice of the multi-objective formulation of the optimization problem is a great advantage, because the final decision of the patch or blend repair design could be made based on the whole Pareto-optimal set and accounting for additional engineering preferences. The weighting between individual optimization objectives could be done a posteriori to the optimization run.

In future work, the computational scheme developed in this work could be extended by several aspects. One aspect corresponds to the damage model introduced in Sect. 3.1. In this contribution the optimization is carried out exemplarily on few damage patterns only. Moreover, this patterns are generated based on engineering experience but do not correspond to measurements. In the future, it would be beneficial to systematically inspect damaged blisk blades with optical measurement equipment. Using this data multiple optimization runs could be performed and a more generalized conclusion could be drawn. Further, the works does not consider aerodynamic requirements or mistuning effects. Since both aspects are important for the functionality of blades, these aspects should be investigated in future research. Similarly, simulation models e.g. welding simulations could be improved to capture effects on the structure in more detail. However, with more complex numerical simulation the optimization becomes more difficult, since numerical costly simulations slow down optimization processes.

Acknowledgements Funded by the Deutsche Forschungsgemeinschaft (DFG, German Research Foundation) CRC 871/3 119193472.

References

Adjei, R. A., Fan, C., Wang, W, and Liu, Y. (2020). Multidisciplinary Design Optimization for Performance Improvement of an Axial Flow Fan Using Free-Form Deformation. Journal of Turbomachinery.

Angus, D. and Woodward, C. (2009). Multiple objective ant colony optimization. Swarm Intelligence.

Beck, J. A., Brown, J. M., and Scott-Emuakpor, O. E. (2017). Probabilistic Study of Integrally Bladed Rotor Blends using Geometric Mistuning Models. Proceedings of the 58th AIAA/ASCE/AHS/ASC Structures, Structural Dynamics, and Materials Conference.

Berger, R., Bruns, M., Ehrmann, A., Haldar, A., Häfele, J., Hofmeister, B., Hübler, C., and Rolfes, R. (2021a). EngiO-Object-oriented framework for engineering optimization. Advances in Engineering Software.

Berger, R., Quaak, G., Hofmeister, B., Gebhardt, C. G, and Rolfes, R. (2021b). Multi-objective Approach Toward Optimized Patch Repairs of Blisk Blades. AIAA Journal.

Bleuer, S., Laumanns, M., Thiele, L., and Zitzler, E. (2003). PISA - A Platform and Programming Language Independent Interface for Search Algorithms. Evolutionary Multi-Criterion Optimization.

Brown, J. M., Slater, J., and Grandhi, R. V. (2003). Probabilistic Analysis of Geometric Uncertainty Effects on Blade Modal Response. Proceedings of the ASME Turbo Expo 2003.

Bussmann, M. and Bayer, E. (2009). Blisk production of the future - technological and logistical aspects of future-oriented construction and manufacturing processes of integrally bladed rotors. Proceedings of the International Symposium on Air Breathing Engines.

Day, W. D., Fiebiger, S. W., and Patel, H. N. (2012). Parametric Evaluation of Compressor Blade Blending. Proceedings of the ASME Turbo Expo 2012.

Deb, K., Agrawal, S., Pratap, A., and Meyarivan, T. (2000). A Fast Elitist Non-dominated Sorting Genetic Algorithm for Multi-objective Optimization. NSGA-II. Lecture Notes in Computer Science.

Dornberger, R., Stoll, P., Büche, D., and Neu, A. (2000). Multidisciplinary turbomachinery blade design optimization. Proceedings of 38th Aerospace Sciences Meeting and Exhibit.

Eberlein, A. (2007). Phases of high-tech repair implementation. Proceedings of 18th International Symposium on Airbreathing Engines, Beijing, 1–8

Heinze, K., Vogeler, K., and Friedl, W.H. (2010). The Impact of Geometric Scatter on High-Cycle-Fatigue of Compressor Blades. Proceedings of the ASME Turbo Expo 2010.

Karger, K. and Bestle, D. (2015). Parametric Blending and FE-Optimization of a Compressor Blisk Test Case. Advances in Evolutionary and Deterministic Methods for Design, Optimization and Control in Engineering and Sciences.

Parejo, J. A., Ruiz-Cortes, A., Lozano, S., and Fernandez, P. (2012). Metaheuristic optimization frameworks. A survey and benchmarking. Soft Computing.

Poloni, C. (1997). Hybrid GA for multi objective aerodynamic shape optimization. Genetic Algorithms in Engineering and Computer Science.

Schoenenborn, H. and Reile, E. (2005). Analytical analysis of the welding and heat treatment of a compressor blisk leading edge repair process. Meeting Proceedings of RDP.

Single Crystalline Laser Welding

Robert Bernhard, Irene Buchbender, Volker Wesling, and Stefan Kaierle

Abstract Rising demands on engines in the aerospace industry require high engine inlet temperatures in modern gas turbines in order to increase the efficiency. Single-crystal nickel-based superalloys were developed in order to meet these rising demands and confer high pressure turbine blades with the necessary wear and oxidation resistance at high temperatures. The blades of the high-pressure turbine are subject to significant wear due to the high mechanical strain under extreme conditions, which appears in the form of cracks in the single-crystal substrate material. There are no approaches for the restoration of the original material properties since the repair of cracks and erosions by polycrystalline laser cladding can be applied only to a limited extent. The aim of the subproject is the restoration of defect, single-crystal high-pressure turbine blades. This objective was achieved by developing a novel two-stage laser metal deposition process based on simulation, laser process development, process monitoring and temperature control.

Keywords Single crystalline · Additive manufacturing · Acoustic emissions · Laser metal deposition

1 Introduction

In order to reduce fuel costs to the airline and comply with environmental legislation on emissions, the optimization of the thermal and propulsive efficiencies of the modern aeroengine has become paramount. Even a small increase in efficiency and

R. Bernhard · I. Buchbender · V. Wesling · S. Kaierle (✉)
Laser Zentrum Hannover e.V., Hollerithallee 8, 30419 Hannover, Germany
e-mail: s.kaierle@lzh.de

R. Bernhard
e-mail: r.bernhard@lzh.de

S. Kaierle
Institute of Transport and Automation Technology, Leibniz University Hannover, An der Universitaet 2, 30823 Garbsen, Germany

© The Author(s) 2025
J. R. Seume et al. (eds.), *Regeneration of Complex Capital Goods*,
https://doi.org/10.1007/978-3-031-51395-4_11

service interval of these engines relates to substantial savings in fuel and maintenance costs over the life of the engine (McNutt 2015).

The development of directionally solidified (DS) castings allowed for the elimination of grain boundaries transverse to the loading axis, while single-crystal (SX) investment castings eliminated grain boundaries almost completely (Donachie and Donachie 2002). This allowed for a huge improvement in creep resistance, since grain boundaries are sites of damage accumulation and crack initiation at high temperatures (Schneibel 2004).

Nickel-based superalloys consist primarily of a two-phase microstructure: A face-centered cubic (FCC) gamma solid solution matrix phase and a uniform distribution of ordered gamma prime precipitates. Advanced single-crystal superalloys contain high volume fractions of gamma prime, typically in the range of 50 to 70 volume percent. The single-crystal nature combined with the high volume fraction of gamma prime precipitates, provide superior high-temperature mechanical properties (Vitek et al. 2002).

High-pressure turbine blades are located in the hot section of aircraft engines and convert gaseous energy exiting the combustion chamber into mechanical energy. These blades are subject to high temperatures, high stress, vibration effects, centrifugal and fluid forces that can result in creep, fracture or yielding fractures (Boyce 2006).

Components in aircraft engines are exposed to extreme conditions. Due to high temperatures and pressure as well as the impact of foreign objects, defects such as plastic deformations, wear by hot gas and CMAS (Ca-Mg–Al–Si) corrosion, impact damages and cracks can occur, resulting in high costs (Alfred et al. 2018).

Due to the high cost of manufacture, repair and replacement, there is a clear financial incentive for more economical manufacturing routes. These result in surface defects as a result of erosion and corrosion, wear of the turbine blade tip and cracking in the base material.

The process of laser metal deposition (LMD) has established itself as a robust process for the repair of parts. Its flexibility with regard to materials used, near net shape of the deposited structures and control of thermal processes make it interesting for specialized applications. Microstructurally, the rapid solidification rates in the melt pool during the LMD process result in small grains (McNutt 2015) with less segregation of alloying elements when compared to casting and conventional arc welding processes. The process also allows for a high degree of control of the heat flow, thereby allowing control over the microstructure (Gäumann et al. 2001).

The Fig. 1 shows the γ and γ' phases of typical Nickel-based superalloys (a). Figure 1b shows the fine microstructure obtained during the deposition of such alloys due to high solidification rates. Using an electron backscatter diffraction (EBSD) analysis of the deposited clad, the absence of grains and high-angle grain boundaries are detected, which is characteristic of single crystal material.

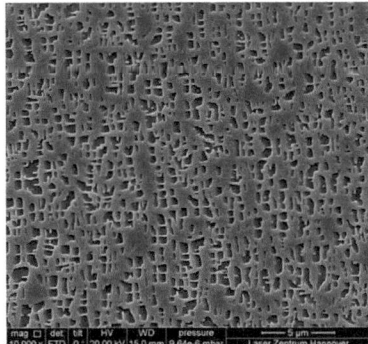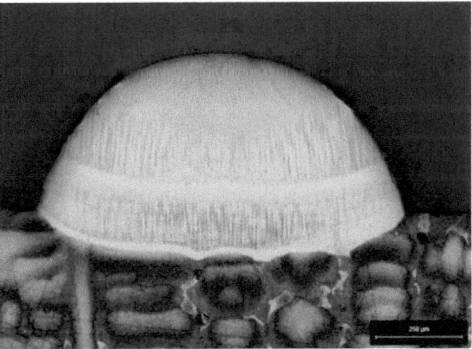

Fig. 1 **a** two-phase microstructure of CMSX-4, **b** micrograph of deposited material

1.1 Material Challenges

While the high performance capabilities of such components result in longer life spans, the repair of these parts still poses a challenge (Rottwinkel et al. 2014a, b). The processing of these Nickel-based Superalloys is challenging due to its low weldability and high susceptibility to cracking (Buchbender et al. 2020), which arises from its complex chemical structure. Many Ni-based superalloys are considered to have low weldability due to the rapid precipitation of the strengthening phase gamma prime (Wahl and Harris 2011). Further- more, an alloy is considered non-weldable if the total Al-Ti content exceeds 4 wt.% (De Luca et al. 2021).

The high aluminum and titanium content of the alloy results in a high gamma fraction, which improves performances at high temperatures, but makes it exceedingly difficult to repair via conventional fusion methods such as welding (McNutt 2015).

When filler metals with high volume fractions of gamma prime are used, weld cracking is common. They are characterized by an abundance of stray grains, which are new grains that are formed during solidification. The high-angle grain boundaries that exist in con- junction with the stray grains are weak links in the microstructure and act as preferred crack propagation paths. Therefore, existing weld technologies must use inferior filter metals with low gamma prime volume fractions (or no gamma prime) in order to produce crack-free welds (Vitek et al. 2002).

In summary, CMSX-4 is characterized as difficult to weld, resulting in material that is highly susceptible to cracking during processing. In addition, to columnar dendritic structure of these parts poses a challenge with respect to reproducing the orientation during additive repair. Balancing these requirements with contour fidelity is difficult, as the parameters that result in high single crystallinity and low cracks, do not necessarily guarantee an optimal shape (Buchbender et al. 2020).

1.2 Scope of the Sub-Project–Single Crystalline Laser Welding

In single crystalline laser welding the heat distribution has a significant impact on the formation of the micro structure. Therefore, the simulation of heat distribution during the laser metal deposition, the influence of an active substrate cooling and preheating on the heat distribution and temperature gradient in the substrate material were carried out. Building on this, a process for the repair of superficial defects using the process of remelting was developed. This allows the regeneration of cracks and defects in the substrate material located parallel and lateral to primary dendrite orientation.

Followed by the deposition of multi-layered structures, single-crystal structures by means of the two-step repair process of deposition and remelting were realized.

This involved the analysis of the influencing process factors on the properties of the deposited material as well as the quantification of the effect of the remelting parameters on the resulting microstructure.

In order to ensure a robust and reproducible process, thermal and acoustic process monitoring methods were implemented. Since the laser beam is the primary source of heat input in the process, the melt pool temperature during deposition and remelting was monitored. Empirically determined optimal temperatures for deposition and remelting were determined for a particular geometry in order to maximize single-crystal height.

The susceptibility of Nickel-based superalloys to cracking as well as the lack of information about crack formation during the process limits the application of repair process by laser metal deposition (Weingärtner et al. 2015).

The analysis of structure-borne emissions was implemented for the detection of cracking during the turbine blade repair process.

With the results from this sub-project B5 it was possible to develop a repair process under conditions determined by industrial applications. The transfer project T5 highlights these successes. Additionally, the turbine blade repair process was then implemented on an industrial machine in the scope of the automated process chain presented in system demonstrator.

2 Materials and Methods

The research was carried out on a 6-axis laser metal deposition machine. In order to melt the material a 680 W laser was used. This diode LDF 400–650 laser by Laserline GmbH emits laser light with wavelengths of 940 and 980 μm. The powder material is conveyed to the process zone using inert argon gas. An additively manufactured processing head with an annular powder deposition ring is used to generate the material deposition. The powder material was deposited coaxially with a powder focus diameter of 2 mm. The CMSX-4 powder particle diameter ranges between 25

Table 1 Chemical composition of CMSX-4 powder and substrate material in wt.% (Buchbender et al. 2020)

	Ni	Cr	Cu	Mo	W	Ta	Re	Al	Ti	Hf
Powder	bal.	6.5	9	0.6	6	6.5	3	5.6	1.0	0.1
Substrate	bal.	6.6	9.7	0.6	6.4	6.5	2.9	5.6	1.0	0.1

and 75 µm. The substrates consist of similar composition to the powder material as shown in the Table 1. Additionally, the angle of primary dendrite orientation is these substrate is specified.

2.1 Process Development

The conditions for the transformation from liquid to solid during solidification processing, such as the temperature gradient and the growth rate, vary from process to process as a function of time and space (Gäumann et al. 2001). This in turn determines the resulting microstructure in the deposited material. By varying the laser power, powder volume flow and the process speed in statistical designed experiments, the laser metal deposition process can be tuned to change the heat distribution and the thermal gradient during the solidification. The foundations for these experiments are based on numerical simulations.

2.2 Simulation

The simulation was carried out in order to analyze the temperature distribution during laser processing of CMSX-4. Ansys 12.1 was used to solve the finite element models in a 10-s transient model (Kirbach 2011).

In comparison to conventional materials like structural steel, the thermal conductivity of the nickel super alloy is lower by a factor of three. In the analyzed process duration of 10 s, the entire volume of the structural steel substrate heats up. In contrast, only 50% of the CMSX-4 substrate is heated up in the same process time. The simulations show that the isotherms in CMSX-4 are aligned in an exponential direction, while, for reference, the isotherms in structural steel are aligned approximately parallel to the components top edge. Thus, in CMSX-4 material a temperature gradient exists in several spatial directions inside the substrate (see Fig. 2). The main one-dimensional temperature gradient through the workpiece is already dictated by the thin and wide geometry of the substrate. The use preheating for the welding process is required. Otherwise, cracks are very likely to occur. With the preheating temperature of 300 °C, the temperature gradient for single- crystal solidification

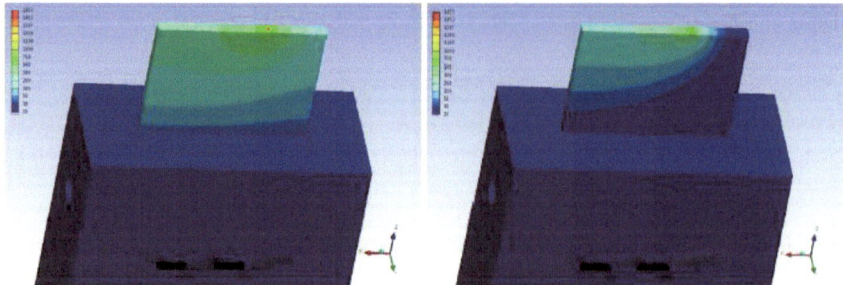

Fig. 2 Thermal simulation of the laser metal deposition of structural steel (left) and CMSX-4 (right)

cannot be guaranteed for the process duration. This conflicts with necessary temperature for stress relief and crack prevention. Using a hot enough temperature field prevents cracks, but the temperature gradient needed for single-crystal solidification would no longer exist.

With an added constant cooling at the bottom edge of the substrate, the CMSX-4 substrate heats up slower and a cold area is formed from below the welding, which provides homogeneous heat dissipation.

Overall, the simulation of preheating of the substrate indicated a loss of the required temperature gradient for the formation of epitaxial structures. This is in line with the recommendation by Gäumann et al. (2001) stating that the temperature of the substrate should be as low as possible.

3   Results

During the project, results have been achieved regarding the deposition of the CMSX-4 material. Material properties were investigated regarding multi-layered deposition and lateral repair. A key aspect was the integrated process monitoring and control.

3.1   Deposition of Single and Multi-Layered Structures

The resulting microstructure of the clad depends on the solidification rate and temperature gradient. The measured temperature gradient during laser metal deposition ranges between 10^5 and 10^7 K/m and the solidification rate is between 10^{-3} and 10^{-1} m/s, as shown in the Fig. 3a.

These thermomechanical rates are in turn primarily determined by the laser power (W), the laser speed (mm/s) and powder feedrate (g/min) as well as the interactions of these factors amongst each other. In order to deposit structures with the

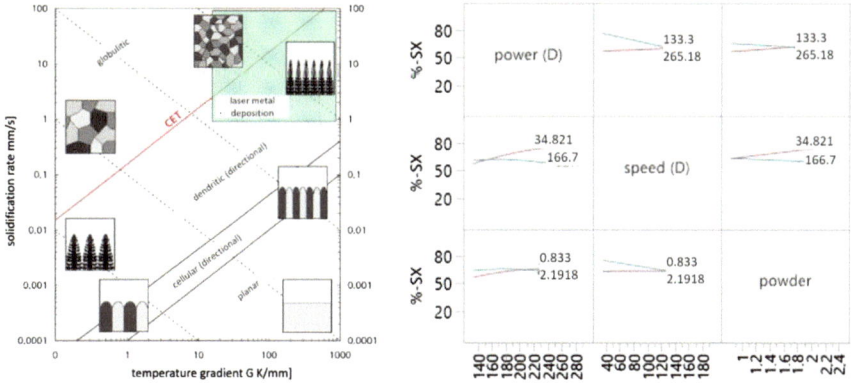

Fig. 3 **a** microstructure as a result of solidification rate and temperature gradient (modified from Burbaum 2011), **b** percentage single-crystallinity (%-SX) as a result of varying significant factors during the deposition process

desired microstructure, an analysis of the effect of the process parameters on single-crystallinity is required. Within the design of experiments, the following range of process parameters was analyzed. The laser power to create the melt pool was varied between 130 and 265 W. The laser speed ranged from 40 to 165 mm/min. CMSX-4 material has been fed with a powder feedrate from 0.8 to 2.15 g/min. The resulting interactions are dis- played in the Fig. 3b. These results indicate that a combination of high power, low laser speed and high powder feedrate result in the highest percentage single-crystallinity. Since literature (Gäumann et al. 2001) and previous work (Buchbender et al. 2020) indicate that the use of high laser power leads to the loss of the temperature gradient, by analyzing the incidence of cracks, microcracks and misorientations, a threshold for these factors can be set. If the parameters are set below this threshold, the microstructural results are desirable.

These findings formed the basis for the deposition of multi-layered structures. In order to further evaluate the effect of parameters during process development, geometrical and microstructural features of the deposited clad were chosen as analysis criteria.

The focus is set on the percentage of single-crystallinity by dividing the single-crystal area by total surface area. Furthermore, the height difference is calculated by subtracting actual height of the weld form the desired metal deposition height. With the help of computer tomography and micro sections the number of cracks, microcracks and misorientations were counted and evaluated in the statistical model.

A key finding during the research is that an additional remelting step involving a single pass of the laser beam allowed for the recrystallization of previously misoriented material within a track. This has been shown by Rottwinkel et al. (2014a, b). With this laser remelting strategy for track single-crystallinity extension has been proven a tool capable of improving and simplifying the formation of large SX-volumes by laser powder cladding. It was expected that upon remelting, a resolidification of the misoriented volume would occur, allowing for the extension of the

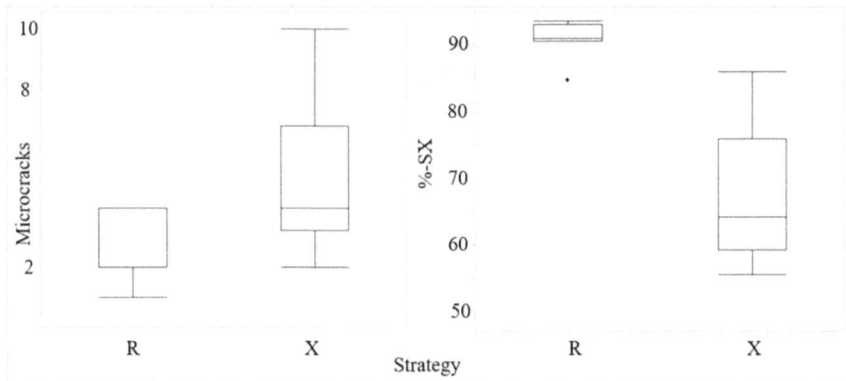

Fig. 4 Effect of remelting strategy (R) and no remelting strategy (X) on microcrack formation and %-SX (Buchbender et al. 2020)

overall single crystalline height (Rottwinkel et al. 2017). A study of builds with and without an additional remelting pass between layers showed that the former resulted in higher %-SX, fewer microcracks and higher deposition height. The largest differences between the strategies of remelting (R) and without remelting (X) were seen in the incidence of microcracks and percentage single crystallinity (%-SX) shown in Fig. 4. Deposits built with the remelting strategy showed fewer microcracks and higher percentage single-crystallinity, which are amongst the most influential factors for epitaxial height extension. Hence, the remelting strategy was chosen for further experiments (Buchbender et al. 2020).

Using optimized parameters and remelting, a study of deposition strategies showed over- all improvement across all five analysis criteria. Percentage single-crystallinity of above 80% was achieved for all parts, with the highest being 97.9% measured by EBSD. Several builds showed zero microcracks, cracks and misorientations with a deposit height of 4480 μm can be seen in Fig. 5 (Buchbender et al. 2020).

3.2 Lateral Repair

With the creation of multi-layered structures, the repair of high pressure turbine blade tips can be targeted. For other repair situations, a lateral repair is needed. This is done by transferring the optimized parameters and strategies to substrates prepared for lateral repair. In addition to the changed thermal conditions of the meltpool, the dendrite orientation in the deposited material is also a function of the altered dendritic orientation in the substrate.

This is especially relevant for the repair of defects located lateral to the direction of primary dendrite orientation. Here, the angle of processing determines the direction of heat flow and therewith the orientation of the microstructure. As soon as the

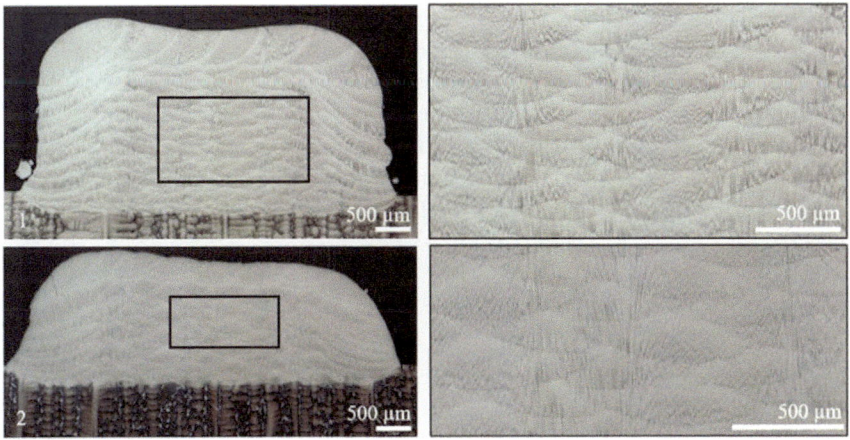

Fig. 5 Selected cross-sectional micrographs of builds using optimized parameters and strategies (Buchbender et al. 2020)

first dendritic solidification starts after the laser process, the solidifying structure takes over the orientation of the crystal structure originated from the base material. This means, that the structure grows epitaxially on the partially melted dendrites inside the base material. If the growth direction equals direction of travel of the laser beam, the orientation matches the primary dendrites. For a predefined direction of travel of the laser beam and a known position and orientation of the primary dendrite stems, the growth direction of the dendrites can be calculated (Burbaum 2011). With this knowledge and in order to develop a repair strategy for lateral defects, it is required to determine the optimal processing angle that results in the highest percentage single-crystallinity. The results of the effect of processing angle on percentage single-crystallinity and the percentage of secondary dendrites are shown in Fig. 6.

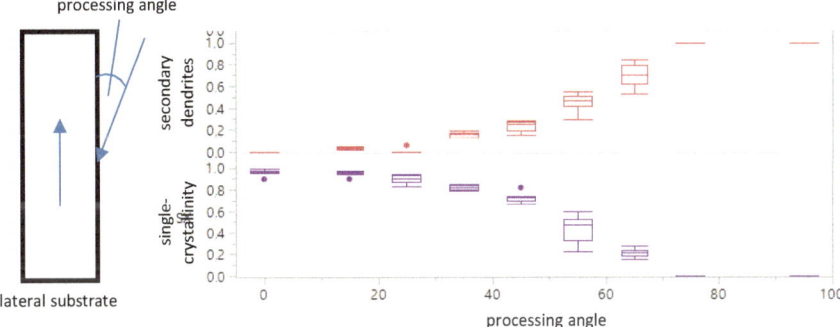

Fig. 6 **a** processing angle and **b** percentage primary and secondary dendrite orientation as a result of processing angle

These results show that at a processing angle of 20° the %-SX gradually declines, while the percentage of secondary dendrite orientation increases. This leads to the decision to advance with a processing angle of 20° for further experiments. In order to be able to process parts along the entire length of its lateral edge, a modification of the processing nozzle was carried out, since the geometry of the current processing head did not allow for flexible processing. The modification ensured the working distance of 9 mm was maintained and allowed the processing of larger parts.

3.3 Process Monitoring

Process monitoring can be employed to increase reproducibility and monitor part quality. Monitoring methods for laser processing are based on the physical phenomena that occur due to laser-material interactions. These can be acoustical, optical, electrical or thermal (Purtonen et al. 2014).

Increasing meltpool temperature and substrate temperature in the higher layers results in process irregularities, the formation of misorientations and cracks as well as the loss of the required temperature gradient for epitaxial deposition. Hence, the applicability of thermal process monitoring in the deposition of epitaxial structures was studied.

A challenge in the current state of the art of process monitoring is the inability of visual and thermal methods to detect cracking within the material during processing. Especially for materials, such as the investigated Nickel-based superalloys, that are prone to cracking and have narrow processing windows. This inhibits the industrial applicability of the process. Challenges in acoustic process monitoring are the inadequate signal to noise ratios and therefore a very high susceptibility to noise, as well as limited literature especially with regard to higher frequency bands. Within the project duration, the uses of structure-borne sound to identify the acoustic signatures of cracks were investigated. This was done by equipping the laser metal deposition process of the difficult to weld Nickel-based superalloy CMSX-4 by applying a time–frequency analysis method.

Thermal process monitoring

Pyrometry allows the measurement of the spectral intensity of a surface within a particular spectral range, which is correlated with a temperature based on the emissivity ε of the material. The applied two-color pyrometer or ratio pyrometer is a type of pyrometer that measures the temperature in two spectral ranges simultaneously and calculates the temperature by converting the ratio of spectral irradiance between the two measured wavelengths. This two-color pyrometers is used to minimize the wavelength-dependent influence of the emissivity of the measuring surface, provided the emissivity changes proportionally for both wavelengths. The schematic setup is shown in Fig. 7a.

It was hypothesized that an empirically determined optimal meltpool temperature range exists for deposition and remelting at a constant deposition speed and powder

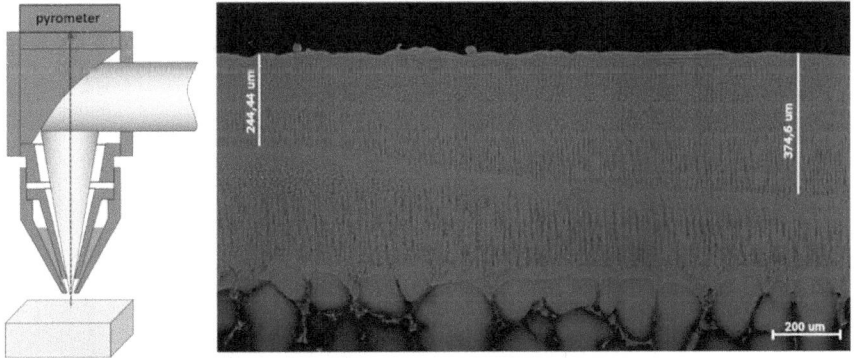

Fig. 7 a schematic diagram of coaxial thermal process monitoring setup and **b** the change in penetration depth as a result of change in laser power during remelting

feed rate. This allows for the maximization of single-crystallinity in the deposited material.

In order to determine the optimal meltpool temperature range, single clads were deposited onto single-crystal substrates. Keeping the powder feed rate and laser speed constant allowed for the quantification of the influence of the laser power on the microstructure. The meltpool temperature was measured with varying laser power across the length of each track for deposition and remelting respectively. Longitudinal micrographs of the clads were prepared and the ratio of SX (single crystallinity) to PX (polycrystallinity) as well as the remelting depth measured.

As shown in the previous section, the choice of remelting parameters influences the %- SX (percentage single-crystallinity) significantly.

The Fig. 7b shows the change in penetration depth during remelting as a result of a change in the laser power. Remelting power of 84 W resulted in a remelting depth of 244 μm, while a power of 140 W resulted in a remelting depth of 375 μm (Honisch 2018). This manifests an adequate remelting depth to allow complete recrystallization of previously misoriented material.

In this setup, the design of experiments concludes an optimal temperature for deposition at 1550 ± 50 °C with a laser speed of 75 mm/min and powder feed rate of 1 g/min. These parameters were determined based on the highest single-crystallinity in longitudinal micrographs of deposited clads.

The optimal temperature for the remelting process lies at 1350 ± 50 °C for a laser speed of 60 mm/min, which was determined, based on the penetration depth required to recrystallize the deposited material.

In order to achieve these temperatures during the deposition and remelting process a custom process control was implemented. By modulating the laser power, the targeted temperatures were controlled (see Fig. 8).

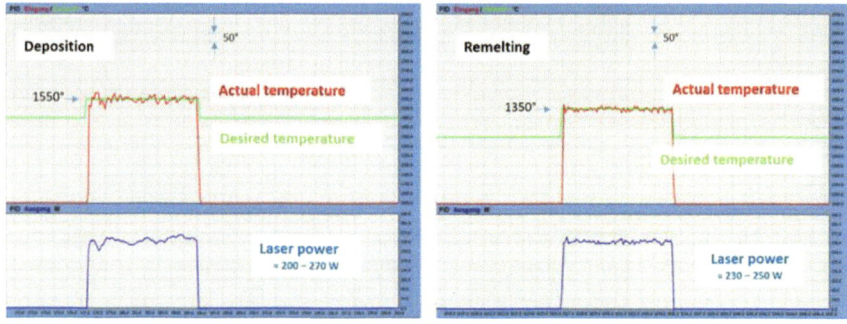

Fig. 8 Thermal process monitoring and control for the deposition and remelting process

Process monitoring by acoustic emission

The analysis of acoustic emissions (AE) have been used in structural health monitoring, to detect phase transformations and cracking. These acoustic emissions are elastic waves or stress waves that propagate through materials in various modes, which are characterized by the vibrations of the particles. In solids, this propagation occurs in the form of longitudinal and transverse waves. Longitudinal waves travel in the direction of sound transmission, causing local regions of compression. Transverse waves, also called shear waves, oscillate perpendicular to the direction of wave motion. The type of mode and propagation depends on the nature and geometry of the medium and are affected by numerous factors and material properties.

In laser material processing, the detection of acoustic emissions has been carried out by monitoring the optical components of the laser welding setup, the work piece or air borne sound off-axially. AE sensors can therefore be structure-borne sensors or airborne acoustic sensors. In order to analyze the emissions detected, a signal processing method must be defined to obtain useful information. Baccar (2015) mentions three approaches for the analysis of AE signals, of which the following two are most relevant to the analysis carried out during the collaborative research center.

First, a parameter-based analysis is investigated, which consists of the extraction of time- based parameters such as events, signal amplitude, rise time, count, duration and energy. Secondly, a time–frequency analysis that analyses features in the frequency domain such as power spectrum, peak frequency and dominant frequency band.

With parameter-based analysis, on one hand, the amplitude of a signal can be correlated with the severity of the occurrence. However, amplitude is affected by attenuation and hence varies with proximity to the sensor. Therefore, amplitude or hit counts based on the amplitude do not convey sufficient information about the source of the acoustic emissions. Time–frequency analyses on the other hand offer more insight. Common methods used for time–frequency analysis are amongst others the Short Time Fourier Transform (STFT), Continuous Wavelet Transform (CWT) and Hilbert Huang Transform (HHT). This study makes use of the STFT method, as is depicted in Fig. 9. In the STFT method of time–frequency analysis, the signal

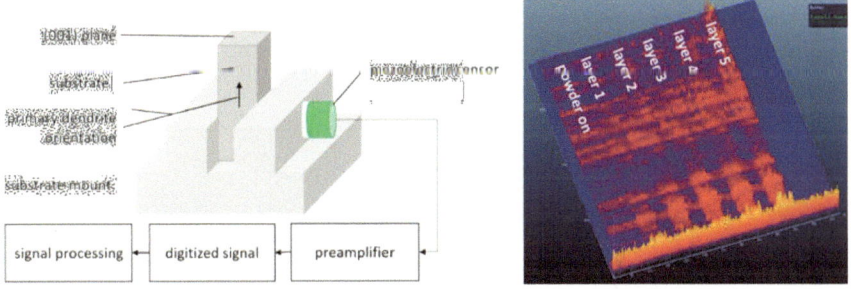

Fig. 9 **a** process set up and signal processing chain, **b** time–frequency analysis of acoustic signal

x(t) is multiplied by a window function g(t-τ), the FFT of the windowed signal is calculated, while the sliding window moves along the time axis, calculating the FFT for each time interval. Common window functions are the Gaussian, Hamming and Hanning windows. The width of the window has a major influence on the resolution of the resulting signal. Large windows ensure high frequency resolution and low time resolution, while small windows provide lower frequency resolution and high time resolution. The choice of window width therefore requires prior knowledge of the process events, as the window size should not exceed the time period of the events to be detected in the signal. STFT allows for the representation of the frequency information of the signal over time, upon which a frequency range can be selected in which the signal to noise ratio is large enough to detect anomalies.

For data acquisition, the commercially available piezoelectric sensor and preamplifier by QASS GmbH were used. Experiments indicated that attaching the piezoelectric sensor to the substrate mount led to the most reproducible results and favorable signal to noise ratio. All measurements were carried out with a sampling frequency of 390 kHz.

The time–frequency analysis was carried out to identify the acoustic characteristics of the laser metal deposition process (Fig. 9b).

This showed that the status of the powder (on/off) can be identified across all frequency bands. Laser on/off status can be identified by higher amplitudes in characteristic frequency bands, in this case around 25 and 100 kHz. Low frequency bands around 10 kHz showed low signal-to-noise ratios. Continuous emissions of low amplitude indicated electrical noise (as seen at 100 kHz). Furthermore, it was observed that amplitude is highly dependent on the geometry of the part, the position of the sensor and the proximity of the process zone to the sensor.

In order to detect cracking during the process of blade repair, the anomalies of the acoustic signal were further analyzed. Formation and propagation of cracks inside the material can be correlated with anomalies. The cracks are characterized by short, broad-banded, high frequency transient signals in the time domain as well as in the frequency domain (Fig. 10a). With the use of signal processing to visualize high amplitude, transient signals in the frequency domain are computed. By setting a

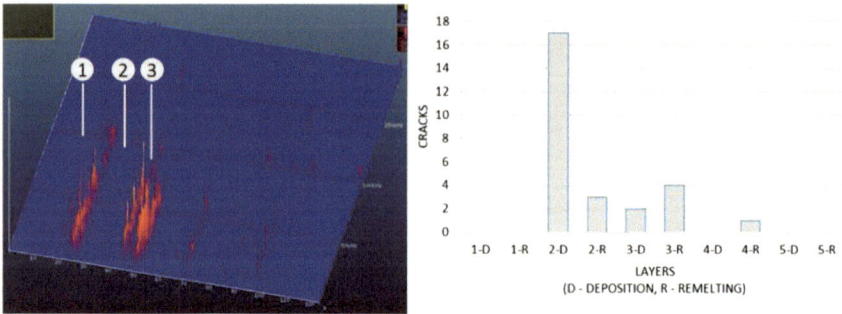

Fig. 10 **a** characteristic signal in the time domain, **b** signal processing for the visualization of high amplitude transient signals

threshold, anomalies are categorized. These classified cracks are counted in each layer (Fig. 10b) and then verified by computer tomography and micrographs.

4 Conclusions

The rising demands on engines and efficiency in the aerospace motivated the repair of high pressure turbine blades. The single-crystal nickel-based superalloys presented challenges during the research due to crack formation and globulitic microstructure. To over- come these difficulties a novel two-step process for a single crystalline repair has been developed during the project duration (Kaierle et al. 2017; Buchbender et al. 2020). It consists of the material deposition followed by a remelting pass of the laser without additional material.

As a result, the single-crystalline repair of the three most important damage patterns is possible. With the help of process monitoring and pyrometric control the microstructure is ensured in all three damage patterns.

Despite the optimized process parameters, small cracks can occur during the laser metal deposition of the CMSX-4 material. By enhancing the process monitoring occurring cracks can be detected by means of structure-borne acoustic measurements while the welding process is carried out.

It has been shown, that Laser metal deposition is a suitable repair process for automated regeneration of the single crystalline turbine blades. In a collaborative effort with direct support of the subprojects A1–thermomechanical properties, B1– combination of brazing and cladding and B2–recontouring a high pressure turbine blade has been refurbished. Therefore, the complex capital goods can be repaired. Furthermore, the findings have been transferred in the transfer project T5, where a deposition of multi-layered structures with single-crystallinity of 98% has been achieved. The integration into the automated process chain of the system demon- strator has been realized. A repaired high pressure turbine blade, its microstructure and the final automated laser metal deposition process can be seen in Fig. 11.

Fig. 11 High pressure turbine blade with material deposition (left), the visible orientated microstructure (right) and the automated laser metal deposition process implemented in the system demonstrator

Acknowledgements Funded by the Deutsche Forschungsgemeinschaft (DFG, German Research Foundation) SFB 871/3 119193472.

References

Alfred, I., Nicolaus, M., Hermsdorf, J., Kaierle, S., Möhwald, K., Maier, H.-J., and Wesling, V. (2018). Advanced high pressure turbine blade repair technologies. In Procedia CIRP (Vol. 74, pp. 214–217). Elsevier BV. https://doi.org/10.1016/j.procir.2018.08.097.

Baccar, Dorra (2015). Development, Implementation, and Validation of an Acoustic Emission-based Structural Health Monitoring System. Online:https://nbn-resolving.org/urn:nbn:de:hbz:464-20151218-165352-0, Dissertation.

Boyce, M. (2006). Gas Turbine Engineering Handbook Elsevier Butterworth- Heinemann, Oxford, UK.

Buchbender, I., Hoff, C., Hermsdorf, J., Wesling, V., and Kaierle, S. (2020). Single-crystal height extension by Laser Metal Deposition of CMSX-4. In Procedia CIRP (Vol. 94, pp. 304–309). Elsevier BV. https://doi.org/10.1016/j.procir.2020.09.057.

Burbaum, B. (2011). Verfahrenstechnische Grundlagen für das Laserstrahl-Umschmel- zen einkristalliner Nickelbasis-Superlegierungen (No. RWTH-CONV-143013). Lehrstuhl für Lasertechnik, Dissertation.

Donachie, MJ. and Donachie SJ. (2002). Superalloys – a technical guide. 2nd ed. ASM International.

De Luca, A., Kenel, C., Griffiths, S., Joglekar, S. S., Leinenbach, C., and Dunand, D. C. (2021). Microstructure and defects in a Ni-Cr-Al-Ti γ/γ' model superalloy processed by laser powder bed fusion. Materials and Design, 201, 109531 (p. 14). https://doi.org/10.1016/j.matdes.2021.109531.

Gäumann M, Bezencon C, Canalis P, and Kurz W. (2001). Single-crystal laser deposition of superalloys: processing-microstructure maps. Acta Materialia 2001;49:1051–1062.

Honisch, R. (2018). Einfluss des Temperaturgradienten auf die Mikrostruktur von Nickel- Basis Superlegierungen während des Laserpulverauftragschweißens, masterthesis, La- ser Zentrum Hannover e.V.

Kaierle, S., Overmeyer, L., Alfred, I., Rottwinkel, B., Hermsdorf, J., Wesling, V., and Weidlich, N. (2017). Single-crystal turbine blade tip repair by laser cladding and remelt- ing. CIRP Journal of Manufacturing Science and Technology, 19, 196–199. https://doi.org/10.1016/j.cirpj.2017.04.001

Kirbach, S. (2011). Prozessauslegung durch Simulation der makroskopischen Wärme- verteilung im Laser-Pulver-Auftragschweißprozess der Nickelbasis-Superlegierung CMSX-4, student project , Laser Zentrum Hannover e.V.

McNutt, P.A. (2015). An investigation of cracking in laser metal deposited nickel super- alloy CM247LC. PhD thesis, University of Birmingham.

Purtonen, T., Kalliosaari, A., and Salminena, A. (2014). Monitoring and adaptive control of laser processes, 8th International Conference on Photonic Technologies LANE.

Rottwinkel, B., Nölke, C., Kaierle, S., and Wesling, V. (2014a). Crack repair of single crys- tal turbine blades using laser cladding technology. In: Procedia CIRP 22 (2014), Nr. 1, S. 263–267. https://doi.org/10.1016/j.procir.2014.06.151.

Rottwinkel, B., Schweitzer, L., Noelke, C., Kaierle, S., and Wesling, V. (2014b). Challenges for single-crystal (SX) crack cladding. In: Physics Procedia 56, S. 301–308. https://doi.org/10.1016/j.phpro.2014.08.175.

Rottwinkel, B., Pereira, A., Alfred, I., Noelke, C., Wesling, V., and Kaierle, S. (2017). Turbine blade tip single crystalline clad deposition with applied remelting passes for well oriented volume extension. In Journal of Laser Applications (Vol. 29, Issue 2, p. 022310). Laser Institute of America. https://doi.org/10.2351/1.4983667.

Schneibel JH (2004). Beyond Nickel-Base Superalloys. In Processing and Fabrication of Advanced Materials XIII: Volume II. Stallion Press, Singapore (2004).

Wahl, J. B. and Harris, K. (2011). Advanced Ni base superalloys for small gas turbines. In Canadian Metallurgical Quarterly (Vol. 50, Issue 3, pp. 207–214). Informa UK Limited. https://doi.org/10.1179/1879139511y.0000000010

Weingärtner, W., Schweizter, S., Rottwinkel, B., Nölke, C., Kaierle, S., and Wesling, S. (2015). Single-Crystal (SX) Laser Cladding of CMSX-4, 8th Brazilian Congress of Manu- facturing Engineering.

J. Vitek, S. David, S. Babu (2002). Welding and weld repair of single-crystal gas turbine alloys. Online: https://www.semanticscholar.org/paper/WELDING-AND-WELD-REPAIR-OF-SINGLE-CRYSTAL-GAS-Vitek-David/ec103c2345abdce80f1e0595cf6bcc72aa6d8a21

Regeneration and Surface Hardening of Titanium Components Using the Example of Titanium Alloy Ti6Al4V

J. Torben Carstensen, Thomas Hassel, and Hans Jürgen Maier

Abstract In this chapter, two different methods for repairing titanium components are presented. These include patch repair and additive repair using metal inert gas welding and wire-and-arc additive manufacturing. The use of inoculant and flux to affect the weld zone is discussed for both repair methods. Using the titanium alloy Ti6Al4V, it is shown that both flux and inoculant can be used to improve the mechanical properties of the overall joint. In addition, local atmospheric nitriding is presented, which is a method to locally harden the surface of titanium components. Surface hardness of more than 800 HV1 were achieved.

Keywords Titanium alloy · Repair technology · Surface hardening · Inoculation · Flux · Maintenance repair overhaul

1 Introduction

In order to reduce environmental impact, it is important to further increase the service life of components. To this end, maintenance, repair and overhaul (MRO) is becoming increasingly important. Especially for components made of expensive materials such as titanium alloys, a cost advantage can arise in addition to the environmental aspect (Wits et al. 2016).

In order to extend the range of application of MRO, this subproject B6 of the Collaborative Research Center (CRC) 871 has researched possibilities to influence the joining zone during the repair of titanium components. In addition to the existing

J. T. Carstensen (✉) · T. Hassel · H. J. Maier
Institut für Werkstoffkunde (Materials Science), Leibniz University Hannover, An der Universitaet 2, 30823 Garbsen, Germany
e-mail: carstensen@iw.uni-hannover.de

T. Hassel
e-mail: hassel@iw.uni-hannover.de

H. J. Maier
e-mail: maier@iw.uni-hannover.de

© The Author(s) 2025
J. R. Seume et al. (eds.), *Regeneration of Complex Capital Goods*,
https://doi.org/10.1007/978-3-031-51395-4_12

patch repair, which was extended by the use of fluxes and inoculants, a new repair method was developed: additive repair using wire-and-arc-additive-manufacturing (WAAM).

Within the CRC 871, the titanium blisk was selected as a component to be repaired. The titanium alloy Ti6Al4V was selected, as it is the most widely used titanium alloy. However, the methods investigated are not limited to the blisk and Ti6Al4V, but can also be applied to other alloys.

There are several repair methods in aerospace, these include patch repair, blend repair and tip repair. In patch repair, the defect is removed. In this process, the focus can be either to create a simple patch geometry or to transfer the seam to a lower-stressed area. A corresponding patch is manufactured for this application. The patch is welded to the component using one of various welding processes. This subproject investigated the repair possibilities of arc processes. After welding, the component is recontoured to restore the original geometry. In blend repair, a crack is not filled, but the crack is removed and a geometry is created that deviates from the initial geometry, but the notch effect of the crack is reduced. In tip repair, a seam is welded onto the tip to fill material and then the blade is recontoured to restore the initial geometry. An extension of the tip repair is additive repair. Here, instead of welding a single seam, the desired geometry is produced by a layer-by-layer technique. In order to be able to select a repair path, it must be clear in advance which properties the component attains for each repair path. With this knowledge the decision can be made if the component is still usable as is, if the part is scrap and needs to be replaced with a new part, or if it can be repaired, cf. Figure 1.

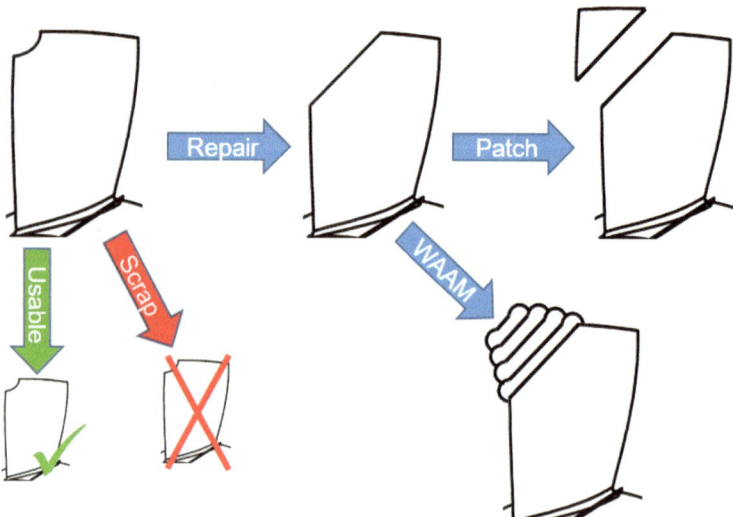

Fig. 1 Possible decision paths for the repair of a highly-stressed component; three basic options are available for the component, i.e., the component can be used as is, it is scrap or it can be repaired

Every repair weld involves an unavoidable heat treatment of the microstructure. In the fusion zone (FZ), a directionally solidified microstructure with isotropic properties is typically formed. In the heat-affected zone (HAZ), grain growth occurs and precipitates possible are dissolved and precipitated again. Welding is accompanied by extreme heating rates and very rapid cooling. For most titanium alloys, including Ti6Al4V, this means a hardening of the HAZ and FZ coupled with a significant decrease in ductility (Schulze 2010).

Usually the base material is homogenized after melting and then deformed. After forming, recrystallization usually takes place for α-alloys. For α-β-alloys, optional recrystallization is followed by an annealing. For β-alloys, either recrystallization annealing, annealing or both are employed. Heat treatment is then completed by (multiple) aging (Derby et al. 2003).

During welding, the homogenized base material is melted and cooled with cooling rates of up to several 100 °K/s (Schulze 2010). The welding of a conventional patch repair ends here. Due to the local heating and cooling, residual stresses are generated (Denkena et al. 2017); due to the lack of a subsequent annealing process, the residual stresses cannot be relieved and have a diminishing effect on the fatigue strength. However, in order to achieve the grain refinement that the base material has, cooling rates of more than 1000 °K/s are necessary or the steps of forming with high degrees of deformation and recrystallization annealing are needed (Derby et al. 2003).

Various investigations have been carried out on the grain refinement of welds. On one hand, there are possibilities to design the process differently. On the other hand, one can add additives whose reactions influences the process.

In welding, heterogeneous nucleation is dominant and heat is mainly dissipated by the substrate. The grain growth progresses in the opposite direction to the heat flow, i.e. from the substrate into the molten pool. For grain refinement, Balasubramanian et al. (2008) pulsed a TIG arc to vibrate the molten pool. They used a frequency of 6 Hz, 40 A base current and 80 A pulse current. They were able to reduce the prior β-grain size to a third of the grain size achieved with the pulse-free arc. The grain size of the reference was 340 μm and the grain size of the pulsed sample was less than 100 μm.

Reisgen et al. and Henckell et al. have investigated active cooling. For this purpose, Reisgen et al. (2020) compared water bath cooling with aerosol cooling and reference welding when welding steel. Water bath cooling showed the greatest increase in hard- ness. Aerosol cooling with water and air also showed an increase in hardness compared to reference welding. Henckell et al. (2017) studied the influence of gas cooling in additive manufacturing of steel. They were able to both increase strength and reduce grain size through the use of a leading gas cooling system. Li et al. (2019) used hot wire in TIG welding of titanium. The use of hot wire allows a reduction in arc energy, which in turn reduces the weld pool temperature. Thus, the weld geometry becomes narrower. At the same time, grain refinement occurs and the mechanical properties in the vertical and horizontal directions equalize. Gou et al. (2020) stimulated the melt pool of Ti6Al4V with ultrasound. This reduced the average length of α-grains from 40 to 20 μm. In addition, the anisotropy of the component was reduced. Hönnige et al. (2017) introduced residual compressive stresses by ultrasonic impact treatment of the weld. By welding the next layer, recrystallization annealing occurs and grain refinement is achieved. Colegrove et al. (2017) rolled over the weld at 75 kN after welding Ti6Al4V, introducing residual compressive stresses. Grain refinement occurred in the HAZ of the next layer. Significant grain refinement was achieved and isotropic mechanical properties were obtained that exceed those of the reference. In the present study, the influence of mechanical deformation was also investigated. Mechanical working of the component significantly restricts the repair process. Forming usually takes place outside the inert gas atmosphere, so the component must be cooled down before treatment. This means that high forces are necessary to generate the required residual stresses in the component. Due to their geometry, components such as the blisk limit the force with which the residual stresses can be introduced without damaging the component. In order to take advantage of the mechanical processing, either a fixture must be manufactured to absorb the forces so that the component is not unintentionally loaded, or the energy input and the associated layer thickness must be reduced.

Another way to influence the process is by in-situ alloying. In the present subproject, the effect of fluxes was investigated. Among the fluxes used (AlCl$_3$, NaCl, FeCl, VCl$_3$, AlF$_3$, CaF$_2$), AlF$_3$ had the greatest effect on the deep welding effect during joint welding. The results of AlF$_3$ are presented below. Yin et al. (2017) also investigated the influence of adding CaF$_2$ in additive manufacturing of Ti6Al4V and observed an effect on the weld geometry and mechanical properties.

The addition of inoculants was also studied in the subproject. With the idea of adding high melting components to the molten bath, which then act as heterogeneous nucleating agents, and thus lead to grain refinement of the FZ. Specifically, Si, SiC, TiC, C, B, TaC were used (Langen 2021). Bermingham et al. (2019) and Mereddy et al. (2017, 2018) have also investigated the grain refinement potential of adding Si, C, and La$_2$O$_3$ to titanium alloys.

2   Materials and Methods

2.1   Materials

For all tests, substrates made of the alloy Ti6Al4V (3.7164) were used. The chemical compositions for the different materials used in the studies are shown in Table 1. The samples were cut by plate shears or by water jet cutting.

Table 1 Chemical analysis of the substrate and welding wire in wt.%

	Substrate flux	Substrate inoculant	WAAM wire	WAAM substrate
Al	5.7	6.15	6.03	6.03
V	3.9	3.8	4.12	3.95
Fe	0.13	0.12	0.1	0.11
C	0.02	0.013	0.008	0.002
B	0.01	0.002	0.007	0.001
H	0.001	0.006	0.005	0.0025
O	0.1	0.12	0.14	0.11
Ti	Balance	Balance	Balance	Balance

2.2 Methods

2.2.1 Shielding Gas Chamber

The shielding gas chamber was filled with argon with purity of 99.998%. Figure 2 shows the chamber used. The workspace is 2 m × 2 m × 1 m. Manual tests can be carried out via eight gloves. A six-axis robot, type KUKA Sixx R900 from KUKA AG, was employed for automated welding.

A gas purification device, type E-Line Gasreinigung from GS GLOVEBOX Systemtechnik GmbH, automatically maintained a positive pressure of 1.5 mbar and allowed tests to be performed at a residual oxygen content of less than 10 ppm and a residual water content of less than 0.5 ppm. Various welding methods could

Fig. 2 Argon welding chamber for processing refractory metals; workspace: 2 m^3; handling system: KUKA Sixx R900

be used in the chamber including: microplasma welding, tungsten inert gas welding (TIG), metal inert gas welding (MIG) and plasma nitriding with non-transferred arc.

2.2.2   Preparation and Welding of the Flux Samples

Specimens (180 mm × 210 mm × 2 mm) were cut from a plate. The specimens were cleaned with acetone. Using an airbrush gun, a solution of ethanol and AlF_3 was applied to the specimens through a spray stencil. Thus, a strip of 20 mm × 150 mm, cf. Figure 3a and a weight of 0.13 ± 0.01 g was applied to the center of the sample. The area density was 0.04 mg/mm². The welding parameters are listed in Table 2.

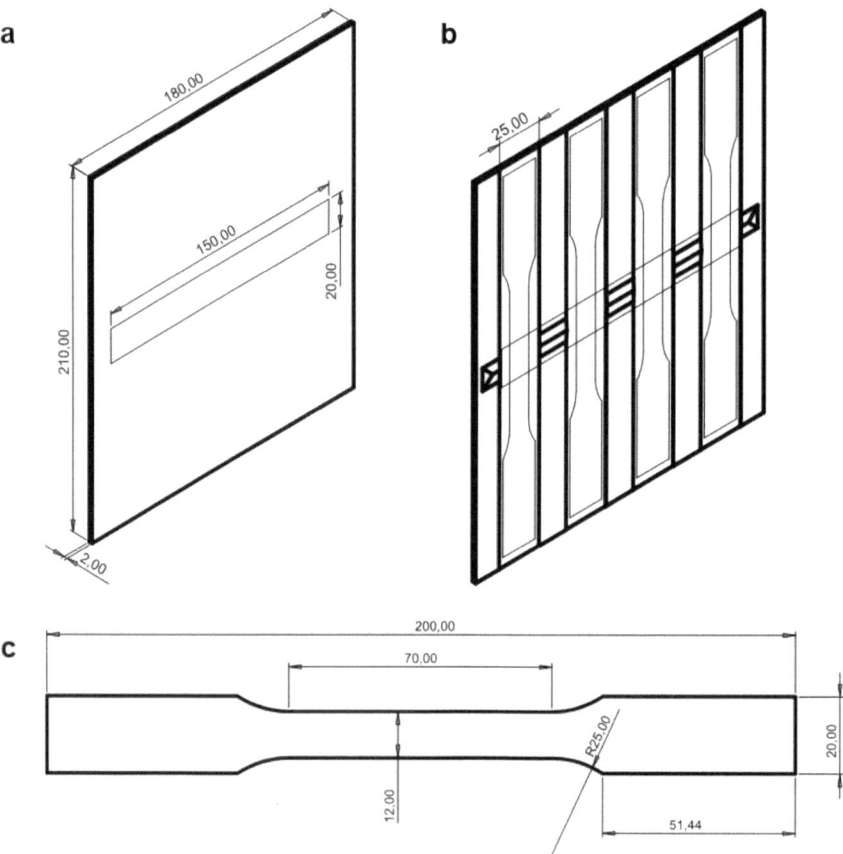

Fig. 3 Process design of the flux samples: **a** representation of specimen geometry with flux application; **b** location of tensile specimens; **c** geometry of the tensile specimens

Table 2 Welding parameters of additive repair and tests with flux or inoculant

	Flux	Inoculant	WAAM TIG1	WAAM TIG2
Current, A	20 and 60	90	80	100
Shielding gas, l/min	6	4	4	4
Electrode distance, mm	3	3	5	5
Travel speed, m/min	0.12	0.12	0.12	0.12
Wire speed, m/min	–	–	1.1–2.3	1.0–1.3
Wire diameter, mm	–	–	1.2	1.2
Dwell time, s	–	–	40	120
Layer height, mm	–	–	1.1	0.8

2.2.3 Preparation and Welding of the Inoculation Samples

The specimens for the tensile tests with inoculant were prepared in the same way as the flux specimens. However, the applied strip was smaller. The width was only 3.3 mm. The weight of the applied inoculant was 0.17 ± 0.02 g. The resulting area density was 0.34 mg/mm^2.

The experiments for the micrographs were performed on samples with a size of 180 mm $\times$ 56 mm $\times$ 2 mm. The area on which inoculant was applied was 3.3 mm $\times$ 150 mm. The weight of the applied inoculant was 0.6 g; the area density was 1.21 mg/mm^2.

As on the flux specimens, the inoculant specimens were welded without filler material in the welding chamber using the same welding system, cf. Figure 3. The welding parameters are listed in Table 2.

After welding, the weld buildup was removed by milling. The tensile specimens had the same design as the flux tensile specimens cut out using wire EDM.

2.2.4 Preparation and Welding of the Wire and Arc Additive Manufactured Samples

Substrate plates measuring 330 mm $\times$ 60 mm $\times$ 5 mm were cut from a titanium Ti6Al4V plate. The chemical composition of the substrate plates and the welding wire used are shown in Table 1. The substrate plates were cleaned with acetone. A single-wall 300 mm $\times$ 300 mm $\times$ 6 mm was welded in the shielding gas chamber. The process parameters are listed in Table 2 (WAAM TIG1). The welding power source used was a Tetrix 350 AC/DC from EWM AG. Waterjet cutting was used to cut out the ground and tensile specimens.

To increase the cooling rate, and thus decrease the prior β-grain size, a new wall was build. The dwell time was increased and the other process parameters had to be adjusted slightly, see Table 2 (WAAM TIG2). The specimen preparation was carried out in the same way as described before.

2.2.5   Determination of the Mechanical Properties

Zwick Z100 and Z250 universal testing machines from Zwick/Roell were used to perform the static tensile tests. The tensile specimens for flux and inoculation testing were designed to meet the standard DIN 50125 H 1.8 × 12 × 50, see Fig. 2c. The tensile specimens of the additive repair were designed to DIN 50125 E 2 × 6 × 20 with enlarged clamping surfaces. After cutting out of the welds, the tensile specimens were milled to final dimensions. Rotary bending tests were performed on round specimens to determine fatigue properties.

2.2.6   Microscopy

The samples were cut by water jet cutting, embedded, ground with SiC abrasive paper of the grit 500, 800, 1200 and 2500. Afterwards polished with an alcoholic diamond suspension (9 μm). Before etching with Kroll's reagent, the samples were finished on a vibratory polisher VibroMet from Buehler. Images were taken using a BX61 incident light microscope with an Olympus XC30 camera.

2.2.7   Vickers Hardness Testing

The Vickers hardness tests were carried out according to DIN EN ISO 6507–1 with a loading time of 13 s. The micro hardness tester used was a Q10 A + from QNESS GmbH. A load of 9.81 N was applied to the test specimen for the flux and inoculation samples. The distance between the indentations was 0.22 mm. For the hardness measurements of the nitrated samples, a load of 0.245 N was applied. The tests were carried out with a Vickers Microhardness Tester FM-310 from Futer-tech. The distance between the indentations differed between 5 and 30 μm.

2.2.8   Local Atmospheric Nitriding with Non-transmitting Arc

Local atmospheric plasma nitriding was applied using various methods. These included TIG and plasma nitriding with non-transmitting arc. Nitrogen was used as the working gas. Figure 4 shows the nitriding process. Nitrogen flows to the arc as the working gas. There, the nitrogen molecules are dissociated in the arc. The nitrogen atoms are then ionized by the arc. The ions either diffuse into the titanium or they recombine at the cold surface to form molecules again. In TIG welding, the surface is melted and the nitrogen diffuses into the molten pool.

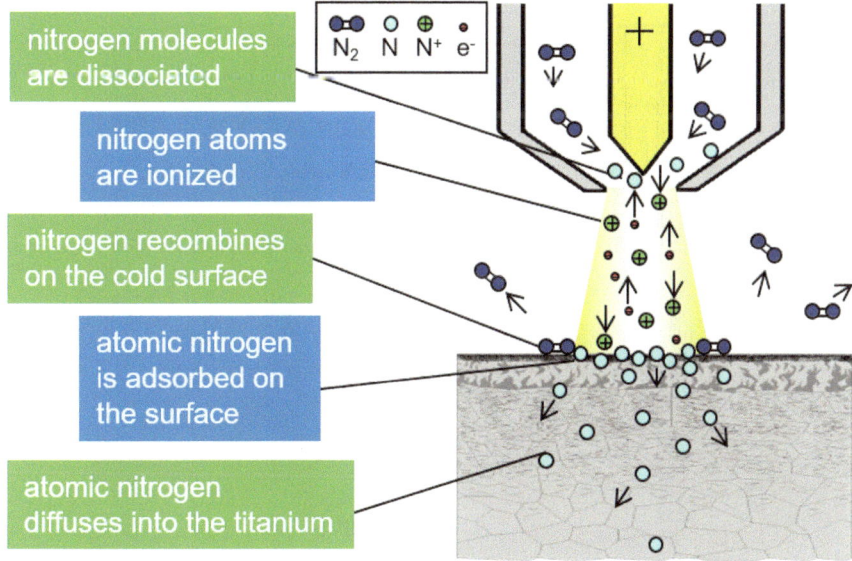

nitrogen molecules are dissociated

nitrogen atoms are ionized

nitrogen recombines on the cold surface

atomic nitrogen is adsorbed on the surface

atomic nitrogen diffuses into the titanium

N_2 N N^+ e^-

Fig. 4 Schematic process of the nitriding with non-transferred arc with the steps dissociation, ionization and diffusion

3 Results and Discussion

3.1 Reduction of Line Energy by Addition of Flux

Without flux, a line energy of at least 163 J/mm (current 60 A) was required to weld through the 2 mm thick specimens. By using 0.04 mg/mm^2 aluminum fluoride, AlF$_3$, as flux, the energy per unit length required for welding could be reduced from 163 to 83 J/mm (current 20 A). The flux influences the gradient of the surface tension dγ/dT, where T is the temperature. With a sufficiently large amount of flux, the gradient becomes positive (Schulze 2010). This means that the surface tension increases with increasing temperature. As a result, the direction of the Marangoni flow is reversed and the melt pool flow direction at the top of the melt pool is inverted. Figure 5a shows the usual melt pool flow for titanium. Figure 5b shows the course of the molten pool flow with a sufficiently large quantity of surface-activating substances (flux) to make the gradient positive.

Figure 6 shows the reference weld and flux weld with 60 A, 40 A and 20 A. Only with 60 A, tests welded trough the 2 mm plate, cf. Figure 6a. With the addition of flux, all tests at 60 A, 40 A and 20 A welded trough, cf. Figure 6b. By reducing the energy required for joint welding, both the heat-affected zone and the FZ become significantly smaller compared to the reference. The smallest width of HAZ of the reference, which welded trough was about 12 mm and the smallest width of the HAZ of the weld with flux which welded trough was about 7 mm.

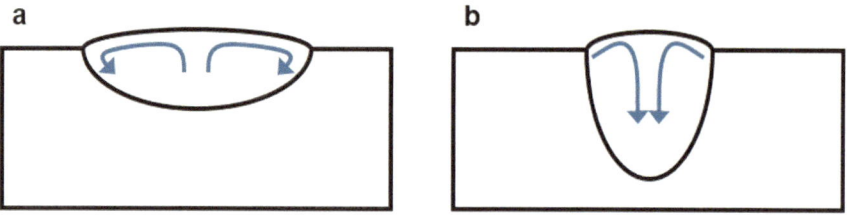

Fig. 5 Melt flow into the liquid weld: **a** shows the usual direction of the melt pool flow for titanium, **b** shows the direction of the flow when $d\gamma/dT$ is >0 due to the influence of the flux (Schulze 2010)

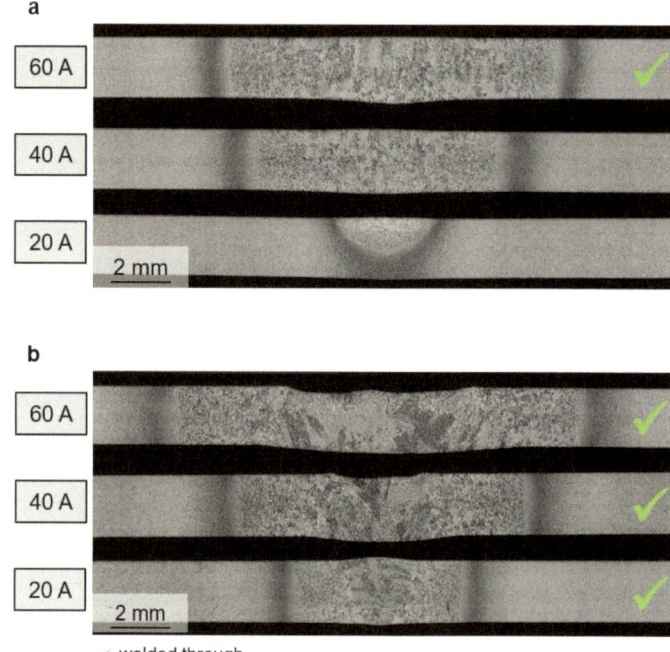

✓ welded through

Fig. 6 Cross-section of bead on plate welds: **a** shows the reference welds with 60 A, 40 A and 20 A, **b** shows the weld with the use of flux with 60 A, 40 A and 20 A

The coarse grain growth in the HAZ can be seen in both welds. The FZ of both specimens shows a martensitic structure. The flux weld shows smaller former β-grains due to the increase cooling rate compared to the reference. The hardness measurement shows a welding-induced hardening of the HAZ and FZ for the reference. The HAZ and FZ have approximately the same hardness at 360 HV1, which is 40 HV1 more than the base material. The hardness measurement of the flux sample shows the same hardness level of 360 HV1 for the HAZ, but the FZ is harder than for the reference. It reaches 400 HV1 and is thus 40 HV1 higher than in the reference. Due to the lower energy input, the meltpool has less overheating and cools down

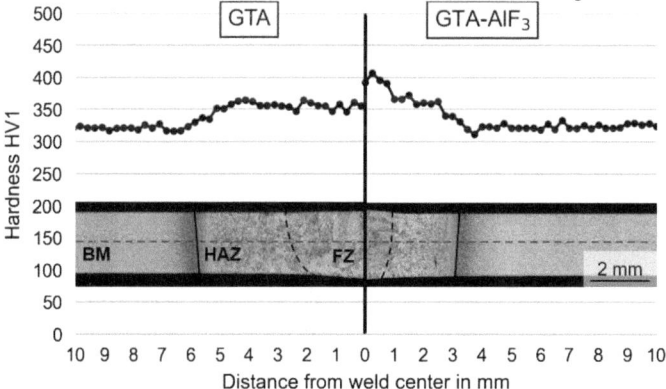

Fig. 7 Comparison of the hardness profile of the reference on the left, the current used is 60 A; the flux-welded specimen is on the right, welded with 20 A.; the dashed red line show the position of the hardness measurements; BM = base material; HAZ = heat affected zone; FZ = fusion zone

faster. Smaller β-grains are formed, but also more elements remain dissolved in the α'-martensite and harden it, cf. Figure 7.

The positive influence of the flux is visible in the mechanical properties of the overall joint, see Fig. 8. The energy reduction reduces the proportion of HAZ plus FZ of the repaired overall joint. Although the strength in the HAZ and FZ increases, the overall connection only reaches the level of the base material, since it breaks there. The worsened elongation due to the unwanted heat treatment can be seen. The weld with flux has visibly changed the microstructure in zone with a width of approx. 7 mm. For the reference the corresponding width was approx. 14 mm. The extensometer used to measure the strain had an initial length of 50 mm. Although less than 20% of the measured area of the flux had been optically changed, the sample only achieved 60% of the elongation. In the case of the reference, one third of the area was optically modified and only just under 35% of the strain of the base material was achieved. This is due to the fact that the HAZ and FZ hinder the movement of the dislocations.

3.2 Grain Refinement of the Prior β-grains with Inoculation

Inoculants are added to the melt during casting, the high-melting compounds act as heterogenic nuclei and stimulate grain formation, resulting in grain refinement (Bermingham et al. 2008). When inoculating welds with SiC, it has been shown that SiC melts and recombines to form TiC and Si (Langen et al. 2018). Thus, the inoculant does not act as an inoculant but as an in-situ alloy. With the SiC inoculant, a significant reduction in β-grain size can be achieved. In addition to the results presented here, grain refinement could also be achieved with TaC, VC, TiC, Si, and

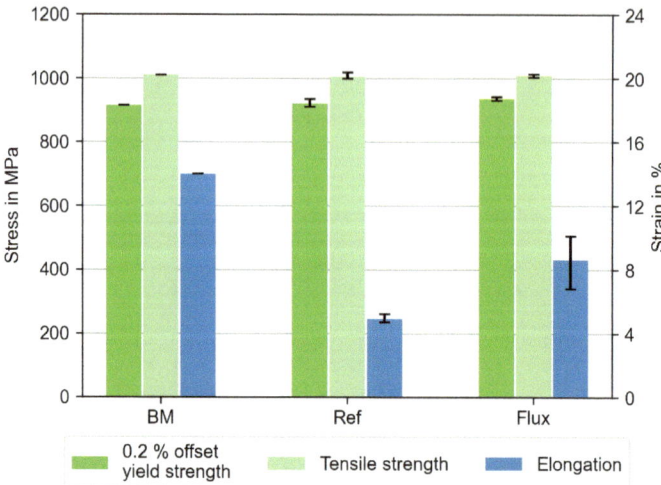

Fig. 8 Yield strength, tensile strength and elongation at break of base material, welded specimens and flux-welded specimens BM means the base material, Ref means the reference weld at 60 A and Flux means the weld at 20 A and the use of flux; error bars show the highest and lowest values obtained from the specimens

C (Langen 2021). In the FZ, the β-grains reached lengths above 1 mm. This is shown in Fig. 9a. Figure 9b shows a transverse section of the inoculated sample. First, the transition from FZ to HAZ is clearly visible, as grain refinement with the inoculant only affects the FZ. In the HAZ, coarse grain growth occurs as in the reference.

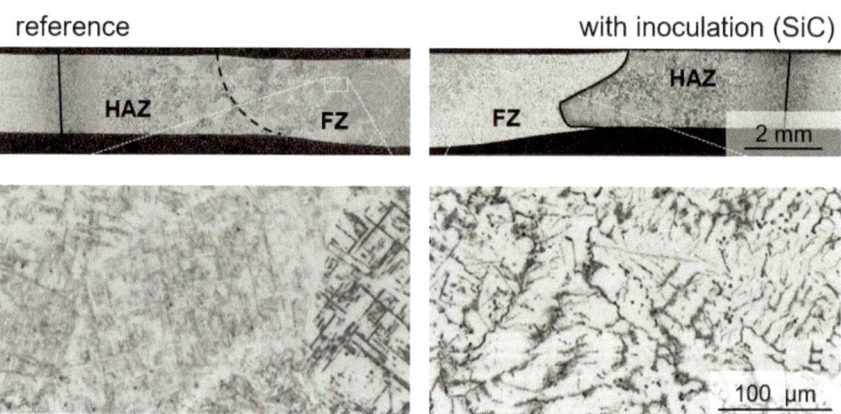

Fig. 9 Cross-section of reference and with SiC inoculated sample both welded at 90 A; on the left reference weld is shown, the prior β-grain size reach a length up to 2 mm; on the right an inoculated sample welded is shown; the prior β-grain size in the FZ is less than 100 μm; the precipitated carbides reach a length of <20 μm; however, in both samples, coarse grain formation occurs in the HAZ

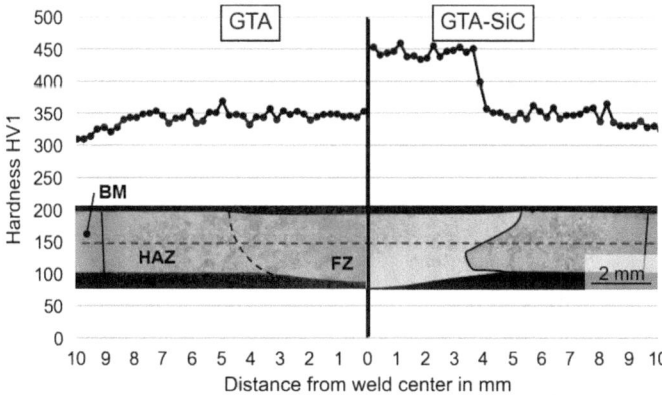

Fig. 10 Comparison of the hardness profile of the reference on the left and the specimen welded with SiC inoculant on the right (Langen 2021)

The hardness measurement on the cross-section shows that the inoculant only acts in the FZ. The coarse grain formation already seen in the microsection in the HAZ shows the same hardness as the HAZ of the reference. However, the addition of SiC results in further hardening of the FZ in the inoculated sample. Detailed EDX studies showed that the SiC dissolves in the melt and subsequently the carbon reacts with the titanium to TiC (Langen 2021). The precipitated titanium carbides further harden the FZ, see Fig. 10.

Tensile tests showed that incolation with an area density of 1.21 mg/mm^2 SiC (1.7 wt.% Si in the FZ) reduced the elongation significantly to less than 0.5% (Langen 2021).

When inoculated with an area density of 0.34 mg/mm^2 SiC (0.8 wt.% Si in the FZ), grain refinement is decreased but elongation at break is improved (Langen et al. 2018). Figure 11 shows the results of the tensile tests for the base material, the reference weld and inoculated specimens with SiC, B and TaC. From the values of the base material and the reference, it is clear that a single weld significantly degrades the mechanical properties of the component. Both strength and ductility are decreased. By using all inoculants, the strength of the weld zone is increased. The inoculated specimens broke outside the weld and showed a typically overmatching of the weld seam. The ductility of the specimens inoculated with B and TaC are similar to the reference weld. Only the specimens inoculated with SiC show better ductility of the overall joint.

Mereddy et al. were also able to reduce the average length of the α-grains from 180 μm to below 40 μm with the addition of 0.41 wt.% carbon to Ti6Al4V. However, the TiC precipitates formed embrittled the material significantly, causing the elongation at break to drop from 10 to 1%. With the addition of 0.1 wt.% carbon, both strength and elongation at break are higher than the reference weld without carbon addition (Mereddy et al. 2018).

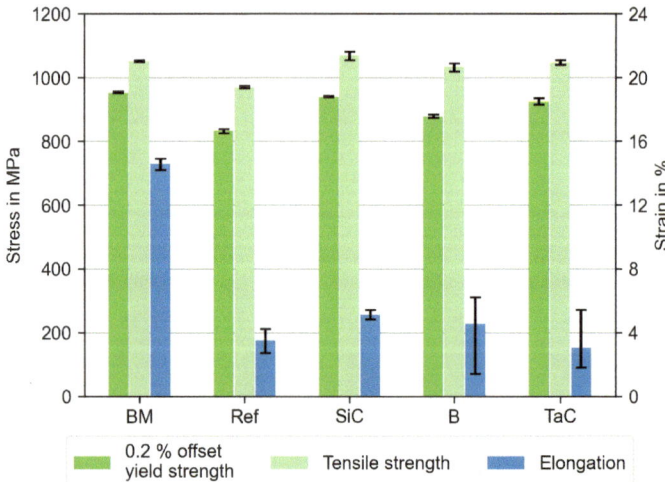

Fig. 11 Yield strength, tensile strength and elongation at break from base material, welded specimens and specimens welded with inoculants; BM is the base material, Ref stands for the reference
and SiC, B and TaC stand for the inoculants used; the current used for all welds was 90 A; error
bars show the highest and lowest value obtained for the specimens (Langen et al. 2018)

3.3 Additive Repair by Wire and Arc Additive Manufacturing

The first test series showed a structure stretching in the build-up direction. The β-
grains partially grew over several layers and reached lengths of up to 15 mm. The
microstructure appeared martensitic, cf. Figure 12. The tensile strength achieved
ranged between 880 and 910 MPa, depending on the direction, cf. Figure 13. The
elongation at fracture was also similar in all directions and, at least 6%, was above
the values of the inoculated specimens and in the range of the specimens welded
with flux. Although the grains grew predominantly in the build-up direction, the
mechanical properties did not show any pronounced anisotropy. This is due to the
fact that the cooling rate was reduced by the high interlayer temperature of 300 °C.
As a result, the proportion of energy dissipated via the substrate decreases and the
proportion dissipated via radiation and convection increases.

In the second investigation, the cooling rate was increased to achieve grain refinement. Therefore the dwell time was increased to reduce the interlayer temperature.
Figure 14 shows the longitudinal and transverse sections. The microstructure shows
the α' martensite. It can be clearly seen that the former β-grains have grown in the
build-up direction. They reach lengths >20 mm. Due to the long cooling periods
between the layers of 120 s, welding was performed with a significantly lower inter-
pass temperature than in the first test. Cooling took place mainly via the substrate, as
is usual in welding. The preferred growth direction of the grains can also be clearly
seen in the mechanical properties, see Fig. 15. In the 45° direction, a tensile strength
of 1050 MPa was achieved. In the other two directions, about 10% less was reached

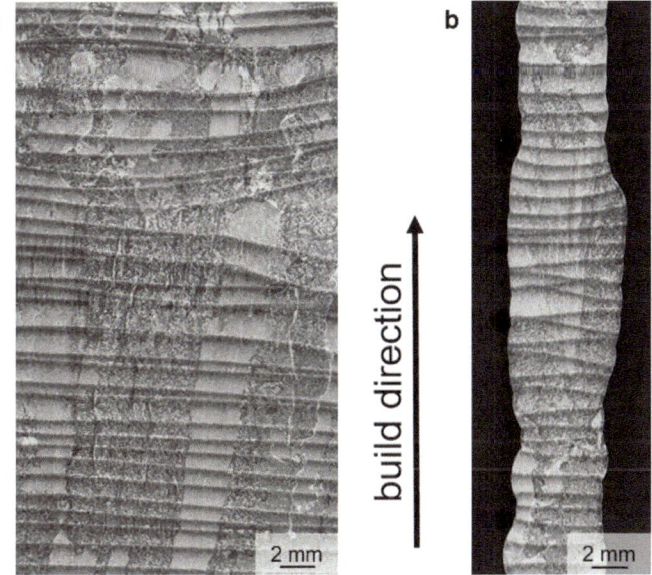

Fig. 12 Micrographs from the first test; build-up direction is upwards; **a** shows the longitudinal micrograph; **b** shows the transverse micrograph

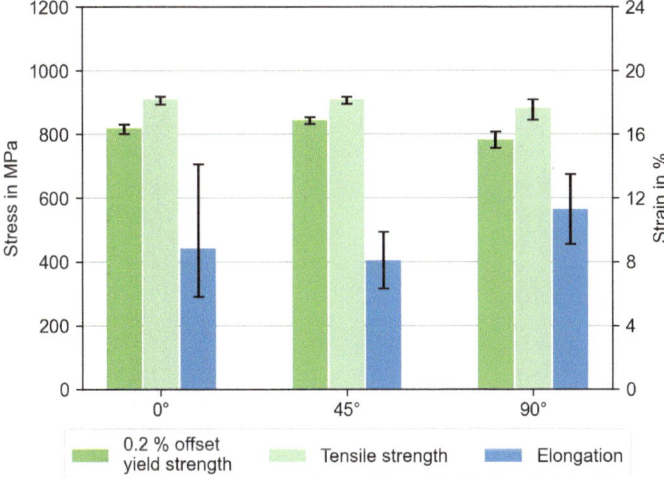

Fig. 13 Yield strength, tensile strength and elongation at break of the first survey; 0° is in the welding direction, 90° in the build-up direction and 45° is the diagonal between the welding direction and the build-up direction; error bars show the highest and lowest value obtained for the specimens

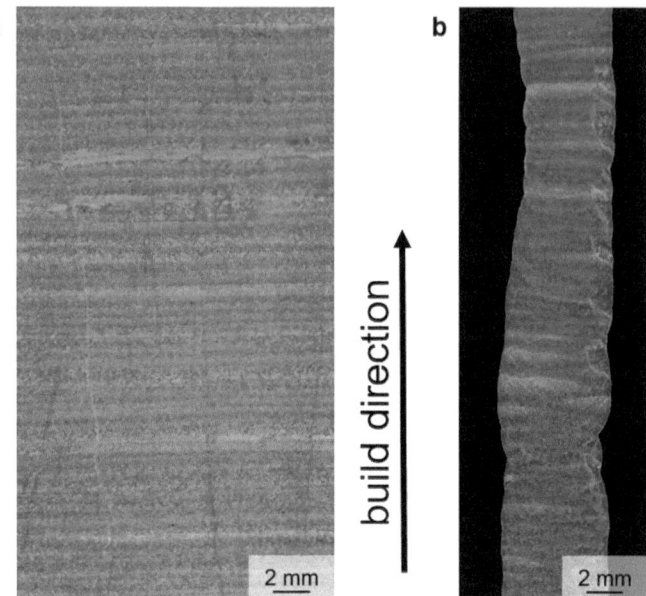

Fig. 14 Micrographs from the second test; **a** shows the longitudinal micrograph; **b** shows the transverse micrograph

(940 MPa). In the 45° direction the ductility was less than in the other two directions, i.e., 7.5% versus 11% and 16%. The higher cooling rate led to smaller α-grains, which increased the strength. Compared to the first series of tests, the increase in cooling rate for the second series of tests increased the strength in all directions above the maximum of the first series of tests. The ductility was similar.

3.4 Local Nitriding

Nitrogen was introduced directly after solidification of the molten pool of a TIG weld. Figure 16 shows the hardness profile and a micrograph of a nitrided sample. A nitride layer of 10 to 20 μm is clearly visible as a white boundary layer. This is because nitrogen promotes the formation of the α-phase of titanium. Hardness values of over 1500 HV0.025 were achieved in the boundary layer. The hardness measurement shows that the diffusion depth was approx. 100 μm. Although the local treatment can achieve a significant hardening of the surface, stress is also induced. These stresses can lead to a reduction in strength or result in distortion. Fatigue tests on locally atmospherically plasma-nitrided specimens have shown that a hardness of over 1000 HV0.025 can be achieved. However, the heating required for this leads to distortion, so that the specimens break prematurely in fatigue tests. To reduce the induced stress that lead to distortion, the energy input can be reduced, resulting

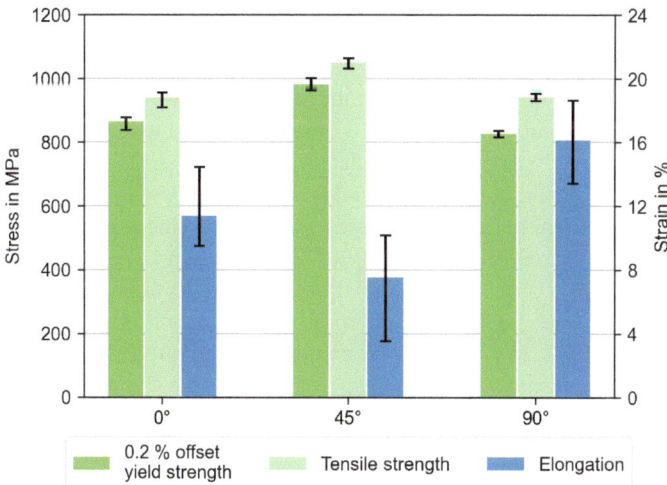

Fig. 15 Yield strength, tensile strength and elongation at break of the second survey; 0° is in the welding direction, 90° in the build-up direction and 45° is the diagonal between the welding direction and the build-up direction; error bars show the highest and lowest value obtained for the specimens

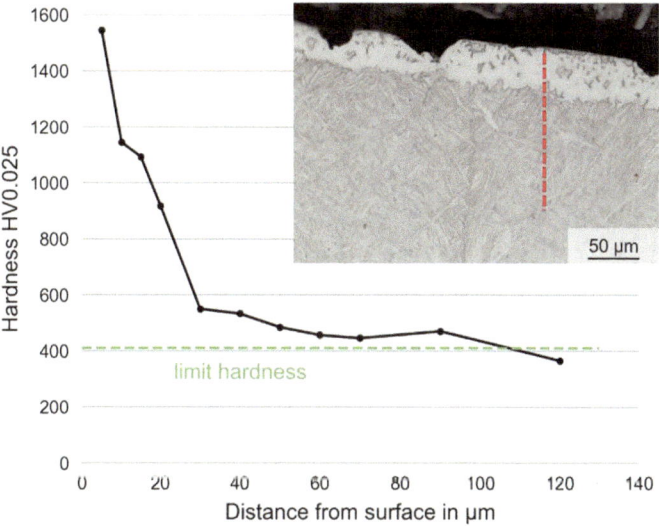

Fig. 16 Hardness profile of a locally atmospherically nitrided sample; limit hardness is 410 MPa (Gushchina et al. 2020); the nitriding hardness depth is approximately 110 μm

in lower surface temperatures. Lower temperatures reduce the nitrogen diffusion resulting in lower surface hardness or longer lead times. With similar fatigue and processing time a surface hardness of 800 HV could be reached.

4   Conclusions

Both inoculants and fluxes can promote the mechanical properties of the overall joint. They can also be combined, see Fig. 17 the narrow seam with small HAZ is clearly visible. The carbide precipitates also clearly distinguish the FZ from the HAZ.

The advantages of the inoculation and flux effect can also be used at least to some extent in additive repair. Grain refinement can also be achieved in additive manufacturing by adding carbon (Mereddy et al. 2018). On one hand, grain refinement helps to improve mechanical properties, and on the other hand, its use reduces the anisotropy of the microstructure (Mereddy et al. 2017). The effect of the flux causes the weld to become narrower. In addition, the average temperature of the molten pool can be reduced without causing bonding defects. The overheated material flows in- and downward rather than outward, preventing bonding defects. Arc stability is reduced by the use of flux. Especially when igniting above the flux layer, misfiring can occur. The arc needs enough energy to melt the flux as well, otherwise the arc jumps and welding defects can occur. The application with spray delivered reproducible results. Small inoculation amounts of 0.1 wt% C as Mereddy had used (Mereddy et al. 2018), are indeed also achievable with spray. But reproducibility is no longer guaranteed. As a result, some areas may reach higher carbon contents, resulting in degradation of mechanical properties. In addition, the spray process for applying inoculant or flux takes place outside the inert gas atmosphere. In order to be able to use the advantages of the inoculant and flux reproducibly with WAAM, the inoculant and flux must be applied differently.

The investigations have shown that the inoculant also can be dissolved during welding. During cooling, the precipitates are formed, which are available as heterogeneous nuclei. This means that no refractory compounds are needed for grain refinement, only the carbon (Langen 2021; Mereddy et al. 2018). Thus, inoculants and fluxes can also be fed directly into the process zone via gases.

Using gas the addition of the inoculant and flux can be carried out internally in the process via the shielding gas in the shielding gas atmosphere. Furthermore the amount of inoculant and flux can be adjusted more precisely and homogeneously.

Local atmospheric nitriding is one way to significantly increase the surface hardness of titanium components. However, one has to consider that the fatigue strengths might be affected by potential distortion. For cyclically loaded components, surface

Fig. 17   Combined inoculation and flux effect

hardening to 800 HV can be achieved without degrading fatigue properties. For structural components not subjected to vibration, hardnesses in excess of 1500 HV can be exploited.

This subproject of CRC 871 is the link between the decision-making process of a possible repair and the repaired recontoured titanium component. The improved repair methods of titanium components developed in this subproject can further increase the mechanical properties of the repaired component. The determination of the achievable mechanical properties can be led back. Thus, they can be taken into account in a decision-making process preceding the repair and prevent unnecessary or inadequate repairs.

Acknowledgements Funded by the Deutsche Forschungsgemeinschaft (DFG, German Research Foundation) SFB 871/3 119193472.

References

Balasubramanian, M., Jayabalan, V., and Balasubramanian, V. (2008). Effect of microstructure on impact toughness of pulsed current GTA welded α–β titanium alloy. In: *Materials Letters* 62 (6–7), S. 1102–1106. https://doi.org/10.1016/j.matlet.2007.07.065.

Bermingham, M. J., StJohn, D. H.; Krynen, J.; Tedman-Jones, S., and Dargusch, M. S. (2019). Promoting the columnar to equiaxed transition and grain refinement of titanium alloys during additive manufacturing. In: *Acta Materialia* 168, S. 261–274. https://doi.org/10.1016/j.actamat.2019.02.020.

Bermingham, M. J., McDonald, S. D., Dargusch, M. S., and StJohn, D. H. (2008). Grain- refinement mechanisms in titanium alloys. In: *J. Mater. Res.* 23 (1), S. 97–104. https://doi.org/10.1557/JMR.2008.0002.

Colegrove, P. A., Donoghue, J., Martina, F., Gu, J., Prangnell, P., and Hönnige, J. (2017). Application of bulk deformation methods for microstructural and material property improvement and residual stress and distortion control in additively manufactured components. In: *Scripta Materialia* 135 (14), S. 111–118. https://doi.org/10.1016/j.scriptamat.2016.10.031.

Denkena, B., Grove, T., Mücke, A., Langen, D., Nespor, D., and Hassel, T. (2017). Residual stress formation after re-contouring of micro-plasma welded Ti−6Al−4 V parts by means of ball end milling. In: *Mat.-wiss. u. Werkstofftech.* 48 (11), S. 1034–1039. https://doi.org/10.1002/mawe.201600743.

Derby, B., Lütjering, G., and Williams, J. C. (2003). Titanium. Berlin, Heidelberg: Springer Berlin Heidelberg.

Gou, J., Wang, Z., Hu, S., Shen, J., Tian, Y., Zhao, G., and Chen, Y. (2020). Effects of ultrasonic peening treatment in three directions on grain refinement and anisotropy of cold metal transfer additive manufactured Ti-6Al-4V thin wall structure. In: *Journal of Manufacturing Processes* 54 (1), S. 148–157. https://doi.org/10.1016/j.jmapro.2020.03.010.

Gushchina, M., Carstensen, T., Maier, H. J., and Hassel, T. (2020). Plasma nitriding Ti-6Al-4V with the aid non-transmitted plasma-arc using different protection atmosphere. In: *Materials Today: Proceedings* 30, S. 694–699. https://doi.org/10.1016/j.matpr.2020.01.524.

Henckell, P., Günther, K., Ali, Y., Bergmann, J. P., Scholz, J., and Forêt, P. (2017). The Influence of Gas Cooling in Context of Wire Arc Additive Manufacturing—A Novel Strategy of Affecting Grain Structure and Size. In: The Minerals Metals &. Materials So TMS (Hg.): TMS 2017 146th Annual Meeting & Exhibition Supplemental Proceedings. Cham: Springer International Publishing (The Minerals, Metals & Materials Series), S. 147–156.

Hönnige, J. R., Colegrove, P., and Williams, S. (2017). Improvement of microstructure and mechanical properties in Wire + Arc Additively Manufactured Ti-6Al-4V with Machine Hammer Peening. In: *Procedia Engineering* 216 (7), S. 8–17. https://doi.org/10.1016/j.proeng.2018.02.083.

Langen, D., Maier, H. J., and Hassel, T. (2018). The Effect of SiC Addition on Micro-structure and Mechanical Properties of Gas Tungsten Arc-Welded Ti-6Al-4V Alloy. In: *J. of Materi Eng and Perform* 27 (1), S. 253–260. https://doi.org/10.1007/s11665-017-3091-y.

Langen, D. (2021). Steuerung des Erstarrungsgefüges der Titanlegierung Ti-6Al-4V beim Wolfram-Inertgas-Schweißen. Dissertation. TEWISS - Technik und Wissen GmbH; Gottfried Wilhelm Leibniz Universität Hannover, Hannover.

Li, Z., Liu, C., Xu, T., Ji, L., Wang, D., and Lu, J. et al. (2019). Reducing arc heat input and obtaining equiaxed grains by hot-wire method during arc additive manufacturing titanium alloy. In: *Materials Science and Engineering: A* 742 (5), S. 287–294. https://doi.org/10.1016/j.msea.2018.11.022.

Mereddy, S., Bermingham, M. J., StJohn, D. H., and Dargusch, M. S. (2017). Grain refinement of wire arc additively manufactured titanium by the addition of silicon. In: *Journal of Alloys and Compounds* 695, S. 2097–2103. https://doi.org/10.1016/j.jallcom.2016.11.049.

Mereddy, S., Bermingham, M. J., Kent, D., Dehghan-Manshadi, A., StJohn, D. H., and Dargusch, M. S. (2018). Trace Carbon Addition to Refine Microstructure and Enhance Properties of Additive-Manufactured Ti-6Al-4V. In: *JOM* 70 (9), S. 1670–1676. https://doi.org/10.1007/s11837-018-2994-x.

Reisgen, U., Sharma, R., Mann, S., and Oster, L. (2020). Increasing the manufacturing efficiency of WAAM by advanced cooling strategies. In: *Weld World* 64 (8), S. 1409–1416. https://doi.org/10.1007/s40194-020-00930-2.

Schulze, G. (2010). Die Metallurgie des Schweißens. Eisenwerkstoffe - Nichteisenme- tallische Werkstoffe. 4., neu bearbeitete Auflage. Berlin, Heidelberg: Springer Berlin Heidelberg (SpringerLink Bücher).

Wits, W. W., García, J. R. R., and Becker, J. M. J, (2016). How Additive Manufacturing Enables more Sustainable End-user Maintenance, Repair and Overhaul (MRO) Strategies. In: *Procedia CIRP* 40 (2), S. 693–698. https://doi.org/10.1016/j.procir.2016.01.156.

Yin, B., Ma, H., Wang, J., Fang, K., Zhao, H., and Liu, Y. (2017): Effect of CaF 2 addition on macro/microstructures and mechanical properties of wire and arc additive manufactured Ti-6Al-4V components. In: *Materials Letters* 190, S. 64–66. https://doi.org/10.1016/j.matlet.2016.12.128.

Consideration of Variability in the Repair Processes and in Material Properties (Project Area C)

Simulation-Based Process Design of Recontouring Technologies

Volker Böß, Berend Denkena, Sven Friebe, and Markus Hein

Abstract Engine manufacturers generate about 50% of their total turnover with maintenance. Damaged parts can either be replaced by spare parts or can be regenerated by e.g. local welding processes. One major step of the manufacturing part of the process chain for regeneration after the material deposition is the removal of excess weld material by cutting, which is called recontouring. Recontouring is often the last process step, which defines the final surface integrity and thus the performance of the repaired parts. Thereby, each component has a batch size of one by reason of individuality. In industrial praxis, the recontouring is done mainly with high manual effort. This results in uncertain and unreproducible repair processes for each component. Hence, a major challenge for recontouring processes is the reduction of the required manual effort by an automated method. In the subproject C1 "simulation-based process design of recontouring technologies", machining investigations and technological simulations were applied in combination with suitable models for a process adaption in order to reach the required workpiece properties. A special focus was the transition from the undamaged area to the deposited material. Furthermore, a method for effective process planning was developed for the generation of 5-axis milling tool paths. This increases the effectiveness of the process planning by adapting them to the individual shape of the component. The present paper gives an overview of the main results of the investigation of methods for the individual process planning of recontouring technologies based on process simulations. This includes the development of an algorithm for the automatic planning of the recontouring process. A Dexel-based simulation method was developed and experimentally validated which allows the prognosis of the geometrical shape of the material deposition. Based on that, the influence of the resulting material allowance

V. Böß · B. Denkena (✉) · S. Friebe · M. Hein
Institute of Production Engineering and Machine Tools, Leibniz University Hannover, An der Universitaet 2, 30823 Garbsen, Germany
e-mail: denkena@ifw.uni-hannover.de
URL: https://www.ifw.uni-hannover.de/de/

V. Böß
e-mail: boess@ifw.uni-hannover.de

S. Friebe
e-mail: friebe@ifw.uni-hannover.de

© The Author(s) 2025
J. R. Seume et al. (eds.), *Regeneration of Complex Capital Goods*,
https://doi.org/10.1007/978-3-031-51395-4_13

on the recontouring was studied. The aim was the generation of a part quality that satisfies the requirements of the functional review. The evaluation of the part quality was conducted by a Dexel-based technological process simulation of the recontouring process. This allowed the prediction of the workpiece surface as well as the research of the influence of the process parameters on the main residual stresses.

Keywords Process planning · Process simulation · Recontouring · Manufacturing technologies · Hybrid manufacturing

1 Introduction

High-value products such as aircraft engines require effective maintenance, repair, and overhaul services to increase economic efficiency and sustainability (Uhlmann et al. 2013). In the case of jet engines, components like blades need specialized repair technologies to maintain or even improve their functional properties throughout the life cycle (Eberlein 2007). Designing these repair processes is a challenging task and differs from new part production (Denkena et al. 2015a, b). Typically, a material deposit is needed to replace missing areas of the damaged part or to attach repair patches or fill cracks. Afterwards, the excess material has to be removed to restore the desired geometrical shape. This removal is referred to as recontouring, where milling processes are commonly used (see Fig. 1). The uniqueness of repair cases, different materials (e.g., base and weld material) and parts prone to vibration complicate the repair. Additionally, 5-axis machining processes are required to restore the complex part geometries and to overcome the limited accessibility. Compared to new part production, it is hardly possible for recontouring processes to test the process with a spare part beforehand. This leads to the necessity of a tailored and precise process planning procedure to ensure high geometric accuracy and surface quality for minimizing scrap and flow loss (Eberlein 2007).

In most cases, the target shape of the blade is not immediately known due to operational deformation processes in the engine, such as thermal creep, so a CAD-technical restoration of the contour is necessary (Gao et al. 2008, Wu et al. 2012). In addition to the measurement of the actual geometry, the planning of the recontouring process

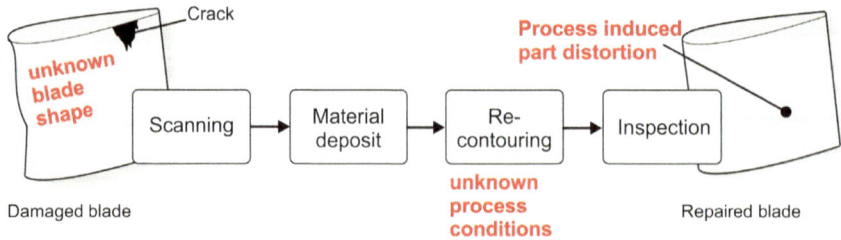

Fig. 1 Repair process chain in terms of manufacturing aspects

requires knowledge about the target geometry. The goal of current research projects is a higher degree of automation since a manual derivation of the target geometry is associated with a high expenditure of time. This is based on measurements of the damaged blade (Rong et al. 2014; Schlobohm et al. 2014). The limits of automation in the repair of engines are largely determined by the complicated component shape combined with a high degree of individualization. For this reason, current research and development projects are concerned with pushing this limit in the direction of optimized and highly efficient recontouring systems. Bremer et al. have developed an automated repair system for turbine components. Here, the CAD surface reconstruction of the unknown target shape (reverse engineering) is taken into account and integrated into an overall process for recontouring the components (Bremer 2006). In particular, the integration of the obtained data into the overall process plays an increasingly important role. Yilmaz developed an automated and integrated process chain for turbine component repair that includes optical 3D measurement, surface reconstruction, and customized milling processes. The aim was to restore the shape of the damaged components within their geometric tolerances with reduced process time (Yilmaz et al. 2010).

For ball end milling, there are numerous studies dealing with process design for new part production. Methods exist to predict process stability, surface topography, and productivity (Markworth 2005; Knobel 2000). They can be categorized into empirical, analytical, and numerical methods. One example of an empirical model was presented by Vakondios et al. (2012), where mathematical regression between the process parameters and the resulting surface parameter Rz is used. Empirical models are very accurate and easy to use but are limited to the experimental scope. Analytical models mathematically describe the shape (Markworth 2005; Knobel 2000) and/or the trajectory of the cutting edges (Arizmendi et al. 2008). For instance, the analytical model of Arizmendi et al. (2008) show the importance of tool runout and its impact on surface topography. However, the analytical equations can get too complex to handle by adding more simulation features, such as vibrations or special tool shapes etc. Therefore, numerical simulations in the time domain are required. They are called material removal simulations (MRS) and use, e.g., Voxel, Dexel or constructive solid geometry (CSG) to discretize the workpiece. In contrast to common analytical approaches, numerical simulation offers advantages regarding the flexibility of their application possibilities and adaptability. The milling kinematics of the machine and a digital model of the workpiece geometry can be used. Thus, it is possible to calculate the cutting conditions engagement parameters even for complex geometries and tools. It has been shown by Liu et al. (2005) that the movement of the cutting edge is a superior approach in terms of accuracy compared to a Boolean subtraction between the workpiece and the rotation body of the milling tool, e.g., a sphere for ball end mills. The resulting surfaces after MRS are perfectly smooth without stochastic influences. This often leads to underestimated surface parameters (Bouzakis et al. 2003; Chen et al. 2005; Liu et al. 2005). Several investigations concerning the interaction of the wall and the near-wall flow have shown that the topography of the wall surface significantly influences the aerodynamic losses that occur on blades or vanes (Abuaf et al. 1998; Bons 2010). Particularly the height and the shape of the surface structure, as well as their alignment with respect to the direction of the flow can increase the overall friction loss (Hohenstein and Seume 2013). Even relatively small structures may have an effect on the friction between the wall and the fluid, as is exemplified by riblets (Boese 2002). Due to their shape and direction on the surface, they reduce the aerodynamic losses. This suggests that every change in the surface topology can have an influence on the aerodynamic loss behavior of an airfoil.

Ball end milling is mainly used for finishing operations and manufacturing of complex parts (Layegh and Lazoglu 2017). Besides high geometric accuracy and low surface roughness, a compressive residual stress state is often required, e.g., in aerospace parts like compressor blades. Compressive residual stresses can be achieved using rounded cutting edges (Denkena et al. 2014b). The disadvantages of using rounded cutting edges are increased process forces because of additional ploughing as well as a possible burr formation (Denkena et al. 2014a, b; Wyen 2011). To reduce process forces causing tool displacements and shape deviations, process parameters can be optimized using process force models. Controlling burr formation, however, is still challenging because the burr formation not only depends on process parameters but also on the workpiece material, tool geometry, and tool path (Link 1992). Burrs deteriorate the surface roughness which is often not acceptable (Aurich et al. 2009). To avoid burrs during the process planning stage, various models for predicting burr are investigated. Sokolowski et al. (1994) apply neural networks and fuzzy logic based on an experimentally determined database. Chu and Dornfeld (2002) uses computed engagement conditions depending on the workpiece geometry, tool path, and cutting parameters as input parameters of a database containing different burr types. Chen et al. (2012) used Finite Element (FE) simulations to predict burr formation in micro ball end milling of Ti-6Al-4 V.

2 Objective

The objective of the subproject C1 of the Collaborative Research Center (CRC) 871 was to establish a cross-process simulation and planning of the machining recontouring process for the functional improvement of the application behavior of the regenerated workpiece. The generated knowledge can be used for common planning, taking into account the interactions between the regeneration processes. The following sub-goals were achieved:

1. Availability of a simulation model for the prediction of the material deposition
2. Digital machine, tool, and workpiece models for efficient process simulations
3. Availability of methods for an optimized surface prediction
4. Knowledge about influences of the material deposition on the recontouring process
5. Method for individual CAM-planning for recontouring processes

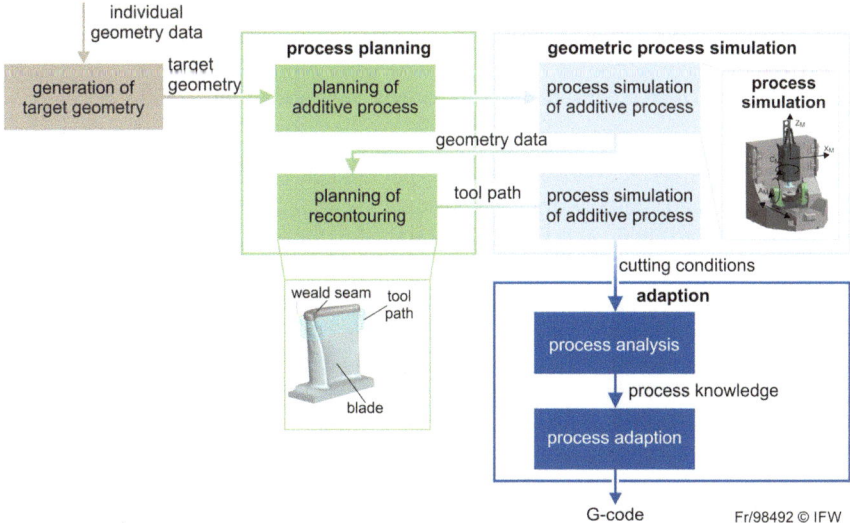

Fig. 2 Concept of adaptive process planning method for blade repair

The results were combined in a holistic concept for adaptive process planning which was established within the sub project C1. The concept of the adaptive process planning method for blade repair is shown in Fig. 2.

In the initial step, based on 3D scan data of the damaged blade, a target geometry model is created (Sect. 3.1). This forms the basis for the subsequent simulation-based process planning of the material deposition process (Sect. 3.2) and the recontouring process (Sect. 3.3). By means of a digital process twin, relevant process variables for the material deposition and recontouring process are predicted during process planning. This includes both technological process data and information about the resulting workpiece geometry. By simulating the material deposition, knowledge is generated about the geometric characteristics of the material deposition, such as the minimum height of the allowance. This knowledge is taken into account in the process design of the recontouring. By simulating the recontouring process, the cutting conditions in the contact zone are calculated. Subsequently, technological process variables and quality data are estimated as part of the process analysis. The data are used to adapt both processes with the aim of increasing process quality and reducing process time. The adaptation of the process parameters as well as the tool path is implemented in the NC code and then transferred to the processing machine.

3 Results

3.1 *Individual Process Planning for Recontouring Processes*

The concept of the individual planning method can be divided into the two sub-methods geometry processing and process planning. The geometry processing method is used to calculate a suitable target model. First, the individual actual geometry is reconstructed on the basis of the 3D scan data. Based on this, a target model is generated. Within the process planning, a 5-axis toolpath is calculated and in combination with the most suitable process parameters used for the generation of the NC code for the tool machine. The method for the individual process planning is shown in Fig. 3.

As discussed in the state of the art, highly stressed components such as turbine blades are subject to continuous operational wear. Macroscopically visible damage such as chipping or cracking is completely removed before the weld preparation and is therefore not relevant for the subsequent repair process chain. In addition, there is typically a slight degeneration in terms of thermal creep effects and abrasive wear of the entire blade. This results in a loss of material, which leads to a significant change in the geometry of the blade. The change in geometry has a particularly strong influence on the repair result and should therefore be taken into account in the entire repair process chain. The resulting deviations from the nominal component

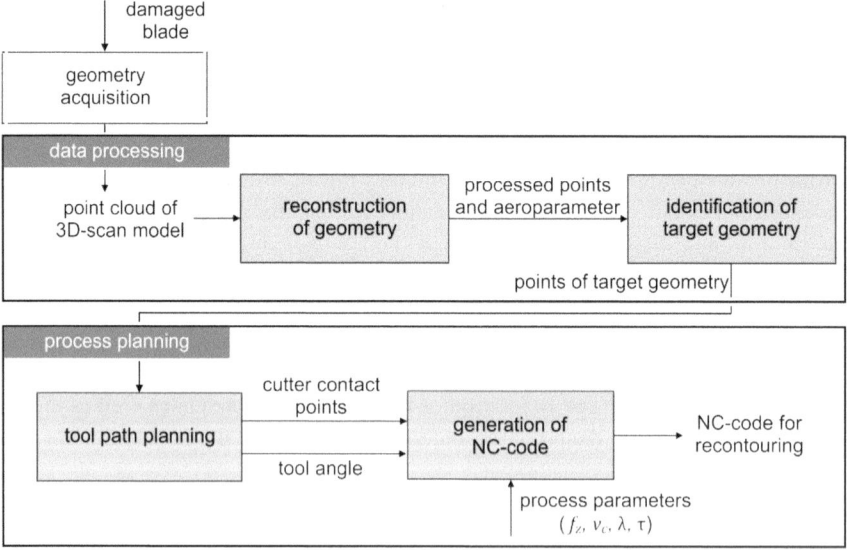

Fig. 3 Adaptable process planning method

geometry (CAD file) necessitate a method for creating a new target geometry for each individual component.

Based on the given geometric information about the current component geometry, the following necessary requirements for the method can be derived:

1. The current component geometrical shape (actual geometry) must be taken into account when creating the nominal geometry.
2. When calculating the nominal geometry, the aim should be to achieve a transition between the actual geometry and the newly calculated geometrical shape (damaged geometry area) that is as smooth as possible in terms of curvature.

The method for the individual computation of a suitable target geometry is carried out in two steps. The calculation of the target geometry requires a continuous surface model without gaps. The approach in praxis for surface reconstruction is done manually in the current state-of-the-art (Bremer 2006, Gao et al. 2008, Rong et al. 2014, Yilmaz et al. 2014). In order to avoid this time-consuming process, first, a new algorithm for the automated reconstruction of the actual geometry for the case of a tip repair was developed. Moreover, by using sliced profile curves the aerodynamical parameters in terms of the position of the leading and trailing edge, the profile thickness as well as the camber line are calculated for each profile. At this stage, the damaged part of the blade is unknown due to the degeneration and thus, requires a restoration of the target model. Therefore, in the second step, the unknown area of the component is calculated by a knowledge-based extrapolation of the reconstructed actual geometry using profile curves and the aerodynamical parameters. Three-dimensional regression functions are built up for the profile thicknesses and the camberlines of all profiles. In order to calculate the geometry of the blade in the unknown area, both functions are extrapolated up to the required target height of the blade. Since the regression functions have a wider space within the planes of the profiles, a trimming at the leading and trailing edge is conducted. This is done by a two-dimensional extrapolation function of the leading and trailing edge points up to the target height of the blade. Then, the information on the profile thicknesses is added to the camber lines in the unknown area. This is done by circles whose diameter corresponds to the thickness and which are positioned along the camber line. The resulting contour in the damaged area is generated by calculating the envelope curve of all the cycles. Finally, the reconstructed actual geometry and the geometry in the damaged area are stitched together. In this way, a suitable target geometry for the blade could be generated.

To validate this approach, a reference model in analogy of a turbine blade of the second stage of the high-pressure turbine (HPT) was designed where the repaired area is known. A damaged model was created by using this reference model and removing the tip area as it is the case for the tip repair. Afterward, a target model was generated on the basis of the damaged model using the developed process planning method. The target model was then digitally machined in a geometric simulation of the recontouring process using the individually and automated generated NC-Code. In this way, the resulting quality of the generated target model could be evaluated by comparing the deviations between the target model and the reference model in the

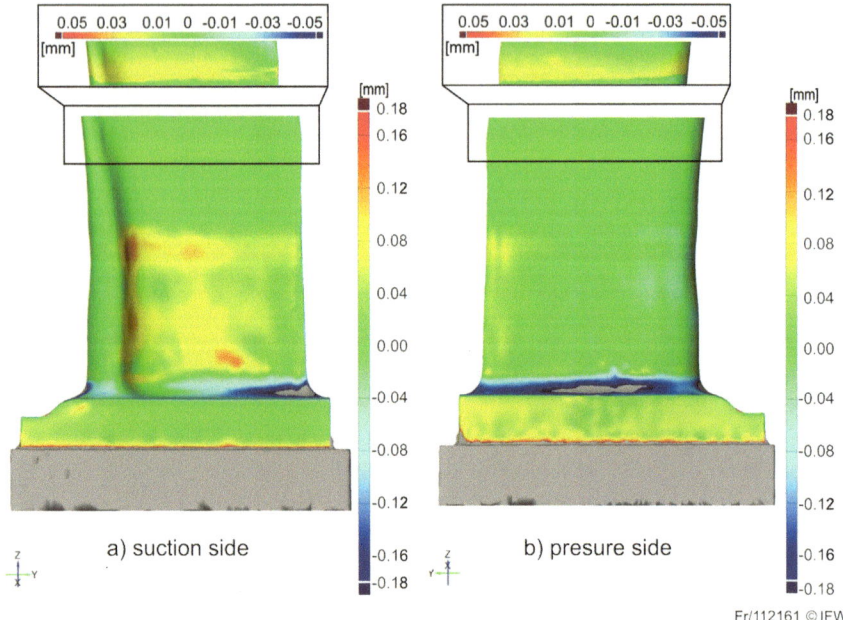

Fig. 4 Surface comparison between generated target model and reference model

tip area. The results of the comparison are shown in Fig. 4 for the pressure side and the suction side.

The criteria for a successful recontouring are a transition from the undamaged to the recontoured area that is as free of offsets and as continuous as possible, while at the same time complying with the shape, dimensional and positional tolerances with regard to the nominal geometry, which includes none damage of the actual geometry. These criteria were proven by the surface comparison. The surface comparison of the repaired blade model shows a good correspondence to the reference model with a maximum deviation of about 0.02 mm. A small step in the transition area can be observed which has a height of 0.02 mm. The deviations decrease with higher z-levels and are nearly negligiblely small at the top of the tip. In addition, it could be ensured that no damage to the actual geometry exists, thus fulfilling the requirements of the repair manual. Thus, the automated planning method for generating the target model can be stated as valid for the considered blade type and repair type.

3.2 Geometric Process Simulation of Additive Process for Blade Repair

To improve the process planning of the recontouring process, knowledge about the geometry of the material deposition is needed. This section introduces a method for the prediction of the geometry of the material deposition after the additive process based on a Dexel-based simulation model. The method was used afterwards to determine the influence of the material deposition on the recontouring process.

An empirical model for the prediction of the geometrical shape of the deposited material according to the process parameters was built. The method is explained in detail in (Böß et al. 2021). The model was parametrized based on experiments by micro-plasma welding as the DED process for titanium alloy Ti6Al4V. The model predicted the width b and the height h of the shape of the deposited material depending on the given process parameters travel velocity v and current I. In the experiments, the material was deposited as a straight line with a length of 70 mm on a titanium substrate with a constant wire feed rate f of 0.3 m/min using argon inert gas. Each experiment was repeated three times.

The cross section of the welding seams in perpendicular direction were measured at five distinct points to calculate the mean weld seam width using a tactile profile measuring device (MarSurf LD 130, Mahr GmbH Göttingen, Germany). In addition, the welding seam were measured in longitudinal direction in the middle of the welding seam to calculate the mean value of the height of the deposited material. By increasing the applied current intensity the height of the deposited material decreased and the width in- creased. This is due to the higher energy input at a higher current. The higher energy input leads to a lower viscosity of the molten material, and thus to a lower ratio between the height and the width of the deposited material. Besides, an increase of the travel velocity led to a decrease of the height and width of the deposited material, which can be explained by the decreasing volume of deposited material per unit length. This resulted in a smaller cross-section profile of the deposited material. Figure 4 shows this effect on the height of the deposited material (Fig. 5).

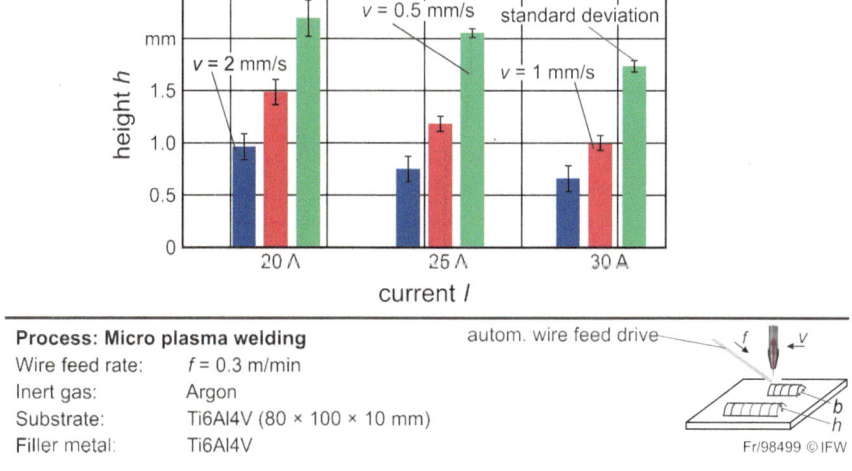

Fig. 5 Measured height for different travel velocity and currents

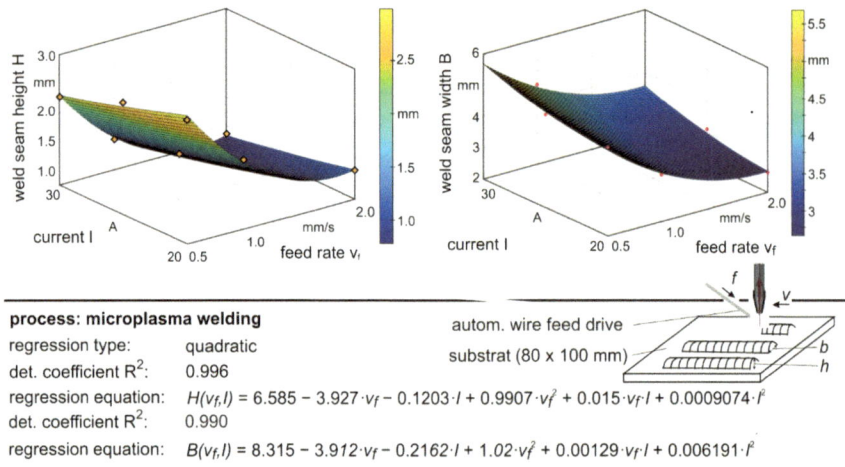

Fig. 6 Regression functions for prediction of welding seam height and width

Based on the experimental results, an empirical regression model was established to predict the height and width of the deposited material based on the process parameters. The quadratic regression functions for the prediction of the width and height as a function of the specified travel velocity and current are shown in Fig. 6. To simulate the shape of deposited material, the developed regression models were implemented in the simulation software IFW CutS (Denkena and BöB 2009). According to the process parameters, the width and height were calculated by means of the regression model, and the digital tool was scaled accordingly. In this way, the geometrical shape of the deposited material could be simulated.

By means of the digital tool, the geometry of the solidified and chilled material deposition was modeled whereby a half-ellipsoid was used. With this simulation method, the geometrical shape of the deposited material can be adjusted by scaling the height and width of the digital tool, which in combination with multi-axis machine kinematics allows the simulation of different weld seam shapes and multi-layer depositions. Furthermore, it enhances the efficient process planning of additive manufacturing processes. The simulation method was used for modeling the material deposition process of blade repair using the scalable digital tool model in combination with a digital workpiece model of the individual damaged blade, which is shown in Fig. 7.

It has become apparent that the resolution of the multi-Dexel model has a high influence on the simulation accuracy. In order to find an appropriate resolution, an experimental setup was conducted in order to show the simulation results of the material deposition simulation for five different Dexel-densities, which are shown in Fig. 8.

On one hand, the results show the influence of the Dexel density on the accuracy of the workpiece model. While at a Dexel density ρ_{XYZ} of $10 \, \text{mm}^{-1}$, the individual layers of the material buildup are weakly pronounced, a continuous increase in the model

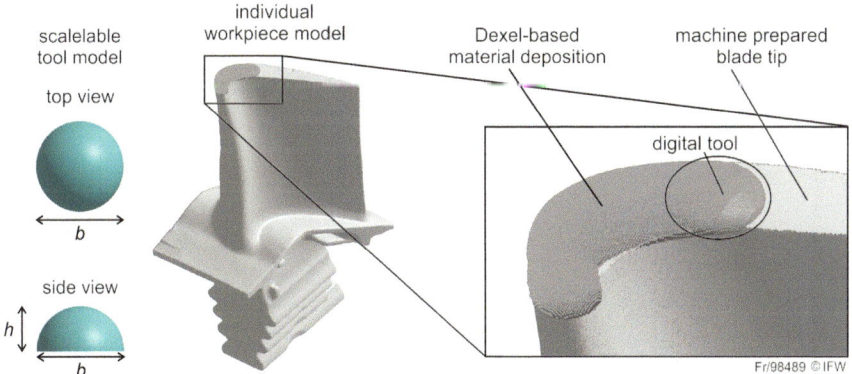

Fig. 7 Geometric simulation of the material deposition process

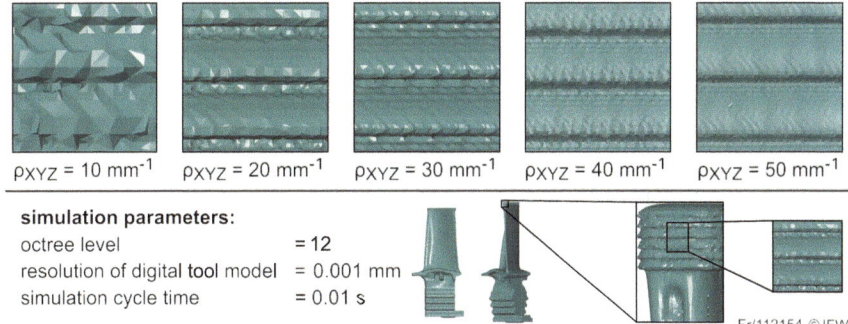

Fig. 8 Simulation accuracy in dependence on the Dexel-resolution

accuracy can be seen when the Dexel density is increased. The highest simulation accuracy was achieved with a Dexel density ρ_{XYZ} of 50 mm^{-1}. On the other hand, the computational effort increased from 15 MB ($\rho_{XYZ} = 10$ mm^{-1}) to 280 MB ($\rho_{XYZ} = 50$ mm^{-1}) of storage space for the Dexel-model.

An important parameter for the process planning of the recontouring process is the material allowance, which is a function of the weld geometry, the shape of the blade profile, and the tool path during the additive process. To plan the weld seam geometry knowledge-based, the influence of the weld seam width on the material allowance has to be investigated. For this purpose, geometric simulations of the material deposition process were carried out on a reference blade geometry with four different weld seam widths b_r of 1.4, 1.8, 2.2, and 2.6 mm. The material allowance was then determined, which is defined as the deviation between the nominal geometry and the actual geometry after the material deposition along the profile curve. The calculated material allowance over the profile length $x_{profile}$ of a blade section is shown in Fig. 9.

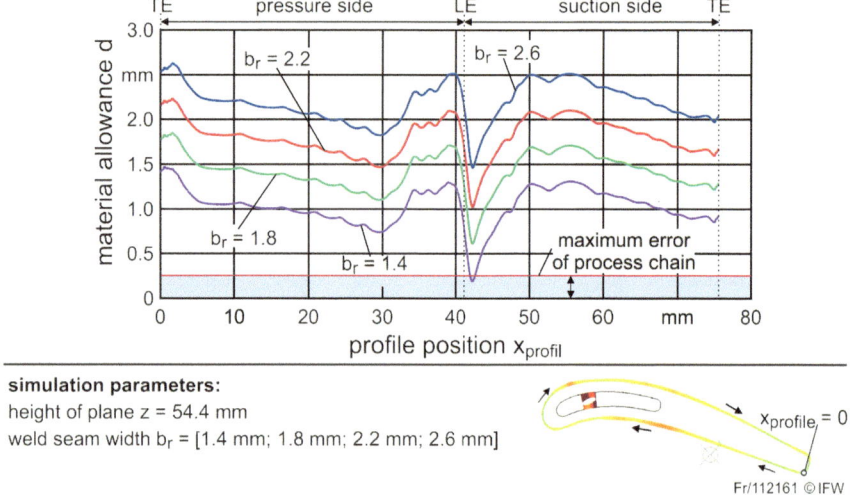

Fig. 9 Influence of weld seam width on the material allowance

The diagram shows a linear relationship between the weld seam width and the material allowance. In addition, the results show a high dependence of the local material allowance on the position along the blade profile. The variation along the length of the curve can be attributed to deviations due to abrupt changes in blade geometry that cannot be taken into account by the material application process, such as the cooling air holes in the area of the pressure side of the blade or high curvatures at leading or trailing edge. In order to achieve the planned target geometry after recontouring, the material allowance along the entire profile curve must not only be at least positive (>0 mm) but also higher as the maximum error of the process chain due to process-induced deviations, which is empirically $\pm 200\ \mu m$. This way, it can be ensured that the planned target geometry can be manufactured in the recontouring process. At the same time, the material allowance should be as small as possible to reduce cutting volume and consequently, tool wear and process time. Therefore, in this case, a weld seam width of 1.8 mm is mostly suitable with regard to an efficient recontouring process without any gaps due to missing material.

To improve the performance of the simulation, a new approach was developed which allows the determination of different material properties of the workpiece. This can be used to distinguish between the undamaged part of the blade and the material deposition (Denkena et al. 2011). Furthermore, a method for a local adaption of the Dexel-density was investigated in the second funding period. This allows the increase of the Dexel resolution within the machined area as well as a decrease within the machined area which leads to lower simulation times (Denkena et al. 2014b). In order to analyze the surface topography of the workpiece at a microscopic level, an abstracted cylinder model for the tool model comes to the limits. Thus, a higher resolution of the tool model is needed which is in conflict with an appropriate simulation time. To solve this problem, a novel simulation method was developed and

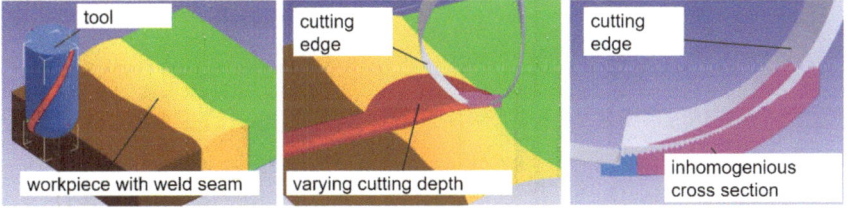

Fig. 10 Simulation with different materials and tool model with distinct cutting edges

researched. This method includes a simplified cylinder model in combination with a detailed tool model with distinct cutting edges. Thereby, the modeling of the cutting edges is only active when the tool comes in contact with the interfering geometry (patch or weld seam). This increases the accuracy with only a slight increase in calculation time (Denkena et al. 2015b) (Fig. 10).

An important advantage of the multi-Dexel-based additive process simulation is the ability to combine this method directly with the subsequent cutting process simulation. Using a coherent discretization model (multi-Dexel-model) for additive and subtractive process simulations allows an integrated digital process chain for hybrid or combined manufacturing (Fig. 11). This makes the approach ideal for the repair process of blades, whereby the influence of material deposition on the recontouring process can be investigated.

In respect to the following recontouring process, taking technological aspects into account, a higher material removal rate generally leads to higher process forces, and thus, to both higher mold defects as a result of displacement and increased tool wear. For the planning of the material deposition process, two goals can be stated which are building a trade-off. On one hand, a minimum material allowance should be planned in order to keep the material removal rate during recontouring as low as possible to reduce process forces (Fig. 11). On the other hand, the material allowance should be kept high enough to ensure a defect free manufacturing of the target geometry (Fig. 9).

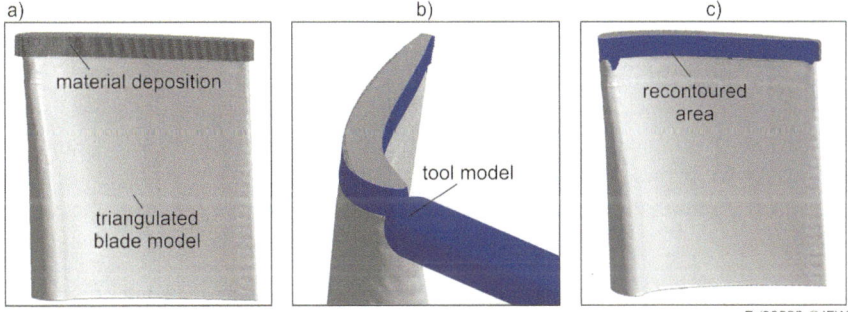

Fig. 11 Simulation results: **a** triangulated blade model with material deposition, **b** geometric simulation of recontouring, **c** recontoured blade model after simulation

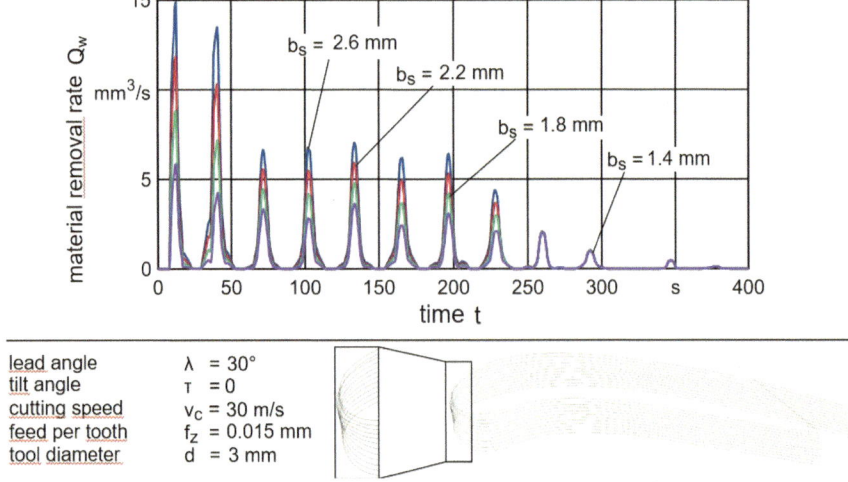

Fig. 12 Material removal rate during the recontouring process for different weld seam widths

For the generation of knowledge about the influence of the material deposition on the machining process, simulation experiments were carried out. The influence of the material deposition geometry on the material removal rate was analyzed for the recontouring of four different weald seam widths (Fig. 12). Thereby, the 5-axis tool paths were generated individually for the considered blade model with the adaptable process planning method with a line width br of 0.3 mm.

The twelve distinct peaks of the material removal rate along the process time are corresponding to the twelve planes of the tool path. The first path curve is at a height $z = 56.1$ mm, and thus slightly higher than the maximum of the material deposition. As a result, the tool is not engaged at the beginning of the first path and only begins to remove material in the area of the leading edge. This leads to a sudden increase in the material removal rate at the beginning. This is also the reason for the high maximum material removal rate of the first and second tool path curves.

In all paths, part of the material of the next paths is removed. This can be traced back to the diameter of the tool $d = 3$ mm. The larger the tool diameter and the larger the depth of cut a_p, the larger the area that is machined. As a result, the remaining material in the following paths decreases which leads to a steady decrease of the material removal rate. It can be observed, that the material removal rate is quite low in the last two paths. The slight irregularities in the material removal rate are due to minor deviations in the creation of the tool path which are resulting from the cooling holes.

Moreover, a clear difference in the material removal rate between the four weld seam widths can be recognized in the first eight paths. The generated knowledge can be used for the determination of the ideal material allowance and enhances the efficient planning of the deposition process.

3.3 Influence of the Process Parameters on the Main Residual Stresses

It is known from the literature that the welding process causes tensile residual stresses on the surface and inside the component due to the cooling or phase transformations (Dattoma et al. 2006; Zain-ul Abdeina et al. 2009). However, the interactions with the introduced residual stresses due to recontouring are unknown or neglected (Dattoma et al. 2006). For this reason, the influence of the process parameters of the recontouring process on the superficial residual stresses was investigated on Ti6Al4V in the first founding period, cf. Figure 13 (Nespor 2015).

The factor with the highest significance on the superficial residual stresses is the cutting edge rounding r_β of the tool, whose effect is about five times higher than for all factors. This means that a cutting edge rounding of $r_\beta = 30$ μm always shifts the residual stresses towards compression by about $\Delta\sigma \approx 200$ MPa, compared to a work-sharp tool. The factors cutting speed vc, depth of cut ap, down and up cut, clearance angle α, and rake angle γ show no significant influence on the residual stresses. Furthermore, the tooth per feed f_z, the step over b_r, and the lead angle λ have an influence on the residual stresses, which is, however, significantly lower than the cutting edge rounding.

For this reason, the effect of the cutting edge rounding is considered in particular below. In Fig. 14a), the cutting edge rounding r_β is plotted against the step over br and in Fig. 14b) against the tooth per feed f_z. The curves of the residual stresses σ1 under variation of f_z, b_r, and r_β are similar. With an increase of the tooth per feed f_z or the step over b_r, the residual stresses close to the surface shift slightly in the direction of tension. Further, when the cutting edge rounding r_β is increased, the compressive residual stresses increase and reach a minimum at about $r_\beta \approx 30$ μm.

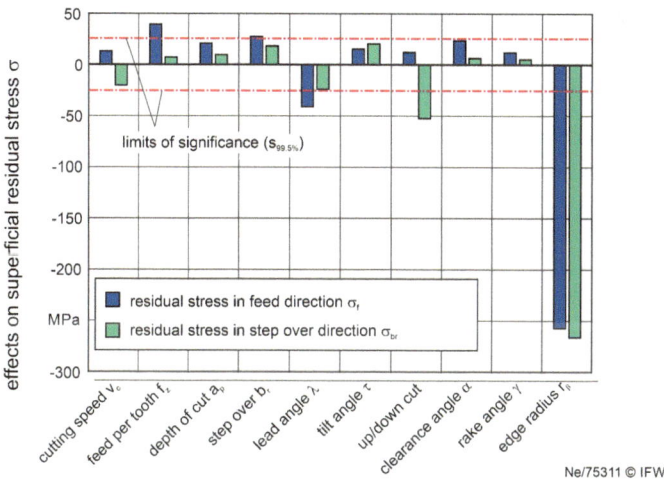

Ne/75311 © IFW

Fig. 13 Influence of the process parameters on the superficial residual stresses (Nespor 2015)

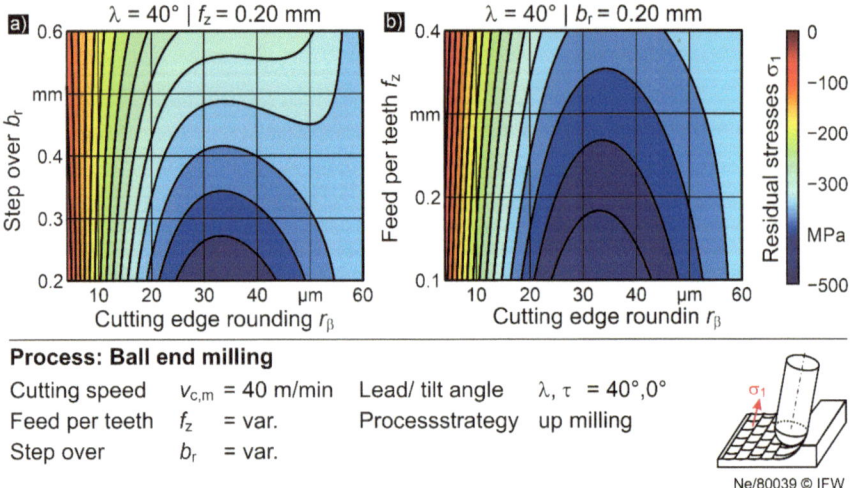

Fig. 14 Influence of the cutting edge rounding on the superficial residual stresses (Nespor 2015)

By further increasing the cutting edge rounding r_β, the residual compressive stresses at the surface decrease again. The residual stress minimum is for the work-sharp cutting edges and rounded tools up to $r_\beta \approx 30\,\mu m$ directly on the workpiece surface. The recommended cutting edge rounding for industrial use is therefore between 30 and 40 μm. The tooth per feed f_z and the step over b_r should be selected as small as possible depending on the required productivity to achieve residual compressive stresses.

The results showed that the process parameters have a significant influence on the residual compressive stresses. With this knowledge, the process can be controlled regarding higher residual compressive stresses which lead to positive material properties (Nespor 2015). In contrast, residual tensile stresses have a negative impact since they can lead to part distortion, especially for thin-walled titanium workpieces such as compressor blades. For this purpose, in the second funding period, a novel method for the compensation of part distortions within process planning for recontouring of thin walled workpieces was developed. The technique includes an optimization of tool angles in 5-axis recontouring. It is based on the recontouring areas as well as the simulated surface-generating cut volume (SGCV). The second part of the method relates to the safe implementation of the tool angles in an automatic tool path modification. The recontouring of two blades shows that the method allows a compensation of part distortion of 21% compared to a conventional process. Currently, the method is based on the equilibrium of internal loads (Böß et al. 2019).

3.4 Prediction of the Workpiece Surface

In addition to the residual stresses, the resulting workpiece surface is also of great significance in recontouring. When machining with ball-end cutters, burr formation also plays a decisive role. For this purpose, a new approach for the prediction of burr formation in 5-axis ball end milling was investigated in the second and third funding periods by applying an MRS considering the micro geometry dependent minimum chip thickness. Therefore, an existing MRS environment, which is able to include dynamic tool deflection based on the tool engagement, was extended by the theory of the plastically deformed volume Q_e, which is penetrated by the cutting edge with an active chip thickness below the minimum chip thickness. Using this theory, multiple cutting processes with different tool orientations and process parameters were investigated and the occurring surface defects are correlated with the previously determined volume Q_e.

To consider the workpiece surface generation effects in process planning, a time-discrete MRS, which uses a multi-Dexel model to discretize the workpiece, was applied. The MRS is described in (Denkena et al. 2019). The Dexel resolution used is approximately 1 μm/Dexel in each direction. The geometries of the rake and flank faces of the tool were derived from a measurement of the tool geometry using the Alicona Infinite Focus G5. The rake and flank faces of the two fluted cemented carbide tools were extracted from the measured geometry and modelled discretely using quadrilaterals with edge lengths of the elements of dS = 0.1 mm (Fig. 15, left). By using the force model of Engin and Altintas (2011) and the consideration of the structural dynamic of the machine tool and workpiece, the simulation is capable of considering the dynamic behaviour for stable and unstable processes as shown in (Denkena et al. 2020). Moreover, the simulation can include the influence of the cutting edge roughness on the surface generation for a more accurate prediction of the milled surface. An example of the surface prediction using the MRS is shown in Fig. 15. Figure 15a) shows the surface of the milling experiment. With the simulation the surface topography can be accurately predicted by taking into account runout error and dynamic displacement of the tool, Fig. 15b). By additionally including the cutting edge roughness in the tool model of the simulation, the characteristic grooves of the workpiece surface can be displayed, Fig. 15c). Overall, it can be stated that the simulation is a suitable tool for the prediction of the milled surface topography of 5-axis ball end milling with arbitrary geometric engagement situations.

Furthermore, burrs can also form on the surface of the workpiece during machining. For the prediction of burr formation, it is important to know which factors have an influence on the development of burr formation and under which conditions burrs may appear. Experimental studies were carried out to investigate burr formation. The tool orientation, the tooth per feed f_z, and the strategy of up and down milling were varied, thus changing the engagement conditions. The resulting generated surfaces were then evaluated by means of video microscopy. The topographies show that the tooth per feed has a significant influence on the formation of the burr on the milling grooves. Thus, at f_z = 0.06 mm, strong burr formation is evident,

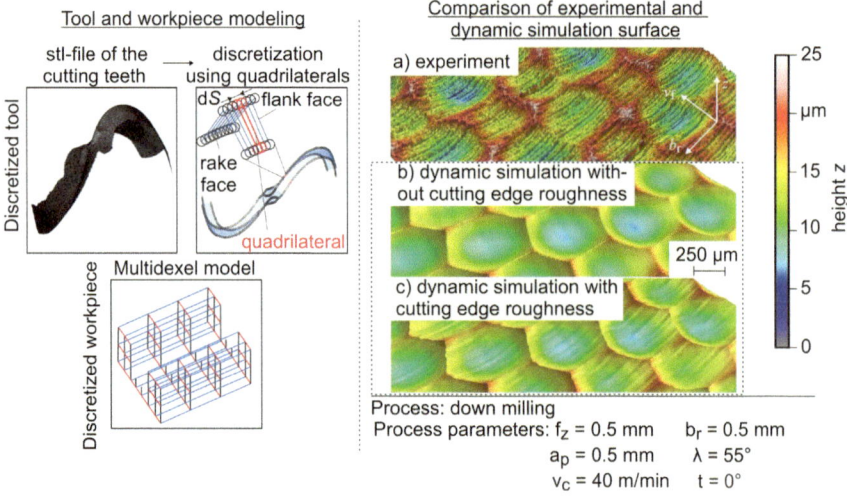

Fig. 15 Dynamic simulation of the workpiece surface

while at $f_z = 0.12$ mm, burr formation is significantly reduced. If a tooth per feed of $f_z = 0.18$ mm is used, burr formation is no longer noticeable (Fig. 16).

Figure 17 shows video microscope images of surface topographies for three different lead angles $\lambda = 15°$, $\lambda = 30°$ and $\lambda = 45°$. For the cutting edge rounding $SS = 15$ μm, a slight burr is visible on the surface at a lead angle of $\lambda = 15°$, despite the use of a larger tooth feed of $f_z = 0.21$ mm compared to the result shown above in Fig. 16 with $f_z = 0.18$ mm. With increasing lead angle, $\lambda = 30°$, no burr can

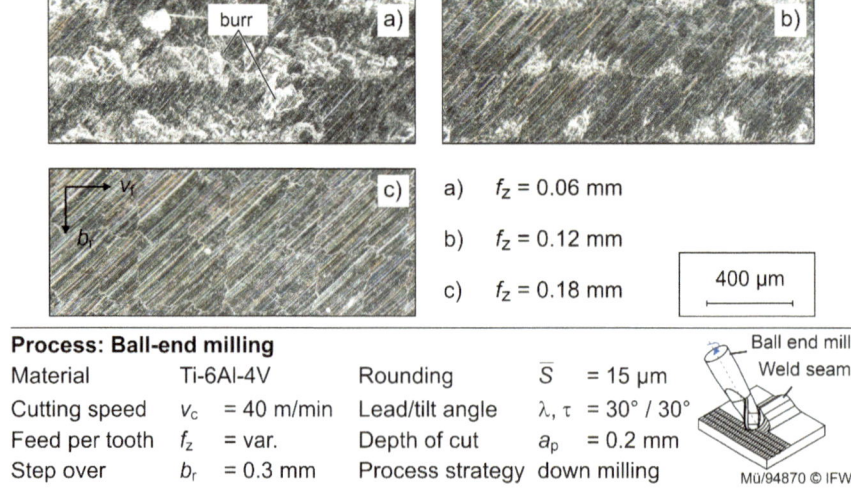

Fig. 16 Burr as a function of the tooth feed (Muecke 2020)

be observed in this example. For $SS = 30$ μm, strong burr formation can be seen at the edges of the milling grooves at both $\lambda = 15°$ and $\lambda = 30°$. If the lead angle is increased to $\lambda = 45°$, the burr can be significantly reduced. After considering all angle combinations and this example, it becomes clear that no universal recommendation can be derived for the choice of tool orientation to reduce burr. With regard to the selected strategy, down or up milling, no general statement can be made either, because there are interactions of the strategy with the tool orientation with regard to the down or up milling parts of the process. At this point, it can only be said that burr formation increases with increased cutting edge rounding and is influenced by the tool orientation and the strategy. Therefore, for the process design to avoid burrs, an individual investigation of the respective process parameter combination is necessary when rounded tools are used. Accordingly, knowledge about the causes of burr formation is required, for which the material removal simulation to be developed is used.

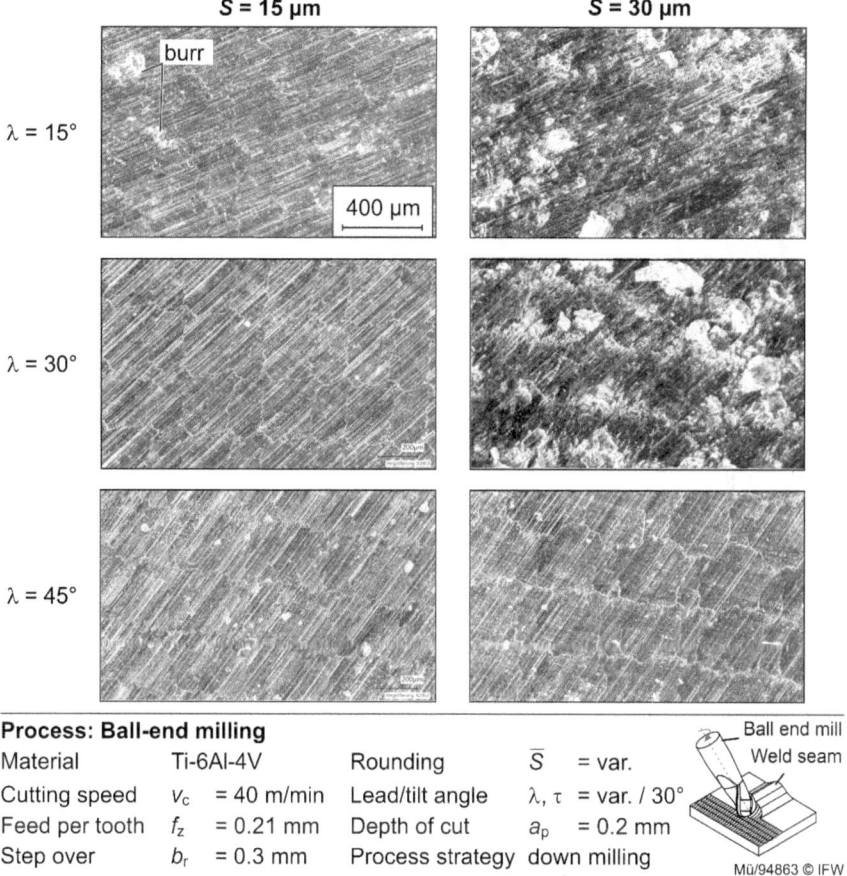

Fig. 17 Burr as a function of the cutting edge rounding and the tool orientation (Muecke 2020)

Plastic surface deformations such as burr formation emerge before the chip formation begins and material is ploughed underneath the cutting edge rounding (Wyen
2011). Therefore, knowledge about the minimum chip thickness depending on the
cutting edge rounding is needed. Further, for ball end milling, it has to be taken into
account that the final surface is generated only at a small portion of the tooth engagement. Consequently, only those periods of the tooth engagement are of interest for
burr formation where the final surface is generated. To determine this specific time
interval, a method was developed that allows the calculation of the surface-generating
periods using the MRS. The method is explained in detail in (Denkena et al. 2021).
With the help of this method it was possible to find out that the time of the final
surface generation is depicted within one tooth engagement and correlated with the
timely progression of the maximum undeformed chip thickness h_{max}. The surface
generation period only depends on b_r and v_c and is independent of f_z. Significant
burr formation is expected to happen from the beginning of the tooth engagement
up to the point where the minimum chip thickness is exceeded and the chip formation begins. The point t_{sp} (Fig. 18) marks the intersection between the maximum
undeformed chip thickness in the process and the minimum chip thickness. It has to
be mentioned that the maximum undeformed chip thickness during the time of the
surface generation is not necessarily reached by a cutting edge element that generates
the final surface. However, for the considered application this simplification is valid
due to the low depth of cut and the orientation of the helix angle.

Therefore, during the time of the surface generation, most of the engaged cutting
elements contribute to the surface generation. For small cutting edge roundings SS,
t_{sp} is reached soon after the tooth entry. With increasing SS and decreasing feed per
tooth f_z the intersection point is reached later in the process. With large cutting edge
roundings, the minimum chip thickness is exceeded after the surface generation.
Therefore, no chip formation occurs during the surface generation and this often
results in burr formation. To evaluate the occurrence of burrs based on a characteristic

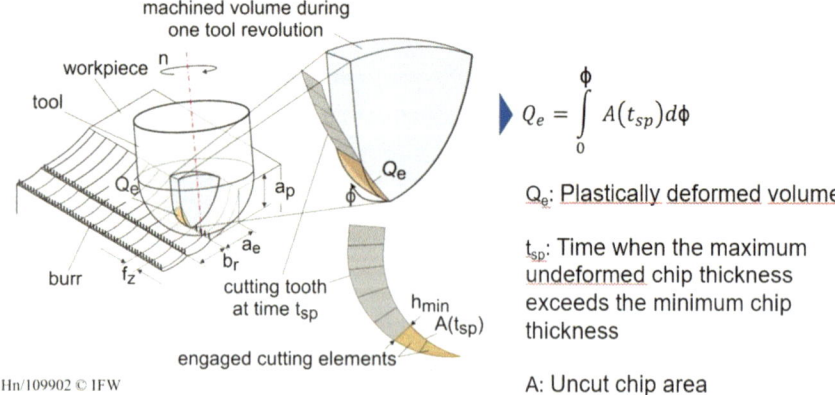

$$Q_e = \int_0^\phi A(t_{sp})\,d\phi$$

Q_e: Plastically deformed volume

t_{sp}: Time when the maximum undeformed chip thickness exceeds the minimum chip thickness

A: Uncut chip area

Fig. 18 Determination of the burr with the help of the material removal simulation (Denkena et al. 2021)

value, the plastically deformed volume Q_e is introduced. This value represents the engagement volume of the tooth up to the beginning of chip formation. As visualized in Fig. 18, Q_e can be calculated by integrating the uncut chip area A from the beginning of surface generation ($t = 0$) to the point, where the minimum chip thickness is reached ($t = t_{sp}$). For this calculation, a MRS can be utilized. A larger value of Q_e correlates with a higher volume portion deformed underneath the flank face.

To verify the use of Q_e for the prediction of plastic surface defects, Q_e was calculated with the MRS for experimental cases. The calculated Q_e and the corresponding experimental surfaces are shown in Fig. 19.

An increase in the feed per tooth results in decreasing values of Q_e because the minimum chip thickness is exceeded earlier during the cut. Furthermore, the applied variation in tool orientation (λ, t) influences Q_e based on the kinematics and the resulting temporal progression of the chip cross-section. The most significant influencing factor is the applied cutting edge rounding size, as it influences the value of the minimum chip thickness itself. The depicted surface plots show a trend of increasing surface defects with increasing Q_e. However, the resulting surface debris, which deviates from the deterministic structure induced by the milling kinematics, is not homogenously distributed over the surface or within one feed track. A reason for the inhomogeneous nature of the remaining burrs on the surface is on one hand the interaction between subsequent milling tracks. Thereby, the milling tool will remove potential burrs from the previous milling track. On the other hand, all specimens were subject to an ultrasonic cleaning bath before the surface investigations. Taking into account this considerations, burrs are also influenced by stochastic effects such as local differences in workpiece microstructure.

4 Conclusions

With regard to the requirements for the resulting workpiece surface, it was possible to determine that residual compressive stresses can be achieved with the use of ball end mills. Here it could also be shown that a medium cutting edge rounding, low feed per tooth, and low step-over are advantageous. In order to achieve a low surface roughness, a large cutting edge rounding, low feed per tooth and low step-over should be used. Furthermore, burr formation should also be avoided. Small cutting edge roundings, high feed per tooth, and larger lead and tilt angle are suitable for this purpose. Here it becomes clear that a compromise must be made for the choice of the process parameters. With the help of the knowledge generated, however, a process with reduced burr formation and good surface quality can be designed, in which high residual compressive stresses are generated.

Furthermore, during the entire funding period, a Dexel-based material simulation method was developed and parametrized by means of welding experiments. With the method, it is possible to predict the geometry of the material deposition process. With the help of the investigation carried out the Dexel-based material deposition simulation allowed the investigation of the influence of the material allowance on the

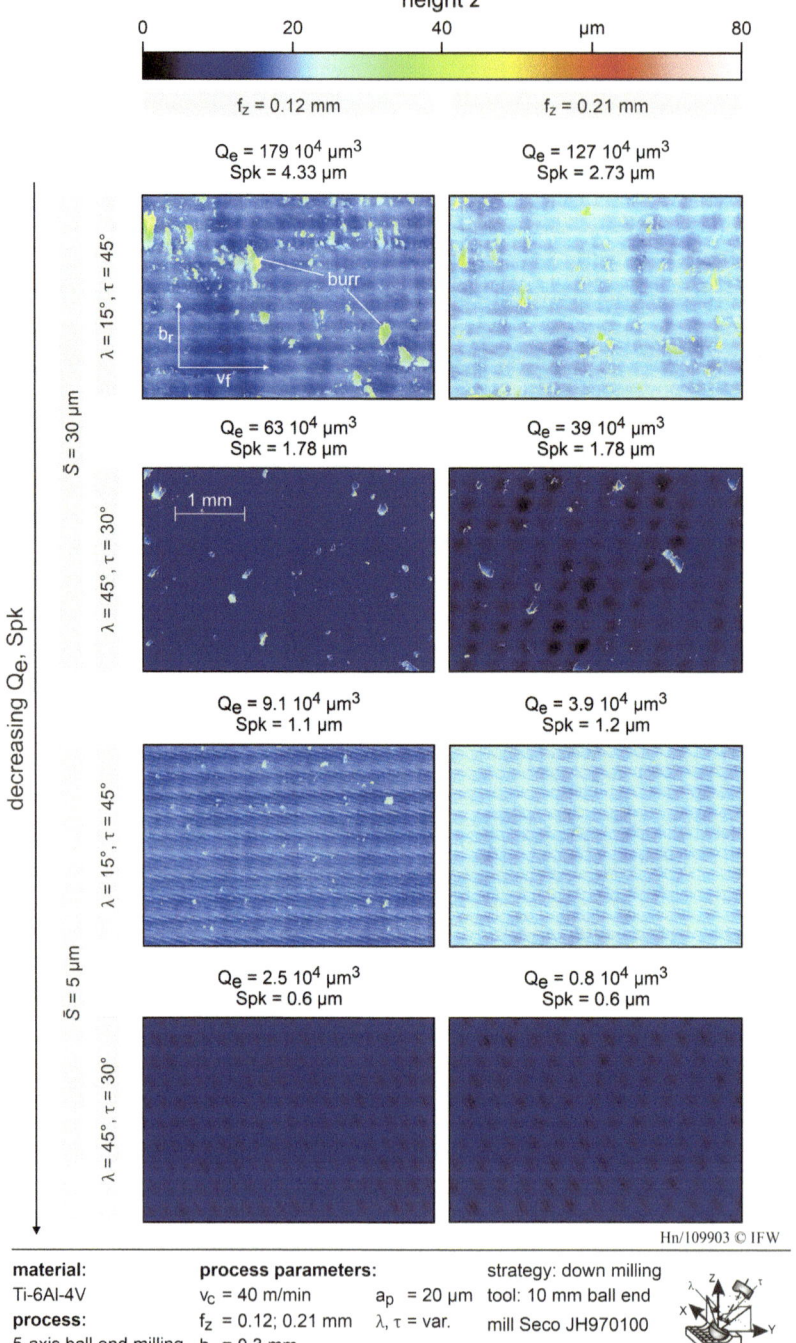

Fig. 19 Machined surfaces and the resulting Q_e (Denkena et al. 2021)

recontouring process in order to identify suitable process parameters. In addition, by means of the adaptive planning method, it is possible to generate the NC-code for the machining individually for each blade and in an automated way.

Acknowledgements Funded by the Deutsche Forschungsgemeinschaft (DFG, German Research Foundation) SFB 871/3 119193472.

References

Abuaf, N., Bunker, R. S., and Lee, C. P. (1998). Effects of Surface Roughness on Heat Transfer and Aerodynamic Performance of Turbine Airfoils, ASME Journal of Turbomachinery; 120, 522–529.

Arizmendi, M., Fernandez, J., Lopesz de Lacalle, L., Lamikiz, A., Gil, A., Sanchez, J., Campa, F., and Veiga, F. (2008). Model development for the prediction of surface topography generated by ball-end mills taking into account the tool parallel axis offset. experimental validation, CIRP Annals - Manufacturing Technology; 57(1), 101–104.

Aurich, J. C., Dornfeld, D., Arrazola, P. J., Franke, V., Leitz, L., and Min S (2009). Burrs – Analysis, control and removal, CIRP Ann Manuf Technol, 58, 519–542.

Boese, M. (2002). Effects of Riblets on the Loss Behaviour of a Highly Loaded Compressor Cascade, Proceedings of ASME Turbo Expo.

Bons, J. P. (2010). A Review of Surface Roughness Effects in Gas Turbines, ASME Journal of Turbomachinery; 132(2).

Böß, V., Rust, F., Dittrich, M., and Denkena, B. (2019). Compensation of part distortion in process design for recontouring processes. Procedia CIRP. 81. 820–825.

Böß, V., Denkena, B., Dittrich, M. A., Malek, T., and Friebe, S. (2021). Dexel-Based Simulation of Directed Energy Deposition Additive Manufacturing. J. Manuf. Mater. Process, 5, 9.

Bouzakis, K. D., Aichouh, P., and Efstathiou, K. (2003). Determination of the chip geometry, cutting force and roughness in free form surfaces finishing milling, with ball end tools, International Journal of Machine Tools & Manufacture 43, 499–514.

Bremer, C. (2006). AROSATEC (Automated Repair and Overhaul System for Aero Turbine Engine Components). Final Report, BCT Alround ISQ Metris MTU Sifco Skytek.

Chen, J., Huang, Y., and Chen, M. (2005). Study of the surface scallop generating mechanism in the ball-end milling process, International Journal of Machine Tools and Manufacture 45, 9, 1077–1084.

Chen, M. J., Ni, H. B., Wang, Z. J., and Jiang, Y. (2012). Research on the modeling of burr formation process in micro-ball end milling operation of Ti-6Al-4V, Int J Adv Manuf Technol, 62, 901–912.

Chu, C., and Dornfeld, D. (2002). Tool path planning for avoiding exit burrs, Journal of Manufacturing Processes, 2, 116–123.

Dattoma, V., De Giorgi, M., and Nobile, R. (2006). On the evolution of welding residual stress after milling and cutting machining. Computers and Structures, 84, 29-30, 1965–1976.

Denkena, B. and Biermann, D. (2014). Cutting edge geometries, CIRP Ann Manuf Technol, 63, 631–653.

Denkena, B., Böß, V., Nespor, D., and Samp, A. (2011). Kinematic and Stochastic Surface Topography of Machined TiAl6V4-Parts by means of Ball Nose End Milling, Procedia Engineering, 19, 81–87.

Denkena, B., Nespor, D., Böß, V., and Köhler, J. (2014a). Residual stresses formation after recontouring of welded Ti-6Al-4V party by means of 5-axis ball nose end milling, CIRP J Manuf Sci Technol, 7, 347–360.

Denkena, B., Rust, F., Böß, V., Nespor, D., and Flöter, F. (2014b). Approaches for improving cutting processes and machine tools in recontouring, Prcedia CIRP, 22, 239–242.

Denkena, B., Böß, V., Nespor, D., Flöter, F., and Rust, F. (2015a). Engine blade regeneration: a literature review on common technologies in terms of machining. International Journal of Advanced Manufacturing Technology, 81, 917–924.

Denkena, B., Böß, V., Nespor, D., and Rust, F. (2015b). Simulation and evaluation of different process strategies in a 5-axis recontouring process, Procedia CIRP, 35, 31–37.

Denkena, B., Grove, T., and Pape, O. (2019). Optimization of complex cutting tools using a multi-dexel based material removal simulation, Procedia CIRP, 82, 379–382.

Denkena, B., Pape, O., Krödel, A., Böß, V., Ellersiek, L., and Mücke, A. (2020). Process design for 5-axis ball end milling using a real-time capable dynamic material removal simulation, Production Engineering, 15, 89–95.

Denkena, B., Krödel, A., Mücke, A., and Ellersiek, L. (2021). Prediction of plastic surface defects for 5-axis ball end milling of Ti-6Al-4V with rounded cutting edges using a material removal simulation, CIRP Annals Manufacturing Technology, 70, 91–94.

Denkena, B., Böß, V. (2009). Technological NC Simulation for Grinding and Cutting Processes Using CutS. In Proceedings of the 12th CIRP Conference on Modelling of Machining Operations, Donostia-San Sebastian, Spain, 7–8, 563–566.

Eberlein, A. (2007). Phases of high-tech repair implementation. 18th International Symposium on Airbreathing Engines (ISABE).

Engin, S., and Altintas, Y. (2011). Mechanics and dynamics of general milling cutters. International Journal of Machine Tools and Manufacture, 41, 2195–2212.

Gao, J., Chen, X., Yilmaz, O., and Gindy, N. (2008). An integrated adaptive repair solution for complex aerospace components through geometry reconstruction. The International Journal of Advanced Manufacturing Technology 36(11-12), 1170–1179.

Hohenstein, S. and Seume, J. (2013). Numerical Investigation on the Influence of Anisotropic Surface Roughness on the Skin Friction, Proceedings of the European Turbomachinery Conference.

Knobel, P. P. (2000). Fräsen von Freiformflächen mit Schleifqualität (milling of free form surfaces with grinding quality), Ph.D. Thesis, Eidgenössische Technische Hochschule Zürich.

Layegh, S. and Lazoglu, I. (2017). 3D surface topography analysis in 5-axis ball-end milling. CIRP Ann Manuf Technol, 66,133–136.

Link, R. (1992). Gratbildung und Strategien zur Gratreduzierung, Dr.-Ing. Dissertation, RWTH Aachen.

Liu, N., Loftus, M., and Whitten, A. (2005). Surface finish visualisation in high speed, ball nose milling applications, International Journal of Machine Tools and Manufacture 45, 10, 1152–1161.

Markworth, L. (2005). Fünfachsige Schlichtfräsbearbeitung von Strömungsflächen aus Ni-basislegierungen (five-axis finishing of nickelbase airfoils), Ph.D. Thesis, RheinischWestfälische Technische Hochschule Aachen.

Muecke, A. (2020) Gestaltabweichungen nach der Rekonturierung reparaturgeschweiß- ter Bauteile. Ph.D. Thesis, Leibniz University Hannover, Hannover.

Nespor, D. (2015). Randzonenbeeinflussung durch die Rekonturierung komplexer Investitionsgüter aus Ti-6Al-4V. Ph.D. Thesis, Leibniz Universität Hannover, Hannover.

Rong, Y., Xu, J., and Sun, Y. (2014). A surface reconstruction strategy based on deformable template for repairing damaged turbine blades. Proceedings of the Institution of Mechanical Engineers, Part G: Journal of Aerospace Engineering 228(12), 2358–2370.

Schlobohm, J., Li, Y., Pösch, A., Langmann, B., Kästner, M., and Reithmeier, E. (2014). Multiscale Optical Inspection Systems for the Regeneration of Complex Capital Goods. Procedia CIRP 22, S. 243–248.

Sokolowski, A., Narayanaswami, R., and Dornfeld, D. (1994). Prediction of Burr Size using Neural Networks and Fuzzy Logic, Proceedings of the Japan-USA Symposium on Flexible Automation, ISCIE, 889–896.

Uhlmann, E., Bilz, M., and Baumgarten, J. (2013). MRO – Challenge and Chance for Sustainable Enterprises. Procedia CIRP, 11, 239–344.

Vakondios, D., Kyratsis, P., Yaldiz, S., Antoniadis, A. (2012). Influence of milling sratcgy on the surface roughness in ball end milling of the aluminium alloy Al7075-T6, Measurement, 45(6), 1480–1488.

Wu, H., Gao, J., Li, S., Zhang, Y., and Zheng, D. (2012). A Review of Geometric Reconstruction Algorithm and Repairing Methodologies for Gas Turbine Components. Indonesian Journal of Electrical Engineering and Computer Science 11, 1609–1618.

Wyen, C. (2011). Rounded cutting edges and their influence in machining titanium, Doctor of Sciences dissertation, ETH Zurich.

Yilmaz, O., Gindy, N., and Gao, J. (2010). A repair and overhaul methodology for aeroengine components. Robotics and Computer-Integrated Manufacturing 26, 190–201.

Zain-ul Abdeina, M., Neliasa, D., Jullien, D., Deloisonb, J. (2009). Prediction of laser beam welding-induced distortions and residual stresses by numerical simulation for aeronautic application. Journal of Materials Processing Technology, 209, 6. 2907–2917.

Fast Measurement of Complex Geometries Using Inverse Fringe Projection

Philipp Middendorf, Markus Kästner, and Eduard Reithmeier

Abstract The inspection in confined spaces, for instance inside aircraft engines, is currently performed manually, since the inspection approaches cannot be sufficiently automated. Using a novel sensor system based on the borescopic fringe projection method, such small installation spaces can be inspected with high precision 3D measurements. This provides a basis for a standardization of the inspection processes during maintenance cycles. In order to automate the inspection process, an approach to plan measurement strategies based on ray tracing simulations of the optical measurement is presented. By taking multiple reflections and the corresponding reconstruction failures into account suitable measurement poses are identified. Finally, an in-situ measurement approach to assess the condition of (aero engine) turbine blades and derived damages is presented.

Keywords Borescopic fringe projection · Turbine blade inspection · In-situ measurement · Ray tracing · Multiple reflections · Virtual fringe projection

1 Introduction

Project C2 is located in the area of early fault detection within the process chain of turbine maintenance of CRC 871. Over the course of the previous two funding periods, research was conducted on a novel type of sensor technology for inspection in confined spaces. The measurement method of fringe projection was transferred to new scales and applications by means of a borescopic structure (Schlobohm et al. 2015, 2016; Pösch et al. 2017). This enables optical 3D measurement in areas that are difficult to access with limited space for movement. In particular, the current advancement in the field of smartphone cameras has made the implementation of miniature sensors possible. The use of these sensors for high-precision geometric

P. Middendorf (✉) · M. Kästner · E. Reithmeier
Institute for Measurement and Automatic Control, Leibniz University Hannover, An der Universitaet 1, 30823 Garbsen, Germany
e-mail: philipp.middendorf@imr.uni-hannover.de
URL: https://imr.uni-hannover.de

© The Author(s) 2025
J. R. Seume et al. (eds.), *Regeneration of Complex Capital Goods*,
https://doi.org/10.1007/978-3-031-51395-4_14

271

component characterization will be demonstrated in this paper using the example of aero engine inspection.

2 Objective

The aim of this subproject is to develop a fast inspection approach for the inspection of complex geometries. For this purpose, the inverse fringe projection method was initially researched and used to perform rapid condition assessments of aircraft engine blades (Pösch et al. 2012; Schlobohm et al. 2017b, a). However, a precise metric derivation of defect sizes is not possible based on a single inverse pattern. Furthermore, precise knowledge of the orientation and geometry of the measured object is required. Due to rapid technical developments in fields of cameras, projectors and computing power, the fringe projection measurement method has caught up with the inverse fringe projection approach for rapid inspection. Today, high-speed fringe projection measuring systems can be built that can perform 3D measurements within one second for high-precision damage analysis. With these, it is even possible to perform handheld data acquisition at reduced accuracy (Matthias et al. 2018).

Significant advancement in the field of miniaturization of camera sensors have also been made due to the needs of the smartphone industry. In addition to the inspection of complex components, this also enables the development of 3D sensors for inspections in confined spaces. In particular, the early fault detection of complex capital goods such as blisks and turbine blades has been identified as an application area. Precise 3D inspection during a maintenance interval on assembled and disassembled aircraft engines is needed to investigate safety-critical aspects and prevent unnecessary and expensive engine disassembly and repair. Accordingly, the objective is to develop miniaturized 3D measuring systems based on the fringe projection method for the purpose of early fault detection. Two different inspection tasks within the maintenance process were defined for this specific application. On the one hand, the particularly challenging inspection of the assembled engine via maintenance openings has to be targeted. On the other hand, blisks and turbine blades of the partially disassembled engine have to be inspected. The two inspection tasks present different challenges for the development of the measuring systems. Therefore measuring head sizes of less than 10 mm in diameter are required and the working distance of the sensor is within the range of 10–20 mm or 40–60 mm. Miniaturized camera sensors in particular are more sensitive to malfunctions than industrial camera sensors and require a suitable measurement strategy with appropriate measurement poses when used within optical sensors. To achieve intelligent measurement pose planning, the reflective properties of the measurement object must be taken into account in addition to the sensor-specific requirements. Especially multiple reflections on highly reflective (shiny) components can lead to faulty reconstructed points within the measurement. For this purpose, a GPU-based simulation approach for determining low-reflection measurement poses and a compensation approach for the error-causing reflections are presented.

This article is structured as follows: First, the concept of a borescopic fringe projection sensor and its challenges during miniaturization are presented. Then a simulation approach to plan suitable measurement strategies is introduced and finally the measurement capabilities and results of the in-situ inspection are demonstrated.

3 Borescopic Fringe Projection Sensor

3.1 Design

For the adaptation of the fringe projection measuring technique to another scale range and the application in confined spaces, the classic camera projector design is adapted. In order to obtain such small measuring heads, miniaturized cameras in the "Chip-on-the-Tip" design are used instead of industrial cameras. To enable the projection of fringe patterns into small installation spaces, a borescope including a lens is used within the projection path. By projecting sinusoidal patterns through a borescope, the sensor head can be spatial separated from the projector and frame grabber board of the camera. Two iterations of borescopic sensors were developed within this subproject. First, the proof of concept for this device class of measurement systems with a measuring head diameter of about 10 mm was designed (Fig. 1 right). As the project progressed, technical advances in the camera sector enabled additional miniaturization of the borescope sensor with a measuring head diameter of about 6.5 mm (Fig. 1 left).

Figure 2 shows a schematic of the measuring head of both sensors. Here the green cone visualizes the field of view of the projector and the pyramid the field of view of the camera. The camera is fixed to the borescope shaft by a customized design of a 3D print clip. This can be flexibly adapted according to the required triangulation basis and thus adjusts the working distance of the measuring system. The two developed measurement systems are based on the components presented in

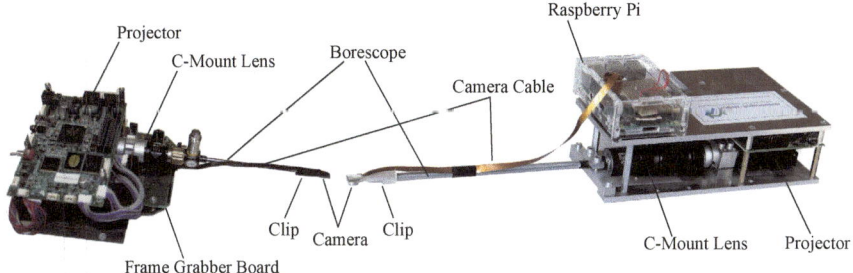

Fig. 1 Borescopic fringe projection systems with a Chip-on-the-Tip camera. Left: 6.5 mm measuring head with a 1/6" sensor and a (ø = 4 mm) borescope, right: 10 mm measuring head with a 1/4" sensor and a (ø = 6.5 mm) borescope

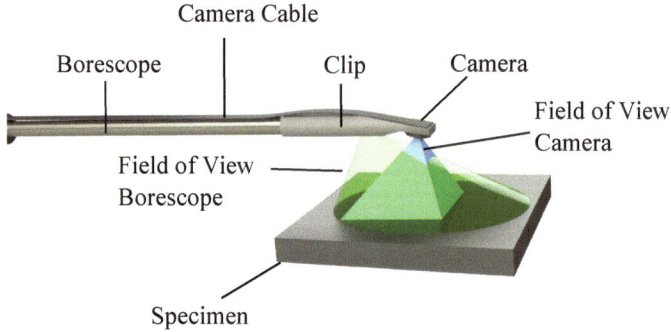

Fig. 2 Essential components and field of view of the measuring head of a borescopic fringe projection sensor. Image according to Middendorf et al. (2022b)

Table 1 Components of both measuring systems. The 6.5 mm measuring head corresponds to the measuring system shown left in Fig. 1

Component	ϕ 6.5 mm measuring head	ϕ 10 mm measuring head
Borescope length	110 mm	200 mm
Borescope ϕ	4 mm	6.5 mm
Camera	OV2740 (1/6")	OV5647 (1/4")
Camera resolution	1928 × 1088	1294 × 972 (binning)
Camera-lens	f = 1.83 mm / F8.0	f = 3.6 mm/F2.9
Camera name	MP-FPC31105-18,350–200	RPI-SPYCAM
Projector	Ti DLP 4500 EVM	Ti DLP 4500 EVM
Borescope-lens	f = 38 mm	f = 38 mm
Borescope view direction	70°	70°
Borescope aperture angle	80°	80°

Table 1 both camera sensors were manufactured by OmniVision Technologies, Inc (Santa Clara, United States). For the projection of the sinusoidal patterns, an evolution module 4500 from Texas Instruments Inc (Dallas, US) was used. The micromirror device forms the fringe pattern by binary tilting each individual micro mirror.

3.2 Miniaturization

In addition to inspection in confined spaces, a miniaturized measuring head has been developed to enable the inspection of maintenance openings in engines. Due to the continuously shrinking size of electrical components and sensors, the quality of the optical measurement also decreases and thus the suitability for use in optical measurement systems. In order to investigate the most relevant properties of the sensors and the influence of miniaturization, analyses on the signal to noise ratio, the edge spread function and spatial frequency response of the cameras are performed below. In addition, the illumination homogeneity, the distortions of the camera-projector pair and the measurement uncertainty of the two measurement systems are presented.

Camera Noise

The signal to noise ratio (SNR) of a camera indicates the ability to differentiate phase information retrieved from the sinusoidal patterns and noise. The noise of an image sensor results from the photon shot noise, sensor read noise, fixed pattern noise, thermal noise, pixel response non-uniformity and quantization noise. In order to determine a corresponding SNR in practical application, Eq. 1 is used to describe the SNR in context of read-and shot-noise. According to the approach of Martinec (2008), the summed intensity of two images ($\mu_{summedImage}$) was related to the standard deviation of the difference of the images ($cr_{differenceImage}$).

$$SNR_{Read,Shot} = \frac{\sqrt{2}\mu_{summedImage}}{2\sigma_{differenceImage}} \tag{1}$$

This test requires the cameras to be out of focus and acquire multiple images under constant illumination conditions. The exposure time is increased incrementally during the test to use the full intensity range of the sensor. In order to create comparative conditions and compensate for the inhomogeneous illumination, regions of interest (ROI) were analyzed in each image. Based on the Fig. 3, it can be shown that miniature sensors (OV5647 and OV2740) with one-piece injection-molded lenses are significantly more susceptible to noise compared to industrial cameras with high-quality lenses (CB120).

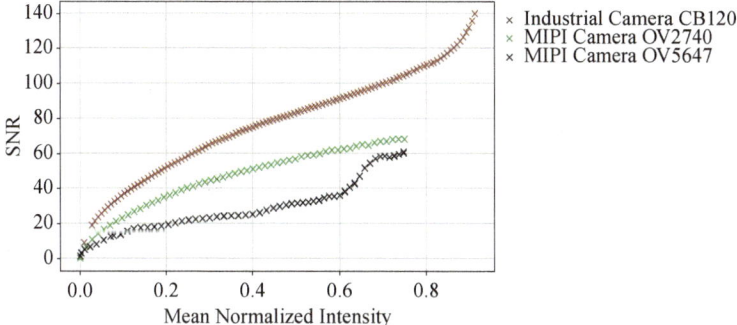

Fig. 3 SNR of both miniaturized image sensors (OV2740, OV5647) and industrial sensor (CB120) for comparison. Results published in Middendorf et al. (2021a)

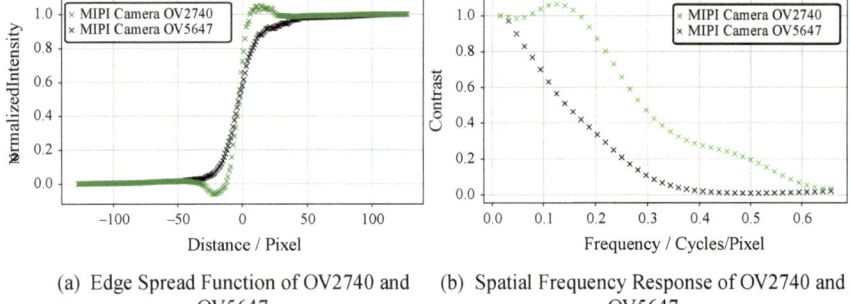

(a) Edge Spread Function of OV2740 and OV5647

(b) Spatial Frequency Response of OV2740 and OV5647

Fig. 4 Edge spread function and spatial frequency response of the miniaturized camera sensors. Results published in Middendorf et al. (2021a)

Spatial Frequency Response

The spatial frequency response (SFR) is one of the most important quality metrics in the camera sector since it quantifies the extent to which a camera and lens system can resolve image details. The slanted edge method (SEM) according to the implementation of van den Bergh (van den Bergh 2019, 2018) is used for this analysis. This approach is robust against distortions, which is particularly relevant for miniaturized systems. For the evaluation according to van den Bergh, the cameras were aligned in their focal plane and the intensity gradients at the edges of black rectangles on white background were examined (test target USAF1951 from *Thorlabs, Inc. (Newton, United States)*). Figure 4a therefore shows the Edge Spread Function (ESF) and Fig. 4b the SFR of the miniaturized camera sensors.

The plot of the ESF depicts the intensity curve over an edge (of a black rectangle) normalized to the average of white and black areas. The sensor OV5647 overshoots and undershoots at the edge, which looks similar to an unsharp masking effect. This behavior is not to be expected, the ESF of the OV2740 and other industrial sensors record a smooth curve and therefore behave approximately ideal. Internal signal processing within the sensor (OV5647) can probably explain this effect. The plot of SFR shows the resulting modular transfer function (MTF) as the contrast over the spatial frequency related to the sensor pixel. The MTF_{50} is usually used to compare optical systems regarding the local frequency to which a sensor images sharply. By the example of sensor OV2740, the MTF_{50} is about $0.15 \, ^{Cycles}$ and the imageable frequency is 288 (pixel count of 1920). In the context of fringe projection, the OV5647 is able to display higher frequencies more accurate than the OV2740, but the OV2740 performs more traceable and can reproduce intensity transitions during fringe projection measurement more accurate.

Homogeneity

The homogeneity of the combined sensor systems is also analyzed, since strong differences in exposure are caused by different lenses, working distances and

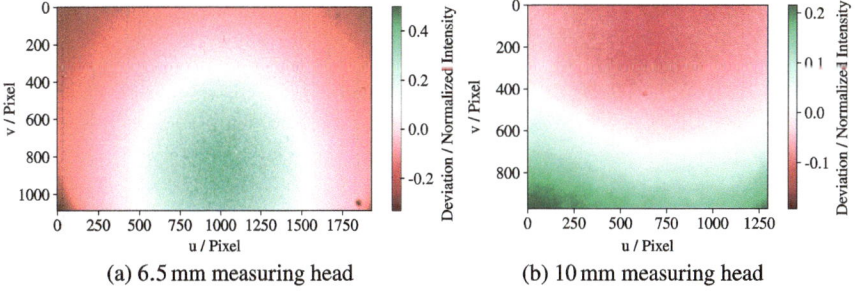

(a) 6.5 mm measuring head   (b) 10 mm measuring head

Fig. 5 Homogeneity of both miniaturized sensors. Results published in Middendorf et al. (2021a)

borescopes. Especially the need of high dynamic range approaches can be evaluated. The homogeneity evaluation is carried out using a white photographic target which is illuminated by a solid field pattern of the projector.

A strong radial intensity gradient can be observed starting from a certain center in Fig. 5. Both sensors show contrary behavior due to a larger measuring volume of the OV2470 and a different alignment of the borescope. When varying the working distance of the sensor, the centers of the rationally symmetrical intensity drops are shifted. In order to precisely adjust the projection of the light into the C-mount lens of the borescope a more flexible design of the digital micro mirror device must be used.

Calibration

In the context of sensor miniaturization, it is meaningful to verify, whether the distortion of the smallest sensor can sufficiently be corrected. Distortion correlates anti proportionally to the aperture size and is not an inherent property of a sensor. Therefore, camera and projector are calibrated according to the pinhole camera approach of Zhang (2000). With respect to radial and tangential distortion of camera and projector, the distortion is modeled via the polynomial approach of Conrady and Brown (Brown 2002). For the determination of the extrinsic system parameters a final stereo calibration of camera and projector is performed. Figure 6 shows the resulting distortion plots. Considering the direction of pixel displacement (arrow directions within the image), it can be concluded that the camera exhibits pincushion distortion, while the projector has a barrel distortion.

Camera and projector are subject to strong radial distortion while the camera also has tangential distortion. Tangential distortion is probably caused by the lens tilting in the thread of the camera package. The corners of the camera image underlie strong image field curvature, so that the features extracted are false and neglected. High pixel displacement within the distortion plot of the projector result from an artificially extrapolated resolution of the projector.

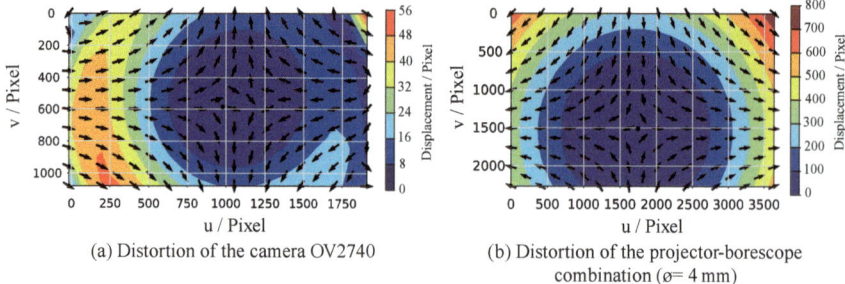

(a) Distortion of the camera OV2740   (b) Distortion of the projector-borescope combination (ø= 4 mm)

Fig. 6 Distortion visualization of the 6.5 mm measuring head. Results published in Middendorf et al. (2021a)

Reconstruction Quality

The probing error with respect to form is often used to classify the reconstruction quality. Using a borescopic sensor with a measuring head at 30 mm working distance it was determined to be 20 μm according to VDI/VDE 2634-2 (Deutsches Institut für Normung e. V. 2012) for a cylindrical feature of a calibrated micro contour standard. Additionally, the probing error with respect to size was calculated following the guideline JCGM 100:2008 (International Organization for Standardization 2008). The probing error with respect to size on this feature is 40 μm within 20 repeated measurements. Please refer to Matthias et al. (2017) for further accuracy and measurement uncertainty investigations and supplementary explanations. These specifications apply to surfaces with good optical cooperativity, for surfaces with limited optical cooperativity, the known physical limits of triangulating optical measurement principles apply.

4 Planing a Measurement Strategy Using Ray Tracing Simulations

During the optical measurement of complex geometries, especially during the measurement of optically non-cooperative surfaces (or glossy surfaces), multiple reflections caused by the shape of the specimen occur frequently. This is critical for fringe projection measurements, since multiple reflections can lead to incorrect phase information which is unwrapped from the camera images. An example of this can be seen in Fig. 7. Here, a fringe projection measurement was performed on the concave surface of a highly reflective compressor blade. Due to multiple reflections, false points are reconstructed outside the actual geometry. The use of such measurement data leads to erroneous damage derivations and prevents the automated data evaluation.

In order to gain a deeper understanding of this problem and to develop a compensation approach for these influences, an optical simulation of the measurement is

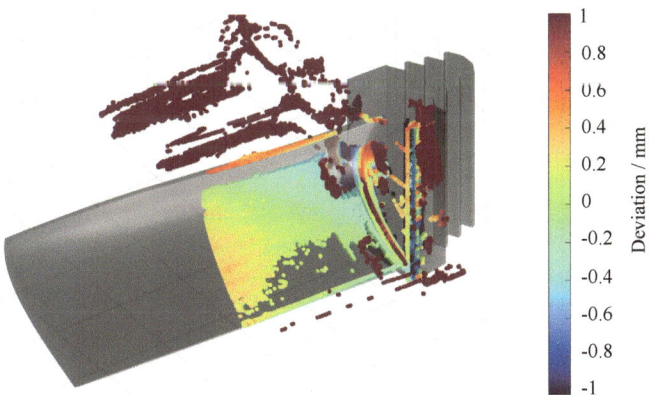

Fig. 7 Reconstructed point cloud of a compressor blade with false reconstructed points due to multiple reflections

performed. Since these kinds of simulations can last up to several days on a CPU basis, a near real-time GPU-based approach has been implemented. Thus, a physically based high resolution simulation of the measurements can be carried out within one second. After explaining the simulation pipeline, an approach to identify low-reflection measurement poses and a method to compensate for erroneous phase information based on the ray tracing simulations is presented.

4.1 GPU-Based Ray Tracing Simulation

Modern ray tracing algorithms rely on the rendering equation of James Kajiya (1986). This equation describes the energy conservation of light rays in space and provides a physically based description of light based on radiometric quantities to simulate an image. To render an image, the following equation has to be solved for each pixel of the image:

$$L_o(\mathbf{p}, \mathbf{w_o}) = \int_{H^2} f(\mathbf{p}, \mathbf{w}_o, \mathbf{w}_i) L_i(\mathbf{p}, \mathbf{w}_i)\cos\theta_i dw_i \qquad (2)$$

The amount of outgoing radiance $L_o(\mathbf{p}, \mathbf{w}_o)$ at a surface point $\mathbf{p}$ is integrated over all incident light ray directions dw_i of a corresponding hemisphere H^2 as a function of the incoming radiance $L_i(\mathbf{p}, \mathbf{w}_i)$ and the reflection properties of the object surface $f(\mathbf{p}, \mathbf{w}_o, \mathbf{w}_i)$. During the simulation of fringe projection measurement cos 0_i describes the angle between the optical axis of the camera $\mathbf{w}_o$ and the optical axis of the projector $\mathbf{w}_i$. The bidirectional reflection distribution function (BRDF) $f(\mathbf{p}, \mathbf{w}_o, \mathbf{w}_i)$ of an object surface describes the distribution of the reflected light. In order to simulate reflections physically based and take the surface roughness into account the BRDF model according to Torrance-Sparrow (Torrance and Sparrow 1967) is applied. The rendering of a gray-scale image of the projection of a sinusoidal pattern of the measurement sequences is depicted in Fig. 8a. For a more detailed mathematical breakdown, Middendorf et al. (2021b) can be referred to. To calculate the occurrence of multiple-reflections and their reflection locations efficiently, an inverse ray tracing approach is used. Since a large number of light rays emitted by the projector are not reflected into the camera, the inverse approach reduces the computational effort significantly. As a consequence, interaction of the light rays and the specimen surface are traced from the camera origin to the projector origin. To further limit the computational effort of tracing multiple reflections, a ray tracing approach according to Whitted is used (Whitted 1980). Starting from the camera origin, a primary ray is traced, and a secondary ray is generated for each intersection of a light ray with an object. This creates a path structure, where the secondary rays are calculated based on the normal direction of the specular reflection of the incident ray. In order to perform ray tracing efficiently, the algorithm was implemented using OptiX a ray tracing engine developed by NVIDIA® (Parker et al. 2010). To parallelize the rendering on the graphics card, the raytracing application is based on NVIDIA's Compute-Unified-Device-Architecture (CUDA) (NVIDIA et al. 2020). This recursive ray tracing is continuously performed until a self-defined reflection depth is reached. In this application, it can be assumed that the influence of a light beam from the 4th reflection on wards has a minor effect on the resulting camera image. Figure 9 shows an exemplary reflection map calculated using the measurement pose from Fig. 8a. In this figure, the maximum reflection depth per camera pixel is color-coded.

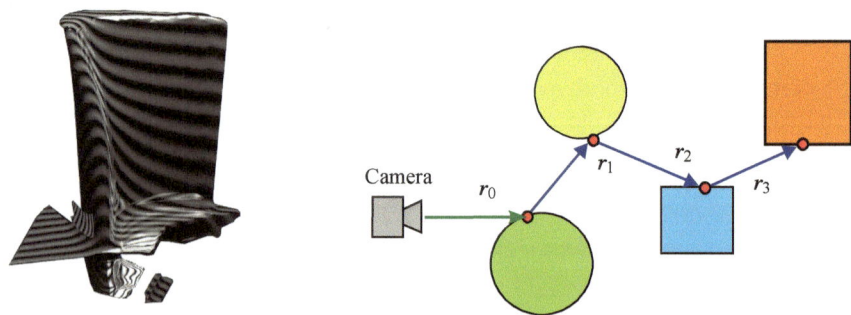

(a) Rendering of a turbine blade

(b) A primary ray $\mathbf{r}_0$ generates secondary rays $\mathbf{r}_1$, $\mathbf{r}_2$ and $\mathbf{r}_3$ while being reflected upon different object surfaces (Suffern and Hu 2014).

Fig. 8 Rendered intensity image of a ray tracing simulation and the functional principle of whitted ray tracing. Results published in Middendorf et al. (2021b)

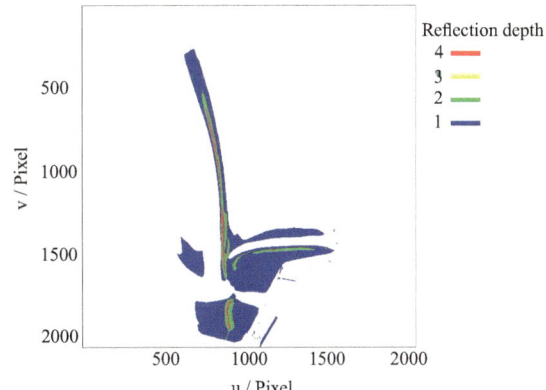

Fig. 9 Calculated reflection depth distribution as result of the predicted reflections of the measurement pose of Fig. 8a. Results published in Middendorf et al. (2021b)

4.2 Evaluation of Suitable Measurement Poses

To identify suitable measurement poses, a consistent evaluation metric is first defined. Based on the sum of all maximum reflection depths per camera pixel, a representative value per measurement pose is determined. To position the sensor according to a targeted field of view and its working distance, a surface point on the specimen is aligned in the focal plane of the camera sensor. This enables the comparison of the reflectivity of a region of interest around the object point. To identify a low-reflectivity measurement pose, a spherical scanning of possible measurement poses around the defined surface point is performed. Figure 10 shows 1024 different measurement poses with respect to the reflectivity of the measurement pose on the sphere. The yellow star shows the center of the sphere and represents the surface point observed. The green star indicates the measurement pose with the lowest reflectivity and the red star represents the measurement pose with the highest reflectivity. The rendering of both measurement poses are shown in Fig. 11a and b. In Fig. 11a, the fringe pattern is projected into the blade and in the direction of the blade root, causing the light to be reflected multiple times. This leads to incorrect phase information recorded by the camera. In contrast, the fringe pattern in Fig. 11b is projected to the blade tip, which avoids reflections.

4.3 Compensation of Multiple Reflections

In order to reduce the influence of the faulty phase information within the reconstruction, a masking approach is presented below. Based on the calculated reflection locations and the reflection depth in the camera image, a binary mask can be created. The masking is applied to the images of the entire measurement sequence so that the areas of multiple reflections are suppressed everywhere. An application of this approach can be seen in Fig. 12. To apply this approach to real measurements and

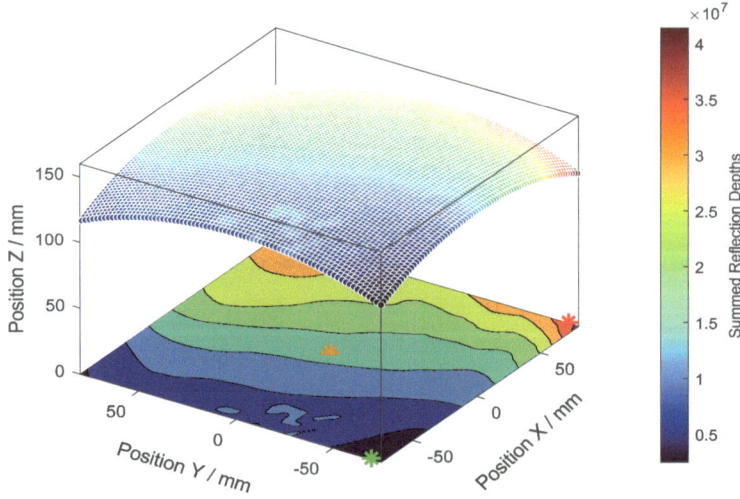

Fig. 10 Spherical scanning of all 1024 measurement poses related to a single surface point. The color indicates the summed reflection depth per pose. Results published in Middendorf et al. (2021b)

(a) Measurement pose with the least reflections (b) Measurement pose with the most reflections

Fig. 11 Comparison of most reflective and least reflective measurement pose. Image **a** has a total reflection count of 41.681.535 while **b** has 2.476.560. Results published in Middendorf et al. (2021b)

to identify reflections in these, a pose estimation of the measurement object has to be performed first. This allows to obtain the relative pose of the specimen in the camera coordinate system. Using the calculated pose, a ray tracing simulation of the measurement can be performed, and a mask can be calculated. The real measurement can be subsequently filtered with the simulated mask. In addition, it is also possible to calculate a mask for the projection, which reduces the reflections during the measurement. To apply this approach for real measurements some assumptions are made. For example, the real position and size of the measured object differs from the simulated one and thus causes a certain uncertainty budget. This is currently taken into account in a simplified manner by performing an erosion of the masking to allow for small errors. In addition, possible defects and other machining operations that

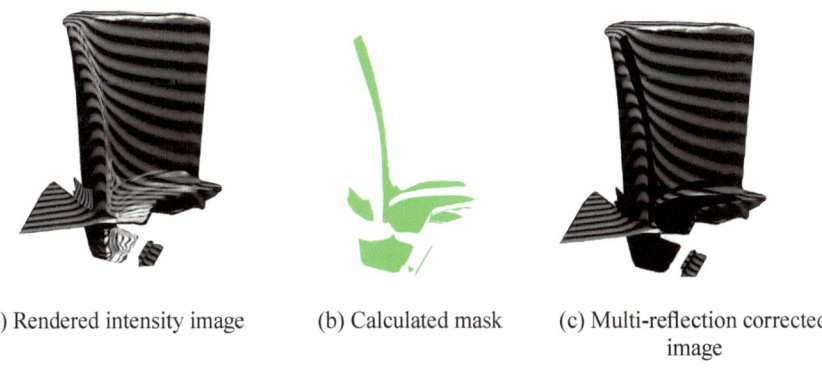

| (a) Rendered intensity image | (b) Calculated mask | (c) Multi-reflection corrected image |

Fig. 12 Example of a masking approach based on the ray traced image and the calculated reflectance map. Results published in Middendorf et al. (2021b)

change the shape and surface of the component cannot be taken into account, resulting in possible uncompensated reflection effects. To validate this simulation approach, three different simulations were performed and reconstructed according to the normal phase shift pipeline used for fringe projection measurement. Subsequently, a deviation analysis of the reconstructed simulation was performed in comparison to the CAD file. Simulation one from Fig. 13a represents an ideal measurement with a reflection depth of one, which avoids erroneous multiple reflection effects. The deviation analysis proves that the simulation was successfully reconstructed despite the optical properties of the measurement system. In the second simulation, a fringe projection measurement was performed with a reflection depth of four. Significant deviations can be seen in the reconstructed point clouds (Fig. 13b). Especially in the area of the leading edge and the blade root, multiple reflections occur. By means of the masking approach, the deviations from the second simulation were compensated. The masked image areas and reduced measurement deviations are shown in Fig. 13c.

5 Inspection of Turbine Blades in Confined Spaces

For the study of measurement in confined spaces, a geometric blade arrangement was reproduced from the mounted aircraft engine. Using five blades in different states of wear, investigations were carried out on damages at the leading edge. Besides the wear at the blade tip, this is one of the main places of wear, as erosion, cracks, nicks, dents, burns, leading edge burn through, coating damages and blocked film-cooling holes occur (IAE. International Aero Engine 2000). Figure 14 shows an image of the borescopic sensor with two blades in different states of wear. The left-hand blade is heavily worn and only slightly visible due to a burnt coating, while the right-hand blade is well visible, as only coating at the leading edge is missing. In addition to the missing coating, a lot of material is missing at the leading edge. An example

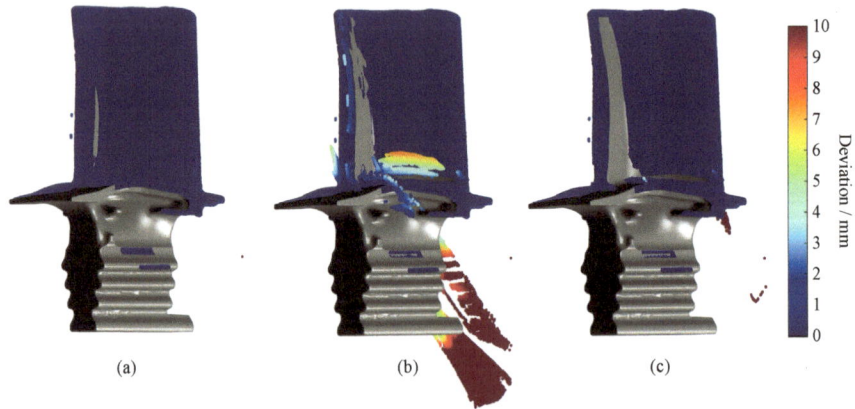

Fig. 13 Deviation analysis of reconstructed raytracing simulations to the CAD geometry. **a** Reconstruction of the simulation with a reflection depth of one. **b** Reconstruction with a reflection depth of four and **c** the reconstruction of the simulation with a reflection depth of four and the multiple reflection mask applied. Results published in Middendorf et al. (2021b)

measurement pose for the borescopic engine inspection is depicted here. The inhomogeneous illumination is caused by the miniaturized measuring system, which is particularly noticeable. Depending on the measurement pose, up to three blades are within the field of view of a borescopic measurement. Due to the rotational degree of freedom of the engine shaft, the position of the turbine blades relative to the measuring system is initially unknown. In order to clearly identify and assign the blades and their position in the engine, the unknown pose must first be determined. The identified measurement pose can then be used to perform wear and deviation analyses. Given the highly variable measurement and wear conditions within the aero engine, a feature segmentation approach is used in addition to the common iterative closest point (ICP) registration approach. The film-cooling holes of the turbine blades have proved to be unique features that can be detected, even after increased wear. These can be segmented both in the two-dimensional image via color gradients as well as in the three-dimensional point cloud via its geometric shaping. To register the measured point cloud to the reference geometry of the CAD models, the film-cooling holes in the CAD reference are first identified. A detailed description of this approach can be found in Middendorf et al. (2022a). Using an equivalent segmentation approach, the film-cooling holes within the measured point cloud are segmented using a clustering approach (DB-Scan). To identify the features, the coordinates of the centroid of the segmented cluster are used. Based on the identified film-cooling holes, a random sample consensus (RANSAC)-based numerical optimization approach is used to find the closest possible match between the set of film-cooling holes from the measurement and the model. The segmented film-cooling holes in the reference geometry are shown in Fig. 15. Based on the estimated pose of the turbine blades, a subsequent fine registration process based on an ICP approach can be used to align the entire measurement to the reference geometry. To evaluate the condition

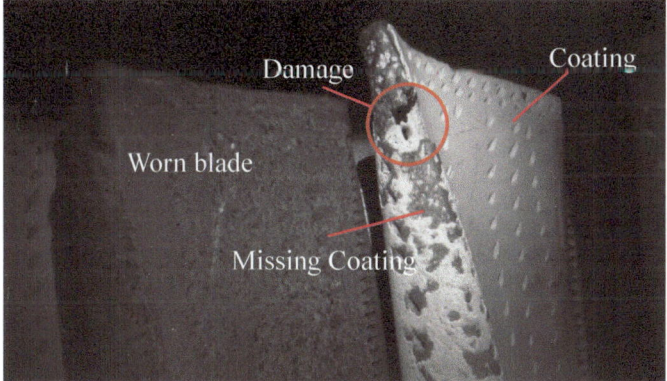

Fig. 1.14 Camera view of an exemplary measurement pose of the borescopic fringe projection sensor

of the measured specimen and derive damages, a surface comparison to the reference geometry is performed. For this purpose, the deviation of the point cloud to the CAD geometry is determined in polygonal normal direction. When assigning the respective (polygon) planes to the corresponding 3D points, a 1-Nearest-Neighbor (NN) classification of the reference point cloud (generated from all polygons) and the reconstructed point cloud is calculated. Subsequently the Euclidean distance of all points is determined to the nearest polygon. The right-hand turbine blade from Fig. 14 is damaged along the leading edge. Deviations of more than 1 mm compared to the reference geometry can be measured, see Fig. 16. Based on the knowledge of the exact geometric shape and location of the damage, the damage can be classified and a disassembly decision for a particular engine can be made. With this metric and high-precision measurement data, the subjective and error-prone assessment of the normal borescope process can be extended (Drury et al. 1997; See 2012; Aust and Pons 2022). In combination with damage classification approaches based on neural networks, such as that implemented by Aust et al. (2021), very fast, efficient and reliable inspection becomes possible. With an appropriate mechanical connection to the engine and a positioning mechanism, an automation of the inspection process based on borescopic fringe projection sensors can be realized in near future.

6 Conclusions

Within this research project, a borescopic inspection approach for confined spaces was developed. Using the example of aero engine inspections, the successful miniaturization and suitability of the measurement system for the intended measurement task could be demonstrated. For the design and development of miniaturized 3D measurement systems, it became evident that especially the camera and the size

Fig. 1.15 Segmented
film-cooling holes plotted on
the reference geometry

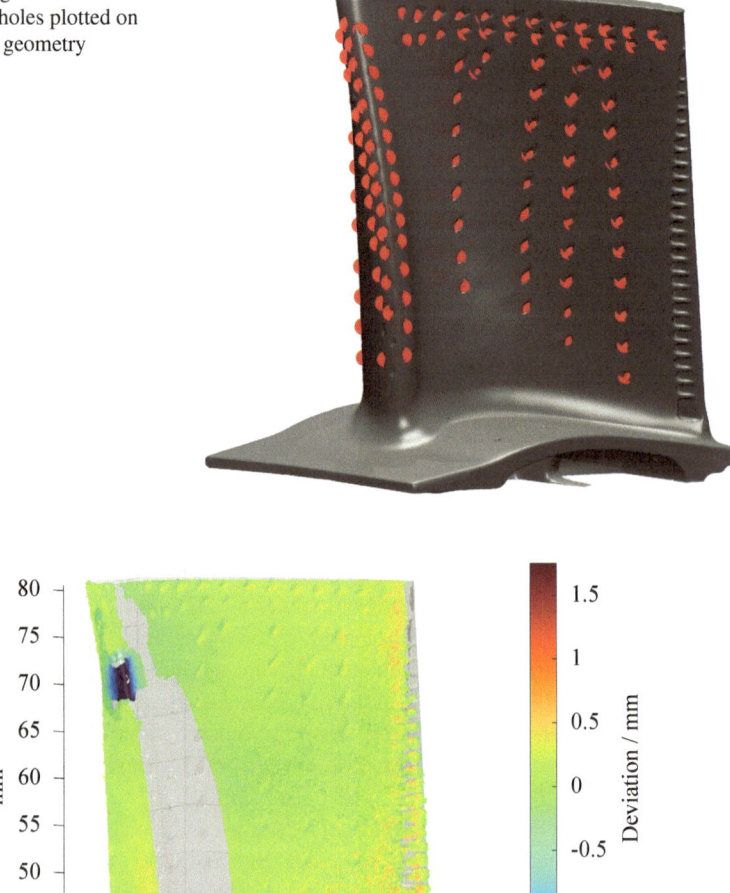

Fig. 1.16 Deviation analysis of an exemplary leading edge damage. Results published in Middendorf et al. (2022a)

of the borescope are decisive. Compared to industrial cameras, strong noise influences, non-linearities in the intensity response, possibly implemented data processing
pipelines on the sensor and strong distortions due to small working distances have to
be considered. Concerning the borescopes, bending effects, which have an influence
on the optical properties of the sensors, have to be taken into account in addition
to the oscillation-sensitive design. A reduction of the borescope diameter also leads

to a loss of intensity within the measurement scene and a drop in intensity within an image. This means high dynamic range measurements must be carried out. By means of a GPU based ray tracing simulation, an approach for automated measurement pose planning could be developed. This approach takes the sensor-specific properties and reflective characteristics of the measurement objects into account. In addition, the measurement pose planning should be extended by further influential factors such as the sensor noise as well as geometry and pose-dependent measurement uncertainties. Furthermore, the influence of production-specific variances of the measurement objects and their effect on the masking and compensation approach should be investigated. For the application of fringe projection measurements on shiny components, a compensation approach for multiple reflections was developed. This enables the examination of highly reflective surfaces, which could previously only be measured with an anti-reflection spray. However, the subsequent compensation of multiple reflections in real measurements requires a successful pose estimation of the measured object and a precise reference geometry. The inspection in confined spaces in an academic environment was also successfully implemented. In particular, navigation and orientation within the aero engine could be addressed with a rigid endoscopic sensor. The pose estimation of the turbine blades within the engine were realized using a feature segmentation approach. In the end, the condition assessment and damage derivation of worn turbine blades could be demonstrated using exemplary damages with impacts at the leading edge. For future tasks, it is possible to bring the developed sensors to a level of industrial maturity to be tested on real aircraft engines outside the laboratory.

Acknowledgements Funded by the Deutsche Forschungsgemeinschaft (DFG, German Research Foundation)—SFB 871/3—119193472.

References

Aust, J. and Pons, D. (2022). Comparative analysis of human operators and advanced technologies in the visual inspection of aero engine blades. *Applied Sciences*, 12(4):2250.

Aust, J., Shankland, S., Pons, D., Mukundan, R., and Mitrovic, A. (2021). Automated defect detection and decision-support in gas turbine blade inspection. *Aerospace*, 8(2):30.

Brown, D. (2002). Close-range camera calibration. *Photogramm. Eng.*, 37.

Deutsches Institut für Normung e. V. (2012). DIN 26343-2 Optical 3-D measuring systems, Optical systems based on area scanning. Beuth Verlag GmbH.

Drury, C. G., Spencer, F. W., and Schurman, D. L. (1997). Measuring human detection performance in aircraft visual inspection. *Proceedings of the Human Factors and Ergonomics Society Annual Meeting*, 41(1):304–308.

IAE. International Aero Engine (2000). V2500 maintenance manual, borescope inspection, standard practices ata 70–00–03 2000. Technical report. Available online: https://www.slideshare.net/RafaelHernandezM/v2500-bsi-issue-01 (accessed on 26 May 2022).

International Organization for Standardization (2008). ISO/IEC GUIDE 98–3:2008, Uncertainty of measurement – Part 3: Guide to the expression of uncertainty in measurement (GUM:1995). Beuth Verlag GmbH.

Kajiya, J. T. (1986). The rendering equation. *ACM SIGGRAPH Computer Graphics*, 20(4):143–150.

Martinec, E. (2008). Noise, dynamic range and bit depth in digital slrs.

Matthias, S., Schlobohm, J., Kästner, M., and Reithmeier, E. (2017). Fringe projection profilometry using rigid and flexible endoscopes. *tm - Technisches Messen*, 84.

Matthias, S., Kästner, M., and Reithmeier, E. (2018). A 3-d measuring endoscope for hand-guided operation. *Measurement Science and Technology*, 29.

Middendorf, P., Hedrich, K., Kästner, M., and Reithmeier, E. (2021a). Miniaturization of borescopic fringe projection systems for the inspection in confined spaces: a methodical analysis. In Ehmke, J. and Lee, B. L., editors, *Emerging Digital Micromirror Device Based Systems and Applications XIII*, volume 11698, pages 151–164. International Society for Optics and Photonics, SPIE.

Middendorf, P., Kern, P., Melchert, N., Kästner, M., and Reithmeier, E. (2021b). A GPU-based ray tracing approach for the prediction of multireflections on measurement objects and the a priori estimation of low-reflection measurement poses. In Beyerer, J. and Heizmann, M., editors, *Automated Visual Inspection and Machine Vision IV*, volume 11787, pages 86–96. International Society for Optics and Photonics, SPIE.

Middendorf, P., Blümel, R., Hinz, L., Raatz, A., Kästner, M., and Reithmeier, E. (2022a). Pose estimation and damage characterization of turbine blades during inspection cycles and component-protective disassembly processes. *Sensors*, 22(14).

Middendorf, P., Rothgänger, M., Peddinghaus, J., Brunotte, K., Uhe, J., Behrens, B. A., Quentin, L., Kästner, M., and Reithmeier, E. (2022b). In situ wear measurement of hot forging dies using robot aided endoscopic fringe projection. *Key Engineering Materials,* 926:1211–1220.

NVIDIA, Vingelmann, P., and Fitzek, F. H. (2020). Cuda, release: 10.2.89.

Parker, S. G., Bigler, J., Dietrich, A., Friedrich, H., Hoberock, J., Luebke, D., McAllister, D., McGuire, M., Morley, K., Robison, A., and Stich, M. (2010). Optix: A general purpose ray tracing engine. *ACM Trans. Graph.*, 29(4).

Pösch, A., Vynnyk, T., and Reithmeier, E. (2012). Using inverse fringe projection to speed up the detection of local and global geometry defects on free-form surfaces. In Bones, P. J., Fiddy, M. A., and Millane, R. P., editors, *Image Reconstruction from Incomplete Data VII*, volume 8500, pages 91–97. International Society for Optics and Photonics, SPIE.

Pösch, A., Schlobohm, J., Matthias, S., and Reithmeier, E. (2017). Rigid and flexible endoscopes for three dimensional measurement of inside machine parts using fringe projection. *Optics and Lasers in Engineering*, 89:178–183. 3DIM-DS 2015: Optical Image Processing in the context of 3D Imaging, Metrology, and Data Security.

Schlobohm, J., Pösch, A., Kästner, M., and Reithmeier, E. (2015). On the development of a low-cost rigid borescopic fringe projection system. In *Photonics, Devices, and Systems VI*. SPIE.

Schlobohm, J., Pösch, A., and Reithmeier, E. (2016). A raspberry pi based portable endoscopic 3d measurement system. *Electronics*, 5(3).

Schlobohm, J., Bruchwald, O., Frackowiak, W., Li, Y., Kästner, M., Pösch, A., Reimche, W., Maier, H. J., and Reithmeier, E. (2017a). Advanced characterization techniques for turbine blade wear and damage. *Procedia CIRP*, 59:83–88. Proceedings of the 5th International Conference in Through-life Engineering Services Cranfield University, 1st and 2nd November 2016.

Schlobohm, J., Li, Y., Pösch, A., Kästner, M., and Reithmeier, E. (2017b). Multiscale measurement of air foils with data fusion of three optical inspection systems. *CIRP Journal of Manufacturing Science and Technology*, 17:32–41. SI: Advanced M&T for TES.

See, J. (2012). Visual inspection : a review of the literature. Technical report. Suffern, K. and Hu, H. H. (2014). *Ray Tracing from the Ground Up*. A. K. Peters, Ltd., Natick, MA, USA, 2nd edition.

Torrance, K. and Sparrow, E. (1967). Theory for off-specular reflection from roughened surfaces. *Journal of the Optical Society of America (JOSA)*, 57(9):1105–1114.

van den Bergh, F. (2018). Deferred slanted-edge analysis: a unified approach to spatial frequency response measurement on distorted images and color filter array subsets. *J. Opt. Soc. Am. A*, 35(3):442–451.

van den Bergh, F. (2019). Robust edge-spread function construction methods to counter poor sample spacing uniformity in the slanted-edge method. *J. Opt. Soc. Am. A*, 36(7):1126–1136.

Whitted, T. (1980). An improved illumination model for shaded display. *Communications of the ACM*, 23(6):343–349.

Zhang, Z. (2000). A flexible new technique for camera calibration. *IEEE Transactions on Pattern Analysis and Machine Intelligence*, 22(11):1330–1334.

Regeneration-Induced Variances of Aeroelastic Properties of Turbine Blades

Lennart Stania and Joerg R. Seume

Abstract Regeneration and wear result in geometric deviations between design intent and reality of turbine blades. These deviations influence the aerodynamic flow field and the aeroelastic behaviour of downstream blades. As an example for the effect of such deviations between modules on blade vibration amplitudes, an experiment is set up to determine the influence of cold streaks, which can occur due to widening of cooling air holes. The vibration amplitude for an off-design point is increased by 20% and for the design point as well. We conclude that the forced response caused by cooling air from the turbine blades must be considered during the design process and for life predictions, especially for higher relative cooling air mass flows. Furthermore, different geometric deviations may occur simultaneously, which can magnify the vibration amplitude even further. A probabilistic process presented here investigates these combined effects. It is shown that a maximum amplitude exists within the given geometric boundaries. Using this information, new safety margins can be set for geometric deviations in the repair or manufacturing process.

Keywords Axial turbine · Aerodynamic · Aeroelasticity · Experiments · Simulations · Parametrisation · Probabilistic

1 Introduction

High aerodynamic, mechanical and thermal loads in axial turbines lead to extensive wear of turbine blades, especially within the high-pressure turbine module. If such a blade is regenerated to save costs and avoid unnecessary scrapping of the blade, the geometry is different compared to its original design intent. Geometric deviations from the design intent influence the aerodynamics across the blade, e.g., they increase the wake deficit. In such a case of higher wake deficit, there is a high possibility of

L. Stania (✉) · J. R. Seume
Institute of Turbomachinery and Fluid Dynamics, Leibniz University Hannover, An der Universitaet 1, 30823 Garbsen, Germany
e-mail: stania@tfd.uni-hannover.de
URL: https://www.tfd.uni-hannover.de/

© The Author(s) 2025
J. R. Seume et al. (eds.), *Regeneration of Complex Capital Goods*,
https://doi.org/10.1007/978-3-031-51395-4_15

changes in aeroelastic properties of a downstream vane or blade, which may increases high-cycle fatigue (HCF) and thus reduce the blade's life. This phenomenon has been investigated in the subproject C4 and is reported below.

A new fifth stage has been designed and manufactured for an existing four-stage axial turbine at the Institute of Turbomachinery and Fluid Dynamics (TFD) within the first funding period. This stage has aeroelastic properties similar to a low-pressure turbine of a real aircraft engine. Furthermore, a computational tool chain is developed for simulating forced response. It is shown experimentally that a stagger angle variation by $1.5°$ increases the vibration amplitude of the downstream rotor by a factor of 4 for a part-load operating point, and up to a factor of 5 for the design point (Aschenbruck and Seume 2015). An additional stage in between the vane row with geometric deviations and the investigated rotor strongly decreases the influence on the vibration amplitude of the blade. For the design point, the vibration amplitude increases by a factor of 1.2. Lastly, the influence of a change in radial tip gap directly within the investigated rotor stage increases the vibration amplitude slightly (Hauptmann et al. 2018). This is due to an increase of the aerodynamic forcing near the blade tip and at the same time reducing damping.

The early work in funding periods (FP) 1 and 2 shows the importance of considering geometric variances in the decision process of blade regeneration based on aerodynamics and aeroelasticity, as well as component life estimation. The component life is heavily influenced by the aerodynamic force of the blade. However, different geometric variances occur simultaneously in reality, which might enhance the vibration amplitude even further and reduce the blade's life. An in-detail study of existing blades and their geometric variances is given in Ernst et al. (2016), where worn low-pressure turbine blades are digitized, parametrised, and investigated regarding their sensitivities to geometric variances. The possibly critical vibrations due to simultaneously occurring variances are analysed in a probabilistic study in the last funding period and reported below.

Next to these geometric variances, there are additional effects like burner malfunction or cooling hole degradation that influence the aeroelasticity of a downstream row. These defects can migrate from one module to another, e.g., from high-pressure turbine to low-pressure turbine, thus accurate prediction of these effects is of importance for product life predictions. This cross-module effect is also investigated in the last funding period 3 through a change in cooling air that leads to cold streaks.

2   Test Setup

The test case for both the experimental setup and the probabilistic study is the 5-stage axial turbine of the TFD. Each stage contains 30 blades and 29 vanes. The numerical domain and different application of geometric variances over the three funding periods are shown in Fig. 1. As previously described, the stagger angle is changed by $1.5°$ relative to the reference cases with the original stagger angle (increase of circumferential velocity at the outlet) for vane stage 5 (V5) in FP1 and

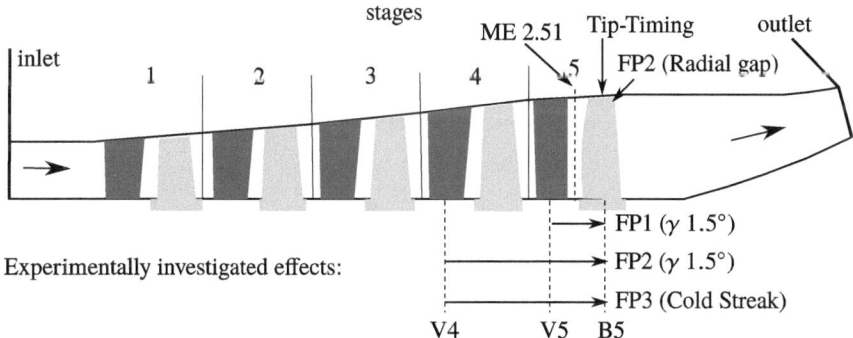

Fig. 1 Overview of different experimental investigations within the three funding periods (FP)

for vane stage 4 (V4) in FP2. In this funding period, V4 is selected for applying the cold streak as an example for cross-module effects.

A tip-timing system is applied to measure the vibration amplitude of the rotor blades in stage 5 (B5) for all cases and an aerodynamic measurement plane over 15° circumferential position between V5 and B5 is used for 5-hole probe measurements. At all times, the condition of the inlet is measured with five 5-hole probes equally distributed around the circumference and five rake probes at the outlet. These measurements ensure that the operating point is maintained between different investigations. Four blades are instrumented with eight thermocouples each, to allow further analysis of the cold streak mixing. The thermocouples are placed on the suction side on two blades and on the pressure side of the other two blades.

During the design process of the fifth stage, three different operating points had been selected based on machine loading and resonance crossings in the Campbell diagram. A crossing with the first structural mode of the blade and engine order (EO) 29 or 30 near the low-load operating point. The same EO has a crossing with mode 2 for the part-load operating point. Additionally, blade mode 1 is excited by EO 15 near the part-load operating pointö, which is shown in Fig. 2. The resulting boundary conditions extracted from experimental data during FP 3 and the crossing for the different operating points are shown in Tab. 1.1.

As described earlier, the cold streak is applied in the fourth stage stator vane, V4. There are five cooling holes with a diameter of 2 mm on the pressure side of each vane near mid-span (see Fig. 3), which allow to investigate the radial migration of cold streaks across the stage. Air is provided by an external compressor and every vane is connected independently with the main air source. Therefore, different engine orders can be investigated for the different operating points. Temperature, pressure, and mass flow of the cold streak are measured continuously. The relative mass flow of the cooling air as a fraction of the total mass flow in the axial turbine and the absolute cooling air temperature are shown in Table 1. For both cases, the reduced mass flow and the rotational speed are held constant for each operating point.

For the computational fluid dynamics simulations (CFD), the unsteady Reynolds-averaged Navier–Stokes equations (URANS) are solved in the pseudo-time domain

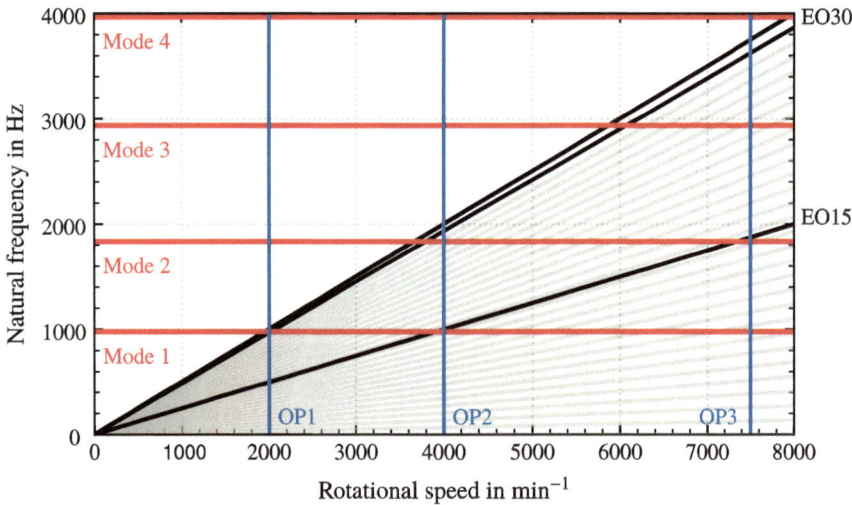

Fig. 2 Campbell diagram

Table 1 Operating points

Operating point	OP1	OP2	OP3
Red. mass flow in kg/s	6.65	7.54	8.5
Red. rotational speed in min^{-1}	2171	4312	7500
Pressure ratio	1.31	1.57	2.45
Inlet temperature in K Outlet temperature in K	356	363	419
	335	325	335
Engine order/mode	29/1	29/2	15/2
		15/1	
rel. cooling air mass flow in % cooling air temperature in K	1.6	1.3	0.6
	305	308	320

with the implicit solver TRACE Nürnberger et al. (2001), developed by the German Aerospace Centre. A finite-volume scheme is used for spatial discretization. The Reynolds-stress is modelled by the Menter SST model (Menter et al. 2003) assuming fully turbulent flow. A stagnation point fix by Kato and Launder (Kato 1993) is used.

All walls are assumed to be adiabatic; the non-slip condition is implemented at the wall. As convergence criterion, the relative change of mass flow has to be below 10^{-4} and the pressure change over one period on different points in the flow field has to be below 10^{-5}. Solving the unsteady equations is necessary due to the mixing of the cold streaks in the flow panel. Therefore, the vanes are scaled to a vane count of 30 for the numerical investigations. This way, the numerical setup can be reduced to single-passage in comparison with full annulus calculations. The forced response and flutter calculations are conducted within the URANS setup. It is assumed that the vibration amplitudes are relatively small, therefore, the aerodynamics has a linear

Aerodynamic measurement between V5 and B5

V5: Tip Timing measurement　　　　　　　　　　　　　　　Flow direction

Cold streak air supply　　　　Cold streak injection

Fig. 3 Close-up of the experimentally investigated stages and cold streak application

relationship with the vibrations. Mode-coupling is neglected from the flow on the structure, because the aluminium blade's mass and stiffness are relatively high in comparison with the influenced surrounding air. However, the impact of structural vibrations on the unsteady flow is taken into account and coupled.

Evaluating the aerodynamic work on the blade, a finite element (FE) simulation with ANSYS Mechanical is used as was done by Pohle et al. (2014). Afterwards, surface displacement is mapped from each FE to a CFD mesh. For this process, the FE mesh is first translated to match the position of the corresponding CFD blade, and then the surface deformations are interpolated to the corresponding CFD cells. This mapping process, described in detail by Voigt et al. (2010), is used to calculate the aerodynamic work W on the blades' surface S by considering the steady pressure p^v and unsteady pressure perturbations $\tilde{p}$ from a CFD by where $\mathbf{n}$ is the normal surface and $\mathbf{x}$ the surface deformation. Finally, the energy method is used and the vibration amplitude can be calculated when forcing W_f and damping work W_d are in equilibrium. For the aeroelastic calculations, the main convergence criterion is constant aerodynamic work within 0.01% over a span of 100 time-steps.

$$W = -i\pi \int_Z \tilde{\mathbf{x}}^H (\tilde{p}\mathbf{n}^0 + p^0\tilde{\mathbf{n}})dS \qquad (1)$$

3 Aerodynamic and Aeroelastic Results

Applying the cold streak (CS) changes some of the main parameters of the axial turbine. The inlet pressure increases, even if the reduced mass flow and rotational speed are kept constant across both cases, resulting in a higher total pressure ratio between inlet and outlet. Additionally, the thermal power output and isentropic efficiency are increasing for the cold streak case. This is due to two effects:

1. Additional mass flow from the cooling air, and
2. potential effect upstream to the machine inlet.

Note that this does not mean, that cooling air will increase an axial turbine's efficiency in all cases, but it is important to investigate the change in the operating point in more detail for further understanding. There are two options: either keeping the reduced mass flow and rotational speed constant or controlling the turbine at constant pressure ratio, thus thermal power output. Due to experimental constraints the first option could be performed only. For aircraft engines the latter would have been closer to reality, where the thrust needs to be constant for the engine to achieve the same aircraft speed with different cooling air mass flows.

This observation in change of axial turbine operating points is confirmed by the pressure profile on V5 at mid-span (see. Fig. 4). For all operating points, a good agreement is achieved between reference and cold streak on the vane's suction side, but a pressure increase on pressure side is observed over the whole length of the vane for the cold streak case, thus resulting in higher blade loading. This increases the wake deficit for the cold streak case in total pressure and circumferential extent. A closer examination of the wakes confirms the change of the wake width on the pressure side, while the suction side remains constant. For OP1, a separation bubble is detected in the experimental data and accurately predicted by CFD at 0–20% chord length on the suction side. The separation bubble originates from the blade's high incidence at low-load operating conditions.

The change in total pressure due to the cold streak is confirmed by the circumferential measurement in traverse plane ME2.51 (cf. Fig. 1) as shown for OP2 in Fig. 5, where the normalized total pressure p_{norm} and temperature T_{norm} are calculated by

$$p_{norm} = \frac{p_t - p_{aus}}{p_{ein} - p_{aus}} \text{ and } T_{norm} = \frac{T_t - T_{aus}}{T_{ein} - T_{aus}} \tag{2}$$

A reduction by 0.25% of the static temperature drop throughout the axial turbine is measured between the reference and the cold streak case. The Mach number is reduced by 0.08 for the cold streak measurement for OP2. The radial temperature distribution is changed between both cases and the temperature is increased near shroud and hub. A minimal temperature at around 40% relative channel height indicates the cold streak. This is confirmed for OP1 and OP2, while for OP3 the influence is within measurement uncertainties.

Repeatability is attained by conducting all tip-timing measurements four times. The final vibration amplitude analysed is the arithmetic mean of all calculations with

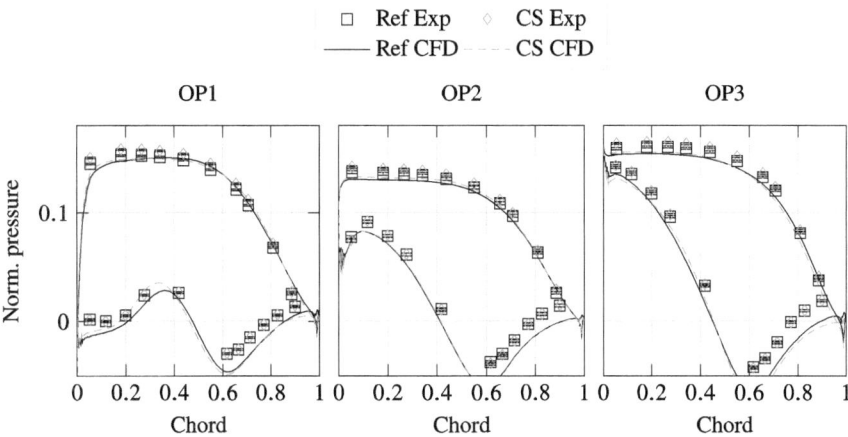

Fig. 4 Pressure profile on V5 for all operating points for reference (Ref) and cold streak (CS). 95% confidence interval in scale of marker

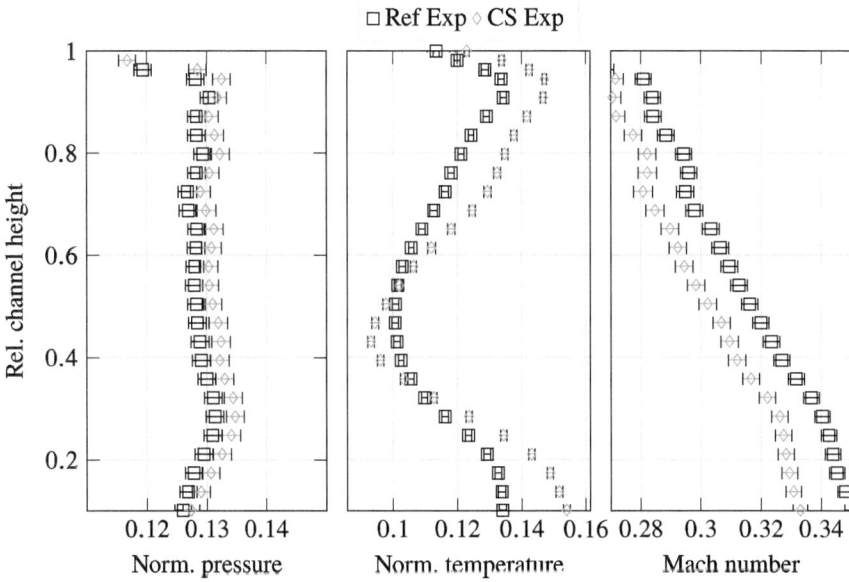

Fig. 5 Area-averaged measurement in plane ME2.51 for OP2. 95% confidence interval in back-to-back measurements

a 95% confidence interval. In Fig. 6 the vibration amplitudes of B5 are shown for the three operating points and all investigated cases over the three funding periods. As described earlier, the vibration amplitude in FP1 and FP2 is increased for OP2 and OP3, while the amplitude is reduced for OP1.

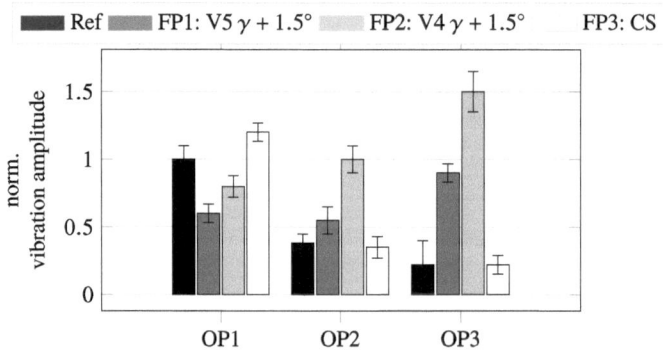

Fig. 6 Vibration amplitude of B5 with 95% confidence interval

The vibration amplitude in increased by 20% in OP1 due to the cold streak and decreases for OP2 and OP3 in a small margin within measurement uncertainties. The EO29 for OP1 has the highest relative cooling mass flow and therefore shows the biggest impact. Especially for OP3 and EO15 for OP2 the cooling mass flow is too low to show a measurable impact. Because of the external compressor in order to supply the cooling air without a cooler system, temperature changes were only possible in small margins with external air blower. These small temperature changes of the cooling air did not have any impact on the vibration amplitude outside the measurement uncertainty.

The unsteady pressure perturbations are increased on the pressure side of the blade's surface as shown in Fig. 7, while the shape between both cases remains the same. On the suction side, the shape is changed. Between 40 and 50% blade length a local reduction occurs of the unsteady pressure for the cold streak case. Outside this area the same behaviour as on the pressure side is observed. These unsteady pressure amplitudes directly impact the force and aerodynamic work on the blade (see Eq. 1).

The change in vibration amplitude is based on three simultaneously occurring effects. First, changes in overall mass flow in the machine due to the change of operating point directly influence the forcing on the blades. This might change the vibration amplitudes if the additional forces have similar forcing frequency as the blade's natural frequency. Second, the additional mass flow due to the cold streak increases the flow potential, which increases the excitation forces similar to the first effect. Finally, the local temperature at a constant total pressure changes, which results in a different density and therefore different mass flow.

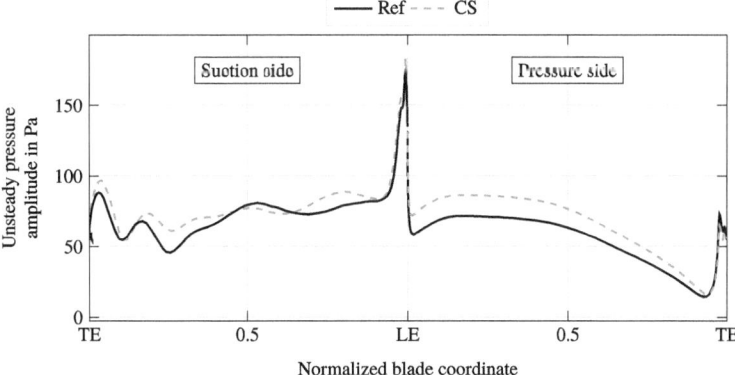

Fig. 7 Unsteady pressure amplitude from CFD results for OP2 at 90% blade height between leading-edge (LE) and trailing-edge (TE)

4   Probabilistic Study

Methodology

Generating a probabilistic data set requires the following steps (see Fig. 8): The first step is creating a parameter model from a given CAD model with 12 characteristic parameters describing a blade section based on the algorithm from Heinze et al. (2014) and Ernst et al. (2016). This parametrisation was conducted on 25 radial blade sections. Using Voronoi diagrams and the Delaunay triangulation Aurenhammer (1991), the camber is approximated, where the smallest distance from the camber to pressure and suction side is determined as profile thickness. The difference between the Delaunay approximation and the true geometry is used as an accuracy criterion and should be below 0.01% thickness deviation for all cases. The leading-edge radius r_{LE} and trailing-edge thickness t_{TE} is determined by fitting a circle into the geometry at the respective position with the Levenberg-Marquardt algorithm (Moré 1978). From the thickness distribution, the value t_{max} and the location $x_{t,max}$ of maximum thickness are determined. Likewise, using the approximated camber, the value c_{max} and positioning $x_{c,max}$ of the maximum camber is calculated. Lastly, the leading-edge angle β_{LE}, the trailing-edge angle β_{TE} and the stagger angle γ are calculated relative to the meridional axis. A summary is shown in Table 2 for all necessary airfoil parameters considered in this work.

Afterwards, based on the geometrical parameters and the percentage deviations of the geometrical parameters obtained by a Latin-hypercube sampling McKay et al. (2000) within prescribed limits, 25 blade sections are reconstructed. No correlation of the input parameters is assumed to exist, and the limits are set with 10% for each parameter to investigate all possible parameter combinations that could occur during the repair process of a blade. This way a wide geometric range is considered without changing the blade completely. Note that this assumption might impact sensitivity

Table 2 Airfoil parameters symbol description

m_{LE}	Axial leading-edge position
$r\Theta_{LE}$	Tangential leading-edge position
l	Chord length
γ	Stagger angle
r_{LE}	Leading-edge radius
β_{LE}	Leading-edge angle
t_{max}	Maximum thickness
$x_{t,max}$	Axial position of maximum thickness
c_{max}	Maximum camber
$x_{c,max}$	Axial position of maximum camber
t_{TE}	Trailing-edge thickness
β_{TE}	Trailing-edge angle

Fig. 8 Probabilistic work flow

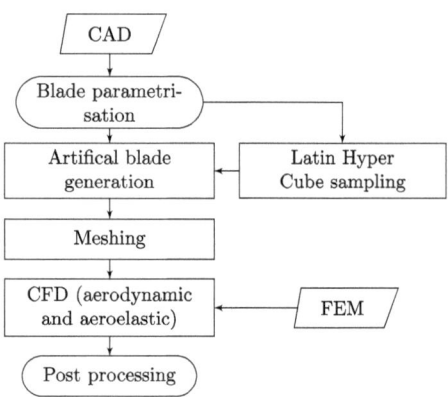

studies, however the final reduced order model can be used for additional sensitivity studies with a correlated input. After applying the reconstruction, the blade sections are transferred to an automatic 3D meshing program. The resulting mesh meets the requirement of a dimensionless distance from the wall $y^+ \leq 1$ for all samples.

The described first step is conducted for the required sample size and is followed by 3D flow simulations. The same solver as described earlier is used, however, only the RANS equations are solved and afterwards flutter and forced response simulations are conducted using the unidirectional time-linearised solver linearTRACE with a frozen gust approach Kersken et al. (2012), saving computational resources in comparison with URANS simulations. A more detailed description of the numerical setup is given in Stania and Seume (2022).

The last stage is isolated for the probabilistic investigation of critical vibration amplitudes. The fifth stator vane V5 is varied and the aerodynamic as well as aeroelastic influence is evaluated for the fifth rotor blade B5. This single stage approach is similar to FP 1. Increasing the simulation domain will increase the required simulation time for each sample, which is not feasible within this project and due to the higher impact of the single stage influence it is not necessary. The inlet boundary condition for the reduced setup is extracted from a 5-stage numerical solution and OP2 is chosen, since the crossing with mode 2 of the blade.

Results

Based on the simulation output, three dimensionless stage parameters and the polytropic efficiency are calculated. The work coefficient Ψ, or stage loading coefficient, is calculated with rotational velocity u_i and the tangential velocity component of the flow $c_{u,i}$ by

$$\Psi = \frac{u_2 c_{u,2} - u_1 C_{u,1}}{u_2^2} \tag{3}$$

where 1 denotes the blade inlet and 2 the outlet. It relates the Euler work over the blade in relation to the rotational speed of the blade. A change of this parameter indicates variations of the loading or extracted work of the blade. The flow coefficient Φ is a dimensionless parameter for the axial or tangential potential of the fluid in comparison to the rotational speed. Variations indicate different flow angles at the trailing edge of the blade. If a variation from the design point is high while keeping the rotational velocity constant, the flow on the blade surface might detach. It is calculated with the axial velocity c_{ax} by

$$\Phi = \frac{c_{ax,2}}{u_2} \tag{4}$$

The reaction R describes the drop of enthalpy h through the blade in relation to the stage. It is calculated by

$$R = \frac{h_2 - h_1}{h_2 - h_0} \tag{5}$$

where the index 0 notes the vane inlet. Decreasing reaction reduces pressure drop through the blade and shifts the velocity triangles to higher velocities within the vane. Note that the definitions of these parameters may vary between different authors. Finally, the polytropic efficiency is calculated by

$$\eta_{poly} = \frac{\kappa}{\kappa - 1} \frac{\log \frac{h_{t2}}{h_{t0}}}{\log \frac{P_{t2}}{P_{t0}}} \tag{6}$$

for the stage with total enthalpy h_t, pressure P_t and heat capacity ratio κ. Another aeroelastic parameter is obtained by calculating the equilibrium between forcing and damping work which leads to the blades' vibration amplitude x. Additionally, the characteristic aeroelastic parameter, the reduced frequency k, is calculated by

$$k = 2\pi f \frac{l}{c} \tag{7}$$

where c is the magnitude of flow velocity, l chord length, and f the characteristic frequency of the blade's mode (eigenfrequency for flutter and vibration frequency for forced response). This parameter is the ratio between the time it takes for the fluid to pass over the blade and the blade's period of vibration, i.e., the reciprocal of the eigenfrequency. Finally, the vibration amplitude x and aerodynamic damping λ are directly obtained by the simulation.

The variance-based sensitivity indices EASI (Effective Algorithm for Computing Global Sensitivity Indices) Plischke et al. (2013) show negligible influence of most geometric input parameters on the output (see Fig. 9). The aerodynamic and flutter calculations show the most significant sensitivity to exist with respect to stagger angle, maximum thickness, and maximum camber. The forced response output shows the highest sensitivity with respect to maximum camber and trailing-edge angle, with negligible impact of all other parameters. This is due to the direct impact on wake deficit of each parameter, which directly influences forced response. Therefore, for future analysis a reduction of the relevant parameters is possible and reduced-order modelling is possible. Reducing the number of independent input parameters reduces the number of required samples until a regression or artificial intelligence model is able to accurately predict the outcome. This is essential within the overall CRC 871 decision making process. Within the performance assessment of the overall regeneration process, the deep learning model is able to predict mass flow, isentropic efficiency and pressure ratio within 1% accuracy.

However, due to the small sensitivity of the vibration amplitude with respect to a few input parameters, the accurate prediction still requires a large sample size. The EASI between aerodynamic output and aeroelastic output shows a similar behaviour. Flutter shows a high sensitivity with respect to the aerodynamics, while the sensitivity

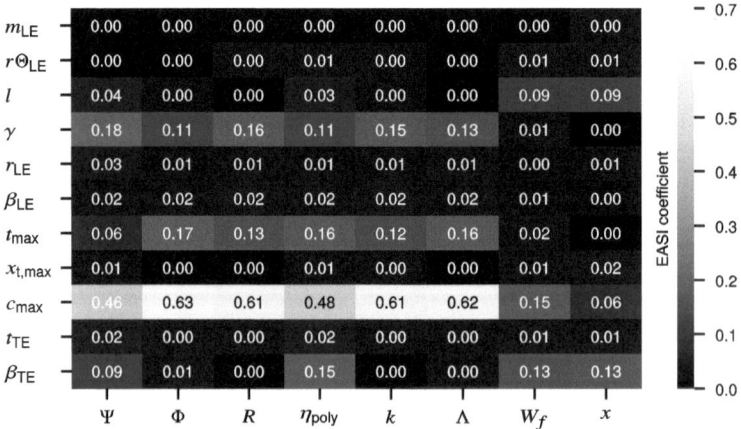

Fig. 9 EASI coefficient between geometric variations and CFD output with an error of $\pm\, 0.02$

of forced response increases compared to the geometric variances first applied but are only up to 0.37. The additional complexity of forced response calculations is due to the frequency and phase dependency between structural modes and aerodynamic periodic force.

For all geometric variances, the maximum increase in blade vibration amplitude is found to be 20%. Note that the maximum stagger angle considered is smaller than the stagger angle variation considered in funding period 1. With this realization, a safety factor can be derived for calculations of the blade's product life and high-cycle fatigue. Furthermore, changes of the maximum thickness, maximum camber, and trailing-edge should be avoided in the repair process to avoid influencing the aerodynamic and aeroelastic behaviour of the downstream blades.

5 Conclusions

Initially, within the first two funding periods, it was shown that geometric variances in a stator vane have a significant impact on the vibration amplitude of a downstream rotor blade. Especially the single-stage influence has to be considered in product life estimation during the design and repair process, because the influence on HCF is eminent. Decisions on vane regeneration need to take these deviations into account, to ensure the aircraft engine's safe operation over many flight cycles. The effect is successfully investigated in an experimental setup and the measurements are used for validating the computational setups, which are able to accurately predict the vibration amplitudes. This approach is applied to a probabilistic study in which a maximum increase of vibration amplitude by 20% is found by simulation for a given parameter range of geometric variances, and important geometrical parameters are identified for changes in the aeroelasticity of the downstream blade. We conclude that reduced order models allow fast estimation of vibration amplitudes by considering only a small amount of parameters. Due to the reduced computational effort in a sensitivity study with less parameters, this is especially important for life predictions, where a data basis has to be created for each blade. The knowledge can be used within the design, repair, and manufacturing process to set new safety margins closer to the calculated worst-case scenario and quickly asses the blades' influence on the next row. However, this knowledge and approach needs to be transferred to further aircraft engines and operating points for applications in the industry.

Using the same experimental setup in FP3 as was earlier used in FP1 and FP2, the cold streak is successfully applied to an axial turbine as an example for cross-module effects. The experimental investigation shows an increase of vibration amplitude by 20% for the low-load operating point, reaffirming the importance of taking into account the cooling air in the design process. If for the sake of simplification, cooling air is ignored during the design process, the product life prediction will be missing this additional forcing mechanism for blade vibrations. Furthermore, the widening of cooling air holes of high-pressure turbine blades and vanes throughout the blade's life needs to be investigated as these changes will continuously influence High Cycle Fatigue.

Acknowledgements Funded by the Deutsche Forschungsgemeinschaft (DFG, German Research Foundation)—SFB 871/3—119193472. The authors would like to thank the DFG for the support. Moreover, the authors would like to acknowledge the substantial contribution of the DLR Institute of Propulsion Technology by providing TRACE code for CFD in turbomachinery. This work was supported by the LUH compute cluster, which is funded by the Leibniz University Hannover, the Lower Saxony Ministry of Science and Culture (MWK) and the DFG. Thus, the authors acknowledge the support of the cluster's system team in the production of this work. Last, we thank Pedro Henrique Calderaro Rodrigues for his help creating the probabilistic chain and an initial data set, as well as Felix Ludeneit for providing substantial contribution on conducting the CFD simulations and data analysis.

References

Aschenbruck, J. and Seume, J. R. (2015). Experimentally verified study of regeneration-induced forced response in axial turbines. *Journal of Turboma- chinery*, 137(3):031006.

Aurenhammer, F. (1991). Voronoi diagrams—a survey of a fundamental geometric data structure. *ACM Computing Surveys (CSUR)*, 23(3):345–405.

Ernst, B., Seume, J. R., and Herbst, F. (2016). Probabilistic cfd-analysis of regeneration-induced geometry variances in a low-pressure turbine. In *52nd AIAA/SAE/ASEE Joint Propulsion Conference*, page 4555.

Hauptmann, T., Meinzer, C. E., and Seume, J. R. (2018). Experimental validation of forced response methods in a multi-stage axial turbine. In *Turbo Expo: Power for Land, Sea, and Air*, volume 51159, page V07CT36A009. American Society of Mechanical Engineers.

Heinze, K., Meyer, M., Scharfenstein, J., Voigt, M., and Vogeler, K. (2014). A parametric model for probabilistic analysis of turbine blades considering real geometric effects. *CEAS Aeronautical Journal*, 5(1):41–51.

Kato, M. (1993). The modelling of turbulent flow around stationary and vibrating square cylinders. *Turbulent Shear Flow*, 1:10–4.

Kersken, H.-P., Frey, C., Voigt, C., and Ashcroft, G. (2012). Time-linearized and time-accurate 3d rans methods for aeroelastic analysis in turbomachinery.

McKay, M. D., Beckman, R. J., and Conover, W. J. (2000). A comparison of three methods for selecting values of input variables in the analysis of output from a computer code. *Technometrics*, 42(1):55–61.

Menter, F. R., Kuntz, M., and Langtry, R. (2003). Ten years of industrial experience with the sst turbulence model. *Turbulence, heat and mass transfer*, 4(1):625–632.

Moré, J. J. (1978). The levenberg-marquardt algorithm: implementation and theory. In *Numerical analysis*, pages 105–116. Springer.

Nürnberger, D., Eulitz, F., Schmitt, S., and Zachcial, A. (2001). Recent progress ini the numerical simulation of unsteady viscous multistage turbomachinery flow.

Plischke, E., Borgonovo, E., and Smith, C. L. (2013). Global sensitivity measures from given data. *European Journal of Operational Research*, 226(3):536–550.

Pohle, L., Panning-von Scheidt, L., Wallaschek, J., Aschenbruck, J., and Seume, J. R. (2014). Dynamic behavior of a mistuned air turbine: comparison between simulations and measurements. In *Turbo Expo: Power for Land, Sea, and Air*, volume 45776, page V07BT33A015. American Society of Mechanical Engineers.

Stania, L. and Seume, R. J. (2022). Robust probabilistic analysis of deterioration- induced aeroe-lasticity in an axial turbine. In *Turbo Expo: Power for Land, Sea, and Air*. American Society of Mechanical Engineers.

Voigt, C., Frey, C., and Kersken, H.-P. (2010). Development of a generic surface mapping algorithm for fluid-structure-interaction simulations in turbomachinery. In *V European Conference on Computational Fluid Dynamics ECCOMAS CFD*, volume 2010.

Prediction of Fatigue Lifetime Using a Wavelet Transformation Induced Multi-time Scaling Method and Xfem

Jian Sun, Stefan Löhnert, and Tengfei Lyu

Abstract Fatigue lifetime prediction due to dynamic crack growth is a significant issue for design and manufacture of engineering components. Damage accumulation in a material on the micro-scale is a main physical mechanism governing crack initiation. For high-cycle fatigue, the number of loading cycles leading to catastrophic fatigue failure can be of the order of millions. The simulation of such processes would be extremely expensive and time-consuming using conventional single time scale methods. In order to overcome this challenging requirement, a wavelet-transformation based multi-time scaling method is successfully adopted in this research to accelerate the prediction of accumulated damage for a large number loading cycles. In this work, the WATMUS technique is coupled with gradient-enhanced damage and the extended finite element method to simulate dynamic crack propagation for a turbine blade.

Keywords Crack growth · Wavelet · Turbine blades · XFEM

1 Introduction

To reduce the operating costs, many captital industrial components, such as turbine blades of aircraft engines, wind turbines or automotive components need regular maintenance and repair. In general, such engineering components are subjected to cyclic loading with high frequencies leading to severe fatigue failure. Fatigue crack initiation and propagation is a main physical mechanism governing the fatigue failure process. To predict the fatigue life of a component, fatigue tests are carried out by applying usually a constant amplitude cyclic load to measure the rate of crack growth.

J. Sun · S. Löhnert (✉)
Institute of Mechanics and Shell Structures, Technische Universität Dresden, August-Bebel-Str. 30, 01219 Dresden, Germany
e-mail: stefan.loehnert@tu-dresden.de

T. Lyu
Institute of Continuum Mechanics, Leibniz University Hannover, An der Universitaet 1, 30823 Garbsen, Germany

© The Author(s) 2025
J. R. Seume et al. (eds.), *Regeneration of Complex Capital Goods*,
https://doi.org/10.1007/978-3-031-51395-4_16

This process is expensive and time-consuming. Therefore in this present research, a numerical method for evaluation of the remaining lifetime of these individual components before and after the regeneration process is developed. This work is established in the collaborative research centre (CRC) 871 with the goal of predicting the functional properties of capital engineering components, like turbine blades. In the previous work, a new multiscale technique has been proposed for the efficient simulation of crack propagation and crack coalescence of macrocracks and microcracks (Holl et al. 2013). Contact formulation is implemented to enforce the non-penetration condition for crack surfaces (Kunin et al. 2017). In a gas turbine engine, air is compressed by the compressor. The temperature of air raises due to the compression and the combustion of fuel in the combustor. The air with high pressure and high temperature passes through the gas turbine. Then the turbine blades of air engines are responsible for transforming the energy of the pressurized exhaust gas to kinetic energy that can be used. The turbine blades are often the limiting components due to the complex and extreme loading conditions. An accurate prediction of the fatigue lifetime for a turbine blade is of high interest due to safety reasons and high replacement costs. The focus of this work lies on accurate and efficient simulation of fatigue failure due to dynamic crack growth under high-cycle loading combined with high thermal loads.

Nowadays, some numerical methods can be used to simulate crack growth. The finite element method (FEM) has been combined with linear elastic fracture mechanics (LEFM) for fatigue failure analysis (Colombo and Giglio 2006; Branco et al. 2015). The boundary element method (BEM) is also an alternative tool for crack simulations (Peng et al. 2017; Portela et al. 1992; Yan 2006). In recent years, phase- field models have gained more attention in fracture modeling. A severe drawback of FEM, as well as the phase-field models for crack simulation is the necessity for severe mesh refinement near the crack surfaces. For FEM, a remeshing process around the crack front is necessary as the crack grows, leading to poor computational efficiency. Hence, dealing with the possible inaccuracy and high computational costs still remains a challenging requirement for crack simulation using FEM. In this work, the extended finite element method (XFEM) (Belytschko and Black 1999; Moës et al. 1999), which was developed by adding proper enrichment functions to the finite element approximations using the partition of unity framework, is adopted. Level set techniques are frequently used to handle the geometric update of crack configurations during crack propagation (Gravouil et al. 2002; Oliver et al. 2004; Sukumar et al. 2003).

The formation of cracks on the macro-scale is a result of damage accumulation in a material on the micro-scale. In continuum damage mechanics, a scalar damage variable is often introduced to describe the loss of material integrity and the reduction of material stiffness. An assumption is that a crack starts to initiate when the accumulated damage in a material reaches a critical value. The results of early finite element computations including a local damage model show severe mesh sensitivity. This can be explained as the damage in material always localizes in the weaker part of the material. To overcome these shortcomings, a gradient-enhanced damage formulation (Peerlings et al. 2000) has been developed for finite element solutions of quasi-brittle and fatigue fractures, and therefore will be adopted in this research. A gradient-enhanced damage model combined with the XFEM is developed to simulate crack growth with the consideration of thermal loading. Dynamics is taken into account by considering inertial effects. The key issue in the prediction of the high-cycle fatigue life is the efficient computation of a large number of loading cycles in a reasonable time. With a conventional single time scale method, each loading cycle is divided into a number of small time steps to ensure an appropriate accuracy. The huge number of time steps in the fatigue life prediction can make the problems nearly unsolvable within a reasonable amount of time. A wavelet transformation induced multi-time scaling (WATMUS) algorithm is applied to improve the computational efficiency for prediction of damage evolution after millions of loading cycles (Joseph et al. 2010; Chakraborty and Ghosh 2013; Yaghmaie et al. 2016; Ghosh and Chakraborty 2013). This method benefits from the split of time into two scales: (i) a fine time scale τ with a rapidly oscillating high-frequency response within a complete cycle, and (ii) a coarse time scale with a low-frequency material response over the entire loading process. The material response in the coarse time scale is a function of the cycle number N, and hence behaves monotonically. The information of fine-scale response of material variables is transformed to the computation and modification of the wavelet coefficients by means of discrete wavelet transformation, and the cycle scale material response is evaluated by appropriate coarse scale rate equations. The coarse scale evolution equation is connected to the fine scale solutions. The contribution presented in this article is the combination of WATMUS, gradient-enhanced damage and XFEM. The evolution of the local damage in the coarse time scale is accurately predicted using the "cycle jump" algorithm in WATMUS. A significant speed-up is obtained comparing the single time scale method.

This paper is organized as follows: In Sect. 2, the elasticity-based damage evolution law for high-cycle fatigue is introduced along with the governing equations and the gradient enhancement of the damage model. A simulation of thermo-mechanical dynamic crack propagation for a turbine blade is demonstrated. In Sect. 3, the mathematical details of the WATMUS method are given including a numerical example to show the high efficiency and accuracy of the WATMUS method for prediction of high-cycle fatigue damage evolution. Finally in Sect. 4 we make conclusions of the research in our subproject within the program of CRC 871, and discuss the possibility of future application of this work.

2 Thermo-mechanical Dynamic Fatigue Crack Growth Using XFEM

Duflot (2008) applied the XFEM for the simulation of steady-state thermo-elastic fractures. This idea can be extended to three-dimensional dynamic crack propagation analysis with transient heat conduction.

2.1 Governing Equations

Due to the high frequency of the external load, the plastic strain can be neglected compared to the elastic part. Hence, a linear-elastic, isotropic material is assumed in this work. Consider a thermally and mechanically loaded, cracked solid in a domain Ω bounded by Γ, the governing equations for damage-based thermo-mechanical dynamic problems with the assumption of small deformation can be defined by

$$\rho c_p \dot{T} + \nabla \cdot \mathbf{q} = 0, \mathbf{q} = -k\nabla T \text{ in } \Omega \tag{1}$$

$$\nabla \cdot \boldsymbol{\sigma} + \mathbf{f} = \rho \ddot{\mathbf{u}} \text{ in } \Omega \tag{2}$$

$$\Phi - c\Delta\Phi = \tilde{D} \text{ in } \Omega \tag{3}$$

$$\boldsymbol{\sigma} = (1 - \Phi)\mathbb{D} : (\boldsymbol{\varepsilon} - \boldsymbol{\varepsilon}_T), \boldsymbol{\varepsilon} = \nabla_s \mathbf{u}, \boldsymbol{\varepsilon}_T = \alpha(T - T_0)\mathbf{I} \tag{4}$$

In these equations, $\boldsymbol{u}$ is the displacement field, $\ddot{\mathbf{u}}$ is the acceleration, T is the temperature field, heat flux vector is given as $\boldsymbol{q}$, stress tensor is $\boldsymbol{\sigma}$, strain tensor is $\boldsymbol{\varepsilon}$, thermal strain tensor is defined by $\boldsymbol{\varepsilon}_T$ with the initial temperature field T_0, local damage is given as $\tilde{D}$, and gradient-enhanced damage field ϕ. There are a variety of damage models to compute the local damage $\tilde{D}$ for brittle and ductile materials. The details about the damage model used in this reseach will be introduced in the next section. The gradient-enhanced damage Eq. 3 proposed in Boggess and Narcowich (2011) is strongly coupled with the displacement and the temperature field through Eq. 4. The material properties include the elastic fourth-order Hooke's tensor $\mathbb{D}$, thermal conductivity k, expansion coefficient α, and c is a nonlocality parameter that can be defined by $c = l^2/2$, where l denotes the internal length of the nonlocality. We assume that there is no body force $\boldsymbol{f} = \boldsymbol{0}$. $\boldsymbol{I}$ is second-order identity tensor and ∇_s denotes the symmetric gradient operator.

2.2 Elasticity Based High-Cycle Fatigue Damage

The cyclic external loading causes a repetitive accumulation of fatigue damage, which leads to initiation, growth and coalescence of cracks at the mesoscale and macroscale levels. The existing experiments of high-cycle fatigue revealed that the damage increases slowly at the beginning of the fatigue process, and then growth of damage accelerates until a sudden loss of stiffness can be observed near the end of the fatigue analysis. To capture this evolving property, the rate equation of high-cycle fatigue damage growth suggested by Peerlings et al. (2000) is used, which can be written as

$$\dot{D} = \begin{cases} g\left(\tilde{D}, \tilde{\varepsilon}\right)\dot{\tilde{\varepsilon}}, & \text{if } f \geq 0, \dot{f} \geq 0 \text{ and } \tilde{D} < 1 \\ 0, & \text{else} \end{cases} \tag{5}$$

with the evolution function expressed by

$$g\left(\tilde{D}, \tilde{\varepsilon}\right) = Ce^{\alpha\tilde{D}}\tilde{\varepsilon}^{\beta} \tag{6}$$

In this equation, C, α and β are material constants, f is a loading function that can be defined by

$$f\left(\tilde{\varepsilon}, \kappa\right) = \tilde{\varepsilon} - \kappa_0 \tag{7}$$

where $\tilde{\varepsilon}$ is the equivalent strain defined in terms of the strain tensor, and κ_0 is given as the threshold value defining the elastic domain. It can be seen that if the equivalent strain during the loading remains smaller than κ_0, the material is always located in the elastic region and no damage development can be observed in all loading cycles. In this case, the fatigue failure will never occur. The condition $f \geq 0$ implies that the damage can increase only when the equivalent strain is outside of the elastic domain. Additionally, the damage variable is allowed to develop only during a loading process, i.e. $\dot{f} \geq 0$.

2.3 Discretization with the Extended Finite Element Method

Based on the discretization scheme in FEM, the displacement field in the XFEM (Belytschko and Black 1999; Gravouil et al. 2002) is described by introducing additional degrees of freedom to the nodes and enrichment functions to capture the discontinuity and singularity of the elements intersected by the cracks. The discretized form of the displacement field in the XFEM can be given as

$$\mathbf{u}^h(\mathbf{x}) = \sum_{I \in \mathcal{I}} N_I(\mathbf{x})\mathbf{u}_I + \sum_{J \in \mathcal{J}} N_J(\mathbf{x})H(\mathbf{x})\mathbf{a}_J + \sum_{K \in \mathcal{K}} \sum_{j=1}^{4} N_K(\mathbf{x})f_j(\mathbf{x})\mathbf{b}_{Kj} \tag{8}$$

The first term is the same as the standard FE displacement approximation and therefore the nodal subset I contains all nodes. The second and third terms are the additional parts enriched by the Heaviside function $H(\mathbf{x})$ and some crack tip enrichment functions $f_j(\mathbf{x})$. Generally, four crack tip enrichment function stemming from linear elastic fracture mechanics can be used:

$$f_{1-4}(\mathbf{x}) = \left\{ \sqrt{r}\sin\left(\frac{\theta}{2}\right), \sqrt{r}\cos\left(\frac{\theta}{2}\right), \sqrt{r}\sin\left(\frac{\theta}{2}\right)\sin\theta, \sqrt{r}\sin\left(\frac{\theta}{2}\right)\cos\theta \right\} \tag{9}$$

where (r, θ) represent the polar coordinate system with a center at the crack tip. The nodal subset $J \subset \mathcal{I}$ in Eq. 8 contains all nodes of the elements that are completely cut by the crack, and the nodal subset $\mathcal{K} \subset \mathcal{I}$ contains all nodes of the elements in which the crack tip or the crack front is located. The Heaviside function is defined by

$$H(\mathbf{x}) = \begin{cases} 1 : \varphi(\mathbf{x}) \geq 0 \\ -1 : \varphi(\mathbf{x}) < 0 \end{cases} \tag{10}$$

where $\varphi(\mathbf{x})$ is the signed distance function from $\mathbf{x}$ to the crack surface. Due to the geometric discontinuity caused by the cracks, the temperature field and the gradient-enhanced damage are also discontinuously distributed across the cracks. Singularity of the temperature gradient at the crack tip or crack front can appear as well. To fulfill these requirements, both the temperature field and the gradient enhanced damage are approximated similar to the displacement field by

$$T^h(\mathbf{x}) = \sum_{I \in \mathcal{I}} \overline{N}_I(\mathbf{x}) T_I + \sum_{J \in \mathcal{J}} \overline{N}_J(\mathbf{x}) H(\mathbf{x}) a_J^T + \sum_{K \in \mathcal{K}} \sum_{j=1}^{4} \overline{N}_K(\mathbf{x}) f_j(\mathbf{x}) b_{Kj}^T \tag{11}$$

$$\Phi^h(\mathbf{x}) = \sum_{I \in \mathcal{I}} \tilde{N}_I(\mathbf{x}) \Phi_I + \sum_{J \in \mathcal{J}} \tilde{N}_J(\mathbf{x}) H(\mathbf{x}) a_J^\Phi + \sum_{K \in \mathcal{K}} \sum_{j=1}^{4} \tilde{N}_K(\mathbf{x}) f_j(\mathbf{x}) b_{Kj}^\Phi \tag{12}$$

If only the first crack tip enrichment function $f_1(\mathbf{x}) = \sqrt{r} \sin(\theta/2)$ is used to reduce the number of unknowns for the coupled problem and significantly decrease the condition number of the resulting coefficient matrices (Legrain et al. 2005; Loehnert et al. 2011), these approximations can be written in a more compact matrix form

$$\mathbf{u}^h(\mathbf{x}) = \sum_I [N_I \ N_I H \ N_I f_1] \begin{bmatrix} \mathbf{u}_I \\ \mathbf{a}_I \\ \mathbf{b}_{I1} \end{bmatrix} = \sum_I \mathbf{N}_I \bar{\mathbf{u}}_I$$

$$T^h(\mathbf{x}) = \sum_I \overline{\mathbf{N}}_I \overline{\mathbf{T}}_I, \ \Phi^h(\mathbf{x}) = \sum_I \tilde{\mathbf{N}}_I \overline{\Phi}_I \tag{13}$$

The displacement field, temperature and gradient enhanced damage are strongly coupled with each other. We consider the weak form of the governing equations for all fields supplemented by the associated boundary conditions. The governing Eqs. (1–3) are multiplied by their corresponding test functions δT, δu and $\delta \phi$, respectively, and subsequently integrated in the domain Ω. Using the divergence theorem, the resulting equation leads to the weak form:

$$\mathbf{R_u} = \int_\Omega \delta\varepsilon \, : \, \sigma d\Omega + \int_\Omega \delta\mathbf{u} \, \cdot \, \rho\ddot{u}d\Omega - \int_\Gamma \delta\mathbf{u} \, \cdot \, \tilde{t}d\Gamma \stackrel{!}{=} 0$$

$$\mathbf{R}_\Phi = \int_\Omega \delta\Phi(\tilde{D} - \Phi)d\Omega - c\int_\Omega \nabla\delta\Phi \cdot \nabla\Phi d\Omega \stackrel{!}{=} 0$$

$$\mathbf{R}_T = \int_\Omega \delta T\rho c_p \dot{T}d\Omega - \int_\Omega (\nabla\delta T) \, \cdot \, \mathbf{q}d\Omega + \int_{\Gamma_q} \delta T\mathbf{q} \cdot \mathbf{n}d\Gamma \stackrel{!}{=} 0 \qquad (14)$$

Applying the Bubnov-Galerkin method and after the complete discretization of $u, \ddot{u}, T$ and ϕ with XFEM ansatz functions, the set of non-linearly coupled equations in each iteration step of the Newton-Raphson procedure can be obtained

$$\begin{bmatrix} \mathbf{M}_{uu} & 0 & 0 \\ 0 & 0 & 0 \\ 0 & 0 & 0 \end{bmatrix} \begin{bmatrix} \Delta\ddot{\overline{u}} \\ \Delta\ddot{\overline{\Phi}} \\ \Delta\ddot{\overline{T}} \end{bmatrix} + \begin{bmatrix} 0 & 0 & 0 \\ 0 & 0 & 0 \\ 0 & 0 & \mathbf{C}_{TT} \end{bmatrix} \begin{bmatrix} \Delta\dot{\overline{u}} \\ \Delta\dot{\overline{\Phi}} \\ \Delta\dot{\overline{T}} \end{bmatrix}$$

$$+ \begin{bmatrix} \mathbf{K}_{uu} & \mathbf{K}_{u\Phi} & \mathbf{K}_{uT} \\ \mathbf{K}_{\Phi u} & \mathbf{K}_{\Phi\Phi} & 0 \\ 0 & 0 & \mathbf{K}_{TT} \end{bmatrix} \begin{bmatrix} \Delta\overline{u} \\ \Delta\overline{\Phi} \\ \Delta\overline{T} \end{bmatrix} = \begin{bmatrix} \mathbf{R_u} \\ \mathbf{R}_\Phi \\ \mathbf{R}_T \end{bmatrix}. \qquad (15)$$

with the definitions of the matrices

$$\mathbf{M}_{uu} = \int_\Omega \mathbf{N}^T \rho \mathbf{N}d\Omega$$

$$\mathbf{C}_{\theta\theta} = \int_\Omega \mathbf{N}\rho c_p \mathbf{N}^T d\Omega$$

$$\mathbf{K}_{uu} = \int_\Omega (1 - \Phi)\mathbf{B}^T \mathbf{DB}d\Omega$$

$$\mathbf{K}_{u\Phi} = -\int_\Omega \mathbf{B}^T \mathbf{D}(\varepsilon - \varepsilon_T)\tilde{\mathbf{N}}^T d\Omega$$

$$\mathbf{K}_{uT} = -\int_\Omega (1 - \Phi)\mathbf{B}^T \mathbf{D}\alpha\tilde{\mathbf{IN}}^T d\Omega$$

$$\mathbf{K}_{\Phi u} = -\int_\Omega \tilde{\mathbf{N}}qs^T \mathbf{B}d\Omega$$

$$\mathbf{K}_{\Phi\Phi} = \int_\Omega \left[\tilde{\mathbf{N}}\tilde{\mathbf{N}}^T + \tilde{\mathbf{B}}c\tilde{\mathbf{B}}^T\right]d\Omega$$

$$\mathbf{K}_{TT} = \int_\Omega \overline{\mathbf{B}}k\overline{\mathbf{B}}^T d\Omega \qquad (16)$$

where q is given by Peerlings et al. (2000)

$$q = \frac{(1-\theta)g(D_n, \varepsilon_n) + \theta g\left(D_{n+1}^{(i)}, \varepsilon_{n+1}^{(i)}\right) + \theta \frac{\partial g}{\partial \varepsilon}|_{n+1}^{(i)}\left(\tilde{\varepsilon}_{n+1}^{(i)} - \tilde{\varepsilon}_n\right)}{1 - \theta \frac{\partial g}{\partial D}|_{n+1}^{(i)}\left(\tilde{\varepsilon}_{n+1}^{(i)} - \tilde{\varepsilon}_n\right)} \tag{17}$$

and the column vector $\mathbf{s}$ is a partial derivate $\mathbf{s} = \partial\tilde{\varepsilon} / \partial\boldsymbol{\varepsilon}$. These equations can be solved using the Newmark-beta method.

2.4 Dynamic Crack Propagation Using the Level Set Methods

Employing the XFEM for crack propagation simulations, the evolving geometry of the crack needs to be tracked. A crack front in the three-dimensional space can be determined by interaction of two orthogonal signed distance functions, also called level sets, that is, the crack surface level set φ and the crack front level set Φ. In this work, the level set method (Gravouil et al. 2002) and a fast marching method (FMM) (Sukumar et al. 2003) are combined for tracking the growth of three-dimensional cracks implicitly. The velocity for crack growth is extended by solving two Hamilton Jacobi equations to steady state in the level set subdomain simultaneously.

2.4.1 Update of ϕ Using a Global Crack Tracking Method

Let $\mathbf{v}$ represent the velocity of crack surface extension, $\mathbf{n}_\phi = \nabla\phi$ is a normal vector perpendicular to the iso-zero surface of the level set ϕ. Two independent vectors T and S on the updated crack surface can be defined by $\mathbf{T} = \mathbf{v}$ and $\mathbf{S} = \frac{\nabla\psi \times \nabla\phi}{\|\nabla\psi \times \nabla\phi\|}$. It can be seen that the vectors T and S are perpendicular to the normal vector $\mathbf{n}_\phi$ of the searched surface. The geometry update of the crack surface is equivalent to the solution of $\mathbf{T} \cdot \nabla\phi = \mathbf{S} \cdot \nabla\phi = 0$. Here we employ the global crack tracking method (Oliver et al. 2004) to solve the extension of velocity more efficiently. Multiplying the equations $\mathbf{T} \cdot \nabla\phi = 0$ and $\mathbf{S} \cdot \nabla\phi = 0$ by T and S, respectively, we obtain:

$$\mathbf{T} \otimes \mathbf{T} \cdot \nabla\phi = \mathbf{S} \otimes \mathbf{S} \cdot \nabla\phi = \mathbf{0} \tag{18}$$

The solution of the following boundary value problem in the level set subdomain satisfies the differential Eq. 18:

$$\nabla \cdot \mathbf{q} = 0, \mathbf{q} = -\mathcal{K} \cdot \nabla\phi \text{ in } \Omega \tag{19}$$

$$\overline{q} \equiv \mathbf{q} \cdot \mathbf{n} \text{ on } \partial\Omega \qquad (20)$$

$$\mathcal{K} = \mathbf{T} \otimes \mathbf{T} + \mathbf{S} \otimes \mathbf{S} + \varepsilon \mathbf{I} \qquad (21)$$

where $\varepsilon \mathbf{I}$ is an artificial treatment to avoid the singularity of the tensor $\mathbf{K}$.

2.4.2   Update of Ψ by Solving Hamilton–Jacobi Equation

To track the motion of crack front, the scheme known as the Hamilton–Jacobi equation of motion can be utilized

$$\frac{\partial \psi}{\partial t} + \mathbf{v} \cdot \nabla \psi = 0 \qquad (22)$$

where $\mathbf{v}$ is the velocity of crack propagation.

2.4.3   Reinitialization of Ψ and ϕ Using FMM

A signed distance function must satisfy the Eikonal equation

$$\|\nabla \phi\| = 1 \qquad (23)$$

When the Hamilton-Jacobi equations are solved numerically, the reinitialization of ψ and ϕ by the FMM (Sukumar et al. 2003) can ensure the Eikonal equation to be almost fulfilled, which can prevent numerical instabilities.

2.4.4   Reorthogonalization of Ψ and ϕ with Embedded Reinitialization

After updating the level set function ψ, it needs to be orthogonalized to ϕ by solving the following Hamilton–Jacobi equation

$$R_e = \frac{\partial \psi}{\partial t} + sign(\phi)\frac{\nabla \phi}{\|\nabla \phi\|} \cdot \nabla \psi = 0 \qquad (24)$$

It can be seen that, the orthogonality condition $\nabla \phi \cdot \nabla \psi = 0$ is fulfilled automatically when the above equation is solved to the steady state. The orthogonalization process by solving the Eq. 24 may destroy the signed distance property of ψ. To ensure the signed distance property, the Eikonal equation $\|\nabla \psi\| = 1$ is embedded into Eq. 24 by means of a weak form. The Hamilton–Jacobi equation can be solved using the Galerkin Least Square (GLS) method (Beese et al. 2018). The potential of the least square part is defined by

$$\Pi^{LS} = \frac{1}{2}\int_\Omega \tau R_e^2 d\Omega \tag{25}$$

The stabilization parameter τ is dependent on the element size. For the determination and physical interpretation of τ the reader is referred to Barth and Sethian (1998) and Beese et al. (2018). The Eikonal equation is enforced by adding a penalty term. The potential of the penalty part is given by

$$\Pi^{PE} = \frac{1}{2}\int_\Omega \gamma(\|\nabla\psi\| - 1)^2 d\Omega \tag{26}$$

where γ is the penalty parameter. Then the weak form can be obtained by

$$\delta\Pi = \delta\Pi^G + \delta\Pi^{LS} + \delta\Pi^{PE} = 0 \tag{27}$$

where

$$\delta\Pi^G = \int_\Omega \delta\psi R_e d\Omega = \int_\Omega \delta\psi\left(\frac{\partial\psi}{\partial\tau} + sign(\psi)\frac{\nabla\phi}{\|\nabla\phi\|}\cdot\nabla\psi\right)d\Omega$$

$$\delta\Pi^{LS} = \int_\Omega \delta R_e \tau R_e d\Omega$$

$$= \int_\Omega \left(\frac{\partial\delta\psi}{\partial\tau} + sign(\phi)\frac{\nabla\phi}{\|\nabla\phi\|}\cdot\nabla\delta\psi\right)\tau\left(\frac{\partial\psi}{\partial\tau} + sign(\phi)\frac{\nabla\phi}{\|\nabla\phi\|}\cdot\nabla\psi\right)d\Omega$$

$$\delta\Pi^{PE} = \int_\Omega \delta(\|\nabla\psi\|)\gamma(\|\nabla\psi\| - 1)d\Omega = \int_\Omega \gamma\left(1 - \frac{1}{\|\nabla\psi\|}\right)\nabla\psi\cdot(\nabla\delta\psi)d\Omega \tag{28}$$

The above formulation can be discretized with the standart finite element procedure.

2.5 Numerical Results of Dynamic Crack Propagation

To demonstrate the applicability of the presented model to dynamic crack propagation problems, a turbine blade with a predefined crack at the front edge, as shown in Fig. 1, is simulated. A turbine blade is subjected to high centrifugal force with thermal loading up to 1400 °C. The dynamic propagation of a predefined crack under high-frequency centrifugal force is investigated with the developed model. Due to the complexity of the geometry, the turbine blade is discretized with ten-node tetrahedral elements with quadratic shape functions. The simulated distribution of the gradient enhanced damage field after a few steps of crack propagation is shown in Fig. 2.

Fig. 1 Mesh of a turbine blade with a predefined edge crack

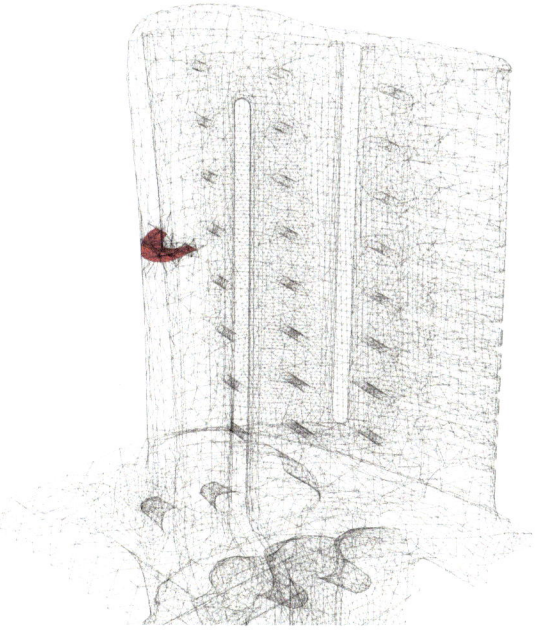

Fig. 2 Simulation results of the gradient enhanced damage field with dynamic crack propagation

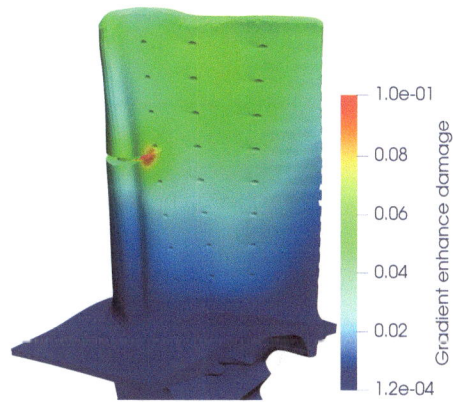

3 Review of the Wavelet Transformation Induced Multi-time Scaling Method (WATMUS)

The wavelet transformation based multi-time scaling (WATMUS) methodology has been successfully developed and implemented to accelerate the time integration in crystal plasticity finite element analysis for a large number of loading cycles Joseph et al. (2010); Chakraborty and Ghosh (2013); Yaghmaie et al. (2016); Ghosh and Chakraborty (2013). In this methodology, the continuous time is split into two scales: a coarse time scale identified with the cycle number N, and a fine time scale τ within each cycle. The value of any state variable α_0 at the beginning of a given cycle N can be thought of as a coarse time scale variable. This state variable will not vary in the fine τ-scale within each cycle and hence, it can be considered as a function of the cycle number N, i.e. $\alpha_0 = \alpha_0(N)$. The objective is to find a coarse time scale (cycle scale) evolution equation in the form

$$\frac{d\alpha_0(N)}{dN} = f\left(\alpha_0(N), u_{i,k}^{\alpha}(N)\right) \tag{29}$$

where u^{α} are the wavelet coefficients for the i-th displacement component of node α. Thus, a two-scale representation of any given variable β can be obtained in terms of the orthogonal wavelet basis functions as

$$\beta(t) = \beta(N, \tau) = \sum_{k=1}^{n} \beta^k(N)\psi_k(\tau) \tag{30}$$

The coefficients β^k depend only on the cycle number N. The wavelets are defined in a finite domain and have compact support. Hence, the sudden change of a material response can be well captured by wavelet-based solutions without the Gibbs phenomena that occurs in the Fourier transformation based solutions. The multiresolution analysis represents a square integrable function in continuous time at different resolutions, and hence an optimal number of coefficients needs to be solved.

3.1 Fundamentals of Wavelets

The wavelet multiresolution analysis (Strang and Nguyen 1997; Boggess and Narcowich 2011) is based on a nested sequence of subspaces at different resolutions

$$\cdots \subset V_{-1} \subset V_0 \subset V_1 \subset V_2 \subset \cdots \subset V_m \subset \cdots \subset L^2(\mathbb{R}) \tag{31}$$

Each subspace V_m is spanned by a set of basis functions constructed from dilations and translations of the scaling function $\phi(\tau)$ defined by:

$$\phi_{m,n}(\tau) = 2^{\frac{m}{2}}\phi(2^m\tau - n) \tag{32}$$

where m denotes the number of resolution and n specifies the number of transition. The multiresolution analysis needs an orthogonal complementary space denoted by W_m. The subspace W_m is spanned by a set of replicas constructed from a single mother wavelet $\psi(\tau)$ at the resolution level m, that is:

$$\psi_{m,n}(\tau) = 2^{\frac{m}{2}}\psi(2^m\tau - n) \tag{33}$$

Any square integrable function $f(\tau)$ can be approximated by its projection onto each subspace V_m and W_m as

$$f_m(\tau) = \sum_n c_{m,n}\phi_{m,n}(\tau), \quad w_m(\tau) = \sum_n d_{m,n}\psi_{m,n}(\tau) \tag{34}$$

Since $V_0 \subset V_1$, the scaling function $\phi(\tau) \in V_0$ can be expanded in terms of the basis that spans V_1 as

$$\phi(\tau) = \sum_k h_k 2^{\frac{1}{2}}\phi(2\tau - k) \tag{35}$$

where h_k corresponds to the low pass filter coefficients. Eq. 35 is known as the dilation or refinement equation. Moreover, W_0 is spanned by the wavelet function $\psi(\tau)$ and its translates $\psi(\tau - n)$. Since $\psi(\tau) \in W_0 \subset V_1$, in analogy to Eq. 35, we have an expansion of $\psi(\tau)$ as

$$\psi(\tau) = \sum_k g_k 2^{\frac{1}{2}}\phi(2\tau - k) \tag{36}$$

where g_k represents the high pass filter coefficients. Filter coefficients for the Daubechies-4 wavelet used in this work (Strang and Nguyen 1997; Boggess and Narcowich 2011) are given as

$$h_0 = \frac{1 + \sqrt{3}}{4\sqrt{2}}, h_1 = \frac{3 + \sqrt{3}}{4\sqrt{2}}, h_2 = \frac{3 - \sqrt{3}}{4\sqrt{2}}, h_3 = \frac{1 - \sqrt{3}}{4\sqrt{2}} \tag{37}$$

$$g_0 = \frac{1 - \sqrt{3}}{4\sqrt{2}}, g_1 = -\frac{3 - \sqrt{3}}{4\sqrt{2}}, g_2 = \frac{3 + \sqrt{3}}{4\sqrt{2}}, g_3 = -\frac{1 + \sqrt{3}}{4\sqrt{2}} \tag{38}$$

Based on the decomposition $V_{m+1} = V_m \oplus W_m$, the approximation of an arbitrary function $f(\tau) \in L^2(\mathbb{R})$ at the resolution of $m + 1$, that is, $f_{m+1}(\tau)$, can be obtained by adding the lower-level approximation $f_m(\tau)$ and its orthogonal complement $w_m(\tau)$.

$$f_{m+1}(\tau) = \sum_l c_{m+1,l}\phi_{m+1,l}(\tau) = \underbrace{\sum_n c_{m,n}\phi_{m,n}(\tau)}_{f_m(\tau)} + \underbrace{\sum_n d_{m,n}\psi_{m,n}(\tau)}_{w_m(\tau)} \quad (39)$$

Using the above equations and defining $a_{m,n} = 2^{\frac{m}{2}} c_{m,n}$ and $b_{m,n} = 2^{\frac{m}{2}} d_{m,n}$, the analysis equation can be obtained

$$\begin{bmatrix} \mathbf{a}_{m-1} \\ \mathbf{b}_{m-1} \end{bmatrix} = \frac{1}{\sqrt{2}} \begin{bmatrix} \mathbf{H}_m \\ \mathbf{G}_m \end{bmatrix} \mathbf{a}_m \quad (40)$$

where

$$\begin{bmatrix} \mathbf{H}_m \\ \mathbf{G}_m \end{bmatrix} = \begin{bmatrix} h_0 & h_1 & h_2 & h_3 & 0 & 0 & \dots & 0 & 0 \\ 0 & 0 & h_0 & h_1 & h_2 & h_3 & \dots & 0 & 0 \\ \vdots & \vdots & \vdots & \vdots & \vdots & \vdots & \vdots & \vdots & \vdots \\ h_2 & h_3 & 0 & 0 & 0 & 0 & \dots & h_1 & h_2 \\ g_0 & g_1 & g_2 & g_3 & 0 & 0 & \dots & 0 & 0 \\ 0 & 0 & g_0 & g_1 & g_2 & g_3 & \dots & 0 & 0 \\ \vdots & \vdots & \vdots & \vdots & \vdots & \vdots & \vdots & \vdots & \vdots \\ g_2 & g_3 & 0 & 0 & 0 & 0 & \dots & g_1 & g_2 \end{bmatrix} \quad (41)$$

This is the essential tool for obtaining the discrete wavelet transformation (DWT). The sampled values of a function $a_m = f = [f_0 f_1 f_2 \dots f_{Np-1}]$, where $N_p = 2^m$, are transformed to all levels of wavelet coefficients and the approximation coefficients of the lowest resolution through the DWT

$$\mathbf{c} = \mathbf{Tf} \quad (42)$$

The vector f with dimension $N = 2^m$ is defined by sampling the function $f(t)$ at N equally spaced points. T is an $N \times N$ orthogonal transformation matrix. Once a DWT has been performed on the given sampled signal, the detail coefficients can be analyzed and modified. The modified signal can be reconstructed through an inverse DWT by

$$\mathbf{f}_{mod} = \mathbf{T}^T \mathbf{c}_{mod} \quad (43)$$

3.2 Solving Coarse Time-Scale Evolution Equations

An implicit two-step second order backward difference scheme (BDF2) has been proposed to integrate the coarse scale variable over a number of cycles (Joseph et al.

2010), for which the residual at the N-th cycle is

$$r_c(\mathbf{x}, N) = (1 + \beta_3 \Delta N) \alpha_0(\mathbf{x}, N) - \beta_1 \alpha_0(\mathbf{x}, N - \Delta N)$$
$$- \beta_2 \alpha_0(\mathbf{x}, N - \Delta N - \Delta N_p) - \beta_3 \Delta N \alpha_0(\mathbf{x}, N + 1) \overset{!}{=} 0 \quad (44)$$

where the parameters are

$$\beta_1 = \frac{(1+r)^2}{(1+r)^2 - 1}, \beta_2 = -\frac{1}{(1+r)^2 - 1}, \beta_3 = \frac{(1+r)^2 - r - 1}{(1+r)^2 - 1} \text{ and } r = \frac{\Delta N_p}{\Delta N} \quad (45)$$

and $\alpha_{Np}(\mathbf{x}, N)$ is the internal variable at the last sampling point of cycle N. ΔN and ΔN_p are cycle jumps of the current and previous steps. Eq. 44 is solved using the Newton-Raphson iterative method for the value of $\alpha_0(N)$. The Jacobian corresponding to the above residual Eq. 44 can be obtained as

$$\frac{\partial r}{\partial \alpha_0}(\mathbf{x}, N) = (1 + \beta_3 \Delta N) - \beta_3 \Delta N \frac{\partial \alpha_0(\mathbf{x}, N + 1)}{\partial \alpha_0(\mathbf{x}, N)}. \quad (46)$$

The partial derivate $\frac{\partial \alpha_0(\mathbf{x}, N+1)}{\partial \alpha_0(\mathbf{x}, N)}$ corresponds to the variation of internal variable at the end of the cycle N with respect to the variation at the beginning. It can be obtained by applying the backward Euler scheme to the internal variable evolution law $\dot{\alpha}(\tau)$ as

$$A_{n+1} = \frac{\partial \alpha_{n+1}(N)}{\partial \alpha_n(N)} = \left[1 - \left(\frac{\partial \dot{\alpha}}{\partial \alpha} \right)_{n+1} \Delta \tau_{n+1} \right]^{-1} \quad (47)$$

Applying the chain rule to Eq. 47 recursively, Eq. 46 can be determined using the result

$$\frac{\partial \alpha_0(\mathbf{x}, N + 1)}{\partial \alpha_0(\mathbf{x}, N)} = \prod_{k=0}^{N_p} A_k \quad (48)$$

where N_p is the number of sampling points in the fine time scale response.

3.3 The WATMUS Method for Three-Dimensional Crack Propagation

In a finite element simulation with a single time scale method, the displacement field is the unknown to be solved. However, applying the WATMUS method, the wavelet coefficients for a complete cycle become the new variables. The iterative update of wavelet coefficients can be achieved by transforming the residual and the unknowns into the corresponding wavelet coefficients. Discretizing and linearizing the weak form of the balance of momentum using the implicit Newmark method yields the equations to compute the update of the unknowns

$$\mathbf{R}^\alpha(t) = \mathbf{K}^{\alpha\beta}(t)\Delta\mathbf{u}^\beta(t) \tag{49}$$

This equation can be expressed by index notation with a two-scale representation

$$R_i^\alpha(N, \tau_l) = K_{ij}^{\alpha\beta}(N, \tau_l)\Delta u_j^\beta(N, \tau_l) \tag{50}$$

where $R_i^\alpha(N, \tau_l)$ is the i-th component of the residual vector corresponding to the node α at the fine time point τ_l within the cycle N, and $u_j^\beta(N, \tau_l)$ represents the j-th component of the nodal displacement of the node β. Each component of the residuum and nodal displacement can be transformed into its corresponding wavelet coefficients by the DWT as described in Eq. 42

$$R_{i,k}^\alpha(N) = T_{kl}R_i^\alpha(N, \tau_l) \tag{51}$$

$$u_{j,k}^\beta(N) = T_{kl}u_j^\beta(N, \tau_l) \tag{52}$$

By applying the inverse DWT to Eq. 52, the iterative update of nodal displacement wavelet coefficients can be obtained

$$\Delta u_j^\beta(N, \tau_l) = T_{lm}\Delta u_{j,m}^\beta(N) \tag{53}$$

Replacing the $R_i^\alpha(N, \tau_l)$ in Eq. 51 by Eq. 50 and substituting into Eq. 52, the corresponding transformed equations are obtained

$$R_{i,k}^\alpha(N) = T_{kl}K_{ij}^{\alpha\beta}(N, \tau_l)T_{lm}\Delta u_{j,m}^\beta(N) \tag{54}$$

from which the update of nodal displacement wavelet coefficients are solved. The nodal wavelet coefficients for cycle N are initialized as the wavelet coefficients of the previous WATMUS cycle $N-\Delta N$ as defined in Eq. 44. For the i-th iteration step in the Newmark-Beta scheme, the sets of wavelet coefficients are computed by

$$\Delta u_{j,m}^\beta(N) = -K_{ijkm}^{\alpha\beta}(N)^{-1}R_{i,k}^\alpha(N)$$

$$u_{j,m}^\beta(N) = u_{j,m}^\beta(N) + \Delta u_{j,m}^\beta(N) \tag{55}$$

with the definition of the WATMUS stiffness matrix as

$$K_{ijkm}^{\alpha\beta}(N) = T_{kl}K_{ij}^{\alpha\beta}(N, \tau_l)T_{lm} \tag{56}$$

3.4 Numerical Example of Fatigue Damage Prediction

The WATMUS method is applied as an efficient and accurate method to predict the fatigue damage evolution in a three-dimensional mode-I fracture problem. A single-edge notched specimen shown in Fig. 3 is fixed to the ground and subjected to an axial cyclic load $f(t) = 10.0 \sin(2.5132 \cdot 10^5 t)$ N with a period of $T_s = 2.5$ x 10^{-5} s. The shape of the structure is 10 mm x 10 mm x 1.25 mm and is discretized with 3017 structured ten-node tetrahedral elements with quadratic shape functions. The material parameters are given in Table 1. Totally 450 cycles are computed so that the computation using single time scale method can be performed within a reasonable time. The damage evolution computed by the single time scale method is compared with results obtained by the WATMUS algorithm in Fig. 4. A speed up of 15–50 times can be achieved. From the comparison it can be concluded that the WATMUS method can accelerate the prediction of fatigue damage with a quite good accuracy.

Fig. 3 A single-edge-notch block subjected to cyclic loading

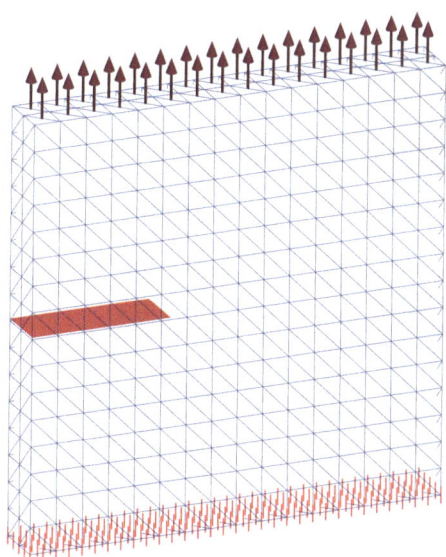

Table 1 Material parameters of the 3D block

α	β	C	E(MPa)	μ	ρ(kg/m^3)
10	1.60	1.37e + 03	2.1e + 05	0.3	7.83e + 03

Fig. 4 Comparison of the predicted coarse scale damage evolution by the WATMUS method and the single time scale method

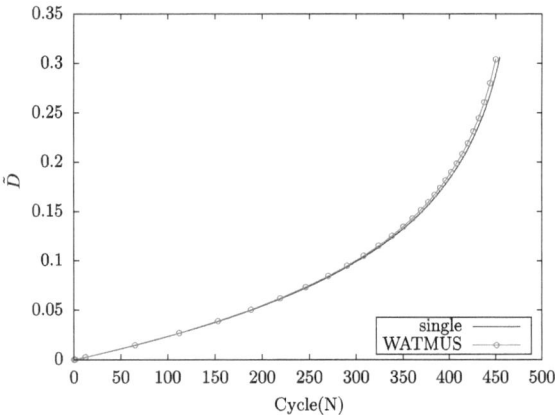

4 Conclusions

The application of a multi-time scaling method is inspired by the high computational cost due to the huge number of time steps for fatigue life prediction using a conventional single time scale FE model. This paper applies the wavelet transformation based multi-time scaling method (WATMUS) for predicting the fatigue damage growth for a huge number of loading cycles. Once the critical value of damage is reached, the crack propagation can be handled by a combination of the XFEM and level set methods. A fatigue damage evolution law, which is suitable for high-cycle fatigue growth is integrated efficiently by employing the WATMUS procedure. The highly oscillating response of the damage variable is retained through the wavelet coefficients obtained by a DWT, and the low frequency damage variable is transformed into a cycle scale variable. The coarse scale damage evolution is performed using the implicit second order backward difference scheme used within the WATMUS method. The performance of the WATMUS method has been investigated by applying it to a three-dimensional mode I problem with predefined planar cracks under cyclic loads. A speed up of 15–50 times can be achieved. The coupling of the XFEM and the WATMUS method can be a powerful tool for modeling three-dimensional fatigue crack growth. An entire fatigue life estimation due to crack initiation and propagation can be conducted in a reasonable computational time.

Acknowledgements Funded by the Deutsche Forschungsgemeinschaft (DFG, German Research Foundation) – SFB 871/3 – 119193472.

References

Barth, T. J. and Sethian, J. A. (1998). Numerical schemes for the Hamilton-Jacobi and level set equations on triangulated domains. *Journal of Computational Physics*, 145(1):1–40.

Beese, S., Löhnert, S., and Wriggers, P. (2018). 3D ductile crack propagation within a polycrystalline microstructure using XFEM. *Computational Mechanics*, 61(1- 2):71–88.

Belytschko, T. and Black, T. (1999). Elastic crack growth in finite elements with minimal remeshing. *International Journal for Numerical Methods in Engineering*, 45(5):601–620.

Boggess, A. and Narcowich, F. (2011). *A First Course in Wavelets with Fourier Analysis*. Wiley.

Branco, R., Antunes, F., and Costa, J. (2015). A review on 3D-FE adaptive remeshing techniques for crack growth modelling. *Engineering Fracture Mechanics*, 141:170–195.

Chakraborty, P. and Ghosh, S. (2013). Accelerating cyclic plasticity simulations using an adaptive wavelet transformation based multitime scaling method. *International Journal for Numerical Methods in Engineering*, 93(13):1425–1454.

Colombo, D. and Giglio, M. (2006). A methodology for automatic crack propagation modelling in planar and shell FE models. *Engineering Fracture Mechanics*, 73(4):490–504.

Duflot, M. (2008). The extended finite element method in thermoelastic fracture mechanics. *International Journal for Numerical Methods in Engineering*, 74(5):827– 847.

Ghosh, S. and Chakraborty, P. (2013). Microstructure and load sensitive fatigue crack nucleation in Ti-6242 using accelerated crystal plasticity FEM simulations. *International Journal of Fatigue*, 48:231–246.

Gravouil, A., Moës, N., and Belytschko, T. (2002). Non-planar 3D crack growth by the extended finite element and level sets-Part II: Level set update. *International Journal for Numerical Methods in Engineering*, 53(11):2569–2586.

Holl, M., Loehnert, S., and Wriggers, P. (2013). An adaptive multiscale method for crack propagation and crack coalescence. *International Journal for Numerical Methods in Engineering*, 93(1):23–51.

Joseph, D. S., Chakraborty, P., and Ghosh, S. (2010). Wavelet transformation based multi-time scaling method for crystal plasticity FE simulations under cyclic loading. *Computer Methods in Applied Mechanics and Engineering*, 199(33):2177– 2194.

Kunin, A. B., Loehnert, S., and Wriggers, P. (2017). Thermo-mechanical modeling of turbine blades taking into account structural defects. *PAMM*, 17(1):519–520.

Legrain, G., Moës, N., and Verron, E. (2005). Stress analysis around crack tips in finite strain problems using the eXtended finite element method. *International Journal for Numerical Methods in Engineering*, 63(2):290–314.

Loehnert, S., Mueller-Hoeppe, D. S., and Wriggers, P. (2011). 3D corrected XFEM approach and extension to finite deformation theory. *International Journal for Numerical Methods in Engineering*, 86(4-5):431–452.

Moës, N., Dolbow, J., and Belytschko, T. (1999). A finite element method for crack growth without remeshing. *International Journal for Numerical Methods in Engineering*, 46(1):131–150.

Oliver, J., Huespe, A. E., Samaniego, E., and Chaves, E. W. V. (2004). Continuum approach to the numerical simulation of material failure in concrete. *International Journal for Numerical and Analytical Methods in Geomechanics*, 28(7-8):609– 632.

Peerlings, R. H. J., Brekelmans, W. A. M., de Borst, R., and Geers, M. G. D. (2000). Gradient-enhanced damage modelling of high-cycle fatigue. *International Journal for Numerical Methods in Engineering*, 49(12):1547–1569.

Peng, X., Atroshchenko, E., Kerfriden, P., and Bordas, S. (2017). Isogeometric boundary element methods for three dimensional static fracture and fatigue crack growth. *Computer Methods in Applied Mechanics and Engineering*, 316:151–185.

Portela, A., Aliabadi, M. H., and Rooke, D. P. (1992). The dual boundary element method: Effective implementation for crack problems. *International Journal forNumerical Methods in Engineering*, 33(6):1269–1287.

Strang, G. and Nguyen, T. Q. (1997). *Wavelets and filter banks*. Wellesley-Cambridge Press.

Sukumar, N., Chopp, D., and Moran, B. (2003). Extended finite element method and fast marching method for three-dimensional fatigue crack propagation. *Engi- neering Fracture Mechanics*, 70(1):29–48.

Yaghmaie, R., Guo, S., and Ghosh, S. (2016). Wavelet transformation induced multi-time scaling (WATMUS) model for coupled transient electro-magnetic and structural dynamics finite element analysis. *Computer Methods in Applied Me- chanics and Engineering*, 303:341–373.

Yan, X. (2006). A boundary element modeling of fatigue crack growth in a plane elastic plate. *Mechanics Research Communications*, 33(4):470–481.

Influence of Regeneration-Induced Mistuning on the Aeroelasticity of Multistage Axial Compressors

Lukas Schwerdt, Niklas Maroldt, Lars Panning-von Scheidt, Joerg Wallaschek, and Joerg R. Seume

Abstract Current developments in turbomachinery favor blade designs which are characterized by low damping. The associated reduction of structural damping, however, creates new problems such as higher vibration amplitudes. Therefore, accurate predictions of these vibration amplitudes and of the fatigue life become increasingly important. Analyzing multistage machinery by only simulating isolated stages can be insufficient. This is particularly the case during repair processes when mistuning, such as that created by wear and repair, magnifies said amplitudes. In this chapter selected current results of the subproject C6 of the Collaborative Research Center (CRC) 871 are presented, building on the earlier research in the subprojects C3 and C6 focussing on mistuning and aeroelasticity, respectively. First, a simulation approach which accounts for structural and aeroelastic interstage coupling under conditions of mistuning is described, followed by an approach for incorporating large mistuning effects into a reduced order model of a single rotor stage. The former model is used to study the effect of intentional mistuning to decrease the sensitivity to additional mistuning due to wear and repair and to allow for smaller safety margins. The aeroelastic models are validated for a $1^{1/2}$-stage axial compressor.

Keywords Reduced order modeling · Mistuning · Aeroelasticity · Multistage turbomachinery · Vibration · Damping

L. Schwerdt · L. Panning-von Scheidt · J. Wallaschek · J. R. Seume
Institute of Dynamics and Vibration Research, Leibniz University Hannover, An der Universitaet 1, 30823 Garbsen, Germany
e-mail: schwerdt@ids.uni-hannover.de

N. Maroldt (✉)
Institute of Turbomachinery and Fluid Dynamics, Leibniz University Hannover, An der Universitaet 1, 30823 Garbsen, Germany
e-mail: maroldt@tfd.uni-hannover.de

J. R. Seume et al. (eds.), *Regeneration of Complex Capital Goods*,
https://doi.org/10.1007/978-3-031-51395-4_17

327

1   Introduction

The design of turbomachinery consistently faces the challenge of increasing the efficiency and, in particular for airplane applications, the power-to-weight-ratio. To achieve this goal, turbomachinery blading is increasingly manufactured as one piece with the rotor disk, as a so called blade integrated disk (blisk). However, eliminating the friction at the blade roots leads to significantly lower structural damping and therefore the blading faces higher vibration responses as well as higher sensitivity to flutter. Further amplitude amplifications are generated by deviations in the cyclic structure of the blisk, called mistuning. This can be a consequence of manufacturing tolerances or balancing procedures, but it also results from wear and regeneration, e.g. from blend and patch repairs. This demands an accurate prediction of the vibrational behavior of such blading in as-new, worn, and regenerated conditions. In particular, the prediction gets more challenging, when multistage effects are considered, where structural and aeroelastic coupling increases the complexity of the problem, as there might be no cyclic symmetry anymore and the model size also increases.

To investigate the influence of blend repairs on the aerodynamic damping a $1^{1/2}$- stage transsonic axial compressor with a realistic nominal rotational speed of 17,100 rpm was designed and manufactured in the second period of funding by Keller et al. (2015); Keller (2021). The compressor features two exchangeable blisks with a blade shape, which, due to its complexity, can only be manufactured as a blisk. Compared to the reference blisk, a second blisk was modified with three blend repairs at different radial heights (Keller et al. 2017), see Fig. 1.1. Each blend repair extends over 17.3 mm of the blade height and is 1.2 mm deep. The modified blades are distributed uniformly over the circumference, with seven nominal blades between each pair of modified blades. Blade vibrations were then excited during operation using an acoustic excitation system (Meinzer and Seume 2020). Measurement of the blade vibrations was conducted using a commercial tip-timing system. Keller (2021) could show experimentally and numerically, that the investigated blend repairs have no significant influence on the aerodynamic damping of the first bending mode.

Measurement data of this compressor is used to validate the aeroelastic numerical models in the following section. Afterwards, the models are extended towards

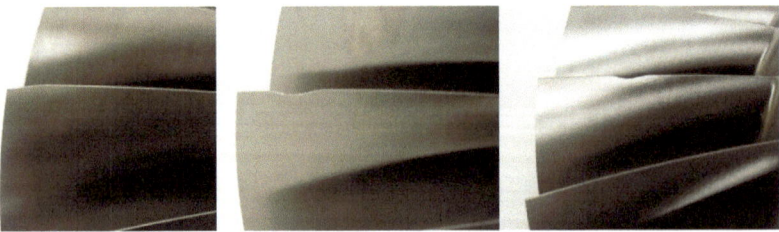

Fig. 1.1 Modified blades with blend repairs of the $1^{1/2}$-stage axial compressor at 100% (left), 80% (middle) and 60% (right) blade height

multistage turbomachinery and applied to a $2^{1/2}$-stage axial compressor test case. Thereafter, a reduced order model for large mistuning is presented.

2 Experimental Validation for a $1^{1/2}$-stage Compressor

The aeroelastic numerical model for the multistage calculations was experimentally validated at TFD's $1^{1/2}$-stage axial compressor. Both flutter and forced response calculations were validated.

For the experimental evaluation of the aerodynamic damping the acoustic excitation system was used to excite different nodal diameters of the first (bending) mode family (Keller 2021). Vibration measurements during operation were taken using a commercial tip-timing system by Agilis. Numerical simulations using Computational Fluid Dynamics (CFD) were obtained using the flow solver TRACE by DLR. The results are shown in Fig. 1.2 left. The numerical results (CFD) mostly agree well with the experiment (Exp.).

For forced response measurements the inlet guide vanes (IGVs) were rotated from a stagger angle of 0° (IGV0) by angles up to 20° (IGV20). The excitation of the third mode family (first torsional mode) by the IGV (engine order 23) was the focus of the investigations. A brief summary of the results is shown in Fig. 1.2. The numerically predicted vibration amplitudes agree with the experimentally measured amplitudes between a stagger angle of 0° and 15°. However, at 20° stagger angle the vibration amplitude is overestimated due to the flow separations at the IGVs, which cannot be predicted accurately by the employed turbulence-models. With these results in mind, it is possible to conclude that the numerical setup is capable of predicting

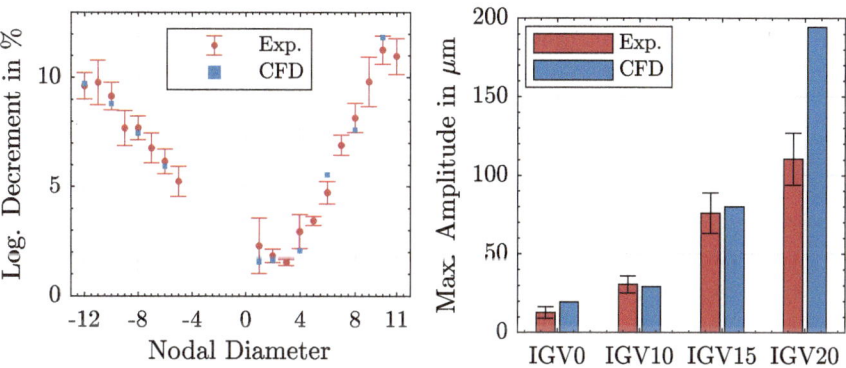

Fig. 1.2 Experimental validation of the numerical setups for the $1^{1/2}$-stage compressor. Left: Aerodynamic damping of first (bending) mode family at 16,245 rpm. Measurement data from Keller (2021). Right: Forced response of third (torsional) mode family at 6660 rpm. Data from Maroldt et al. (2022a)

forced response at operating points with a sufficiently large stall margin, where no significant separated flow regimes occur.

The investigations were supported by flow field measurements using 5-hole probes to evaluate the simulation results regarding the IGVs' wake shape and therefore the excitation intensity. Additionally, in order to mimic multistage effects a second engine order was imposed by circumferentially varying the IGVs' stagger angles during forced response measurements. However, as observed for IGV20 also in this case large flow separations occur, which deteriorates the prediction quality. Details can be found in Amer et al. (2020) and Maroldt et al. (2022a).

3 Multistage Computations

The multistage calculations are performed in a simulation procedure, applying a reduced order model and a CFD harmonic balance approach. The procedure is shown in Fig. 1.3. The full structural model is initially created in Ansys Mechanical and then exported to the Matlab-based application RAMBO (Reduced Analysis of Mistuned Blisks in Operation). Based on a cyclic modal analysis of the separate stages and displacements, generated by certain interface basis functions, the degrees of freedom (DOFs) of the ROM are created. The deformations of the ROM's DOFs are then used in flutter and forced response simulations to calculate the aeroelastic coefficients (aerodynamic damping and stiffness) and the aerodynamic forcing. The results are fed back to the ROM to calculate the system modes including the aeroelastic contributions and the frequency response. Additionally, mistuning can be added to the structure, for example to incorporate regeneration-induced variances or to conduct Monte-Carlo simulations to define limits of manufacturing tolerances. This approach was numerically applied by Maroldt et al. (2022b) to a $2^{1/2}$-stage axial compressor, which is an extension of the existing $1^{1/2}$-stage compressor. A summary of the approach and the results will be given below.

3.1 Structural Reduced Order Model

The goal of the reduced order model is the efficient calculation of the eigenfrequencies and forced response amplitudes of multistage rotors, including aeroelastic coupling and mistuning effects. Therefore, the final ROM must provide accurate results with

Fig. 1.3 Simulation sequence for the calculation of the vibrational behavior of multistage turbomachinery according to Maroldt et al. (2022b)

short calculation times and small memory requirements. But the computational effort for the creation of the ROM cannot be prohibitive, especially considering the large model size due to the presence of multiple stages and the necessary CFD simulations. With these aspects in mind, the ROM is designed to need only sector-level structural calculations and to minimize the necessary aeroelastic calculations. As it is designed to capture only small mistuning the reduction basis is derived from the tuned structure.

The ROM is based on a combination of substructuring methods for the stages, while at the same time making use of the cyclic symmetry of each stage. Building on the Fourier Constraint Mode (FCM) method (Song et al. 2005), the interface reduction is improved to increase the efficiency of the reduction (Schwerdt et al. 2020). In both methods, each stage is transformed into cyclic coordinates. Each harmonic index of each stage is then treated as a single substructure and reduced using the Craig-Bampton reduction method (Craig et al. 1968) with interface reduction. The interfaces in this context are the surfaces connecting adjacent stages. The interior degrees of freedom are reduced with the classic fixed- interface modes, which are modes of each substructure with the interface DOFs held fixed. Constraint modes displace the interface DOF and contain the static deflection of the rest of the structure given the interface displacement.

To facilitate the coupling of stages with non-matching interface meshes, the interface reduction is performed a-priori using predefined interface basis functions. These basis functions are matched to the circular or ring-shaped geometry of the interstage interfaces. In circumferential direction a Fourier basis is used. This matches the FCM method, where the interface is split into rings of nodes, which are reduced individually using Fourier bases. Note that the DOFs of each node corresponding to the three displacement directions are reduced individually. The approach is extended in the improved Polynomial Fourier Constraint Modes method by using polynomial basis functions for the interface reduction in the radial direction, as shown in Fig. 1.5. This makes the reduction basis totally independent from the FE mesh resolution, although the maximum Fourier harmonic to include and the polynomial degree is limited theoretically by the FE mesh to ensure the number of interface DOFs does not increase by introducing the transformation using the basis functions.

To assemble the substructures, the corresponding interface DOFs are matched and displacement compatibility is enforced. Here, the different number of sectors of adjacent stages must be taken into account. This is due to the aliasing of the higher Fourier harmonics, which may belong to different cyclic symmetry harmonics in the stages. A simple example of this fact is a rotor with two stages with ten and 20 blades. The interface harmonic with ten nodal diameters corresponds to alternating displacements of the twenty sectors of the second stage, but is aliased to harmonic index zero in the first stage, because each spatial harmonic period encompasses exactly a single sector.

To increase the computational efficiency, real-valued cyclic modes and interface harmonics are used instead of complex traveling wave coordinates. This is due to the influence of mistuning on the matrix structure. While the different coordinates lead to the same computational effort for cyclic modal analyses, introducing mistuning leads to fully populated matrices of the same size in either case, which means more effort is required when using complex arithmetic.

Different DOFs of the ROM of the $2^{1/2}$-stage compressor are pictured in Fig. 1.4. In the top row, examples of the fixed interface modes are shown, while two interface DOFs are shown on the bottom. Here, the main advantage of the Fourier basis is obvious: The lowest Fourier harmonics are most important, as they are coupled with large blade deflections in the constraint modes, which also highlights their importance for the subsequent aeroelastic simulations. The constraint modes of the higher harmonics are localized and can be omitted from the reduced order model. The same holds true for the polynomial coefficients. Additionally, the influence of the choice of the interface location is evident. Here, the interface is close to the first stage. This leads to large blade deflections of the first stage's blades in the interface DOFs.

To demonstrate the effect of the additional polynomial basis functions on the efficiency of the interface reduction, results from Schwerdt et al. (2020) are shown in Fig. 1.6. Here the maximum eigenfrequency error of the first mode family is

Fig. 1.4 Degrees of freedom of the ROM. Top: fixed interface modes. Bottom: interface DOFs

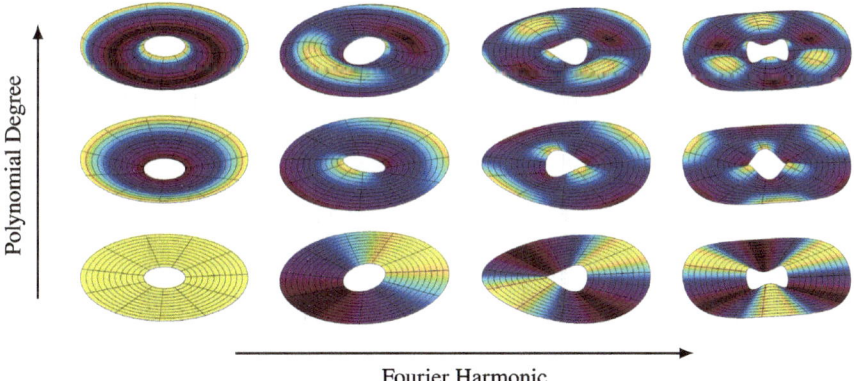

Fig. 1.5 Polynomial Fourier Constraint Modes (PFCM) interface basis functions

plotted against the number of DOFs for a simplified two stage rotor. Only the interface is reduced to isolate the effect of the interface reduction. For each graph the maximum interface polynomial degree is fixed (from 0 to 4) and the number of Fourier harmonics increases from left to right. The interface between the stage is ring shaped and consists of five concentric rings of nodes in the FE model. Therefore, a polynomial degree of four represents the maximum possible, and yields the same results as the FCM method. The results show that including the highest polynomial degree is optimal only if a large number of Fourier harmonics is included as well, which corresponds to little reduction of the interface DOFs. For reasonable reduction levels, a balanced truncation of the polynomial and Fourier basis is optimal, and best-case accuracy improvements of more than a factor of 100 are possible using PFCM compared to FCM for this system. For further discussion on this topic, including the possibility to truncate the individual DOFs instead of using a fixed global limit of the polynomial degree and maximum Fourier harmonic, see Maroldt et al. (2022c).

As with all substructuring-based ROMs the reduction basis contains more degrees of freedom than necessary. Therefore, it can be further reduced using a modal analysis of the reduced model to get the modes of the tuned system (Bladh et al. 2001). Here it is beneficial due to the presence of multiple stages, but for single stage rotors the tuned system modes can be calculated directly, and substructuring is usually not advantageous if it is used only to calculate the tuned modes. To include mistuning into the ROM multiple approaches are possible depending on the required accuracy and targeted calculation times. This is similar for all model order reduction methods based only on the tuned structure. First, the mistuning can be projected directly onto the reduced basis. This is the most accurate but most expensive option. Second, the structural changes can be limited to projection modes selected beforehand. This allows for some calculations to be performed in advance and thus reduces the computational effort. The most popular of these methods is the CMM (Component Mode Mistuning) method (Lim et al. 2007), where the changes of blade-alone modes of each blade are used as mistuning parameters.

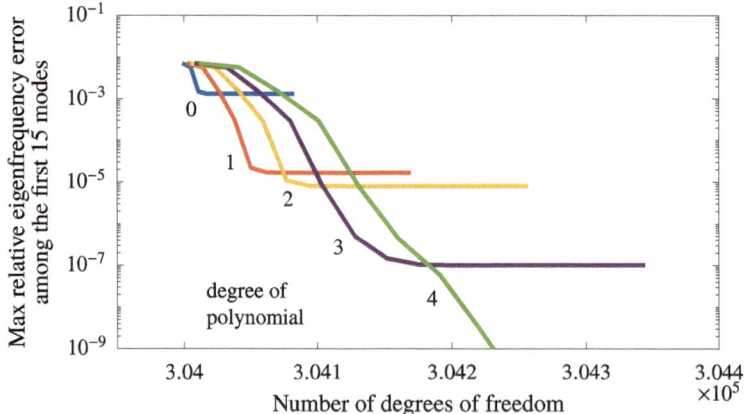

Fig. 1.6 Eigenfrequency error of the first mode family depending on the amount of interface reduction for a simplified two-stage geometry (Schwerdt et al. 2020). Only the interface is reduced (©SEM)

To include aeroelastic effects, CFD simulations are performed which are detailed in the following Sect. 1.3.2. The results are incorporated into the ROM as stiffness and damping terms representing the linearized aeroelastic contributions to the system dynamics. These aerodynamic coefficients are calculated for all DOFs of the ROM. And by calculating the aerodynamic effect of a displacement of one DOF on another, aeroelastic intermodal and interstage coupling is included. Due to the comparatively large computational effort of the CFD calculations, it is paramount to reduce the number of DOFs as much as possible while keeping the monoharmonic nature of the displacements of each DOF. To achieve this, the benefits of a secondary interface reduction were investigated in Maroldt et al. (2022c).

3.2 Aeroelastic Model

The computation of the aeroelastic coefficients follows the procedure described in Willeke et al. (2017) for single-stage simulations. However, compared to the calculation on a single or isolated stage-basis in multistage calculations two challenges arise: First, aeroacoustic scattering of the exciting engine order occurs, leading to the excitation of multiple nodal diameters. These effects need to be considered in the forcing calculations. Second, the unsteady pressure distributions on a blade, generated by a vibrating blade of another stage lead to additional aerodynamic damping and stiffness. Therefore, the computational domain of the flutter simulations needs to incorporate at least the neighboring stages, which highly increases the computational effort.

To solve this problem efficiently, a harmonic balance approach, implemented in the flow solver TRACE, is used. The method allows calculating one passage using phase-

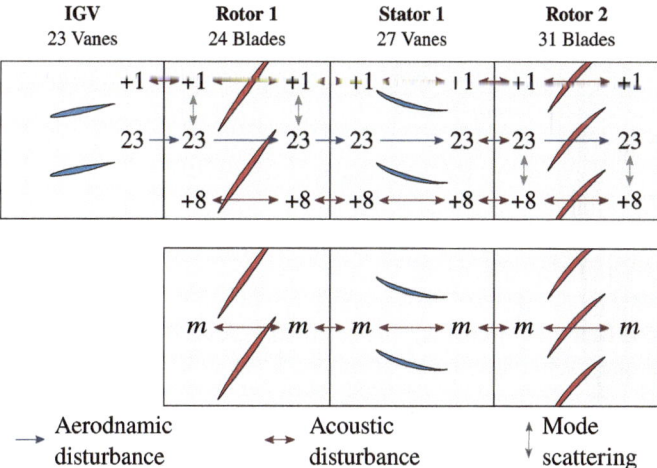

Fig. 1.7 Calculated circumferential modes m in the harmonic balance forcing and flutter calculation. Used with permission of American Society of Mechanical Engineers ASME, from Maroldt et al. (2022b); permission conveyed through Copyright Clearance Center, Inc

shifted periodic boundary conditions. Calculations were performed using the k-log (ω) (Müller and Morsbach 2018) version of the Menter-SST (Menter et al. 2003) turbulence model with the Kato & Launder (Kato and Launder 1993) stagnation anomaly fix. This is the same setup as used for the calculations of the $1^{1/2}$-stage compressor described and validated in Sect. 1.2. Both, aerodynamic excitation and aerodynamic coefficients, are calculated based on the aerodynamic work done on rotor 1 and 2 for the DOFs of each nodal diameter. In each flutter calculation one DOF is set to vibrate at the frequency of interest and the aerodynamic work done on all DOFs of the same nodal diameter is calculated. To account for interstage coupling the disturbance created by the vibration is calculated in the stator and both rotor domains. For example, a vibrating DOF of the first rotor can create unsteady pressure on the second rotor and therefore couples DOFs of both stages. The investigated vibrational response is excited by EO23, which is created by the IGV. To address mode scattering in the forcing calculation, all acoustic cut-on modes are calculated, see Fig. 1.7. The modes are generated by scattering of the EO23, which is scattered at rotor 1 to a forward traveling wave with nodal diameter 1, and at rotor 2 to a forward traveling wave with nodal diameter 8.

3.3 Tuned Results

The results on stage and nodal diameter basis are shown in Fig. 1.8. The frequency response is separated into the excitation of stage 1 and stage 2 by ND1 and ND8. The response is dominated by the excitation of stage 1 with ND1, since it is located just downstream of the exciting IGV. Modes localized in stage 1 (approx. 110 Hz), stage 2 (approx. 132 Hz) and in both stages (e.g. approx. 124 Hz) are visible. Especially the latter mode is created by structural interstage coupling and would not appear in an isolated stage modeling. Modes which are at least partly localized in the second stage also respond, when exciting the second stage with ND1. In contrast, only one mode, solely localized in the second stage, is visible when exciting stage 2 with ND8. The reason for this is that stage-coupled modes usually occur at lower nodal diameters. Lastly, the response, when exciting stage 1 with ND8 only plays a minor role. The overall response, created by the superposition of the individual responses, is shown in Fig. 1.9. Due to the superposition of multiple nodal diameters created by the mode scattering in the forcing calculation, variations of the individual blade amplitudes are visible, leading to an amplitude magnification.

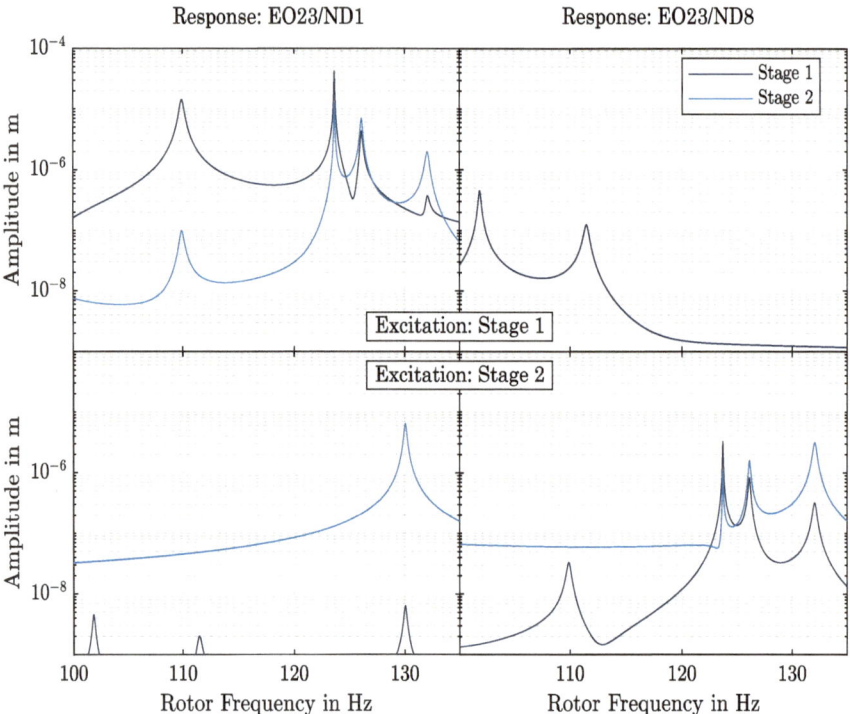

Fig. 1.8 Frequency response of stage 1 and stage 2 when exciting an individual stage with a certain nodal diameter

Fig. 1.9 Overall frequency response of the individual blades of stage 1 and stage 2. Used with permission of American Society of Mechanical Engineers ASME, from Maroldt et al. (2022b); permission conveyed through Copyright Clearance Center, Inc

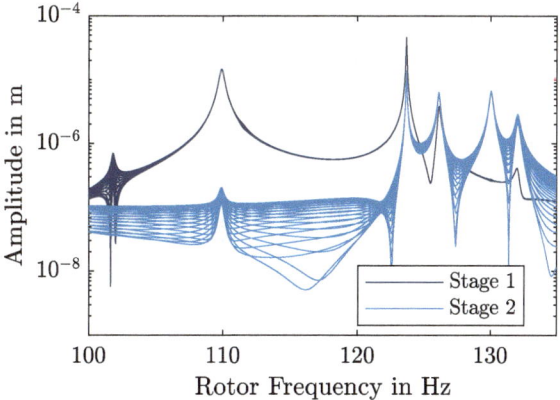

3.4 Mistuned Results

In addition to the forced response calculations using the tuned rotor, Monte-CarloSimulations were done to assess the influence of the multistage aeroelastic coupling on the mistuning amplitude magnification (Maroldt et al. 2022b). The results are pictured in Fig. 1.10. Here, the amplitude magnification due to mistuning are compared for two different damping cases. First, the full aeroelastic damping including interstage coupling (fullA) is used. In the simulations to produce the second graph, only the aerodynamic coefficients within each stage were kept, while the coefficients responsible for the coupling between stages were set to zero (singleStageA). The structural damping was omitted in both cases, due to its comparatively small magnitude for blisks. For the investigated resonance crossing of this rotor and the chosen mistuning level, the mistuning sensitivity is underestimated approximately by ten percent when omitting the aeroelastic interstage coupling. When looking at the resulting stresses and therefore, the estimated fatigue life, it is useful to perform evaluations based on the amplitude frequency strength (af-strength) (Hanschke et al. 2017), which is the ratio of actual and allowable loading. The results with full aerocoupling show lower amplitudes, which is reflected by an increase of the af-strength of 17% (Maroldt et al. 2022b).

Similar to the way intentional mistuning can mitigate the effects of unwanted random mistuning for single stages, it can be applied to rotors with multiple stages. For modes with participation of more than one stage, it is possible to introduce the intentional mistuning in one stage to reduce the vibration amplitudes in another stage. This is analogous to the possibility of affecting the blade vibration with intentional mistuning of the disk (Schwerdt et al. 2019), which was previously researched in the subproject C3 of the CRC 871. This is demonstrated for the $2^{1/2}$-stage compressor in Fig. 1.11, where the splitting of the double mode due to intentional mistuning as well as the distribution of expected amplitude magnifications with and without intentional mistuning (MT) are shown. In this case the median of the amplitude magnification can be reduced from 1.202 to 1.076.

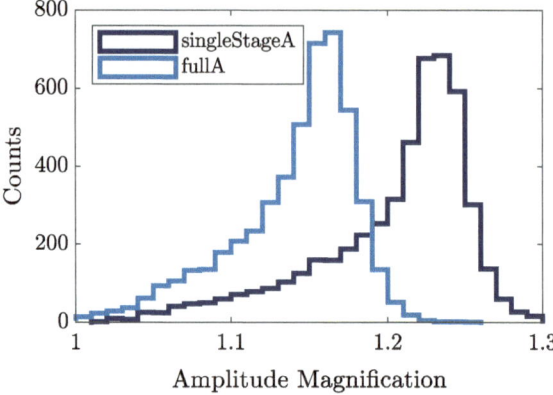

Fig. 1.10 Amplitude magnification of the first stage including full aeroelastic coupling (fullA) and aeroelastic coupling only within each stage (singleStageA). Both cases use 5000 identical samples of Gaussian stiffness mistuning of the sectors with a standard deviation of 0.002. Used with permission of American Society of Mechanical Engineers ASME, from Maroldt et al. (2022b); permission conveyed through Copyright Clearance Center, Inc

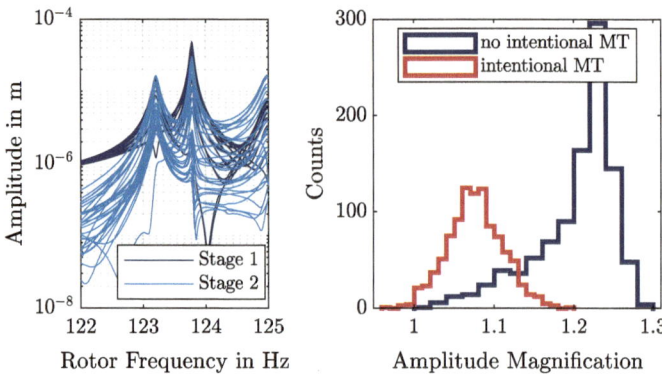

Fig. 1.11 Left: Frequency response of the $2^{1/2}$-stage compressor with intentional mistuning of the second stage. Right: Amplitude magnification of the first stage with and without intentional mistuning (MT) of the second stage

4   Reduced Order Modelling for Large Mistuning

Reduced order models based on the tuned structure, such as the one discussed in Sect. 1.3.1, are most efficient for structures with small mistuning. Different methods are needed if accurate modeling of larger structure changes is necessary. This may be the case when investigating large damages, or the milling process during a patch repair of a blisk. In these cases, the large structural change is confined to a single blade.

In literature, multiple methods are available to deal with these cases with large mistuning. Here, it is important to distinguish between methods that are restricted by their reliance on the structure of the finite element mesh, and those which can capture arbitrary changes of the blades. Due to the constraints on the FE-mesh the former methods (see for example Lupini and Epureanu (2019); Sinha (2009)) can be more efficient, and should therefore be preferred if applicable. But this section deals with the latter methods focusing on large arbitrary geometric mistuning. There, different variants of generally applicable substructuring algorithms have been known for decades (Craig et al. 1968). Specializing on bladed disks with large mistuning, the PRIME (Pristine Rogue Interface Modal Expansion) method (Madden et al. 2012) takes advantage of the cyclic symmetry of the base system. This method was used as a reference when evaluating the newly developed method PRISM (Schwerdt et al. 2021), which is discussed in this section. Only single stage rotors are considered here, although the methods can be extended to multistage applications (Kurstak and D'Souza 2018). The DOFs of the rotor are partitioned into those of nominal (or pristine) sectors, those belonging to modified (or rogue) sectors and the DOFs of the shared interface between these sector types, shown in Fig. 1.12. For the PRIME method, cyclic modal analyses are performed for full blisks consisting of only pristine and rogue sectors, respectively. The full blisk is then assembled in the reduced order space where the interface DOFs are reduced using the pristine modes. The key insight is to use independent sets of DOFs for the pristine sectors, the rogue sector, and the interface to reduce the influence of non-matching displacement subspaces of the interface in the different cyclic modes. The resulting ROM is reduced further to avoid unnecessary DOFs and numerical issues with a badly conditioned reduction basis.

The newly developed PRISM method works by building the eponymous partially reduced intermediate system model of the whole blisk. First, the cyclic modes of the

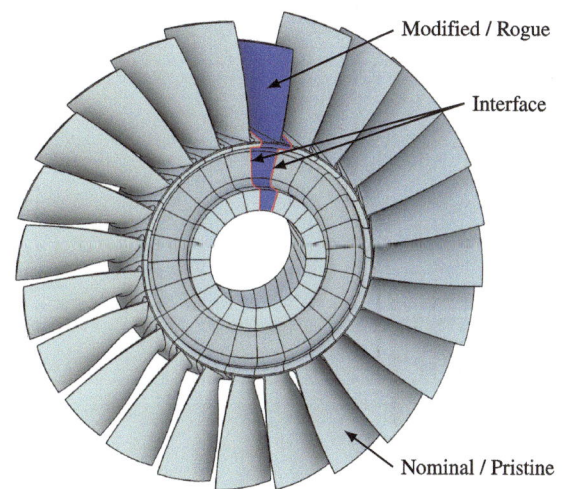

Fig. 1.12 Mistuned compressor blisk split into pristine, interface and rogue degrees of freedom. Used with permission of American Society of Mechanical Engineers ASME, from Schwerdt et al. (2021); permission conveyed through Copyright Clearance Center, Inc

Modified / Rogue

Interface

Nominal / Pristine

rotor with only nominal sectors are calculated. These are used to reduce the nominal sectors while the full FE-model of the modified sector is kept. To assemble the intermediate system model, the interface DOFs of the modified sector are reduced using the modes of the nominal sector. Thereby, displacement compatibility is ensured between the reduced nominal sectors and the non-reduced modified sector in the partially reduced intermediate system model. A modal analysis of this model results in the final reduction basis for the whole blisk.

The key advantage of the PRISM method is the reduced calculation time for the modal analysis of the intermediate system model compared to a full cyclic modal analysis of the modified sector type. This reduces the computational effort for the reduction procedure compared to the PRIME method, with a calculation time reduction of 88, and 45% when including the time taken for the modal analysis of the nominal sector type, as reported in Schwerdt et al. (2021). It should be noted however, that smaller gains are expected when using a fully iterative modal analysis solver, as opposed to the block Lanczos used in Schwerdt et al. (2021), or when calculating more modes. This is due to the time taken for the decomposition of the stiffness matrices for each harmonic index being the dominant part of the calculation time. In addition to the reduced time for the model reduction, compared to PRIME the accuracy is improved as well. By using the modified sector without reduction in the intermediate model, the modes of the complete system are captured slightly better compared to using a limited reduction basis of cyclic modes. This is shown in Fig. 1.13, where the accuracy in predicting the eigenfrequencies of both methods is compared using ten modes per sector for the blisk shown in Fig. 1.12.

Fig. 1.13 Relative eigenfrequency errors of PRIME and PRISM ROMs. Used with permission of American Society of Mechanical Engineers ASME, from Schwerdt et al. (2021); permission conveyed through Copyright Clearance Center, Inc

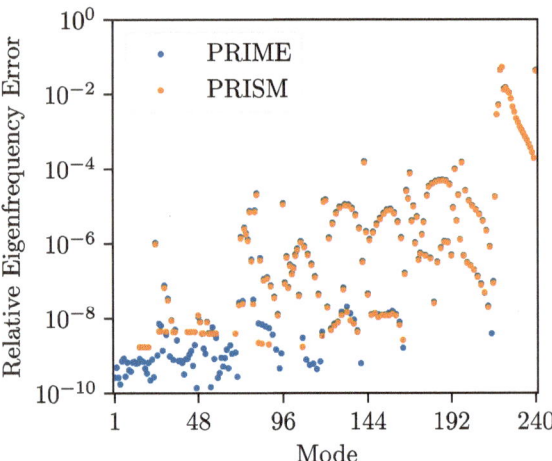

5　Conclusions

In this chapter efficient and accurate reduced order models (ROMs) for the calculation of the vibrational behavior of realistic turbomachinery was presented. The models account for structural mistuning due to realistic geometries, which are shaped by wear and regeneration. If a regeneration process of one blade leads to large geometric changes, the developed approach PRISM is able to calculate the influence on the vibrational behavior with slightly better accuracy in approximately half the time compared to the state of the art method for the analyzed example.

Furthermore, a ROM for the calculation of multistage turbomachinery including structural and aeroelastic interstage coupling was developed. It was shown that the interstage coupling can influence the vibration amplitudes in multistage turbomachinery. For the test case investigated, the predicted amplitude frequency-strength increases by 17% when interstage coupling is included, allowing a more accurate fatigue life prediction. However, the coupling effects depend on various parameters such as number of stages, stiffness of the rotor, and the eigenfrequencies of the isolated stages. Therefore, the influence of interstage coupling may be less severe in other cases.

Since for multistage calculations a large amount of aeroelastic flutter calculations needs to be performed, the ROM was extended by an additional a posteriori interface reduction to reduce the number of aeroelastic simulations necessary for a given accuracy level of the ROM. On this basis more complex investigations, such as using speed-dependent aeroelastic coefficients, can be conducted in the future.

Acknowledgements　Funded by the Deutsche Forschungsgemeinschaft (DFG, German Research Foundation) – SFB 871/3 – 119193472. The authors would like to thank the DFG for the support. Moreover, the authors would like to acknowledge the substantial contribution of the DLR Institute of Propulsion Technology and MTU Aero Engines AG by providing the TRACE code for CFD in turbomachinery.

References

Amer, M., Maroldt, N., and Seume, J. (2020). Investigation of multiharmonic effects in a single stage high speed compressor.

Bladh, R., Castanier, M. P., and Pierre, C. (2001). Component-mode-based reduced order modeling techniques for mistuned bladed disks, part i: Theoretical models. *Journal of Engineering for Gas Turbines and Power*, 123(1):89–99.

Craig, JR., Roy. R. and Bampton, M. C. C. (1968). Coupling of substructures for dynamic analyses. *AIAA Journal*, 6(7):1313–1319.

Hanschke, B., Klauke, T., and Kühhorn, A. (2017). The effect of foreign object damage on compressor blade high cycle fatigue strength. In *Proceedings of the ASME Turbo Exp*, page V07AT31A005, New York, N.Y. ASME.

Kato, M. and Launder, B. E. (1993). The modeling of turbulent flow around stationary and vibrating square cylinders. In *9th Symposium on Turbulent Shear Flows*, pages 10.4.1–10.4.6.

Keller, C. (2021). Einfluss regenerationsbedingter varianzen auf die aeroelastik von verdichterblisks.

Keller, C., Kellersmann, A., Friedrichs, J., and Seume, J. R. (2017). Influence of geometric imperfections on aerodynamic and aeroelastic behavior of a compressor blisk. In *ASME Turbo Expo 2017*. ASME.

Keller, C., Willeke, T., Burrafato, S., and Seume, J. (2015). Design process of a 1.5-stage axial compressor for experimental flutter investigations. In *International Gas Turbine Congress 2015 Tokyo*.

Kurstak, E. and D'Souza, K. (2018). Multistage blisk and large mistuning modeling using fourier constraint modes and prime. *Journal of Engineering for Gas Turbines and Power*, 140(7).

Lim, S.-H., Bladh, R., Castanier, M. P., and Pierre, C. (2007). Compact, generalized component mode mistuning representation for modeling bladed disk vibration. *AIAA Journal*, 45(9):2285–2298.

Lupini, A. and Epureanu, B. I. (2019). On the use of mesh morphing techniques in reduced order models for the structural dynamics of geometrically mistuned blisks. *Mechanical Systems and Signal Processing*, 127:262–275.

Madden, A., Epureanu, B. I., and Filippi, S. (2012). Reduced-order modeling approach for blisks with large mass, stiffness, and geometric mistuning. *AIAA Journal*, 50(2):366–374.

Maroldt, N., Amer, M., and Seume, J. R. (2022a). Forced response due to vane stagger angle variation in an axial compressor. *Journal of Turbomachinery*, 144(8).

Maroldt, N., Schwerdt, L., Berger, R., Panning-von Scheidt, L., Rolfes, R., Wallaschek, J., and Seume, J. R. (2022b). Reduced order modeling of forced response in a multistage compressor under mistuning and aerocoupling. In *ASME Turbo Expo 2022*. ASME.

Maroldt, N., Schwerdt, L., Panning-von Scheidt, L., Wallaschek, J., and Seume, J. R. (2022c). An improved reduced order model for bladed disks including multistage aeroelastic and structural coupling. In *Proceedings of Global Power and Propulsion Society - GPPS Chania22*. GPPS.

Meinzer, C. E. and Seume, J. R. (2020). Experimental and numerical quantification of the aerodynamic damping of a turbine blisk. *Journal of Turbomachinery*, 142(12).

Menter, F. R., Kuntz, M., and Langtry, R. (2003). Ten years of industrial experience with the sst turbulence model. In Hanjalic, K., Nagano, Y., and Tummers, M. J., editors, *Turbulence, Heat and Mass Transfer 4*.

Müller, M. and Morsbach, C. (2018). A logarithmic w-equation formulation for turbulence models in harmonic balance solvers. In *7th European Conference on Computational Fluid Dynamics (ECFD 7)*.

Schwerdt, L., Panning-von Scheidt, L., and Wallaschek, J. (2021). A model reduction method for bladed disks with large geometric mistuning using a partially reduced intermediate system model. *Journal of Engineering for Gas Turbines and Power*, 143(7).

Schwerdt, L., Scheidt, L. P.-v., and Wallaschek, J. (2020). A priori interface re- duction for substructuring of multistage bladed disks. In Linderholt, A., Allen,

M. S., Mayes, R. L., and Rixen, D., editors, *Dynamic Substructures, Volume 4*, Conference Proceedings of the Society for Experimental Mechanics Series, pages 13–21. Springer International Publishing, Cham.

Schwerdt, L., Willeke, S., Panning-von Scheidt, L., and Wallaschek, J. (2019). Reduced-order modeling of bladed disks considering small mistuning of the disk sectors. *Journal of Engineering for Gas Turbines and Power*, 141(5):384.

Sinha, A. (2009). Reduced-order model of a bladed rotor with geometric mistuning. *Journal of Turbomachinery*, 131(3):031007.

Song, S. H., Castanier, M. P., and Pierre, C. (2005). Multi-stage modeling of turbine engine rotor vibration. In *Volume 1: 20th Biennial Conference on Mechanical Vibration and Noise, Parts A, B, and C*, pages 1533–1543. ASME.

Willeke, S., Keller, C., Panning-von Scheidt, L., Seume, J. R., and Wallaschek, J. (2017). Reduced order modeling of mistuned bladed disks considering aerodynamic coupling and mode family interaction. In *Reduced Order Modeling of Mistuned Bladed Disks considering Aerodynamic Coupling and Mode Fam- ily Interaction*, European Conference on Turbomachinery Fluid Dynamics and hermodynamics. European Turbomachinery Society.

Comprehensive Control
of the Regeneration Processes (Project Area D)

Modeling, Configuration and Assessment of Regeneration Supply Chains

Torben Lucht, Tammo Heuer, Thorben Kuprat, Steffen C. Eickemeyer, and Peter Nyhuis

Abstract In contrast to the conventional manufacturing industry, the regeneration of complex capital goods is subject to additional information uncertainty and disturbances throughout the order processing phase. To achieve a high logistics performance despite these constraints, supply chain configurations and appropriate tools are required to address the resulting challenges. Along a roadmap, this report presents measures and tools that support the reduction of information uncertainty, the description of effects of uncertainty within regeneration planning and control and in regeneration logistics as well as a superordinate assessment and configuration of regeneration supply chains. For forecasting a Bayesian Network based approach is presented. The planning quality that is achieved based on these forecasts is assessed using a descriptive model based on the classic milestone trend analysis. To support logistics oriented regeneration supply chain assessment and configuration an extensive simulation model as well as a model for pool stock dimensioning is presented. These tools and models represent the results of the subproject D1 of the CRC871 that had the overall goal to empower regeneration service providers to establish robust regeneration supply chains.

Keywords Supply chains · Modeling · Forecasting · Spare parts · Pooling · Simulation

1 Introduction

In recent years, a growing social, economic and political interest and striving for the sustainable use of resources and products in the sense of a circular economy can be observed worldwide (Cervelló-Royo et al. 2020; Liu et al. 2021). A prominent

T. Lucht · T. Heuer · T. Kuprat · S. C. Eickemeyer · P. Nyhuis (✉)
Institute of Production Systems and Logistics (IFA), Leibniz University Hannover, An der Universitaet 2, 30823 Garbsen, Germany
e-mail: office@ifa.uni-hannover.de

S. C. Eickemeyer
Constructor University gGmbH, Campus Ring 1, 28759 Bremen, Germany

J. R. Seume et al. (eds.), *Regeneration of Complex Capital Goods*,
https://doi.org/10.1007/978-3-031-51395-4_18

example of this kind of circular economy represents the regeneration of complex capital goods. The regeneration of complex capital goods, such as jet engines, stationary gas turbines or wind turbines, comprises maintenance, repair and overhaul (MRO) activities and serves to generate additional "life cycles" at the end of a use phase of the products (Guide et al. 1997). The terms "regeneration" and "MRO" will be used as synonyms below. Starting from an initial diagnosis, the generic regeneration process can be separated into the disassembly of the complex capital good, an inspection, the repair or replacement of damaged components, the subsequent reassembly as well as a final inspection and quality assurance (see Fig. 1) (Eickemeyer et al. 2011; Guide et al. 1997; Lucht et al. 2019; Lund 1984).

The diagnosis performed before physical induction forms the basis for the definition of the required depth of disassembly (full or partial) and the initial scheduling of the regeneration process (Eickemeyer et al. 2011; Seitz et al. 2020). Disassembly can be performed either on flow lines or in so-called docks (Burmeister et al. 2004). Flow line organization tends to offer advantages in efficiency, by allowing to synchronize processing of the regeneration orders along the supply chain. Disassembly in docks allows for a (limited) interchange of disassembly sequences—e.g., sorted by earliest repair start date of single components that have already been identified as damaged in the diagnosis. Further discussion on this can be found in Lucht et al. (2020).

The disassembled components then undergo a detailed inspection to gather reliable information about the repair measures required for regeneration and spare parts demands. Here, discrepancies between the predicted damage pattern and the final inspection result can occur. If this is not taken into account in supply chain design and planning, there is a risk of considerable turbulence in the form of schedule deviation or long waiting times for missing material or capacities (Dombrowski and Sendler 2016). The same may result if (additional) rework is required during repair, should damage only be detected or even caused there. Instead of repairing unserviceable components, the components required for reassembly can be provided from spare parts pools (Kuprat et al. 2016).

Both a pool containing serviceable (SA) components and a pool of repairable (RA) components are established in many MRO supply chains (Lucht et al. 2019). The pools offer the possibility to cushion uncertainties about spare parts stock levels and alternative supply paths (Kuprat 2018; Lucht et al. 2021a). However, these serviceable components are often very expensive, so spare parts inventories must be kept

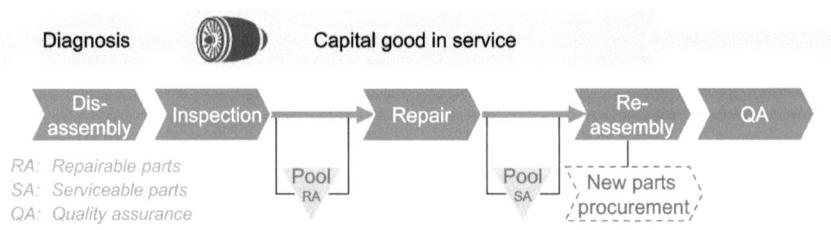

Fig. 1 Regeneration supply chain structure (based on Eickemeyer et al. 2011; Guide et al. 1997)

as low as possible. An order-specific provision via new parts procurement is also possible but mostly comes at very high costs and extensive delivery lead times. Once all components are ready for installation, reassembly and quality assurance can be performed. The capital good can then be handed back to the owner to continue service in another life cycle (Guide et al. 1997; Guide 2000; Lund 1984).

During regeneration, the capital goods are not available for the provision of services (e.g., the transport of passengers) and thus the generation of revenue. Hence, lead time extensions of the regeneration are directly linked to losses in revenue. It should be noted, however, that in most cases the capital goods could not be used without the MRO event either, as components have reached their operational limits or are no longer serviceable due to other influences. Nonetheless, the turnaround time (TAT) forms the basis for the operators' operational planning. Using the example of an aircraft, this means that the aircraft is assigned to specific flights. If there are deviations from the planned TAT, this not only limits the ability to generate revenue, but at the same time leads to a major need for rescheduling by the operators (Herde 2013). Consequently, deviations from promised TAT or delivery dates are usually tied to strong penalties (Eickemeyer et al. 2013). As a result, achieving the highest possible planning reliability and accuracy is of primary importance while also ensuring short TAT. This is countered by a great uncertainty in information regarding the timing and scope of a regeneration order due to the variety of possible damage patterns in conjunction with a large number of components as well as the highly dynamic market and production environment (Reményi and Staudacher 2014).

The actual damage of the individual components can usually only be precisely determined during the inspection following disassembly or even during the repair process—for example, after coatings have been removed. However, unforeseen faults can still occur within the advanced regeneration process. To avoid negative impacts on the regeneration service providers' logistics performance suitable measures are required to ensure the regeneration order can still be completed on time. In other words, regeneration service providers must design their production systems to be as robust as possible, while robustness in here refers to "stability against different varying conditions" (Stricker and Lanza 2014).

To support regeneration service providers, various tools and approaches have been developed in subproject D1 of the Collaborative Research Center 871. These are organized into three major fields of action. They also provide the structure of the remainder of this paper. Section 2 gives an overview of these fields of action that simultaneously represents the overall roadmap of the subproject. Based on this, the fields of action are presented separately in Sects. 3 (improving forecast quality), 4 (assessing planning quality) and 5 (assessing and configuring of regeneration supply chains). Section 6 sums up the outline of this paper and provides an outlook to further research and transfer activities.

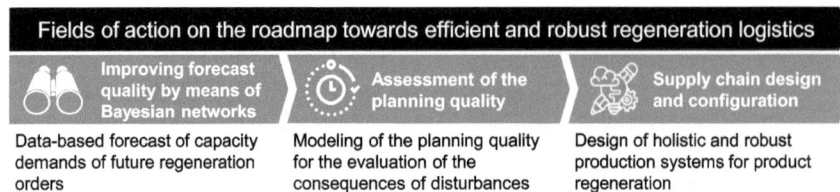

Fig. 2 Roadmap and modular structure of subproject D1

2 Objective and Structure of the Fields of Action

Subproject D1 of CRC 871 focuses on the design of an efficient and robust regeneration logistics system through three major fields of action (see Fig. 2). These also form the overall structure of the following sections.

The first field of action addresses the uncertainty at the beginning of the regeneration process. The aim is to minimize this uncertainty to improve the planning basis for the subsequent regeneration process.

The second field of action focuses on increasing transparency of the logistics consequences of this uncertainty within the regeneration process. In particular the focus lies on assessing the planning quality of the planning processes as critical tasks both for the internal processes as well as for communication and coordination with customers, partners and suppliers and thus making them accessible for improvement measures.

Since random disruptions can still occur even with complete information availability, the third field of action aims to compensate for the remaining disruptive influences as efficiently as possible by means of a suitable design of the supply chain and configuration of production planning and control. The supply chain design and the configuration of production planning and control (subsequently subsumed as "production configuration") must take these challenges into account and provide suitable tools to reduce or compensate for any turbulence to ensure a high level of logistics efficiency.

3 Reducing Uncertainty Using Data-Based Forecasting

The first of the three fields of action focuses on the forecast of the capacity load expected from a regeneration order upstream of the actual regeneration process. This is of essential importance for the planning of the regeneration process, as it supports the definition of realistic delivery times and at the same time the demand-oriented dimensioning of capacities and material stock and allows a higher precision (Reményi et al. 2011). However, due to the unique nature of regeneration orders, conventional, time series-based forecasting approaches are unsuitable for making a forecast with acceptable accuracy for practical applications. Instead, order-specific forecasting approaches allow to take into account the information available at a specific point in time from a specific usage profile of a regeneration object and knowledge from past regeneration orders for the forecast. These order-specific forecasts can then be used to aggregate the total load resulting from the order spectrum. A schematic representation of the described prognosis method developed within the subproject shown in Fig. 3.

In order to make the data of different types and origins required for this purpose usable for forecasting, they can be merged into a higher-level database using data mining methods, as shown in Fig. 3 (Eickemeyer et al. 2013). In this context, the usage profile of the capital goods is characterized, among other things, by information regarding the place as well as the time and type of operation. Following the example

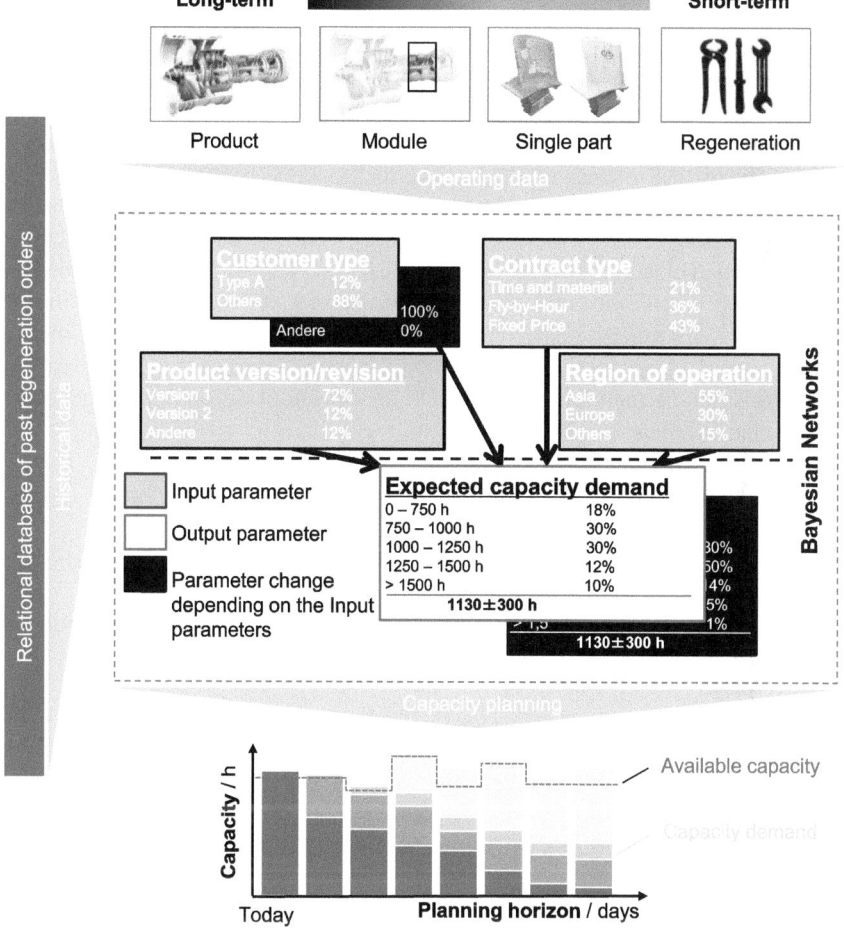

Fig. 3 Data-based forecasting of capacity demand (based on (Eickemeyer et al. 2013; Eickemeyer 2014))

of an aircraft engine, the number of flown cycles (i.e. completed cycles consisting of take-off, flight phase, and landing) can be used for this purpose (Seitz et al. 2020). Potential further information is provided by temperature, pressure and speed curves (Weiss et al. 2022). Information from past regeneration runs, however, can include customer data, schematics, component information, and detailed data on regeneration paths (required repair processes, setup times, processing times etc.).

The damage library compiled in this way makes it possible to predict the existing but not yet identified damage to a capital good depending on the specific usage profile. The actual damage for individual usage profiles can be inferred by using databased fore-casting methods. Additionally, such database can also predict the resulting workload in the event of a need for regeneration. The knowledge stored in corresponding databases can be extracted using Bayesian networks (Berkholz 2012; Eickemeyer et al. 2013; Seitz et al. 2019). With these, a high forecasting quality can be achieved in practical applications. It has been shown that the application of Bayesian networks is useful for forecasting the workload of entire complex capital goods, individual assemblies and components as well as the loads on the workstation level in the regeneration of complex capital goods (Eickemeyer et al. 2013). Furthermore, beyond capacity planning and coordination, e.g., for dimensioning pool inventories, Bayesian networks have the potential to make helpful forecasts and increase planning reliability. A typical example is the decision to accept new orders (Eickemeyer and Herde 2012). The improved prediction of capacity demands of regeneration objects that are still in the utilization phase allows a better estimation of the arising efforts and costs of regeneration service providers. Thus, it forms the basis for a more precise (capacity) planning (Berkholz 2012) and a higher logistics efficiency. Improved throughput or delivery time forecasts are also possible as well taking into account information about the current workload in the regeneration supply chain (Hiller et al. 2021).

4   Assessment of Planning Quality in Regeneration

Even if a significant improvement of the information accuracy can be achieved using the increasing data availability described above (see dotted line in Fig. 4), uncertainties remain, which lead to turbulences in the regeneration process and thus to deviations from planned states and dates. Due to the regeneration-specific uncertainties in information and data basis, which only diminishes within the regeneration process (see Fig. 4), frequent adjustments of planned dates are required in practice to reflect the changing information within these.

Although adjustments to the plans are necessary in practice to provide the operational order processing with the most up-to-date information, this also leads to "concealing" planning errors that—in consequence—cannot be assessed in an a posteriori performance controlling anymore (Lucht et al. 2021b). However, corresponding planning iterations can be systematically described using a so-called plan history diagram (PHD) and thus made accessible for analyses. For this purpose, the temporal distance

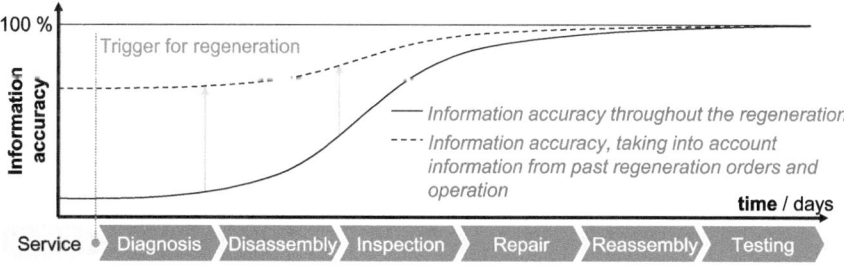

Fig. 4 Qualitative development of information accuracy along the regeneration process (based on (Eickemeyer et al. 2012))

between a planned date and the observation period is plotted over time (see Fig. 5). If a plan remains unchanged over time, this results in a straight line that continuously approaches the horizontal axis. For example, the remaining throughput time until the planned realization of a date is reduced by one day with each additional day that passes. This describes the ideal process by one day scheduling without further schedule adjustments during order processing. Shifts in dates are expressed accordingly as swings in a positive direction (shifting a date into the future) or in a negative direction (bringing a date earlier).

Figure 5 shows the corresponding planned date curves of four orders of a real use case (regeneration service provider) in a PHD. Here, the planned date curves of the

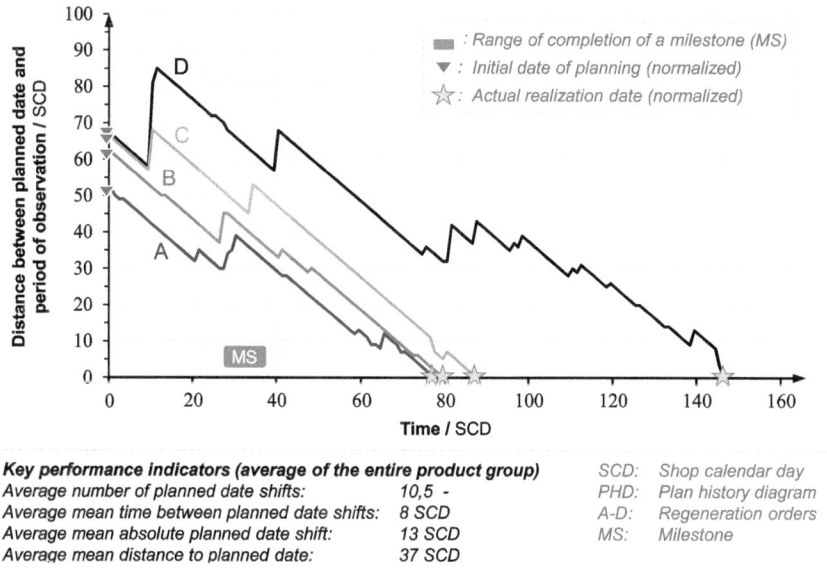

Fig. 5 Normalized plan history diagram of four regeneration orders with characteristic rescheduling patterns (based on (Lucht et al. 2021b))

regeneration orders of one exemplary product group are normalized to a common start date. In addition to a comparison of different planned date curves, characteristic patterns in the planned date curves can also be identified. Characteristic planned date adjustments are recognizable within the time of reaching the marked milestone. In this example, this milestone represents the completion of the inspection of all disassembled components. At this time, for the first time, there are more or less reliable insights into the actual damage pattern of the regeneration good. The recurring planned date adjustments identified in this way can thus be made accessible to a root cause analysis and thus potential measures such as the adjustment of the scheduling algorithm on which these dates are based can be initiated. Typical causes found for recurring schedule delays at the completion of this milestone were found to be systematically underestimated workload of regeneration orders or recurringly required customer clarifications not considered in the planning.

Overall, the PHD allows for a systematic description and investigation of planning processes in turbulent production environments that require frequent planning iterations. In this way, schedule shifts can be described and the causal disruptive influences as well as weak points in the production configuration can be identified. The production configuration must be fundamentally suitable for achieving the required logistics performance and, accordingly, for compensating for external and internal disruptive influences. The definition of a suitable production configuration is an extremely complex task due to the large number of influencing factors to be taken into account and their mutual dependencies, and is therefore discussed in detail in the following section.

5 Configuration of Regeneration Supply Chains

Depending on the selected configuration of the supply chain, regeneration service providers are able to offer a certain logistics performance on the market. As described above, the customer perceives this via key figures for on-time delivery and the delivery time offered. For high profitability of the regeneration process, regeneration service providers must therefore realize a high on-time delivery and short delivery times with the lowest possible logistics costs. Even if the configuration is defined to a certain degree by the capital goods themselves and by the technologies required for regeneration, there remain degrees of freedom on the production logistics side with regard to the provision of capacity, scheduling and lot sizing procedures, and control and processing strategies. The design of these degrees of freedom directly impacts the logistics performance and the logistics costs of all individual processes. The parameterization of the individual processes directly influences the overall performance of the supply chain and thus on the performance perceived by the customer due to the existing interactions within the regeneration supply chain. The production configuration can be divided into four successive levels (see Fig. 6).

At the strategic levels 1 and 2, a distinction can be made between the definition of the goal and the structural configuration of the supply chain. The configuration

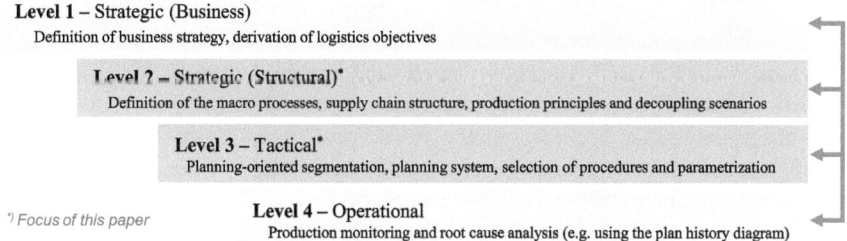

Level 1 – Strategic (Business)
Definition of business strategy, derivation of logistics objectives

Level 2 – Strategic (Structural)*
Definition of the macro processes, supply chain structure, production principles and decoupling scenarios

Level 3 – Tactical*
Planning-oriented segmentation, planning system, selection of procedures and parametrization

*) Focus of this paper   **Level 4** – Operational
Production monitoring and root cause analysis (e.g. using the plan history diagram)

Fig. 6 levels of production configuration (based on (Mütze et al. 2022))

of planning and control within the structure-forming macro processes represents the tactical level 3 of production configuration. Finally, the fourth level addresses production monitoring and root cause analysis for potential deviations as a starting point for adjustments to the existing production configuration. Among other things, the PHD already presented above can be used for this purpose (Mütze et al. 2022). The key configuration decisions are made at levels 2 and 3 so that these are the focus of the following subsections.

For this purpose, a simulation-based evaluation tool for the comparison of different supply chain configurations is presented (Sect. 5.1). On this basis, general cause-effect relationships with a focus on pooling as a regeneration-specific configuration option are modeled (Sect. 5.2) and approaches for the operational implementation of dynamic capacity and material planning are presented (Sect. 5.3).

6 Simulation-Based Support in Strategic and Tactical Configuration of Regeneration Supply Chains

The multitude of influencing factors, configuration options as well as their interactions along the supply chain (Lucht et al. 2019) makes production configuration and especially the selection of appropriate procedures for production planning and control (PPC) extremely complex. Often, decisions on the design of the supply chain structure cannot be made in isolation from configuration decisions such as the selection of specific order release or sequencing procedures at individual work systems (Mundt et al. 2020). At the same time, these usually are also dependent on boundary conditions and input variables that cannot be influenced or are difficult to influence, making it extremely difficult or even impossible to apply common logistics models to evaluate all possible configurations.

Therefore, effective support for regeneration service providers in this respect lies in making configuration decisions assessable and thus manageable. To allow corresponding evaluations without having to run simulations during operation, a model was developed to easily assess and compare potential configurations and scenarios.

Simulation serves as a suitable tool for investigating various configurations in combination with variable boundary conditions and evaluating their logistics performance (Nyhuis et al. 2005). This section presents the simulation model for assessing the logistics effects of various production configurations and MRO-specific disturbances and uncertainties. The focus lies on the overall structure of the model, the design and configuration options modeled herein, and the aggregation of the simulation results within an assessment model, which serves to improve accessibility of the simulation results. This allows the comparison of different configurations and thus supports the selection of a suitable configuration depending on the respective boundary conditions. Thus, both structural design options, which lie in the regeneration in the introduction of one- or two-stage pooling, and parameterizable planning and control procedures of individual processes are made accessible to an evaluation.

6.1 Model and Supply Chain Structure, Disturbances and Uncertainties

To analyze the interdependencies and the logistics performance achievable with various supply chain and PPC configurations the simulation model is implemented as a time discrete model using *Siemens Tecnomatix PlantSimulation v15.3*. The basic supply chain structure follows the generic process model of regeneration presented in Sect. 1. At the top level, it comprises disassembly, inspection, repair, reassembly and a final test (see Fig. 7). The models features arranged below this macroscopic structure are described in the following section. Here, the focus is on the configuration options available.

The capital goods to be regenerated are modeled as modular products consisting of six modules with defined precedence relationships for disassembly and reassembly. This structure is adapted from the modular design of an aircraft engine, which can be seen as a typical example of a capital good to be regenerated. The intermediate arrival time of the orders is assumed to be normally distributed and thus fluctuates around a definable mean value.

When generated, one of three different order configurations is assigned to each of them. These configurations are the "full disassembly" and two variants of "partial disassembly". In the case of "full disassembly", all modules are disassembled, inspected, repaired/replaced and reassembled. The partial disassembly represents a regeneration of only 3 of the 6 modules. The first variant of partial disassembly corresponds to a reduced workscope of the regeneration order that leaves the inner three modules assembled and untouched. In contrast, the second variant assumes the first three modules to be removed fully assembled to access the three remaining modules. While all modules are removed in this configuration, only the three inner modules are fully disassembled, inspected and repaired (if necessary).

Whether a module needs to be scrapped, is serviceable or repairable as well as its specific workscope is determined by a pre-defined damage pattern. This

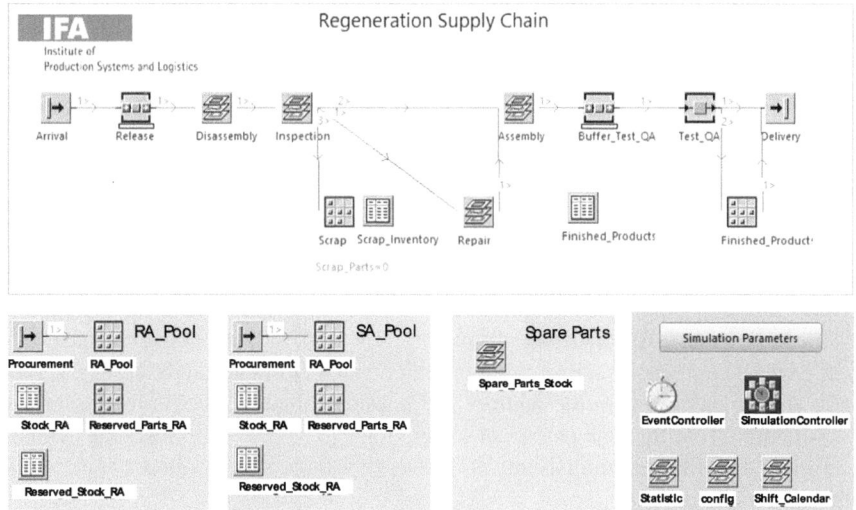

Fig. 7 Macro level of the overall simulation model implemented in PlantSimulation

represents a damage diagnosis or prognosis performed prior to disassembly. This damage pattern information is again randomly assigned to the disassembled modules during order generation. For this purpose, five specific damage patterns (no damage, light, medium, heavy damage and scrap/irreparable) are distinguished. Based on the average damage pattern, the orders are assigned to one of three customer delivery time classes (short, regular, long) corresponding to an overall project workscope. Based on the regular delivery time that also represents a configuration parameter of the simulation model, 10% of the regular delivery time is deducted or given as a premium depending on the order-specific delivery time class. As an additional option, a defined proportion (0–100%) of the orders is randomly generated as a rush order which corresponds to a reduction of delivery time by 10 days.

For disassembly, the line principle as well as a construction-site principle is implemented. While the line principle leads to a fixed disassembly sequence of the modules, the construction-site principle allows for due date-oriented sequencing of the modules. The disassembled modules are then inspected in module-specific work systems. Here, information uncertainty regarding the initial workscope can be modeled by randomly inducing normally distributed deviations of the damage pattern. The initial damage pattern forms the mean value of this distribution. Modules found to be irreparable are removed from the system and must be replaced with a replacement component.

For this purpose, a new parts inventory is implemented, which is supplied by a new parts procurement. This is assumed to be inventory-controlled and allows setting the initial stock level, reorder point as well as the replenishment time and lot size for each module. The same applies to the RA and SA pools, which can be switched on and off separately and filled via procurement or by the regeneration process. Both pool stages can provide modules as scrap replacements or allow for rotating modules between the pool inventories and the capital goods to be regenerated. This requires to remove the link between the modules to be repaired and the original regeneration good. Removing this link opens up the possibility of rotating the modules whenever a delivery date cannot be met. The assignment of pool modules to a specific regeneration order is done when a specific demand is identified in the reassembly process.

In addition, there is a configuration option that allows SA or RA pool modules to be reserved as soon as irreparability is determined during inspection. The following repair shop forms the most complex section within the regeneration supply chain. It is modeled as an area consisting of four different workstations. These can have up to six parallel stations/capacities, each of which can process all modules. The respective work plans differ in the number and duration of work steps to be performed, depending on the damage pattern of the modules to be repaired. Since the compensation of disturbances and failures of the work systems are not subject of this work, they are not explained here.

When the repair is completed, demand for rework can be randomly assigned to a defined number of modules (0–100%). Modules that require rework are routed back to inspection and need to pass the repair process again following the work plan of the next higher damage pattern. Reassembly is modeled based on the construction-site principle and can also be scaled to up to six parallel stations. The final test represents a single work system testing the reassembled good.

6.2   Configuration Options in PPC

Along the supply chain structure, different configuration options as well as regeneration-specific disturbances and uncertainties are implemented. An aggregated overview of the main configuration parameters and options in the model is given in Fig. 8.

The PPC procedures implemented cover the global order management, order scheduling, completion check, capacity control, order release and sequencing. The overall scheduling is implemented as a backward-oriented scheduling based on the order-specific arrival date and delivery time. The planned throughput times are determined on the basis of the predicted damage pattern respectively the corresponding work plans. To determine the planned inter-operation times, a multiple of the work system-specific processing time is defined. This allows for a much simpler and well automatable execution of different configurations but is of limited validity from a production logistic perspective in case of highly varying processing times (Schäfers 2020). Since the resource-specific processing times in this simulation are set comparatively homogeneous, this should only have a minor impact on the overall results.

The global order release, and thus the release into disassembly, can be triggered immediately, on the basis of the planned start date, or by means of the ConWIP (Hopp

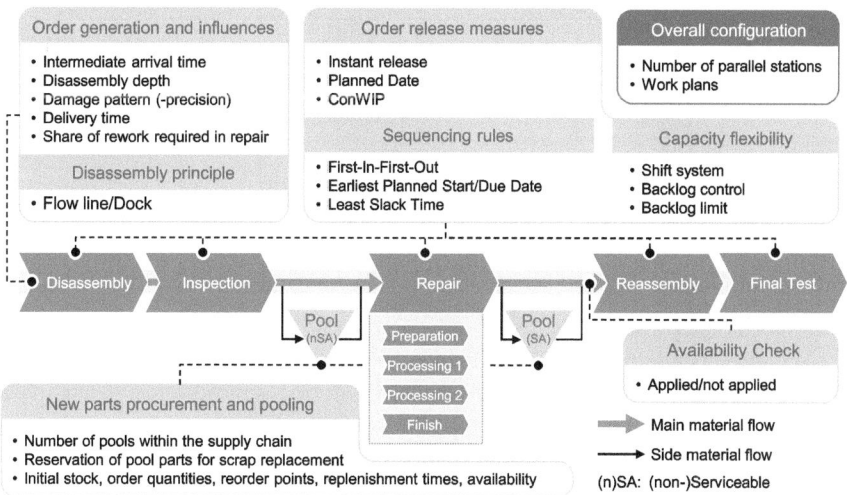

Fig. 8 Schematic overview of the configuration options implemented in the simulation model

and Spearman 2000) procedure. For the latter, the maximum number of orders to be processed in the system at the same time can be set. Order release procedures in inspection, repair and reassembly can be set to the procedures already listed for the global order release. The procedure selected applies for all parallel or sequential work stations within the respective area. For the configuration of order sequencing, both the first-in-first-out (FIFO) principle and due date-oriented principles (comprising the earliest due date or least slack procedure) can be applied in each area. They are also applied for all parallel or sequential work stations within an area. A static completeness check can be set up at reassembly to ensure that reassembly is not started until all the required modules have been provided. If this is switched off, the simulation does not limit the order release to orders which already have the material completely physically ready for reassembly.

Overall, a two-shift system is applied for all work systems and stations, allowing capacity flexibility in additional shifts. These can be activated by a backlog control when exceeding a defined backlog limit per work system. This is reset either after a defined period or after the backlog falls below the defined limit.

The variety of configuration options for results in an almost infinite number of possible combinations and cause-effect relationships. Consequently, a detailed and comprehensive logistics modeling of interdependencies within the overall configurations does not appear to be reasonable. Instead, an assessment model is to be used to support and facilitate easy analysis and comparison of the logistics performance that can be achieved with a particular overall configurations.

6.3   Aggregation of Simulation Data and Comparison of Different Configurations

The assessment model visualizes the logistics performance of the entire supply chain as well as that of the individual work systems incorporating established logistics performance indicators at the most relevant measurement points along the supply chain. While the primary focus is on the punctuality of the reassembly as the crucial point for logistics performance within the supply chain, all the other processes are also covered. This allows to analyse customer-oriented indicators like punctuality and indicators relevant to the MRO service provider like capacity utilization.

While the simulation is implemented in *Siemens Tecnomatix PlantSimulation* the analysis is implemented in *KNIME Analytics Platform* to allow for (semi-)automated analyses of the simulation data. The results are structured in a *Microsoft Excel* spreadsheet that forms the database/lookup table for the presentation and comparison of the logistics perfomance achieved with different configurations in the assessment tool. Since a complete presentation of all simulation results is not possible within the limited space of this paper, an exemplary selection of key simulation results for four configurations is shown in Fig. 9 for discussion instead. This summarizes essential key performance indicators available within the evaluation tool as well as their presentation.

Because of the different scales of the incorporated performance indicators, a comparison is made based on a relative basis compared to a reference configuration. This also allows for a better comparison of the logistics effects of a change in configuration or disturbing influences that can be simulated using the simulation model. To do so, configuration CR is set as a reference. This configuration is chosen based on the findings and conclusions of former simulation and case studies on production configuration within MRO shops (besides others Georgiadis 2017; Guide, JR. 2000; Hoffmann et al. 2017; Liu et al. 2017; Reményi et al. 2011; Sendler 2020). It applies due date oriented sequencing and order release in all supply chain areas except the inspection and final test. Instead sequencing is done based on the FIFO-principle in these processes, since these are neither of particular relevance regarding the overall research goal nor do they significantly influence the logistic performance (due to the low material flow complexity within these). Order release is done immediately after the order's arrival at the respective work station. In addition, configuration CR uses a disassembly based on the line principle and applies a static completeness check in reassembly. Configuration CR outperformes this configuration in nearly every aspect. This reconfirms the results of the previous simulation studies that can be found in scientific literature (e.g. Denkena et al. 2017; Reményi et al. 2011). Since configuration CR forms the basis for assessing the logistics-oriented impact of the disturbances and pooling mechanisms described above, these are not included in these configurations. Instead, these represent the two key differentiators to the other configuration shown in Fig. 9. Configuration C2 equals reference configuration CR but is subject to information uncertainty regarding the actual damage pattern of the capital goods respectively their modules. As a counter measure it also incorporates a

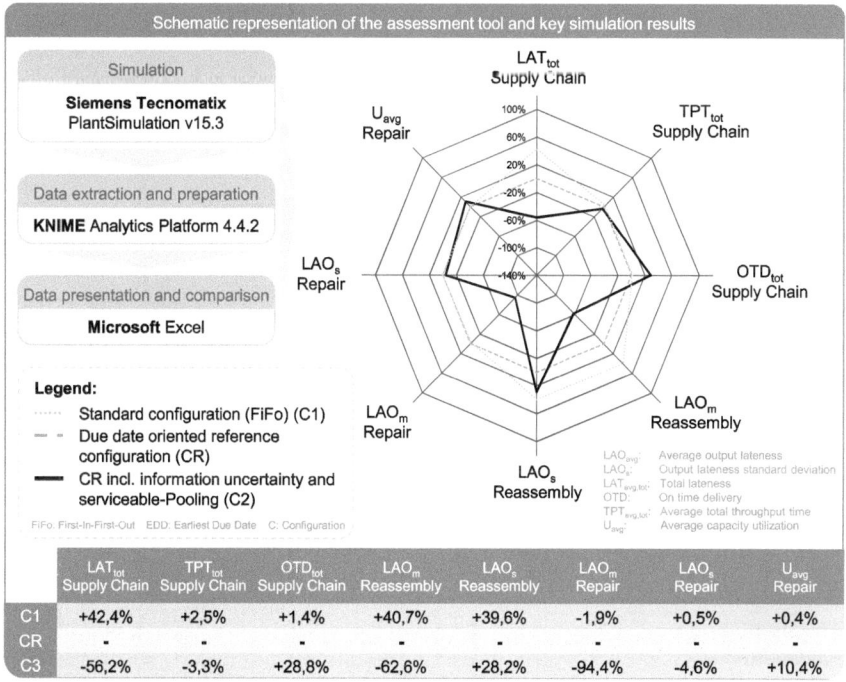

Fig. 9 Schematic representation of the assessment model

SA pool stage, that is added upstream of the reassembly. The use of the components contained in this pool follows the descriptions in sections above.

The introduction of the disturbances in configuration C2 leads to a significantly increased standard deviation of the lateness parameters. Also, the mean work in progress (WIP) in reassembly increases due to the higher disturbed WIP waiting for components. Similarly, there is a drop in the repair capacity utilization, which can be attributed to the partially lower workload induced by the repair jobs. The mean value and standard deviation of the output lateness in repair and reassembly is significantly reduced introducing single-stage SA-components pooling. In addition, a considerable increase in repair capacity utilization can be achieved with a simultaneous reduction of the mean WIP. This can be explained by the due date-oriented provision of the pool components and the thus improved controllability of the workload within the repair.

Overall, the simulations confirm the potential of various configuration options, which are in line with the results, some of which have already been reported in earlier, separate studies. The evaluation model makes it easy to assess the effects of design or control changes on the logistics performance and cost indicators. The derivation of target-oriented measures for the configuration of regeneration supply chains is thus significantly simplified. It serves as decision support for the individual and requirement-oriented design and configuration of regeneration supply chains.

In particular, the logistics-oriented potential of pooling is confirmed in terms of the opportunity to improve logistics performance and compensate for disruptive influences. Thus, the following section focuses on modeling the effects of pooling on logistics performance and costs.

7 Modeling of the Effects of Pooling on the Reassembly Process

The modeling of pooling as a regeneration-specific design and configuration option concerns both the dimensioning of corresponding pool stocks and the use of the available components in operational regeneration processes. This is addressed in the following sub-sections, building on each other.

7.1 Dimensioning of Pool Components

The simulation study confirms that the pooling of spare parts can positively influence on keeping due-dates in reassembly and the resulting missing parts situation (Kuprat et al. 2016). In order to be able to describe the missing parts situation in reassembly, which is supplied by several parallel supply processes (disassembly, repair, procurement, and pools), the supply diagram according to Beck 2013 was enhanced accordingly. This allows several supply processes to be taken into account simultaneously based on their respective schedule deviation distributions in the outgoing material flow (Kuprat et al. 2016; Kuprat 2018). This alternative improved approach of the supply diagram is thus suitable for evaluating the synchronicity of material flows in regeneration. For example, as an indicator for the evaluation, the disturbed WIP can be determined with the supply diagram. This forms the basis for describing general correlations identified in the simulation studies between the mean stock of SA pool components and the missing parts situation, expressed via the disturbed WIP. In principle, a high mean stock level of SA pool components can positively influence the disturbed WIP as well as the adherence to duedates in reassembly. However, the positive influence follows a digressive course, so that above a certain stock level no additional positive influence on the defective part situation can be achieved (see Fig. 10). This point can be determined mathematically under the assumption of ideal process conditions (this includes, in particular, deviations in entry and exit of the pools). In real environments, the boundary conditions on which this calculation is based are usually not given, so that a higher SA pool stock level is required to achieve the identical effect. This leads to a deviating curve (dotted line in Fig. 10).

The choice of components to be included in the pooling also significantly influences the missing parts situation. In view of the logistics performance achieved, components that lie on the critical path of order should be provided from the pool

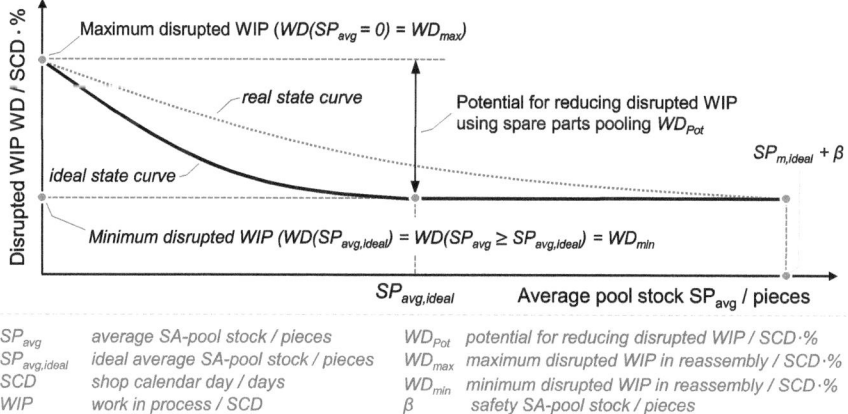

Fig. 10 Effects of pooling on the disrupted WIP in reassembly (based on Kuprat 2018)

and thus tend to cause a schedule deviation. A corresponding procedure for the selection of suitable pool components as well as further explanations of the modeling presented can be found in Kuprat 2018. To fully exploit the logistics potential bound up in the pool stocks built up in this way, they must be integrated into the regeneration operations in the best possible way. This requires integration and consideration of the interactions with the other processes of the regeneration supply chain.

7.2 Integrated Planning of Repair and Two-Stage Spare Parts Pooling

Mathematical optimization models offer a possibility to represent the resulting multitude of parameters and influences to be considered in the decision-making process. While only one SA pool level has been represented in the modeling in the previous section so far, both the SA pool level and—if available—an additional RA pool level upstream of the repair must be taken into account in an integrated manner to leverage the entire logistics potential of pooling. A corresponding mathematical optimization model is presented in Lucht et al. 2021a. This model couples classical repair scheduling, as presented in Hoffmann et al. 2017 with flexibility options opened up by pool stocks. This means that delays in the provision of materials caused, for example, by disruptions or deviations from the plan can be responded dynamically by demand-oriented replacing the original components to be repaired with pool components. While pool components from the SA pool in serviceable condition can be used for this purpose without throughput time, the repairable components held in the RA pool first require repair before they can be reinstalled in capital good. While they are, therefore, less flexible in terms of logistics, they usually have a significantly

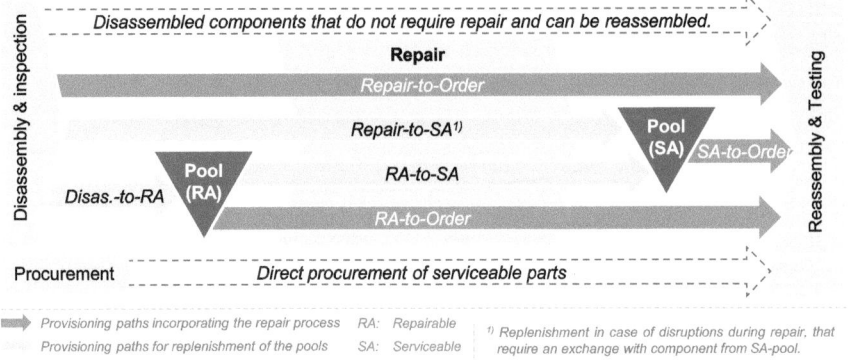

Fig. 11 Potential material provisioning paths with two-stage pooling (based on Heuer et al. 2020; Lucht et al. 2021a)

lower value than new or already repaired components [35] and thus also a significantly reduced capital tie-up. All in all, this results in various supply paths in Fig. 11 (Berkholz 2012; Heuer et al. 2020), that needs to be dynamically allocated to material demands from regeneration orders, which represents the elementary purpose of the model.

This approach thus forms a decision support system that allows integrated and time dynamic order and material coordination while simultaneously taking into account the multitude of relevant interactions in regeneration. Input variables of the model are the available pool stocks, their damages and values, if applicable, and the information about the order spectrum. On one hand, this includes the known date, planned and actual arrival date as well as the delivery date of the orders. On the other hand, additional information can be taken into account, including whether the use of pooling is permissible in the respective order, whether rework will be required at the end of the repair, and the damage class used as a basis for planning and the actual damage class determined in the findings.

This information can be used to map typical regeneration disturbances, which the model then takes into account in the decision-making process. Both predefined, completed scenarios and continuously updated data from practical applications can be processed. The objective function stored in the model minimizes the total costs, which consist of penalty costs for schedule deviations (caused by disruptive influences) in the reassembly as well as costs for the use of pool components. As a result, the model provides minimum-cost repair scheduling for the complete (but not necessarily punctual) servicing of material demand in reassembly.

A generalizable dependence of the cumulative delay on the stock level in the two pools was identified in simulation studies carried out with the model using synthetically generated scenarios. Thus, it can be seen that a reduction of the cumulative delay can be achieved with increasing stock levels in the respective pools. However, increasing the stock level in the SA pool shows a significantly higher impact—regarding schedule reliability but also capital commitment—than increasing the available stock level in the RA pool. This is due to the differences in the flexibility with which the respective components can be provided in the reassembly. In addition, however, it is also shown that the combined use of heterogeneous pool inventories makes it possible to achieve an identical cumulative delay with reduced use of new parts (Lucht et al. 2021a). The reduction in the number of new parts required to provide the desired logistics service that can be achieved in this way offers potential for a reduction in regeneration costs and a better evaluation and an increase in the reliability of regeneration service providers. In the long term, this also contributes to support a more sustainable and at the same time competitive use of resources.

8 Conclusions and Future Research

This section sums up the outline of this paper and gives an overview of potential future research activities. Future research focuses on the extension and transfer of the scientific results presented in this paper. The transfer addresses both the utilization of the scientific results in industry and the transfer of the results to other applications and industries.

9 Summary and Conclusions

The regeneration of complex capital goods is faced with both high customer demands on the logistics performance to be achieved and a simultaneous high degree of information uncertainty. Therefore, it is a highly demanding industry in terms of logistics, and the special boundary conditions prevailing here must be taken into account in its logistics design. In this respect, this article presents three fields of action that build upon each other to support regeneration service providers. These address the reduction of information uncertainty, the creation of transparency about the effects of interference and information uncertainty on essential planning tasks, and the design and configuration of efficient and robust regeneration supply chains.

Information uncertainty represents an essential perturbation impact on regeneration logistics. This concerns in particular the knowledge of the workload to be expected from a regeneration order, which is the basis for short-, medium- and long-term planning. To reduce this information uncertainty (field of action 1), Bayesian networks can be used to forecast the damage of future regeneration orders based on data from past regeneration orders enriched with information from the operation of the capital goods to be regenerated and structured in a damage library. Thus, the capacity load to be expected from these can also be determined.

In addition to the design of the capacity structure, this has a particular influence on the timing (scheduling) of the regeneration orders. Precise and stable scheduling is of great relevance for operational order processing and coordination and communication with customers. However, unclear information, internal and external disturbances, and other planning errors often lead to deviations from this planning so that planned dates are adjusted accordingly. A model for describing planning iterations was presented in the second field of action to make this planning behavior transparent and to uncover inefficiencies potentially hidden by planning iterations. This model serves both to visualize and to analyze the progress of planning due-dates so that, for example, recurring patterns in planning behavior can be identified and thus made accessible for the derivation of suitable countermeasures.

The third field of action addresses the support of regeneration service providers in the design and configuration of their supply chains. This can be supported by the presented simulation model and the evaluation model based on it. In particular, the pooling of spare parts was identified as a suitable measure to improve material supply. It was thus possible to model a generalizable influence of the average pool stock of SA components ready for installation on the disturbed WIP as a key figure for the missing parts situation in the reassembly. The use of the available pool components in the operative regeneration process requires the simultaneous consideration of many influencing factors and (mutual) dependencies. This is especially true when adding an additional pool stage with repairable RA components. The presented mathematical optimization model allows to consider these qualitatively heterogeneous pool conditions within the regeneration planning and thus to exploit their logistics potential in the best possible way. In the process, a further, potentially globalizable interdependency between the inventories of the two pool levels and the achievable adherence to delivery dates could be identified. This represents a central starting point for further research work.

10   Future Research

Future research focuses on the extension and transfer of the scientific results presented in this paper. While the developed optimization model addresses load adjustment in the form of pooling, its combination with the capacity adjustment modeling in Eickemeyer et al. 2014 offers the possibility to best exploit the joint flexibility or robustness potential of capacity flexibility and multi-level spares pooling. Further potential is promised by extending pooling as a regeneration-specific option for supply chain structure design. While the influence of single-stage pooling on disturbed WIP could already be modeled in a generally valid way, an extension of the modeling to two-stage pool structures as well as their influence on the robustness of material supply promises the potential to harness the joint logistics potential along supply chains of distributed inventories. This is of increasing importance, especially in the context of increasing supply uncertainty due to events that are difficult to predict, such as conflicts or natural catastrophes, to assess the actual material availability in the event of supply failures or to identify unnecessarily high (intermediate) material inventories. The application is not limited to regeneration service providers and promises great potential in the manufacturing industry. With regard to regeneration service providers, the increasing availability of operating data of capital goods opens up additional potential for a drastic reduction of information uncertainty about the material demand to be expected from a regeneration order and thus more precise input information for material management. A particular challenge here will be to allocate the forecast requirements to the respective available pool levels within the regeneration supply chain. For this purpose, the total demand must be broken down with regard to further requirements such as the available time for material provision per demand or the available stock of used but repairable components.

Acknowledgements Funded by the Deutsche Forschungsgemeinschaft (DFG, German Research Foundation)—SFB 871/3 119193472.

References

Beck, S. (2013). Modellgestütztes Logistikcontrolling konvergierender Materialflüsse. Dissertation. Berichte aus dem IFA, vol 2013,3. PZH-Verl. TEWISS—Technik und Wissen GmbH, Garbsen

Berkholz, DA. (2012). Grundmodell zur Kapazitäts- und Belastungsabstimmung eines Arbeitssystems in der Regeneration. Berichte aus dem IFA, vol 2012,2, Hannover

Burmeister, R., Exler, F., Petrick, T., Siedow, H.-J., and Winkler, L. (2004). Method of assembly/disassembly of aviation engines: European Patent (WO 2004/097179)

Cervelló-Royo, R., Moya-Clemente, I., Perelló-Marín, MR., and Ribes-Giner, G. (2020). Sustainable development, economic and financial factors, that influence the opportunity-driven entrepreneurship. An fsQCA approach. Journal of Business Research 115:393–402. https://doi.org/https://doi.org/10.1016/j.jbusres.2019.10.031

Denkena, B., Dittrich, M.-A., and Georgiadis, A. (2017). Combining in-house Pooling and Sequencing for Product Regeneration by Means of Event-driven Simulation. Procedia CIRP 62:153–158. https://doi.org/https://doi.org/10.1016/j.procir.2016.06.005

Dombrowski, U. nad Sendler, M. (2016). Planung und Steuerung in Instandhaltungs- werkstätten. ZWF 111:622–625. https://doi.org/https://doi.org/10.3139/104.111601

Eickemeyer, S.C. and Herde, F. (2012) Regeneration komplexer Investitionsgüter. Zeit- schrift für wirtschaftlichen Fabrikbetrieb 107:761–765. https://doi.org/https://doi.org/10.3139/104.110836

Eickemeyer, S.C., Doroudian, S., Schäfer, S., and Nyhuis, P. (2011). Ein generisches Prozessmodell für die Regeneration komplexer Investitionsgüter. ZWF 106:861–865

Eickemeyer, S.C., Goßmann, D., Wesebaum, S., and Nyhuis, P. (2012). Entwicklung einer Schadensbibliothek für die Regeneration komplexer Investitionsgüter. Industrie Management 28:59–62

Eickemeyer, S.C., Borcherding, T., Schäfer, S., and Nyhuis, P. (2013). Validation of data fusion as a method for forecasting the regeneration workload for complex capital goods. Prod. Eng. Res. Devel. 7:131–139. https://doi.org/https://doi.org/10.1007/s11740-013-0444-8

Eickemeyer, S.C., Herde, F., Irudayaraj, P., Nyhuis, P. (2014). Decision models for capacity planning in a regeneration environment. International Journal of Production Research 52:7007–7026. https://doi.org/https://doi.org/10.1080/00207543.2014.923122

Georgiadis, A. (2017). Ersatzteildisposition und Reihenfolgebildung in der Produktregeneration. Dissertation, Gottfried Wilhelm Leibniz Universität Hannover; TEWISS—Technik und Wissen GmbH

Guide VDR, JR. (2000). Production planning and control for remanufacturing: industry practice and research needs. Journal of Operations Management 18:467–483. https://doi.org/https://doi.org/10.1016/S0272-6963(00)00034-6

Guide VDR, JR., Kraus, M.E., and Srivastava, R. (1997). Scheduling policies for remanufacturing. International Journal of Production Economics 48:187–204. https://doi.org/https://doi.org/10.1016/S0925-5273(96)00091-6

Herde, F. (2013). Rahmenbedingungen der industriellen Regeneration von zivilen Flugzeugtriebwerken. Books on Demand, Norderstedt

Heuer T, Lucht T, Nyhuis P (2020) Material Disposition and Scheduling in Regeneration Processes using Prognostic Data Mining. Procedia Manufacturing 43:208–214. https://doi.org/https://doi.org/10.1016/j.promfg.2020.02.138

Hiller, T., Lucht, T., Kämpfer, T., Vinke, L., Holtsch, P., and Nyhuis, P. (2021). Hybride Lieferzeitprognose: Verbesserte Termin- und Auftragsplanung im volatilen MRO-Umfeld. ZWF 116:882–888. https://doi.org/https://doi.org/10.1515/zwf-2021-0197

Hoffmann, L.-S., Kuprat, T., Kellenbrink, C., Schmidt, M., and Nyhuis, P. (2017). Priority Rule-based Planning Approaches for Regeneration Processes. Procedia CIRP 59:89–94. https://doi.org/https://doi.org/10.1016/j.procir.2016.09.028

Hopp, W.J. and Spearman, M.L. (2000). Factory physics: Foundations of manufacturing management, 2nd edn. Irwin/McGraw-Hill, Boston

Kuprat, T. (2018). Modellgestütztes Ersatzteilmanagement in der Regeneration komplexer Investitionsgüter. Dissertation, Leibniz Universität Hannover; TEWISS Verlag—Technik und Wissen GmbH

Kuprat, T., Schmidt, M., and Nyhuis, P. (2016). Model-based Analysis of Reassembly Processes within the Regeneration of Complex Capital Goods. Procedia CIRP 55:206–211

Liu P., Zhang, X., Shi, Z., Huang, Z. (2017). Simulation Optimization for MRO Systems Operations. Asia Pac. J. Oper. Res. 34:1750003. https://doi.org/https://doi.org/10.1142/S0217595917500038

Liu, F., Lai, K., and Cai, W. (2021). Responsible Production for Sustainability: Concept Analysis and Bibliometric Review. Sustainability 13:1275. https://doi.org/https://doi.org/10.3390/su13031275

Lucht, T., Kämpfer, T., and Nyhuis, P. (2019). Characterization of supply chains in the regeneration of complex capital goods. In: Dimitrov D, Hagedorn-Hansen D, Leipzig K von (eds) International Conference on Competitive Manufacturing (COMA 19), 31 January-2 February 2019, Stellenbosch, South Africa. Department of Industrial Engineering, Stellenbosch University, Stellenbosch, pp 444–449

Lucht, T., Heuer, T., Nyhuis, P. (2020). Disassembly sequencing in the regeneration of complex capital goods. In: Nyhuis P, Herberger D, Hübner M (eds) Proceedings of the 1st Conference on Production Systems and Logistics (CPSL 2020). publish-Ing, pp 12–20

Lucht, T., Wojcik, A., Nyhuis, and P. (2021a). Integrated Repair Shop Scheduling and Spare Parts Pooling for Robust Product Regeneration. In: 2021 IEEE International Conference on Industrial Engineering and Engineering Management (IEEM). IEEE, 200–206

Lucht, T., Mütze, A., Kämpfer, T., and Nyhuis, P. (2021b). Model-Based Approach for Assessing Planning Quality in Production Logistics. IEEE Access 9:115077–115089. https://doi.org/https://doi.org/10.1109/ACCESS.2021.3104717

Lund, R.T. (1984). Remanufacturing: The experience of the United States and implications for developing countries, 1st edn. World Bank Technical Paper, vol 31. Banco Mundial, Washington

Mundt, C., Winter, M., Heuer, T., Hübner, M., Seitz, M., Schmidhuber, M., Maibaum, J., Bank, L., Roth, S., Scherwitz, P., and Theumer, P. (2020). PPS-Report 2019: Studienergebnisse, 1st edn. TEWISS, Garbsen

Mütze, A., Lucht, T., and Nyhuis, P. (2022). Logistics-oriented Production Configuration Using the Example of MRO Service Providers. IEEE Access:1. https://doi.org/10.1109/ACCESS.2022.3146420

Nyhuis, P., Cieminski, G. von, Fischer, A., and Feldmann, K. (2005). Applying Simulation and Analytical Models for Logistic Performance Prediction. CIRP Annals 54:417–422. https://doi.org/https://doi.org/10.1016/S0007-8506(07)60135-8

Reményi, C. and Staudacher, S. (2014). Systematic simulation based approach for the identification and implementation of a scheduling rule in the aircraft engine maintenance. International Journal of Production Economics 147:94–107. https://doi.org/https://doi.org/10.1016/j.ijpe.2012.10.022

Reményi, C., Staudacher, S., Becker, H., Dinc, S., Manejev, R., and Fichtelmann, R. (2011). Simulation of the Maintenance Process in an Aircraft Engine Maintenance Company. In: Duffie, N.A., DeVries, M.F. (ed) Proceedings of the 44th CIRP Conference on Manufacturing Systems, Madison, Madison, Wisconsin/USA

Schäfers, P. (2020). Modellbasierte Untersuchung der Wirkung von Planungs- und Steuerungsverfahren auf die Termintreue einer Produktion, 1st edn. Berichte aus dem IFA, 02/2020. TEWISS, Garbsen

Seitz, M., Sobotta, M., and Nyhuis, P. (2019). A Data Mining Approach to Support Capacity Planning for the Regeneration of Complex Capital Goods. In: Ameri F, Stecke KE, Cieminski G von, Kiritsis D (eds) Advances in Production Management Systems. Towards Smart Production Management Systems, vol 567. Springer International Publishing, Cham, pp 583–590

Seitz, M., Lucht, T., Keller, C., Ludwig, C., Strobelt, R., and Nyhuis, P. (2020). Improving MRO order processing by means of advanced technological diagnostics and data mining approaches. Procedia Manufacturing 43:688–695. https://doi.org/https://doi.org/10.1016/j.promfg.2020.02.121

Sendler, M. (2020). Ganzheitliche Konfiguration der PPS in der Instandhaltung hoch-wertiger Investitionsgüter, 1st edn. Schriftenreihe des IFU, vol 36. Shaker, Düren Stricker, N., Lanza, G. (2014). The Concept of Robustness in Production Systems and its Correlation to Disturbances. Procedia CIRP 19:87–92. https://doi.org/10.1016/j.procir.2014.04.078

Weiss, M., Staudacher, S., Becchio, D., Keller, and C., Mathes, J. (2022). Steady-State Fault Detection with Full-Flight Data. Machines 10:140. https://doi.org/https://doi.org/10.3390/machines10020140

Selection of Efficient Regeneration Modes for the Regeneration of Complex Capital Goods

Carolin Kellenbrink, André Schnabel, Marleen Hoppmann, Jan Niklas Woidtke, and Stefan Helber

Abstract In contrast to the production of new parts, for the regeneration of complex capital goods, various modes of regeneration are often available. They reflect, e.g., different repair technologies and/or different personal qualifications. In this paper, we describe solution approaches for the selection of efficient regeneration modes. Thereby, we simultaneously schedule maintenance tasks as they influence mode selection. Using the example of turbine blades of aircraft engines, we explain the problem setting and the need to consider the customer's business model. For immobile capital goods such as wind turbines, the selection of efficient regeneration modes requires additional decisions concerning the transportation of personnel and material. We explain this adjacent problem setting and solution approaches. In this context, we include stochastic service times and weather conditions as well as the uncertain condition of the good.

Keywords Maintenance, repair, and overhaul · Maintenance scheduling · Multi-modal routing · Aircraft engines · Wind turbines

1 Introduction

The maintenance, repair, and overhaul (MRO) of used products is an important topic for ecological and economic reasons. Particular challenges arise for planning the so-called regeneration processes of complex capital goods such as aircraft engines or wind turbines. Due to the high remaining value, decisions must be made on how to efficiently regenerate objects with unique deterioration. The specific condition of the good and the customer's business model are important for the selection of such an efficient regeneration mode and the scheduling of associated activities. For example, a mainly time-sensitive customer has different requirements on the regeneration process and its outcome than a mainly quality-sensitive customer.

C. Kellenbrink · A. Schnabel · M. Hoppmann · J. N. Woidtke · S. Helber (✉)
Institute of Production Management, Leibniz University Hannover, Koenigsworther Platz 1, 30167 Hannover, Germany
e-mail: stefan.helber@prod.uni-hannover.de

© The Author(s) 2025
J. R. Seume et al. (eds.), *Regeneration of Complex Capital Goods*,
https://doi.org/10.1007/978-3-031-51395-4_19

The regeneration of *immobile* capital goods, such as wind turbines, imposes additional requirements on the process planning as the teams and the tools need to be transported to the location of the capital good. Therefore, transportation times and costs must be considered when scheduling the tasks. There are also different modes available, as the implementation of the tasks depends on the qualification and on available tooling of the assigned team. For example, the rotor blade of a wind turbine can be repaired either on the rope or by using a hydraulic lifting platform. Planning is further complicated by the fact that bad weather can delay or even prevent the implementation of certain tasks. The start and duration of the repair is, therefore, subject to uncertainty. Again, the condition of the good must be taken into account. In Sect. 2 of this article, we focus on the selection of efficient regeneration modes in the context of aircraft engines to describe the problem setting in detail and to set the foundation for the rest of the paper. Section 3 considers the additional challenges for the regeneration of immobile capital goods such as wind turbines. We extend the problem setting by considering the condition of the good in Sect. 4 and by additionally considering the uncertainty of service times and weather in Sect. 5.

2 Fundamentals: Selection of Efficient Regeneration Paths for Turbine Blades of Aircraft Engines

For the regeneration of complex capital goods, such as turbine blades of aircraft engines, there are multiple technically permissible options available. These different regeneration modes lead to different process parameters, such as makespan and cost but also to different functional features of the regenerated good concerning performance and remaining lifetime.

The different regeneration paths for a turbine blade are shown in Fig. 1.1. Depending on the former condition of the good, there are up to seven different regeneration paths available. We assume that for one model turbine blade and one model customer, paths 1, 2, and 3 are not applicable due to the condition of the blade. The resulting model attributes of the remaining paths 4 to 7 are shown in Table 1.1. Regeneration path 4 leads to the highest possible remaining lifetime and a very good performance, so that the customer assigns the result with a quality level A. However, it has high costs of 4,000 e and a quite long duration of 674 minutes.

When choosing an efficient regeneration mode, the customer business model needs to be explicitly considered. The customer does not always demand or is willing to pay for the best functional state that can be achieved. For a quality- sensitive customer, path 4 would be the optimal path with a high quality and lower costs compared to a new part (path 7). In contrast, for a time- and quality-sensitive customer, path 5 may be a better choice with a much shorter duration compensating for the lower quality. This shows that the trade-off among functional condition, cost and duration must be considered to find a profit-maximizing solution for the specific customer. The customer's business model is reflected by the time- and quality-dependent revenue.

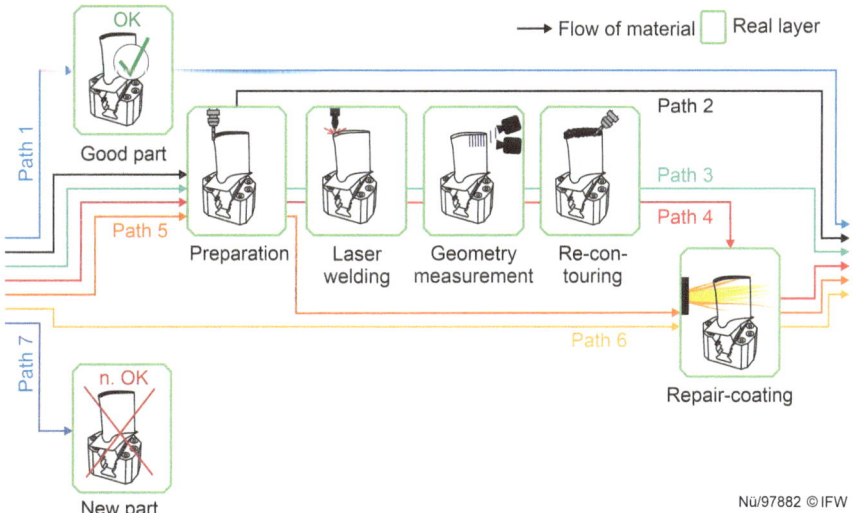

Fig. 1.1 Different regeneration paths

Table 1.1 Properties of the possible project structures

Path	Description	Lifetime Flightcycles	Performance K	Quality level	Duration min	Costs €
4	Welding & Coating	20,761	1.45	A	674	4,000
5	Preparation & Coating	19,986	3.92	B	417	2,000
6	Coating	18,325	5.48	C	392	1,500
7	New part	20,761	1.43	A	62	10,000

In addition, the capacity utilisation on the shop floor and the limitation of the mobile handling system must be considered as different regeneration projects are coupled via those common capacities. Assume, for example, that path 6 is chosen for a time-sensitive customer leading to a quality level C. If the repair-coating, as shown in Fig. 1.1, is temporarily occupied, the resulting waiting time could be used for the preparation step of path 5 without a temporal extension of the whole process. Due to the higher quality level of path 5 compared to path 6, customer satisfaction could be improved by choosing this path. This shows that the optimal regeneration path is influenced by the resource competition of different regeneration orders.

To model the corresponding decision problem, our approach for scheduling flexible projects, as presented in Kellenbrink and Helber (2015), Kellenbrink and Helber (2016) and Hoffmann et al. (2020), can be used. In such projects, there are mandatory activities and optional activities reflected by decisions and caused activities, as shown in Fig. 1.2. The goal is to find a profit-maximising schedule including the

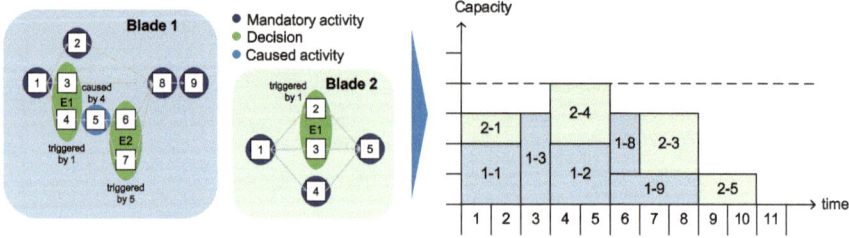

Fig. 1.2 Scheduling multiple flexible projects

decision on the implemented activities. Using heuristic solution approaches, such as genetic algorithms or simulated annealing, it is possible to efficiently schedule multiple regeneration projects. The approach is not limited to aircraft engines but can be applied to different mobile capital goods.

The abovementioned research has been carried out in subproject "D3: Selection of efficient regeneration modes" in the German Collaborative Research Centre (CRC) 871 "Product-Regeneration". This research network aims to transform experience-based decisions on MRO processes into knowledge-based decisions using a virtual twin. To prove the operability of the concept, a physical system demonstrator was built up using the example of the aircraft engine and, in particular, the turbine blade. The described decision on an efficient regeneration path and on the scheduling of the process steps plays a central role in the system demonstrator. It supports the consolidation and coordination of the results from the other subprojects for the purpose of a holistic view of the regeneration processes. In Kellenbrink et al. (2022), we show that our approach can successfully be applied to select efficient regeneration modes in the context of this system demonstrator.

3 Selection of Efficient Regeneration Modes for Maintenance Actions at Multiple Distributed Wind Farms

3.1 Introduction

The question of how to select regeneration modes occurs not only for mobile complex capital goods such as aircraft engines but also for the regeneration of immobile capital goods such as wind turbines. Wind turbines are subject to wear and tear both during operation and downtime. In this case, the regeneration modes can also differ. They can be especially distinguished due to organizational aspects. For example, blade damage can be repaired by either climbing the blade using a rope or by getting access to the blade using a lifting ramp. Both options require specific skills from the team and special equipment to be placed inside the vehicle used by the team. In addition, the trade-off between lower costs, if a team is not equipped with all essential spare parts and the latest tools, and a longer processing time due to delivery times or slower processing is considered.

In addition to the selection of an efficient regeneration mode, which is reflected by the assignment of one specifically qualified and equipped team, an operator responsible for the maintenance of multiple wind parks must simultaneously schedule those activities by assigning suitable starting times to each activity. To regenerate all goods, a tour must be found in which all spatially distributed locations are visited.

The problem outlined is very similar to the established models and procedures from the literature on routing problems. However, the routing problem for the regeneration of immobile capital goods becomes challenging and cannot be solved by known methods as the different team-specific regeneration modes become available. In this situation, an interesting combination of routing and project scheduling arises.

3.2 Problem Setting

Different wind turbines must undergo necessary maintenance activities depending on the lifetimes of different expendable parts and on condition monitoring, which utilizes the data collected by multiple sensors at or near the turbine. The locations of those turbines are given. The driving times required by the teams to travel from the location of one wind turbine to the location of another turbine cannot be disregarded. The operator of the wind turbines must ensure that the combined availability and, thus, energy production of the turbines in a wind park is above a specific level. When a repair occurs too late, the turbine might have a significantly decreased power output or even cease to operate. Contrariwise, if the team attempts to execute an operation too early, it may interfere with the fixed interval of inspections and cause unnecessary costs. This leads to individual time windows for each operation.

As shown in Fig. 1.3, each team starts and finishes its tour at the depot of the operator, reflected by the dark blue dot. In the depot, the vehicles and equipment are stored when not being used. There are no turbines and therefore no tasks to be carried out at the depot. The qualifications of the teams are heterogeneous, and each team travels with a specific selection of tools and equipment in their vehicle. Therefore, the duration of maintenance operations and hourly wages differ between the teams. Based on historical data and information from wind turbine condition monitoring, a rough estimate is made of the expected scope of work in the different modes, i.e., for the different teams. Information from previous regeneration events can also be included if, for example, an uncritical error has already been identified but has not yet been remedied. Initially, it is assumed that all values are known in advance.

A solution is characterized by an assignment of a starting time (within its time window) and a responsible team for each maintenance activity at each turbine; thus, the team specific tours can be derived. In addition, the start and end working time of each team needs to be defined to compute the total working hours of each team. A model solution is shown in Fig. 1.4. The activities on turbines 2, 3 and 4 are executed

Fig. 1.3 Example instance
and possible tours for two
teams

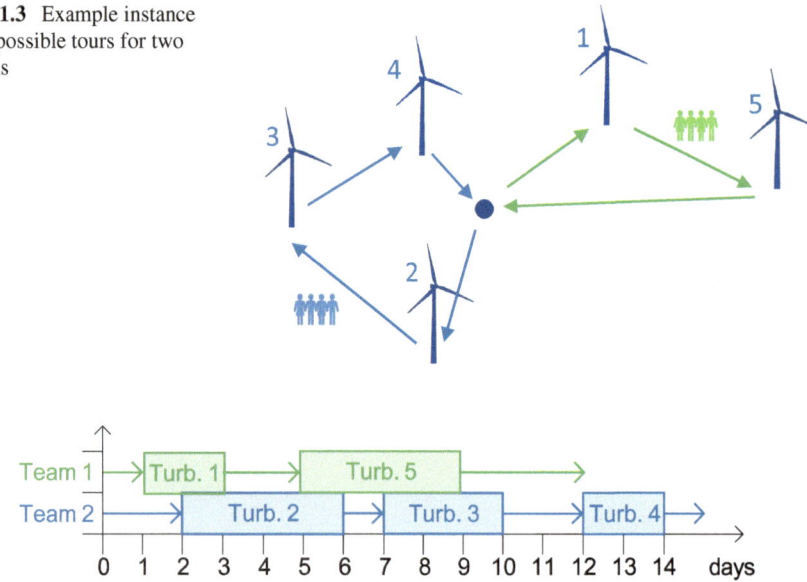

Fig. 1.4 Feasible solution

by blue team 2. For example, the maintenance process on turbine 2 starts at day 2, and the activity on turbine 3 starts at day 7 after a transfer of this team from the location of turbine 2 to the location of turbine 3. The overall working hours of both teams in this schedule are 12 and 15 days, respectively. This results in personnel costs of 39 monetary units, assuming one day of team 2 costs 2 monetary units and team 1 is cheaper with 1 monetary unit per day.

We formally defined the introduced problem setting as a mixed-integer linear model. The first numerical results when solving instances to proven optimality using a commercial standard solver show an almost exponential increase in the required CPU times. This, in combination with the fact that multiple-hard problems are subproblems of this new problem, makes it very likely that no exact solution procedure will be able to identify the optimal solutions of this problem in an acceptable time. Therefore, it seems promising to develop problem-specific heuristic solution procedures for this problem to quickly obtain good solutions. Since both local search procedures and column-generation approaches have been applied successfully to similar problems, we now adapt these two inexact solution methods and evaluate their efficiency at solving this problem in a heuristic fashion.

3.3 Related Literature

The interest in renewable energy sources in general and wind energy in particular has been growing in recent years due to regulatory incentives, which are motivated by public concerns regarding climate change. Therefore, the number of publications dealing with issues related to wind turbines has significantly increased in recent years. Irawan et al. (2017), Dai et al. (2015), and Stålhane et al. (2015) investigated routing and scheduling for offshore windfarms. Turbines located in offshore parks are typically harder to maintain due to the weather-dependent accessibility of the sites. Kovács et al. (2011) integrate their scheduling with condition monitoring and error diagnosis. Sinha and Steel (2015) are performing a risk-based analysis of turbine outages. Byon et al. (2010) consider weather-related issues and Tautz-Weinert et al. (2019) the relevance of financial aspects. Salo et al. (2018) use text mining on historical data of turbines. None of the cited sources considers integrated routing and scheduling with heterogeneously qualified teams and time windows.

Multiple publications deal with the technical aspects of wind turbine repairs. A review of different methods for the repair of wind turbines can be found in Mishnaevsky Jr (2019). Ozturk and Fthenakis (2020) predict the frequency, repair times and costs for wind turbine failures using Bayesian updating. Colone et al. (2019) present predictive repair scheduling of wind turbine drive-train components based on machine learning.

In addition to the obvious proximity to routing problems, this problem is related to machine scheduling, in which orders are assigned to different machines. Travel times can be interpreted as sequence-dependent setup times, as shown by Brucker and Knust (2012).

3.4 Heuristic Solution Approaches

3.4.1 Local Search with Indirect Encoding

The problem setting can be interpreted as an integrated partitioning and sequencing problem. The maintenance tasks are partitioned into subsets whereby each subset corresponds to one team. The order of the wind turbines in such a set then defines the sequence of the maintenance actions. Decoding these sequences into an overall schedule is possible by using a schedule generation scheme, which derives the starting times for all maintenance activities.

The schedule is derived from that sequence by letting the team start its tour before the beginning of the first activity in its sequence so that it arrives exactly at the beginning of the time window belonging to the maintenance action at the turbine. The team then finishes processing and drives to the next activity in the sequence, again having a look at the feasible time window and so on. If the processing of one activity ends after the right-hand-side of the time window, the procedure aborts

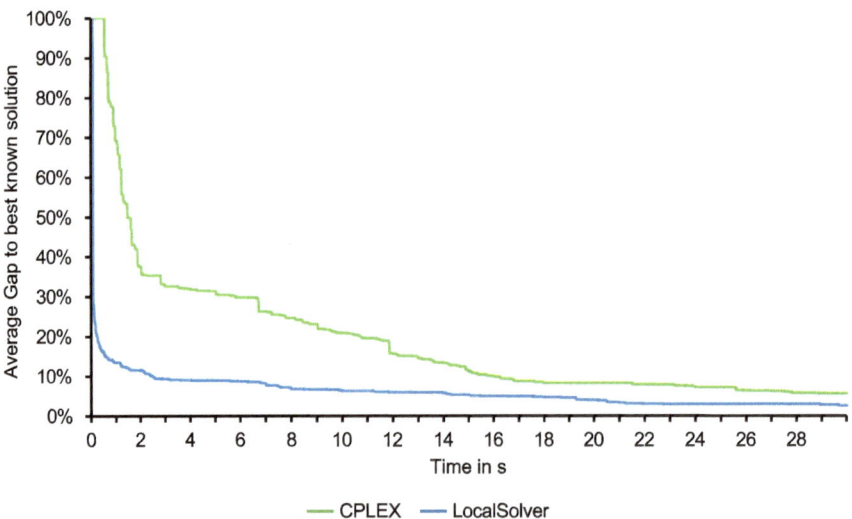

Fig. 1.5 Solution quality LocalSolver vs. CPLEX

and sets the total working hours of team k to the number of teams K multiplied with planning horizon T as a penalty. After all activities are scheduled, the activities except the latest one are aligned to the right to ensure that the total working hours are not unnecessarily increased by the previous left alignment of activities. For example, the solution presented in Fig. 1.4 would be represented by the sequence 1, 5 for green team 1 and the sequence 2, 3, 4 for blue team 2.

As an indirect encoding, the shown representation was implemented in a problem-tailored heuristic solution approach using the commercial software LocalSolver, cf. www.localsolver.com. We compare this approach with solving the mixed-integer linear program (MIP) using the standard solver CPLEX for an instance with 24 tasks and 4 teams. Figure 1.5 contains a diagram showing the progression of the average gap to the best-known solution over time up to a time limit of 30 seconds. Both solution procedures start with a gap of 100%. LocalSolver is able to quickly improve upon the initial candidate solution and identifies solutions with a gap of under 10% after just 3 seconds of running time. The MIP solver CPLEX is significantly slower in finding better solutions. CPLEX is not able to overtake LocalSolver before reaching the time limit.

3.4.2 Column-Generation Based Problem Size Reduction

After the assignment of maintenance tasks to teams is finalized, there are no dependencies between the teams. Therefore, it seems promising to decompose the underlying problem into a set of subproblems dealing with the scheduling of the different teams. Of course, the choice of the actually implemented tasks can also be part of the subproblem of creating a schedule for a team. This decomposition can be used in a column-generation scheme, where the master problem consists of choosing a set of possible team schedules such that each task is covered by at least one chosen schedule and for each team, no more than one schedule is selected.

To generate new schedules with potentially reduced costs in the subproblems, the shadow prices depicted from the LP-relaxed version of the restricted master problem (RMP) are included in the objective function of the subproblem for each team. For example, they reflect the attraction of including a certain location in the solution of the subproblem. The objective computes the reduced costs by considering the actual costs of the newly generated schedule and subtracting the shadow prices π_i and σ_k of the constraints for all covered tasks i and the selected team k, respectively. If the objective function value of any schedule is strictly negative, this schedule is added to the restricted master problem, and the loop of iteratively solving the master and the subproblem is started again.

The column-generation approach does not lead to feasible solutions, as only a relaxation of the master problem is solved. For this reason, the mixed-integer linear programming formulation of the original model is solved afterwards. The advantage of our approach is that the problem size can be heuristically reduced by fixing some variable bounds. Team-/task-assignments, which never occur in any selected schedule in the final master RMP solution, are forbidden by setting the upper bounds of the corresponding decision variables to zero.

3.4.3 Numerical Results

Figure 1.6 shows the required time and reached gap to the solution obtained with a MIP solver with a time limit of 3,600 seconds and an average computational time of 1,445.83 seconds for instances with 6 teams, 4 locations, and 12 tasks for three different solution methods. The first solution method uses LocalSolver and the integrated partition- and sequencing-problem approach to solve the problem monolithically. LocalSolver is able to reach a deviation of less than 1% in only approximately 7 seconds. The second solution method is a combination of mathematical programming for column generation and a MIP solver for the resulting reduced problem with forbidden team-task associations. This approach requires slightly less time but results in a lower solution quality. As a third solution method, the final reduced problem is solved using LocalSolver. Here, the CPU times are higher, and the solution quality is lower.

The results show that all three methods are much faster than the MIP solver with less than 1% of its computational time and maintaining a very high quality. This shows that these approaches are useful for efficiently scheduling maintenance tasks on immobile capital goods. Using LocalSolver to solve the base problem seems the most promising. The column-generation approach can be used to obtain lower bounds in a Branch-and-Price approach to compute exact solutions. This is an interesting aspect for future research.

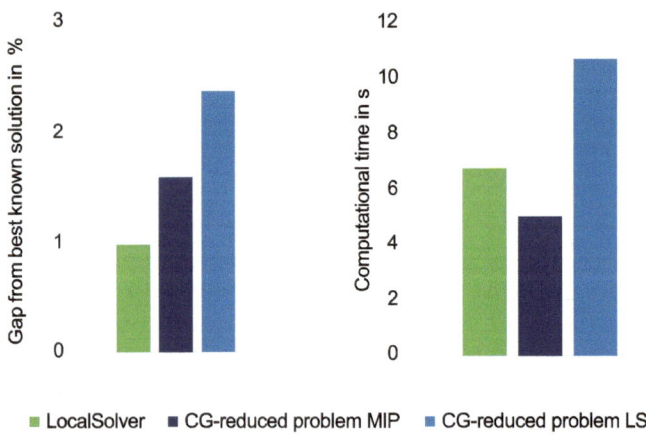

Fig. 1.6 Numerical results

4 Condition-Based Planning of Maintenance Actions

4.1 Problem Setting

In the previous section, it was assumed that each turbine must be maintained within a defined time interval reflecting a certain service level for energy production. However, in reality, such hard time windows do not exist. Instead, there is a desired point in time for the maintenance of a certain wind turbine, and deviations from this time are feasible but should be avoided. On one hand, if a turbine is maintained too early, in long-term maintenance actions, the corresponding costs occur too often. On the other hand, if the desired time is exceeded due to the limited availability of the teams and requirements of other turbines, the turbine may fail. To avoid both situations, we penalize delays and earliness in the objective function. Thus, our approach allows for condition-based scheduling without limiting the possible solutions using hard time windows.

The desired maintenance time can be determined, for example, by evaluating automatically transmitted turbine condition data, which can be used to predict the failure time of a component. Nevertheless, in practice, the many uncertain factors, such as different wear of individual components due to different environmental conditions or production qualities, make an exact forecast of the desired maintenance time difficult. The desired maintenance time of each turbine is, therefore, considered a stochastic variable.

4.2 Related Literature

In Ahmad and Kamaruddin (2012b), Ahmad and Kamaruddin (2012a) as well as Bousdekis et al. (2015), general reviews on condition-based maintenance decision-making and its application in industry are given. An overview of condition-based maintenance strategies in the context of offshore wind energy with a focus on condition monitoring, fault diagnosis and prognosis, and maintenance optimization is given by Kang et al. (2019).

Zhang et al. (2015) presented a framework for the condition-based maintenance and operation of wind turbines. Zhou and Yin (2019) report a reduction of annual maintenance cost by 30–40% due to an opportunistic condition-based maintenance strategy for offshore wind farms.

Song et al. (2018) define a condition-based maintenance policy with periodic inspection intervals. Therefore, probabilistic models are built for stochastic wind speeds and directions. The preventive maintenance scheduling problem with interval costs is introduced by Bangalore and Patriksson (2018). It considers both age-based and condition-based models to provide more flexible maintenance schedules than the constant-interval approach. Likewise, the rolling horizon stochastic optimization model on a daily basis by Besnard et al. (2011) is able to minimize service maintenance costs compared to performing the maintenance during a fixed period.

Wang et al. (2021) analyse the cost advantages of grouped maintenance actions for a wind farm compared to the loss of revenue to do waiting times for single failed turbines. A genetic algorithm to solve the optimisation problem is presented. The combination of condition-based maintenance and time-based preventive maintenance can be used to find a cost-optimal degradation threshold and to avoid overuse or underuse of maintenance tasks, cf. Dao et al. (2021).

Ghamlouch et al. (2019) propose a decision-making procedure for condition-based maintenance planning in a dynamic environment and/or variable working conditions leading to stochastic deterioration and production processes. Tian et al. (2011) develop an approach to determine an optimal maintenance schedule given the failure probabilities for components and the system. A case study of a wind turbine gearbox is discussed in Byon and Ding (2010). The consideration of seasonal weather effects in the context of condition-based maintenance is able to improve the cost structure.

4.3 Scenario-Based Compensation Model

The stochastic model arising can be interpreted as a two-stage compensation model, cf. Gendreau et al. (1996) and Scholl (2001). In such a model, a distinction is made between deterministic first-stage decisions and stochastic second-stage decisions. In the first stage, the schedule is defined. In the second stage, the deviation between actual and (beforehand unknown) desired maintenance time is compensated by the

Fig. 1.7 Simple random sampling

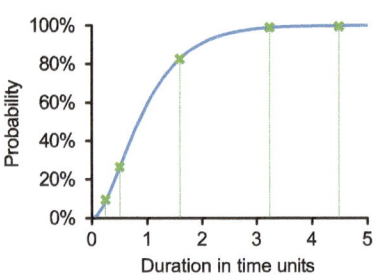

penalty costs for earliness or tardiness. Please note that the second stage only is introduced as a fictitious stage to consider the stochastic nature of the problem setting. As the optimal desired maintenance time of a turbine is not observable, the penalties are interpreted as opportunity costs.

To include the penalty costs for the compensation measure on the first stage and to be able to formulate a deterministic substitute model solvable with standard methods, a scenario approach is used. In this method, many different fictitious scenarios with their associated probability of occurrence are considered even in the first-stage decision. The scenario approach makes it possible to work with several deterministic substitute values for the original stochastic value instead of only using the mean value. The objective function is no longer calculated for a single scenario, but a sum is minimised over all scenarios multiplied by the probability of occurrence of the scenario.

The generation of the individual scenarios is very important for the quality of the scenario approach. On one hand, the more scenarios that are generated, the more robust the final solution becomes. On the other hand, the computational effort also increases considerably with an increasing number of scenarios. In simple random sampling, in which a random sample is drawn independently from the given distribution, a useful distribution of the scenarios is not ensured, as shown in Fig. 1.7. In the descriptive sampling by Saliby (1990), the distribution function is divided into different quantiles, as shown by the dotted black lines in Fig. 1.8. As many quantiles of the same size are formed as scenarios are desired. Then, for each quantile, the value of the distribution function at the position of the medium value of the quantile is chosen; see green lines. In this way, a better representation of the stochastic distribution is achieved. For this reason, descriptive sampling is used to create the scenarios.

4.4 Numerical Analysis

The effect of the condition-based approach on the optimal maintenance time is examined for different ratios of the deviation costs for earliness and tardiness. To focus on the general behaviour, only one turbine that can be maintained by one team is

Fig. 1.8 Descriptive
sampling

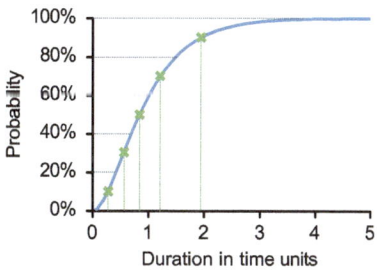

considered. The only decision to be made is about the cost-optimal starting time over
all scenarios.

The (unknown) desired maintenance times for the different scenarios are generated
in the interval [3, 5]. As a fixed cost rate, the delay costs are set to $c_{delay} = 20$ monetary
units. The earliness costs are drawn from the interval $c_{pre} = [1, 500]$ monetary units.
Thus, ratios in the range c_{pre}/c_{delay} [0, 25] are considered. The optimal maintenance
times are plotted against this cost ratio c_{pre}/c_{delay}; see Fig. 1.9. With a cost ratio of
1, the optimal maintenance time with a value of 4 reflects as expected the mean
value of the linearly distributed scenarios. For ratios of $c_{pre}/c_{delay} < 1$, it is attractive
to maintain the turbine earlier than at the mean value, and for ratios above 1, the
processing should start later than the mean value. For a model ratio $c_{pre}/c_{delay} = 5$, it
is five times more costly to regenerate the good one period too early than to delay the
regeneration event by one period. In this case, the optimal regeneration time would be
4.5. The analysis shows that the optimal starting time cannot be determined linearly
from the cost ratio, which justifies the usage of a scenario approach.

Of course, this analysis only focuses on the regeneration of one turbine. In the
case of multiple spatially distributed wind parks and different teams, the selection
of an optimal start of the regeneration process is substantially more complex but
follows the same approach. It is done by integrating the described approach into the
model reflecting the problem setting in Sect. 3.

5 Consideration of Stochastic Service Times and Weather-Dependent Availability

5.1 Problem Setting

The assumption of deterministic team-specific durations for the maintenance activities does not hold in practice since manual operations and unknown conditions
may increase the actual time required for any maintenance activity. Therefore, the
formerly deterministic parameter for the service time of a given team at a given
location is now modelled as a random variable. A gamma distribution is able to
represent stochastic service times very well. Figure 1.10 shows the density function

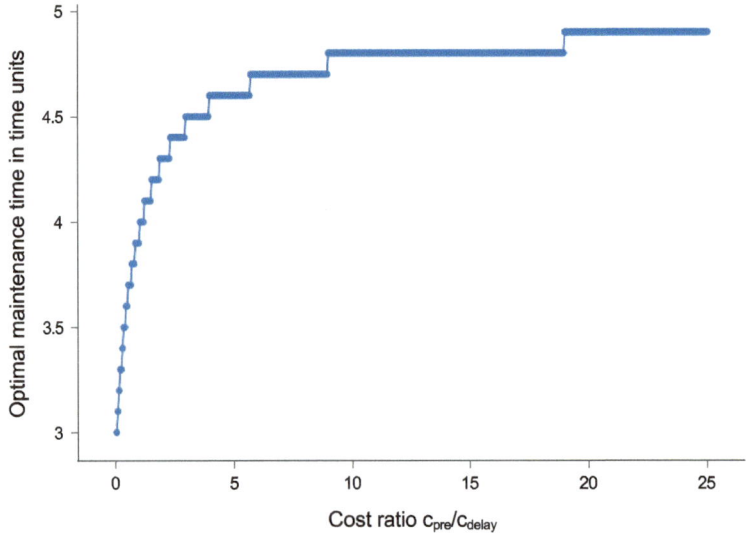

Fig. 1.9 Course of the optimal maintenance time for different cost ratios c_{pre}/c_{delay}

of a gamma distribution, which is right-skewed. This means that small deviations from the service time are very likely, while very strong deviations are possible but rather unlikely.

In addition, we now explicitly model the weather situation. If at a certain location on a certain day the weather is so bad that it does not allow maintenance work, this location cannot be included in the tour on this day. To be able to take the single days into account, we assume a finer time structure with an exact modelling of the workdays with a maximum daily working time and a maximum weekly working time. In this situation, neither time windows nor penalties for earliness or tardiness are considered, as it is assumed that for this shorter planning horizon, all included turbines are schedulable and that the deviation from the desired starting time is no longer decision-relevant.

Again, we use a two-stage approach in a scenario-based decomposition model. The sequence of the locations as well as the team assignments are independent of the scenarios and reflect the decisions on the first stage. They are determined in advance

Fig. 1.10 Density function of the gamma distribution

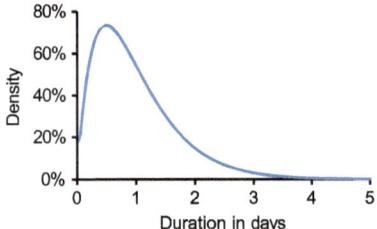

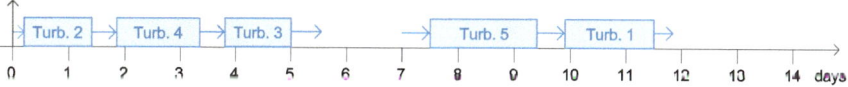

Fig. 1.11 Schedule in scenario 1

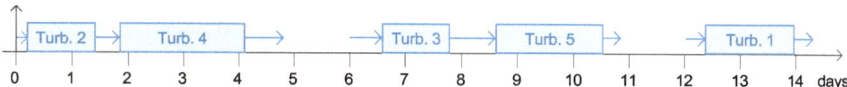

Fig. 1.12 Schedule in scenario 2

before the service times are realised. For example, for one team, the sequence of locations could be defined as (2, 4, 3, 5, 1). In the second stage, the maintenance actions in the given sequence are assigned to specific days. Please note that the scenarios are no longer fictitious as for the conditions-based approach but that the realized service times can be observed ex-post.

Possible solutions for different scenarios are presented in Figs. 1.11 and 1.12. In scenario 1, the team travels from turbine 2 to turbine 4 and then to turbine 3 in the first tour, but in scenario 2, due to a longer duration of the maintenance actions on turbine 4, the team only visits turbines 2 and 4 in the first tour.

As a compensation measure to ensure that the decision on the first stage leads to feasible schedules for the second stage, we introduce the possibility of not visiting a location at all being associated with penalty costs.

5.2 Related Literature

In practical vehicle routing problems, the parameters are often not known or are subject to certain fluctuations. Most likely, the most commonly assumed source of uncertainty, according to Gendreau et al. (1996), is stochastic demand. In this case, the level of demand at the customer location is uncertain and may fluctuate. Other approaches include stochastic customers. In this case, the demand at each customer location is deterministic, but there may be a certain probability that customers are not present. Bertsimas (1992) consider both stochastic customers and stochastic demands, which consist of two stages. In the first stage, several routes are planned, with each of the routes starting and ending at the depot. In the second step, the customers' actual absence and realized demand are considered. Han et al. (2013) present a tour planning model where travel times are uncertain. They also use a two-stage model, whereby the routes are determined in the first stage. In the second stage, the exact travel times are known, and penalty costs are calculated for the excessive duration of a route.

5.3 Heuristic Solution Approaches

5.3.1 Schedule Generation Scheme

Even for a very small instance with only twelve locations, the MIP solver Gurobi could not find an optimal solution within a given time limit of 3,600 seconds. This shows that again, a heuristic solution approach is necessary. The solution representation remains the same as described in Sect. 3.4.1, but there are small adjustments in the schedule generation scheme necessary. The scheduling takes place for each scenario. The maintenance tasks are scheduled one after the other in the given sequence until the maximum daily and/or weekly working time for the current day is exceeded. The remaining locations are then visited on the following day. Locations that cannot be visited due to bad weather are first put on hold, and an attempt is made to schedule them again the next day. For all locations that are not visited at all, penalty costs are added to the service and travel costs.

5.3.2 Genetic Algorithm

Genetic algorithms have been widely studied, cf., e.g., Holland (1992), and are also used to solve a route planning problem with stochastic elements, cf., e.g., Ando and Taniguchi (2006). In this approach, each individual represents a single solution. A population consists of multiple individuals and can be created by randomly assigning locations to teams for each individual. This ensures that all teams have a roughly identical number of locations.

To generate offspring, two parents are randomly selected from the population. Two offspring are produced using the Partially Mapped Crossover according to Mak and Guo (2004). The generated offspring can mutate with a certain probability. In the selection, the best individuals are selected from all parent and child individuals of the current generation to form the parents of the next generation. Once a maximum number of generations is reached, the algorithm is terminated, and the individual with the best objective function value represents the solution to the problem.

5.3.3 Tabu Search

According to Oyola and Arntzen (2017), tabu search is the most commonly used search strategy among local search strategy options. Approaches in this area are also able to find very good solutions to routing problems, cf., e.g., Gendreau et al. (1994). Therefore, tabu search was chosen as an alternative heuristic.

Tabu search is a local search method. Starting from a previously created initial solution, new neighbouring solutions are generated using the cross-exchange operator by Taillard et al. (1997). The order of the locations of two randomly chosen routes from different teams are exchanged with each other, resulting in a new solution. For each neighbouring solution, it is checked whether this solution is on the tabu list. If this is the case, it is deleted to prevent solutions from being visited repeatedly and the procedure from getting stuck in a local optimum. The solution with the best objective function value is adopted as the new solution and is added to the tabu list. This is independent of whether the new neighbouring solution is better or worse than the original solution so that a larger solution space is searched, cf. Scholl and Domschke (2010). When the maximum tabu length is reached, the first solution is deleted from the list. The next iteration starts until a termination criterion is reached.

5.3.4 Numerical Results

The numerical results are shown in Fig. 1.13. For the first instance with 12 wind turbines, the genetic algorithm (GA) and the tabu search (TS) were able to find a solution that is slightly worse than the solution of the MIP solver Gurobi. However, they took only approximately 5 seconds. For the second instance with 20 turbines, the genetic algorithm was able to find a solution with a better objective function value in a very short time of 11 seconds. The tabu search performed slightly worse than the genetic algorithm but was still able to find a better solution than the Gurobi model in only 5 seconds.

For the two larger instances, Gurobi could only find a solution in which no location is visited in the given time limit of one hour. The genetic algorithm and the tabu search were able to find remarkably better solutions. The solutions of the genetic algorithm are slightly better than those of the tabu search, but the genetic algorithm has slightly longer solution times of 22 and 35 seconds, respectively, while the tabu search is again terminated after five seconds.

In summary, it can be stated that with the deterministic replacement model, it is almost impossible to find an optimal solution due to the size of the model. The use

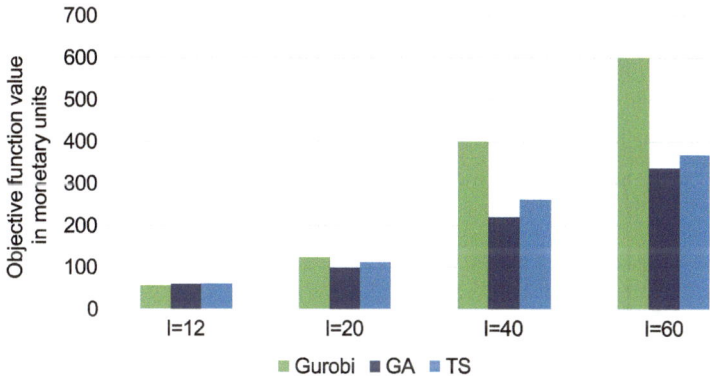

Fig. 1.13 Objective function value for the solution approaches for instances of different sizes

of a heuristic solution procedure is significantly less time-consuming. The solutions of the genetic algorithm were slightly better than those of the tabu search. This shows that both approaches are very promising for solving the problem of scheduling maintenance tasks on distributed wind farms considering uncertainty in a reasonable amount of time.

6 Conclusions

In this paper, we addressed several variants of the selection of efficient modes for the regeneration of complex capital goods. For this efficient design of regeneration processes, the scarce capacities of the regeneration system and the teams need to be explicitly considered in the planning process. Using heuristic operations research methods such as simulated annealing, tabu search and genetic algorithms, it is possible to choose an efficient mode and to schedule the corresponding tasks quickly and with good solution quality.

We have demonstrated the practical applicability of our approach for the regeneration of turbine blades of aircraft engines by embedding it in the system demonstrator of CRC 871 with the customer business model as an important factor for determining the most efficient regeneration mode.

The general considerations and operations research methods are transferable to maintenance actions on immobile capital goods such as wind turbines. As the distances between the goods need to be considered, aspects of routing problems were included. In addition, the uncertainty arising in this context was explicitly included in a stochastic two-stage approach.

Future research should consider the maintenance processes on more complex capital goods with their additional challenges. For example, for a network of waste incineration plants in which domestic waste is thermally utilized to generate process steam, district heating, and electrical energy, specific and local waste input commitments and energy output commitments are given. They must be fulfilled even if a plant is regenerated. Therefore, balancing processes between the different plants must be considered. Again, with operation research methods, a decision support system can help select and schedule the associated operations.

Acknowledgements Funded by the Deutsche Forschungsgemeinschaft (DFG, German Research Founda- tion) – SFB 871/3 – 119193472.

References

Ahmad, R. and Kamaruddin, S. (2012a). An overview of time-based and condition-based maintenance in industrial application. *Computers & Industrial Engineering*, 63(1):135–149.

Ahmad, R. and Kamaruddin, S. (2012b). A review of condition-based maintenance decision-making. *European Journal of Industrial Engineering*, 6(5):519–541.

Ando, N. and Taniguchi, E. (2006). Travel time reliability in vehicle routing and scheduling with time windows. *Networks and Spatial Economics*, 6:293–311.

Bangalore, P. and Patriksson, M. (2018). Analysis of SCADA data for early fault detection, with application to the maintenance management of wind turbines. *Renewable Energy*, 115:521–532.

Bertsimas, D. (1992). A vehicle routing problem with stochastic demand. *Operations Research*, 40(3):574–585.

Besnard, F., Patriksson, M., Strömberg, A.-B., Wojciechowski, A., Fischer, K., and Bertling, L. (2011). A stochastic model for opportunistic maintenance planning of offshore wind farms. In *2011 IEEE Trondheim PowerTech*, pages 1–8.

Bousdekis, A., Magoutas, B., Apostolou, D., and Mentzas, G. (2015). A proactive decision making framework for condition-based maintenance. *Industrial Management & Data Systems*, 115(7):1225– 1250.

Brucker, P. and Knust, S. (2012). *Complex scheduling*. GOR-Publications. Springer, Berlin Heidelberg.

Byon, E. and Ding, Y. (2010). Season-dependent condition-based maintenance for a wind turbine using a partially observed markov decision process. *IEEE Transactions on Power Systems*, 25(4):1823–1834.

Byon, E., Ntaimo, L., and Ding, Y. (2010). Optimal maintenance strategies for wind turbine systems under stochastic weather conditions. *IEEE Transactions on Reliability*, 59(2):393–404.

Colone, L., Dimitrov, N., and Straub, D. (2019). Predictive repair scheduling of wind turbine drive-train components based on machine learning. *Wind Energy*, 22(9):1230–1242.

Dai, L., Stålhane, M., and Utne, I. B. (2015). Routing and scheduling of maintenance fleet for offshore wind farms. *Wind Engineering*, 39(1):15–30.

Dao, C. D., Kazemtabrizi, B., Crabtree, C. J., and Tavner, P. J. (2021). Integrated condition-based maintenance modelling and optimisation for offshore wind turbines. *Wind Energy*, 24(11):1180–1198.

Gendreau, M., Hertz, A., and Laporte, G. (1994). A tabu search heuristic for the vehicle routing problem. *Management Science*, 40(10):1276–1290.

Gendreau, M., Laporte, G., and Seguin, R. (1996). Stochastic vehicle routing. *European Journal of Operational Research*, 88(1):3–12.

Ghamlouch, H., Fouladirad, M., and Grall, A. (2019). The use of real option in condition-based maintenance scheduling for wind turbines with production and deterioration uncertainties. *Reliability Engineering & System Safety*, 188:614–623.

Han, J., Lee, C., and Park, S. (2013). A robust scenario approach for the vehicle routing problem with uncertain travel times. *Transportation Science*, 48(3):373–390.

Hoffmann, L.-S., Kellenbrink, C., and Helber, S. (2020). Simultaneous structuring and scheduling of multiple projects with flexible project structures. *Journal of Business Economics*, 90(5):679–711.

Holland, J. (1992). Genetic algorithms. *Scientific American*, 267(1):66–72.

Irawan, C. A., Ouelhadj, D., Jones, D., Stålhane, M., and Sperstad, I. B. (2017). Optimisation of maintenance routing and scheduling for offshore wind farms. *European Journal of Operational Research*, 256(1):76–89.

Kang, J., Sobral, J., and Guedes Soares, C. (2019). Review of condition-based maintenance strategies for offshore wind energy. *Journal of Marine Science and Application*, 18:1–16.

Kellenbrink, C. and Helber, S. (2015). Scheduling resource-constrained projects with a flexible project structure. *European Journal of Operational Research*, 246(2):379–391.

Kellenbrink, C. and Helber, S. (2016). Quality- and profit-oriented scheduling of resource-constrained projects with flexible project structure via a genetic algorithm. *European Journal of Industrial Engineering*, 10(5):574–595.

Kellenbrink, C., Nübel, N., Schnabel, A., Gilge, P., Seume, J. R., Denkena, B., and Helber, S. (2022). A regeneration process chain with an in- tegrated decision support system for individual regeneration processes based on a virtual twin. *International Journal of Production Research*, https://doi.org/https://doi.org/10.1080/00207543.2022.2051089.

Kovács, A., Erdős, G., Viharos, Z. J., and Monostori, L. (2011). A system for the detailed scheduling of wind farm maintenance. *CIRP Annals*, 60(1):497–501.

Mak, K. and Guo, Z. (2004). A genetic algorithm for vehicle routing problems with stochastic demand and soft time windows. *Proceedings of the 2004 IEEE Systems and Information Engineering Design Symposium*, pages 183–190.

Mishnaevsky Jr, L. (2019). Repair of wind turbine blades: Review of methods and related computational mechanics problems. *Renewable energy*, 140:828–839.

Oyola, J. and Arntzen, H. und Woodruff, D. (2017). The stochastic vehicle rout- ing problem, a literature review, part II: solution methods. *EURO Journal on Transportation and Logistics*, 6(4):349–388.

Ozturk, S. and Fthenakis, V. (2020). Predicting frequency, time-to-repair and costs of wind turbine failures. *Energies*, 13(5):1149.

Saliby, E. (1990). Descriptive sampling: A better approach to monte carlo simulation. *The Journal of the Operational Research Society*, 41:1133–1142.

Salo, E., McMillan, D., and Connor, R. (2018). Value from free-text maintenance records: converting wind farm work orders into quantifiable, actionable information using text mining. *Analysis of Operating Wind Farms 2018*.

Scholl, A. (2001). *Robuste Planung und Optimierung: Grundlagen - Konzepte und Methoden - experimentelle Untersuchungen*. Physica-Verlag, Heidelberg.

Scholl, A. and Domschke, W. (2010). *Logistik: Rundreisen und Touren*. Oldenbourg Wissenschaftsverlag GmbH, München.

Sinha, Y. and Steel, J. A. (2015). A progressive study into offshore wind farm maintenance optimisation using risk based failure analysis. *Renewable and Sustainable Energy Reviews*, 42:735–742.

Song, S., Li, Q., Felder, F. A., Wang, H., and Coit, D. W. (2018). Integrated optimization of offshore wind farm layout design and turbine opportunistic condition-based maintenance. *Computers & Industrial Engineering*, 120:288–297.

Stålhane, M., Hvattum, L. M., and Skaar, V. (2015). Optimization of routing and scheduling of vessels to perform maintenance at offshore wind farms. *Energy Procedia*, 80:92–99.

Taillard, E., Badeau, P., Gendreau, M., Guertin, F., and Potvin, J. (1997). A tabu search heuristic for the vehicle routing problem with soft time windows. *Transportation science*, 31(2):170–186.

Tautz-Weinert, J., Yürüşen, N. Y., Melero, J. J., and Watson, S. J. (2019). Sensitivity study of a wind farm maintenance decision-a performance and revenue analysis. *Renewable energy*, 132:93–105.

Tian, Z., Jin, T., Wu, B., and Ding, F. (2011). Condition based maintenance optimiza- tion for wind power generation systems under continuous monitoring. *Renewable Energy*, 36(5):1502–1509.

Wang, J., Zhang, X., and Zeng, J. (2021). Optimal group maintenance decision for a wind farm based on condition-based maintenance. *Wind Energy*, 24(12):1517–1535.

Zhang, T., Dwight, R., and El-Akruti, K. (2015). Condition based maintenance and operation of wind turbines. In Tse, P. W., Mathew, J., Wong, K., Lam, R., and Ko, C., editors, *Engineering Asset Management - Systems, Professional Practices and Certification*, pages 1013–1025, Cham. Springer International Publishing.

Zhou, P. and Yin, P. (2019). An opportunistic condition-based maintenance strategy for offshore wind farm based on predictive analytics. *Renewable and Sustainable Energy Reviews*, 109:1–9.

Resilience-Based Decision Criteria for Optimal Regeneration

Julian Salomon, Matteo Broggi, and Michael Beer

Abstract Complex capital goods, such as jet engines, are critical to the functioning of modern societies. These sys- tems are exposed to various threats that cannot be prevented entirely. Thus, the concept of resilience – encompassing reliability as well as robustness and recovery in the presence of a disruptive event – is combined with efficient reliability methods to support decision making for complex capital goods. As fundamental step, the current work addresses the generation of a functional model from a physical model based on sensitivity analyses. The developed resilience analysis framework is applied to this model in order to derive conclusions supporting decision maker while incorporating monetary and technical aspects. A combination with the concept of survival signature allows efficient reliability analysis in repeated model evaluations. A novel methodology is developed by amalgamating the non-intrusive stochastic simulation method and the concept of survival signature leading to an significant reduction of the computational effort when considering mixed uncertainties.

Keywords Resilience optimization · Reliability analysis · Uncertainty quantification · Sensitivity analysis · Decision making

J. Salomon (✉) · M. Broggi · M. Beer
Institute for Risk and Reliability, Leibniz University Hannover, Callinstraße 34, 30167 Hannover, Germany
e-mail: salomon@irz.uni-hannover.de

M. Beer
Department of Civil and Environmental Engineering, University of Liverpool, Liverpool L69 3BX, United Kingdom

International Joint Research Center for Resilient Infrastructure & International Joint Research Center for Engineering Reliability and Stochastic Mechanics, Tongji University, Shanghai 200070, China

© The Author(s) 2025
J. R. Seume et al. (eds.), *Regeneration of Complex Capital Goods*,
https://doi.org/10.1007/978-3-031-51395-4_20

1 Motivation

Modern societies are highly dependent on a broad variety of complex capital goods including aircraft engines, industrial plants, and infrastructure systems, Salomon et al. (2020). Aircraft engines, for example, are of paramount importance for both private mobility and the industrial transportation sector of these societies. For economic and safety reasons, it is vital that such complex systems are as reliable as possible Salomon et al. (2021b). To ensure this efficiently and sustainably, the Collaborative Research Center "Regeneration of Complex Capital Goods" (CRC 871) investigates scientific fundamentals for the maintenance, repair and overhaul (MRO) of these complex capital goods, especially in the field of civil aviation, as proposed in Denkena et al. (2019). In fact, these systems are exposed to various threats and it is extremely challenging to identify all possible critical impacts and prevent them accordingly. Therefore, recent developments focus not only on enhancing the reliability and robustness of these systems, but on increasing their recoverability as well. This has led to the concept of resilience that comprises all of these aspects, cf. Salomon et al. (2020).

Information required as basis in the design, maintenance, and repair of systems are commonly governed by uncertainties, Salomon et al. (2021b). Thus, it is critical for decision making in such processes to have tools capable of efficiently performing resilience and reliability analyses of complex systems, taking into account precisely these uncertainties comprehensively. An additional major concern in MRO processes is not only the identification of direct influences of individual components, but more importantly of complex and elusive interaction effects among multiple components and their impact on the key performance measures of the capital good under investigation Salomon et al. (2021a). Global sensitivity analyses are a well-established tool in this specific context.

The current work addresses the development of a computationally efficient theoretical and algorithmic framework for evaluating the resilience and reliability of complex capital goods under consideration of uncertainty to support decision making in MRO processes. Correspondingly, the following guiding principles were defined as:

- guarantee of resilience before, during and after regeneration, and in particular on the functionality of the complex capital good;
- consideration of monetary and technical constraints;
- quantification of uncertainties during regeneration;
- identification of regeneration paths improving resilience such that technical and economic risks are minimized.

Typically, complex physical models are employed to derive conclusions in system engineering. A large number of model evaluations are required to analyize and optimize the performance of a system and its continuous guarantee. However, the utilization of a physical model for repeated evaluations with varying model parameters is often accompanied by an enormous computational burden. Thus, a derivation of a

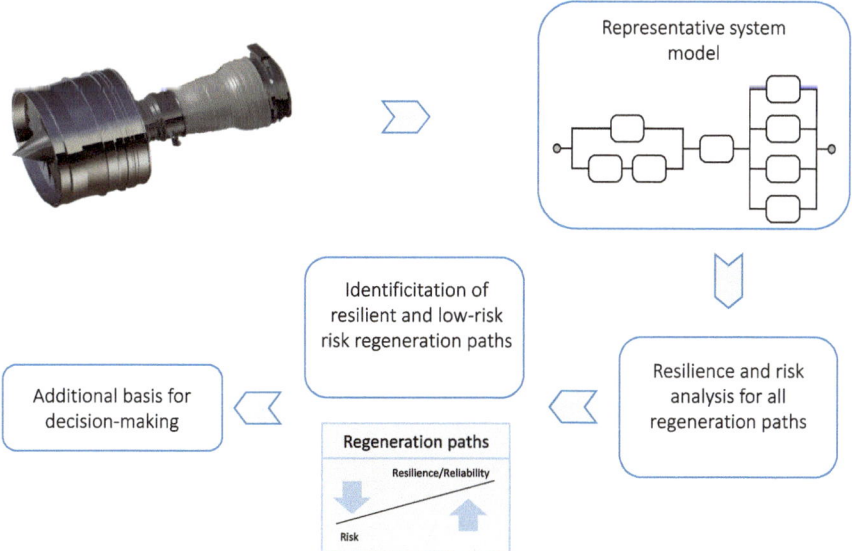

Fig. 1 Objectives and corresponding work flow

function-based model from the complex physical model, mapping the core properties of interest and thus reducing computational effort, is proposed in the current work, as illustrated in Fig. 1. After the generation of such a functional model, the developed resilience assessment framework is applied to derive additional information for the decision making process.

2 Scope of the Paper

Given the principles above, four key objectives are formulated and addressed in the subproject *D*5 "Resilience-based Decision Criteria for Optimal Regeneration" of the CRC 871:

1. the establishment of a comprehensive function-based modular system modeling approach of the overall engine for resilience and reliability assessment;
2. efficient dynamic system modeling in dependence of operating states due to the concept of survival signature for enhanced computational efficiency;
3. the development of models for mixed – aleatoric and epistemic – uncertainty and the utilization of simulation methods reducing computational effort for sampling;
4. the identification of resilient regeneration paths.

More precisely, this means that at first a representative system model is extracted from a physical simulation model, e.g., of an aircraft engine, by the utilization of a sensitivity analysis. The corresponding findings are presented in Sect. 3. As illustrated in Fig. 1, the resulting functional model is basis for further in-depth analysis and investigation.

Given the functional model of an arbitrary, complex capital good, the comprehensive resilience analysis includes the parts illustrated in Fig. 2. At the top level a resilience analysis forms the fundamental frame for the resilience assessment of complex capital goods, evaluating all possible regeneration paths. Subsequently, limiting technical and monetary constraints are taken into account and a reduced set of acceptable resilient and low-risk regeneration paths is identified.

The reliability analysis based on the concept survival signature, introduced in Coolen and Coolen-Maturi (2013), for enhanced computational efficiency, especially

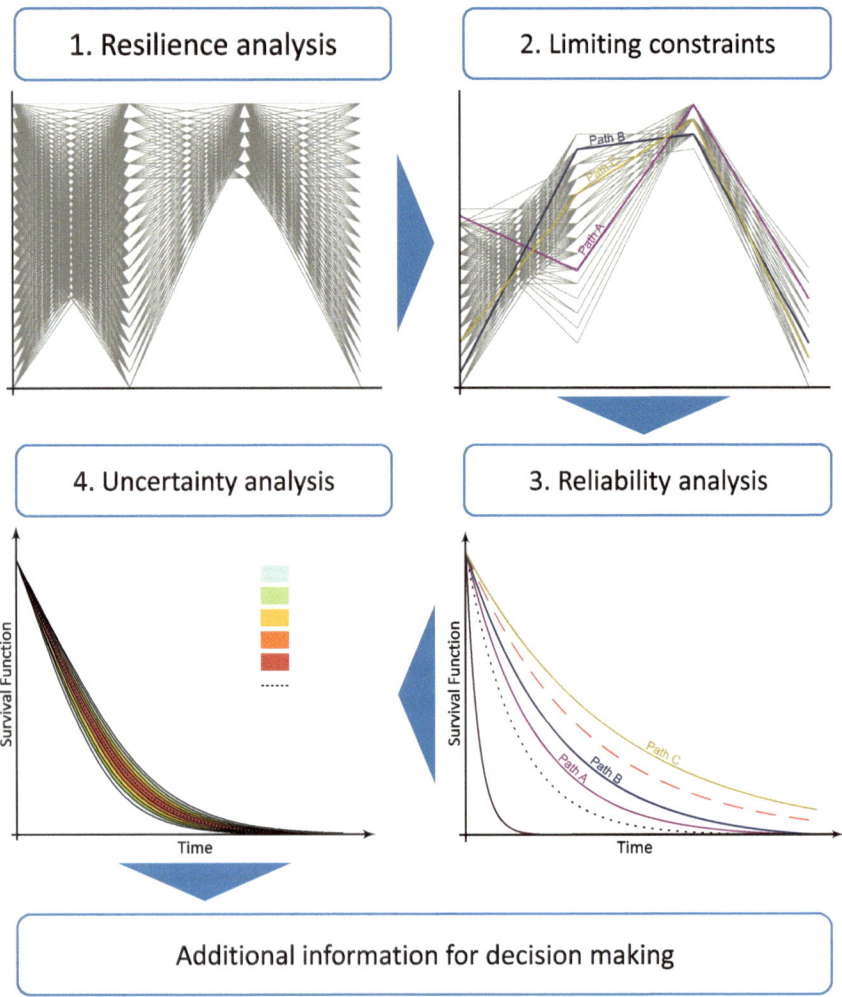

Fig. 2 Work flow in the analysis framework

in case of repeated model evaluations, is integrated into the resilience analysis for each regeneration path considered. This leads to a significantly reduced computational effort during resilience analyses of considered complex capital goods. The additional uncertainty analysis allows the consideration of diverse uncertainties utilizing novel developed, highly efficient algorithms, see Salomon et al. (2021b), reduce the sample size tremendously.

The resilience analysis is introduced in Sect. 4 and the consideration of monetary constraints is demonstrated in Sect. 5. Further, an efficient approach for the integrated reliability analysis is proposed in Sect. 6. The uncertainty analysis considered in Sect. 7 forms the last part and allows for a computationally efficient uncertainty quantification when it comes to mixed uncertainties. As a result, an additional basis for decision making in the virtual level of the regeneration process management is obtained taking into account systemic interactions and uncertain data.

3 Functional Modeling Approach

In the current section, developments concerning various functional models and their generation, as illustrated in Fig. 1, are presented. In the context of the CRC 871, a fundamental procedure was established to generate a functional model based on sensitivity analysis utilizing Sobol indices, see Miro et al. (2019). However, these developments focused on binary-state systems. In the current work, the approach proposed in Miro et al. (2019) is further developed for the consideration of multi-state systems. In addition, an alternative sensitivity measure is considered, allowing for the incorporation of interdependencies between various input parameters, enabling for a more comprehensive and realistic system modeling. Once derived from the physical model, the functional model is investigated in the analysis framework that was outlined in Sect. 2 and is presented in subsequent sections.

3.1 Extraction of Structure Functions Based on Sensitivity Analyses

As fundamental step, the methodology to derive a functional model from a physical simulation is presented within the CRC 871, Miro et al. proposed in Miro et al. (2019) a procedure to extract a functional model from a performance model of an multistage axial compressor. A multistage compressor combines multiple rotor and stator blade rows in an alternating series of connected stages. It was shown that various performance measures are dependent on the blade roughness. According to Miro et al., the blade surface roughness is considered as input variable for further analysis. The four-stage high-speed axial compressor of the Institute of Turbomachinery and

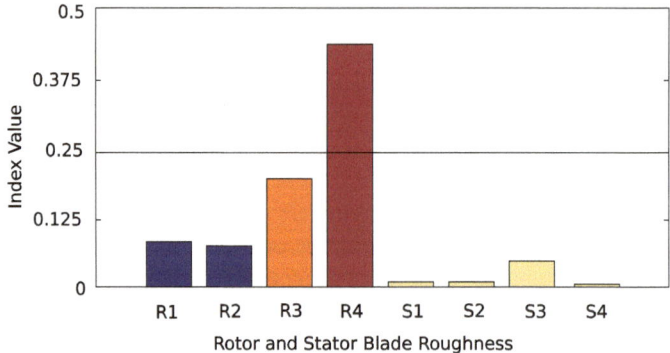

Fig. 3 Component importance measure with threshold of 25%, adopted from Miro et al. (2019)

Fluid Dynamics at Leibniz University Hannover is the baseline compressor of this study, consisting of four stator rows S1 - S4 and four rotor rows R1 - R4.

Miro et al. established a functional model based on results of a sensitivity analysis considering Sobol indices, see Sobol (2001), of an one-dimensional aerodynamic simulation model of that axial compressor. They chose a variance threshold of 25% based on expert knowledge, see Fig. 3. Correspondingly, the system is considered to fail due to roughness related effects if a 25% total variation of the system performance measure, estimated via Monte Carlo Simulation (MCS), is reached.

The functional model developed by Miro et al. describes the dependence of the overall compressor performance, i.e., the total-to-total isentropic efficiency, on the roughness of the rotor and stator blades as binary-state structure function in the form of a Reliability Block Diagram (RBD), see Fig. 4. According to the concept of RBDs, the system functions if there exists a connection between start and end node and fails if this connection is interrupted, corresponding with a performance variation of at least 25%.

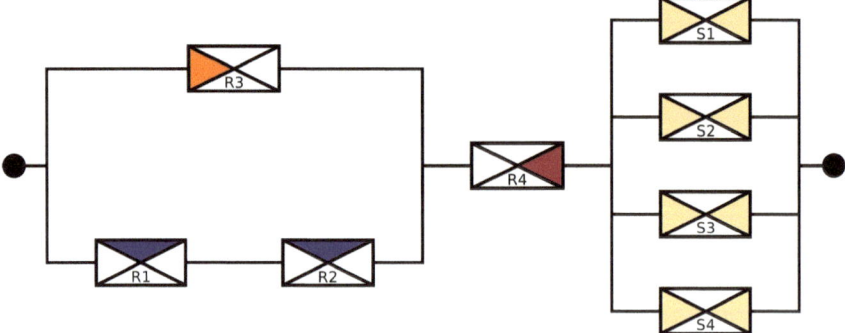

Fig. 4 Functional model of the multistage high-speed axial compressor

In the approach proposed by Miro et al., a certain row is specified as one of four component types ci for i 1, ... , 4. Components of the same type are prone to identical distributions describing degradation while being independent from each other. The classification of the component type and the arrangement of the components in the RBD is chosen based on the sensitivity of the component blade roughness affecting the total-to-total isentropic efficiency. Figure 3 shows the sensitivity results that are based on Sobol indices. Components with similar sensitivity values are determined to be of one component type. In the current work, the component type allocation suggested by Miro et al. is adopted. Correspondingly, the stator and rotor rows are assigned as $(R1, c_1)$, $(R2, c_1)$, $(R3, c_2)$, $(R4, c3)$, $(S1, c4)$, $(S2, c4)$, $(S3, c4)$, $(S4, c4)$.

The arragement of the components is established as follows: If the sensitivity value of a single component exceeds the threshold, it is set in series with other components going beyond the threshold due to significant importance to the overall system performance; if a sum of sensitivity values exceeds the threshold the corresponding components are set in parallel and then linked in series. For example, component R4 goes beyond the threshold alone and therefore is considered as the most important component. Thus, the system should fail if component R4 fails, i.e., R4 exceeds a critical roughness and due to that the roughness-related performance variation of the system exceeds the threshold. Further, R1 or R2 only go beyond the threshold in sum with R3. Correspondingly, the functioning of this subsystem is described as $(R1 \lor R3) \land (R2 \lor R3) = (R1 \land R2) \lor R3$, as shown in Fig. 4, where $R1$, $R2, R3 \in \{0, 1\}$. Following this idea all stators should be arranged in parallel as each of them has rather a small impact. This parallel block $S = S1 \lor S2 \lor S3 \lor S4$ is again in series with the R1, R2, R3 block as well as with R4. Miro et al. argued that the parallel stator block should be allocated in series as shown in Fig. 4 based on expert knowledge, even though the sum of their sensitivity values doesn't reach the threshold. To summarize, the entire system and its functional state is described by $F = (R1 \land R2) \lor R3) \land R4 \land (S1 \lor S2 \lor S3 \lor S4)$ with $F \in \{0, 1\}$ and the system components $R1, R2, R3, R4, S1, S2, S3, S4 \in \{0, 1\}$.

In the current work, this approach is adapted for a multi-state system multi-state component consideration to prove the applicability of the functional modeling approach in the context of partial functionality. Thereby, suppose the system is functioning in the state j or above if the j-th rule is satisfied for components in state j or above with $j = 1, ... J$. For illustrative purposes, four rules are defined as structure functions represented by RBDs and thus $J = 4$. Correspondingly, four thresholds are determined as basis to generate four structure functions corresponding to four levels, see Fig. 5.

Figure 6a shows the structure function via an RBD for the system state of perfect functioning $j = J = 4$. The corresponding threshold is set to 2.5%. The components R1, R2, R3, R4 and S3 exceed this threshold, while components S1, S2 and S4 only go beyond the threshold if summed up. Thus, R1, R2, R3, R4 and S3 are connected in series, while S1, S2 and S4 are connected in parallel and then set in series. This seems reasonable as the components S1, S2 and S4 do not have a critical impact, while the components R1, R2, R3, R4 and S3 definitely harm the perfect state due to their significant influence on the system performance variation.

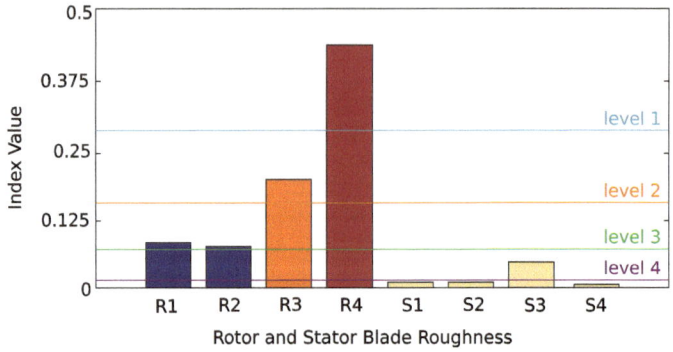

Fig. 5 Component importance measure with thresholds of 2.5, 7.5, 15 and 30%

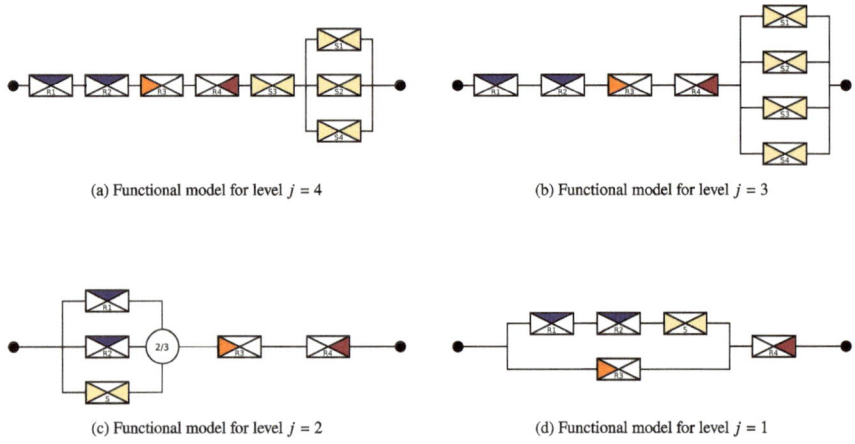

(a) Functional model for level $j = 4$ (b) Functional model for level $j = 3$

(c) Functional model for level $j = 2$ (d) Functional model for level $j = 1$

Fig. 6 RBDs for different performance levels

Figure 6b shows the structure function via an RBD for the intermediate system state $j = 3$. The corresponding threshold is set to 7.5%. The components R1, R2, R3 and R4 exceed this threshold, while components S1, S2, S3 and S4 only go beyond the threshold if summed up. Thus, R1, R2, R3 and R4 are connected in series, while S1, S2, S3 and S4 are connected in parallel and then set in series.

Figure 6c shows the structure function via an RBD for the intermediate system state $j = 2$. The corresponding threshold is set to 15%. The components R3 and R4 go beyond this threshold. Thus, R3 and R4 are connected in series. In contrast, the components R1, R2 as well as the parallel stator block S only exceed the threshold if at least two of those are summed up. Based on the series connection of R3 and R4, the system functions in state $j = 2$ if at least $R1\ R2$, $R1\ S$ and $S\ R2$ take a value of 1, i.e., function in state $j = 2$ or above. As a consequence, the latter relationship is modeled via an at-least-2-out-of-3-connection.

Figure 6d shows the structure function via an RBD for the last system state $j = 1$ before complete failure. The corresponding threshold is set to 30%. Note that all thresholds are set arbitrarily and only for illustrative purpose. The component R4 goes beyond this threshold and is connected in series. The components R1, R2 and S only exceed this threshold if summed up with R3. In case of functioning it holds that $(R1 \lor R3) \land (R2 \lor R3) \land (S \lor R3) = (R1 \land R2 \land S) \lor R3$.

The obtained binary-state structure functions represented in Figs. 4 and 6 are utilized for multi-state system reliability analysis. The corresponding findings are presented in Sect. 6.3.

3.2 Kucherenko Indices

Typically, Sobol indices are utilized for conducting sensitivity analysis as, e.g., proposed in Miro et al. (2019). These variance-based indices display effects of single input variables on output variables (first-order effect indices), and interaction effects between several input variables and their impact on the output variables (total-effect indices). The Sobol indices, as well as most other sensitivity analysis tools, are based on the assumption that all input variables are independent of each other. However, this assumption rarely applies in reality, and in various engineering fields, input variables are correlated, see e.g., Jacques et al. (2006); Keitel and Dimmig-Osburg (2010). Therefore, in this work, a sensitivity analysis of the above mentioned steady-state performance model for an aircraft engine is conducted by applying a generalized form of the Sobol indices according to Kucherenko et al. (2012), hereinafter referred to as Kucherenko indices. These indices are capable of taking into account dependencies between input variables and are therefore more suitable for addressing real world problems.

In practice, an analytical determination of the Kucherenko indices is often not feasible. Therefore, Kucherenko et al. presented Monte Carlo estimators for their indices in their work, Kucherenko et al. (2012). Both estimators require a conditional sampling. Conditional sampling, however, might be tedious or even impossible for some models due to computational demand, such as for the jet engine iteration matching model, considered in this work. Therefore, in Marelli et al. (2019), Marelli et al. provide sample-based Monte Carlo estimators for both Kucherenko indices.

As an example, consider the V2500-A1 jet engine that is a two-spool turbofan with a fan, low-pressure compressor (LPC), high-pressure compressor (HPC), high-pressure turbine (HPT), low-pressure turbine (LPT) and a common thrust nozzle. This jet engine was considered in Salomon et al. (2021a) to show the applicability of a sophisticated sensitivity measure for an entire jet engine. In cooperation with the subproject D6 of the CRC 871 concerning module interactions and the overall system behavior, the first-order effects for the six considered output quantities with respect to the five varying input efficiencies of the main turbomachines, are determined by utilizing the sample-based Monte Carlo estimators. The corresponding results are shown in Fig. 7. It can be seen that among all of the direct effects on output

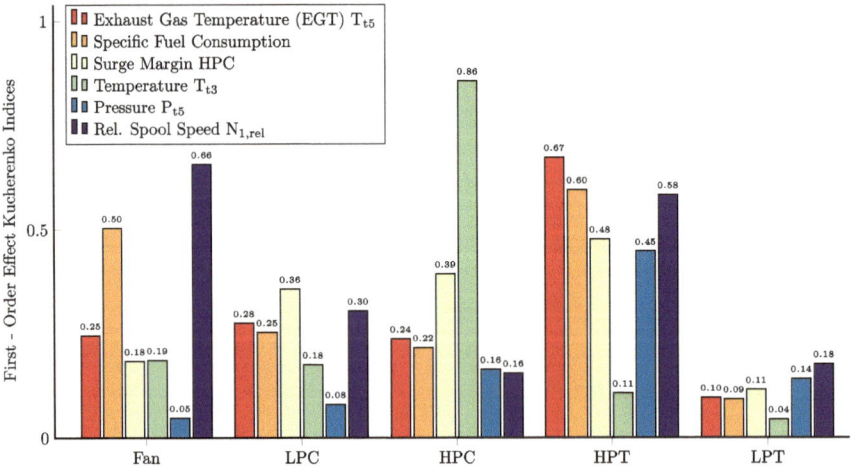

Fig. 7 First-order effect Kucherenko indices

variances, the variance of the HPT efficiency is the dominant factor for four to five out of all six output quantities. This is the fact as the HPT exhibits the highest index value and therefore constitutes the main influence on the system performance. In this manner the approach is applicable for the assessment of an entire jet engine under consideration of interdependencies between engine components.

In Salomon et al. (2021a), Salomon et al. additionally compute the total-effect Kucherenko indices and both results are discussed in detail. To summarize, the study shows that simple correlations are not sufficient to explain the influence of combined module variances and find the causes of deterioration. Therefore, sensitivity analyses under consideration of dependent variables by means of Kucherenko indices and digital performance twins are powerful tools to determine the influence on a scientific basis. For an overall view, however, the change in capacity and work must also be examined at different operating points with an engine pressure ratio regulation. It shall be noted, that these results can be utilized as a basis for a detailed reliability analysis by developing a functional model according to Miro et al. (2019) and Eryilmaz and Tuncel (2016) of the V2500-A1 aircraft engine performance model.

4   Resilience Analysis

The resilience analysis, developed in Salomon et al. (2020), forms the first phase and basis of the analysis framework, illustrated in Fig. 2, and is presented in the current section. Therefore, a fundamental notion of resilience and a corresponding metric is suggested. Subsequently, a resilience decision making framework is developed consisting of two key ingredients, an adapted systemic risk measure and a sophisticated resilience metric, enabling for systematic computation of the resilience for various endowment configurations. These endowment configurations can be intepreted in a variety of ways, e.g., as different regeneration paths of the considered complex capital good. Finally, the grid search algorithm and its advantageous proper ties in terms of computational efficiency are presented. For illustrative purposes, the developed algorithmic framework is than applied to the functional model established in Sect. 3.1, whereby it is not limited to this particular use case, but can be utilized to a variety of system models. Illustrative results are presented in combination with the second phase of the analysis framework in Sect. 5.

4.1 Resilience Metric

Given a system being exposed to a disruptive event and recovering its functionality afterwards, three essential phases can be defined classifing the system states, as illustrated in Fig. 8: (i) The original stable state, whose duration relates to the reliability of the system, forms the first phase. (ii) The second phase is the loss of performance after the occurrence of a disruptive event. This loss depends on the vulnerarbility or robustness of the system; the robustness of the system is interpreted as the resistance to a loss of performance. (iii) The disrupted state of the system and its recovery to a new stable state is the last phase and governed by the recoverability. In general, the new stable state may differ from the original state and, accordingly, its performance may be higher or lower. The majority of resilience metrics available in the current literature is based on system performance, i.e., on the three states and their transitions shown in Fig. 8. Consequently, a quantitative measure of resilience depends on the specific choice and definition of system performance, see e.g., Ayyub (2015). Performance-based approaches may be ratio-based, integral-based, or both.

In the current work, the probabilistic resilience metric by Ouyang et al. Ouyang et al. (2012) is utilized. The metric is denoted by *Res* and defines the expected ratio of the integral of the system performance (t) over the time interval $[0, T]$ and the integral of the target system performance T $Q(t)$ over the same time interval:

Fig. 8 The three resilience phases before and after a distruptive event; adapted from Henry and Ramirez-Marquez (2012)

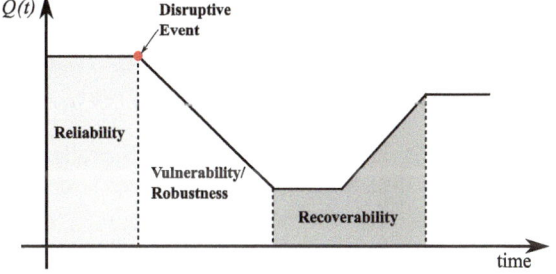

$$Res = E[Y], \quad \text{where } Y = \frac{\int_0^T Q(t)dt}{\int_0^T TQ(t)dt}. \tag{1}$$

The system performance $Q(t)$ is described as a stochastic process. In general, $TQ(t)$ might be considered as a stochastic process as well, but for expediency it is assumed to be a non-random constant TQ in this work. The resilience metric takes values between 0 and 1 when limiting the recovered performance at maximum equivalent to the original performance. The value $Res = 1$ indicates a system performance corresponding to the target performance, while $Res = 0$ captures that the system is not working during the considered time period at all.

4.2 Adapted Systemic Risk Measure

As proposed in Salomon et al. (2020), the resilience metric presented in Sect. 4.1 is integrated into an adapted systemic risk measure, enabling the systematical assessment of various system configurations, that might be, e.g., regeneration paths. In particular, technical systems for which a meaningful system performance $Q(t)$ can be determined are considered.

Assume that the system encompasses l system components. Each component is characterized by its type and n relevant properties that influence the overall system performance. For convenience, apply matrix notation. A component $i \in \{1, ..., l\}$ can be characterized by a row vector

$$(a_i; j_i) = (\eta_{i1}, \eta_{i2}, \ldots, \eta_{in}; j_i) \in R^{(1 \times n)} \times \mathbb{N} \tag{2}$$

where $\eta_{i1}, \eta_{i2}, ..., \eta_{in}$ represent the numerical values of the n properties and $j_i \in \{1, 2, ..., b\} \subseteq \mathbb{N}$ defines its type. The system is described by a pair including the matrix $A \in \mathbb{R}^{(l \times n)}$ and the column vector $z \in \mathbb{N}^l$ that captures the types of the components:

$$(A; z) = \begin{pmatrix} \eta_{11} & \eta_{12} & \cdots & \eta_{n1}; & z_1 \\ \eta_{21} & \eta_{22} & \cdots & \eta_{2n}; & z_2 \\ \vdots & \vdots & & \vdots & \vdots \\ \eta_{l1} & \eta_{l2} & \cdots & \eta_{ln}; & z_l \end{pmatrix} \tag{3}$$

The input-output model $Y = Y A$; is evaluated for these pairs. Below, a corresponding adapted systemic risk measure is constructed as follows. As a specific example, choose the acceptance set

$$\mathbb{A} = \{X \in \mathbb{X} \mid E[X] \geq \alpha\} \text{ with } \alpha \in [0, 1] \tag{4}$$

The risk measure is defined as

$$R(Y; \ K) \ = \ R(Y; (K; \ z)) \ = \ \{A \ \in \mathbb{R}^{l \times n} \mid Y_{(K+A;z)} \ \in \ \mathbb{A}\}, \tag{5}$$

that is the set of all allocations of modified system properties A that are added to the base properties K for which the altered system $(K+A; z)$ exhibits a resilience greater or equal to α. Without loss of generality but to keep the notation simple, set $K = 0$, and $R(Y; 0)$ is written as $R(Y)$.

Practical applications might require to impose restrictions for the structure of the matrix in Eq. (3). For instance, components of a specific type might require an equivalent configuration, i.e., the corresponding row vectors ai must possess equal values. Following Feinstein et al. (2017), such constraints can be captured by functions : $\mathbb{R}^p \to \mathbb{R}^{(l \times n)}$ that are monotonously increasing with $a' \to (A; z)$, where $z \in \mathbb{R}^l$ indicates the types of the components. Such a function maps a lower-dimensional set of parameters $a' \in \mathbb{R}^p$ to the system description.

4.3 Grid Search Algorithm

In accordance with Feinstein et al. (2017), a set-valued systemic risk measures as presented in Sect. 4.2 can be computed via a combination of the so-called grid search algorithm and stochastic simulation. In two dimensions, a box-shaped subset of endowment properties is subdivided by a grid of equidistant points.

The algorithm proceeds as follows. The search starts at the origin of the search space; assume that the origin is outside of $R(Y)$. In a successive manner, the acceptance criterion is evaluated for each adjacent grid point on the grid diagonal along the direction $(1, 1, \ldots , 1)^T$. Typically, in each evaluation stochastic simulation is performed. The search along the diagonal terminates as soon as a grid point that meets the acceptance criterion is identified. Given the monotonicity of the input-output model and the properties associated with the acceptance criterion (cf. Feinstein et al. (2017)), all grid point configurations in the box-shaped subset with the first accepted one as the bottom left corner are acceptable as well and consequently belong to $R(Y)$. Analogously, all endowments in the box-shaped subset with the first accepted one as the top right corner are rejected. Thus, these points belong to $R(Y)^C$ that is the complement of the systemic risk measure. Precisely this monotonicity property makes the algorithm efficient.

Each neighboring pair of diagonally adjacent points with one of these points meeting the requirements and the other not, defines a sub-box. In the next step, the algorithm checks the remaining corners of this sub-box, assigning a status to dominating and dominated endowments, respectively. Subsequently, the next neighboring pairs of points can be determined. The algorithm terminates as soon as all points on the grid have an assigned acceptance status. Finally, risk measure $R(Y)$ is determined as a discrete grid-approximation. This algorithm, combined with the methods proposed in Sects. 4.1 and 4.2, allows decision-making to be made regarding the optimal trade-off between resileince-enhancing endowments for complex capital goods.

5   Constrained Resilience Analysis of an Axial Compressor

In the current section, the methodology presented in Sect. 4 is demonstrated for illustrative scenarios, while the procedure for considering monetary constraints is elaborated, see phase one and two in Fig. 2. The method can be applied to assess a variety of complex capital goods. In Salomon et al. (2019) and Salomon et al. (2020), Salomon et al. proved the applicability of the proposed approach for a wide range of complex systems, e.g., flow networks, an axial compressor and the Berlin metro network.

5.1   Resilience Analysis Setting

In the context of the CRC 871, consider the functional model illustrated in Fig. 4 of the axial compressor developed in Miro et al. (2019) presented in Sect. 3.1. Again, an interruption between start and end represents system failure, i.e., a roughness-realted performance variation of the physical system of at least 25%. The system functionality is utilized as meaningful system performance $Q(t)$ that was claimed in Sect. 4.2 for the subsequent application of the resilience decision making procedure. The system performance is evaluated at each point in time th and equals 1 if there is a connection from start to end and is 0 if the connection is interrupted.

Components $i \in \{1, \dots, 8\}$ of the functional model represent stator blade rows and rotor blade rows. In this example, each of them is assumed to have the same component type, i.e., it holds that $ji = 1 \; \forall i \in \{1, \dots, 8\}$. For simplicity, denote $(a_i; j_i) = (a_i ; 1) = a_i \; \forall i \in \{1, \dots, 8\}$. Suppose that each row, i.e., each component, is characterized by two endowment properties, namely, a roughness resistance re and a recovery improvement r^*. Then, the component is described by $a_i = (re_i , r^*_i)$. It holds true that $re_i = re_i^*$, r_i^*, if $j_i = j_i'$, consequently, the endowment pair (re_i, r_i^*) has equal numerical values for all components. Each of these configuration pairs might represent a particular regeneration path.

After evaluating the system performance in a previous time step t_h, each component can fail randomly. A failed component is removed from the model and no longer contributes to the system performance at time t_{h+1}. The component remains in the failed state until its full recovery. Assume that the failure probability of the component i is assumed to be constant in the time interval (t_h, t_{h+1}). For illustrative purposes, it is given by

$$P\{\text{Component i fails during } (t_\text{h}, t_\text{h+1})\} = t \; \cdot \lambda_i \tag{6}$$

with

$$\lambda_i = 0.8 - 0.03 \cdot re_i, \tag{7}$$

where λi is the time-independent failure rate. This single-step failure model corresponds to a simple approach for considering reliability and robustness. A consideration of system reliability in multiple states, where the system passes through several intermediate states before failure, as presented in a subsequent section, is one possibility for a more comprehensive modeling approach.

Suppose that a failed component instantly recovers to the original performance level after a certain number of time steps passed. Then, the component recovery is described by

$$r_i = r_{\max} - r_i^* \text{ with } r_i^* < r_{\max}, \tag{8}$$

where $rmax$ denotes the maximum number of time steps required for recovery and r_i^* is a reduction depending on the current endowment of the component i. Since each time step is of the length $t = \frac{T}{u}$, with T denoting the investigated duration and u the amount of considered time steps, the duration of the recovery process is $r_i \cdot \frac{T}{u}$. In accordance with Ayyub (2014) and Ayyub (2015), this simple recovery model corresponds to a one step recovery profile; however, various other characteristic profiles of recovery in time are conceivable.

Note that in this setting increasing the roughness resistance of a blade row, i.e., a component i, mitigates the degradation of the surface, i.e., counteracts the roughning process, and correspondingly reduces the failure rate λ_i. If the component i fails, its functionality is fully recovered after ri time steps specified via Eq. (8).

5.2 Costs of Endowment Properties

A certain endowment relates to the property quality of one or more components. In general, a higher quality of components results in a more resilient system. However, an increase in quality is typically associated with an increase in costs. Consequently, it is essential to take into account monetary aspects for an expedient decision making procedure. In accordance to Mettas (2000), assume that increasing the reliability of components in complex systems corresponds to an exponential increase in their costs.

Assume that the cost associated with improving the endowment property roughness resistance is given by

$$cost^{re} = \sum_{i=1}^{8} price^{re} \cdot 1.2^{(re_i - 1)}, \tag{9}$$

where re^i is the roughness resistance value of component i. Further, $price^{re}$ is a common basic price independent of i in the current case study. Analogously, assume an exponential relationship for the costs associated with the recovery improvement r_i^*:

$$cost^{r*} = \sum_{i=1}^{8} price^* \cdot 1.2^{(r_i^* - 1)}. \tag{10}$$

The total cost of an endowment results from the sum of these costs:

$$cost = cost^{re} + cost^{r*}. \tag{11}$$

This cost function shown is subsequently utilized to determine the cost of a certain endowment. Consequently, the endowment pair with minimum cost can be identified. The combination of the adapted systemic risk measure developed in Sect. 4.2, including the corresponding acceptance set, with the cost function allows the evaluation of optimal endowment pairs regarding resilience and monetary constraints.

5.3   Scenario and Numerical Results in a Two-Dimensional Setting

Below, the decision making method for identifying resilience-enhancing endowments under consideration of monetary constraints is demonstrated for the multistage high-speed axial compressor presented in Fig. 4 in Sect. 3. For illustrative purposes, the model parameters and simulation parameter values, shown in Table 1, are considered.

Assume an resilience acceptance threshold of $\alpha = 0.8$, an arbitrarily selected number of $u = 200$ time steps, a constant failure rate of $\lambda = 0.8$ as well as an arbitrarily selected time step length of $t = 0.05$. The first step in the analysis is to determine the set of all acceptable endowments that correspond to a resilience value of at least $Res = 0.8$ over the time period under consideration. In practice, any improvement of the axial compressor blades is associated with costs. Consequently, the second step is to identify the least expensive acceptable endowment, denoted by $\hat{A}$. The grid search algorithm described in Sect. 4.3 explores the roughness resistance re and the recovery improvement r^* over $re_i \in \{1, ..., 20\}, r_i^* \in \{1, ..., 20\} \, \forall_l \in \{1, ..., l\}$. Increasing a value of the properties of a component i is interpreted as increasing its the quality level. The roughness resistance values are interpreted as various quality levels of coatings applied to the blades. In terms of recovery, the quality increasing leads to a reduced recovery time for the components taking values from a maximum of 20 time steps (for $r_i^* = 1$) to a minimum of one time step (for $r_i^* = 20$) given $r_{max} = 21$.

Table 1 Parameter values for the resilience decision making method for the functional model of the multi-stage high-speed axial compressor

Parameter	Index	Value in scenario
Number of rotor/stator blade rows	l	8
Acceptance threshold	α	0.8
Number of time steps	u	200
Length of a time step	t	0.05
Failure rate	λ	0.8
Maximum recovery time	r_{max}	21
Recovery improvement	$r*$	$r_i^* \in \{1, \ldots, 20\}$
Roughness resistance	re	$re_i \in \{1, \ldots, 20\}$
Recovery improvement price	$Price^{r*}$	600 €
Roughness resistance price	$Price^{re}$	500 €

Figure 9 shows the results of the grid search algorithm. The acceptable pairs of component properties, i.e., roughness resistance and recovery improvement, are depicted as blue, filled dots. Clearly, the quality of recovery improvement and the quality of the blade coatings can be compared regarding their impact on the system resilience. For instance, given recovery improvement values with $r_i* \geq 15$, the minimum roughness resistance value of $re_i = 1$ is already sufficient to achieve the desired level of resilience.

For the determination of $R(Y)$ only about 10% of all possible endowment pairs had to be evaluated due to the grid search algorithm presented in Sect. 4.3. More precisely, the number of endowment pairs on the diagonal plus the number of pairs equivalent to the size of the set of pairs with minimum acceptable resilience, i.e., the boundary or pareto front of $R(Y)$, had to be evaluated. Taking into account the

Fig. 9 Numerical results of the grid search algorithm for the functional model of the axial compressor with explored roughness resistance/recovery improvement values

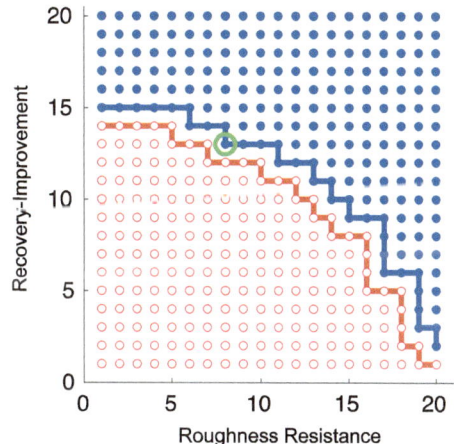

base prices in Table 1, the most cost-efficient endowment i among the boundary set is characterized by a roughness resistance of $re_i = 8$ and a recovery improvement of $r_i* = 13$ for each of the eight components. In Fig. 9 the corresponding pair is highlighted by a green circle. According to Eqn. (11) the total cost equals 136 930€. Based on these results, the decision maker is advised to realize $re_i = 8$ and $r_i* = 13$.

Note that in case of analyzing regeneration paths, resilience applies to the regeneration paths in two ways: 1. as a part of the overall performance over the entire life cycle of the complex capital good, and 2. as a resilient regeneration path in itself. Clusters are formed or identified of similar, equally acceptable regeneration paths to which, in the event of a problem, it is possible to switch without great effort.

5.4 Scenario and Numerical Results in Multiple Dimensions

As shown in Salomon et al. (2019), the methodology developed in Salomon et al. (2020) can be utlized in the multidimensional case as well. The current subsection shall prove the applicability in a four dimensional setting given the model of the multistage high-speed axial compressor. The model parameter and simulation parameter values shown in Table 2 are considered. Assume the recovery improvement $r*$ to be fixed for all components, regardless of their type, $r_i* = 11\ 1, \ldots l$, while the roughness resistance re is explored over $re_i \in \{1, \ldots, 20\}\ \forall\ I \in \{1, \ldots, l\}$. Again, the roughness resistance values can be interpreted as increasing quality levels of coatings. In this scenario, the four component types suggested in Sect. 3.1 are adopted. Correspondingly, the first and second rotor blade rows are assigned as c_1), the third and fourth as c_2 and c_3, respectively, while all stator blade rows are assigned as c_4. The set of all acceptable endowments leading to a system resilience value of at least $Res = 0.85$ over the time period under consideration is determined via the grid search algorithm. Then, the most cost-efficient endowment denoted by $\hat{A}$ is identified (Table 2).

Figure 10 shows the corresponding results. In Fig. 10a, all combinations with a satisfactory system resilience of at least $Res = 0.85$ are depicted, corresponding to phase one in the analysis framework, see Fig. 2. This is the set of roughness resistance endowment pairs contained in. In fact, the roughness resistance of the fourth rotor blade row (c_3) has the highest impact on the system resilience compared to other rows. This can be concluded as only pairs with a high roughness resistance quality for this type are acceptable. Regardless of the endowment property values of all other component types $c_i \in \{1, \ldots, 4\}$, the endowment pairs with coating qualities of $re_i \leq 15$ for c_3 are not sufficient to provide an acceptable level of system resilience. In contrast, the roughness resistance of the four stators (c_4) has minor influence on the system resilience compared to all other types. Even endowments with $(re_i, 4) = 1$, i.e., a minimum coating quality level, are sufficient to achieve acceptable resilience values. The same holds true for the rotors of of type c_1 and c_2. Although, in comparison to the stators, components of types c_1 and c_2 require significantly

Table 2 Parameter values for the resilience decision making method for the functional model of the multistage high-speed axial compressor

Parameter	Index	Value in scenario
Number of blade rows	l	8
Acceptance threshold	α	0.85
Number of time steps	u	200
Length of a time step	t	0.05
Failure rate	λ	0.8
Maximum recovery time	r_{max}	21
Recovery improvement	r^*	11
Roughness resistance	r^e	$re_i \in \{1, \ldots, 20\}$
Recovery improvement price	$Price^*_{(r^*; j_i)}$	600 €
Roughness resistance price	$Price^{re}_{(re_i; j_i)}$	800 € $j_i \in \{1, 2, 3\}$ 500 € $\forall j_i = 4$

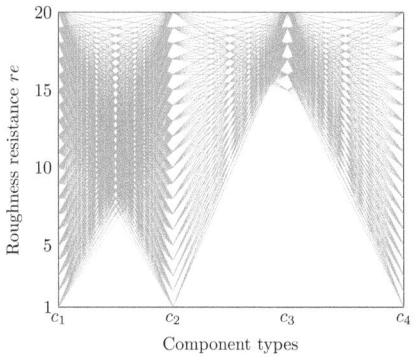 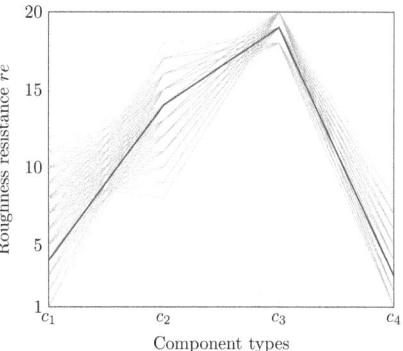

(a) Numerical results with explored roughness resistance values   (b) Numerical results considering a budget threshold

Fig. 10 Numerical results for multi-dimensional setting

higher level of coating quality to compensate small, i.e., values other than maxium values of roughness resistance for c_3.

For decision making, it is crucial to be able to take into account monetary constraints. Therefore, Fig. 10b shows the endowment pairs contained in $R(Y)$ that lead to a satisfying system resilience of $Res = 0.85$ considering a budget threshold that is set to $cost^{re}_{max} = 50\ 000$€ for illustrative purposes, corresponding to phase two in the analysis framework, see Fig. 2.

The results illustrated in Fig. 10b show that only configurations with low coating quality levels for all stators (c_4) are below the cost limit. Firstly, this is the case due to their low influence on system resilience, and secondly, to the high costs for increasing the quality levels for the stators caused by their amount and exponential cost-quality behavior. In contrast, only configurations that provide the highest quality levels of $(re_i, 3) \geq 17$ for the rotor of type 3 are acceptable and below the price limit simultaneously. The roughness resistance of the rotor of type c_3 has a critical influence on the system resilience. Consequently, the compensation of lower quality levels for c_3 by higher quality levels of the remaining blade rows exceed the given budget treshold. Even though the roughness resistance of the rotor of c_2 has a lower influence on the system resilience than that of c_3, mininum quality levels for c_2 cannot be compensated by high qualities of the other components either. Correspondingly, at least $(re_i, 2) = 4$ is required to meet the acceptance criterion.

Considering the base prices in Table 2, the most cost-efficient endowment is characterized by the pair with roughness resistances of $(re_i, 1) = 4$, $(re_i, 2) = 14$, $(re_i, 3) = 19$ and $(re_i, 4) = 3$. The corresponding configuration is highlighted in blue in Fig. 10b. Via Eqn. 11 the total cost is obtained as $cost_{(\hat{\lambda};z)} = cost^{re} + cost^* = 42\ 604€ + 35\ 664€ = 78\ 268€$.

The numerical effort for the computation of $R(Y)$ was reduced by about 98% due to the grid search algorithm compared to a naive evaluation of the search space. Correspondingly, only 2% of all possible combinations of roughness resistance values had to be evaluated. Note that the application of this methodology to higher-dimensional problems is only limited by constraints of computational memory and time.

6   Reliability Analysis

The reliability analysis follows the resilience analysis and the reduction of all, in terms of system resilience, acceptable system configurations respectively regeneration paths due to technical and monetary restrictions. It thus forms the third phase in the analysis framework, see Fig. 2.

6.1   Repeated Evaluation of the Survival Function

For all remaining endowment pairs of interest for decision makers, a system reliability analysis is conducted to evaluate the system failure probability at given time t. The system reliability is evaluated based on the stochastic properties of the system components represented as probability distribution functions that describe the event probability of failure due to degradation or a disruptive event. Each model evaluation leads to the so-called survival function, i.e., the probability that the system is still functioning at time point t, performing its predefined task. In Fig. 11 three different survival functions are shown as Path A, B and C for illustrative purposes.

Different endowment pairs evaluated in the grid search algorithm correspond to various component properties due to different regeneration paths and lead to different survival functions. The three survival functions in Fig. 11 correspond to three different regeneration paths, i.e., endowment pairs. In addition, there might exist minimum and maximum requirements of system reliability due to practical experiences, customer

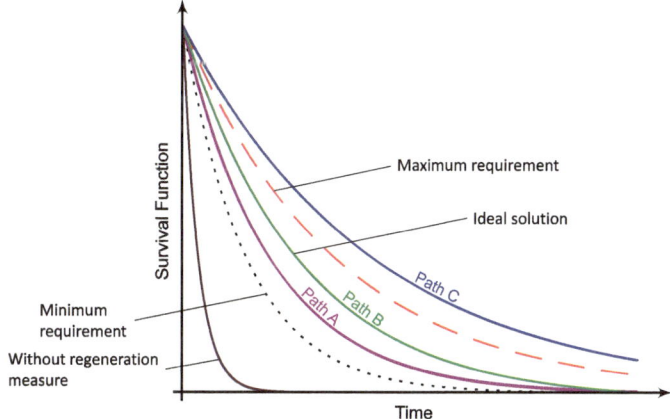

Fig. 11 Reliability analysis based on the concept of survival signature considering multiple regeneration paths

requirements, e.g., budget limitations, or other circumstances. An efficient proce-dure is required to realize the large number of repeated model evaluations, i.e., computations of survival functions, with changing component properties.

6.2 Concept of Binary-State Survival Signature

Figure 12 illustrates the concept of survival signature introduced in Coolen and Coolen-Maturi (2013). The most beneficial attribute of this approach is its separa-tion property. That means that the system structure is separated from the probability structure of the system describing the component failure behavior. This leads to a significant reduction of the computational effort, since once the typical costly to determine system structure has been computed, any possible characterization of the probabilistic part can be tested with no need to recompute the structure. This means that any number of system configurations, i.e. regeneration paths, can be simulated and analyzed, since these only affect the probability structure and usually not the system structure. At the same time, the survival signature radically condenses infor-mation on the topological reliability for systems with multiple component types with K being the maximum number of component types. The failure times of compo-nents of one type are claimed to be independent and identically distributed (*iid*) or exchangeable. For more information on claimed exchangebility in practice, see Coolen and Coolen-Maturi (2016) and Salomon et al. (2021b).

For a deeper understanding of this concept, consider a coherent system with a given binary-state structure function defining the system state to be either 0 or 1 for a binary-state vector out of the set of all possible state vector. The binary-state

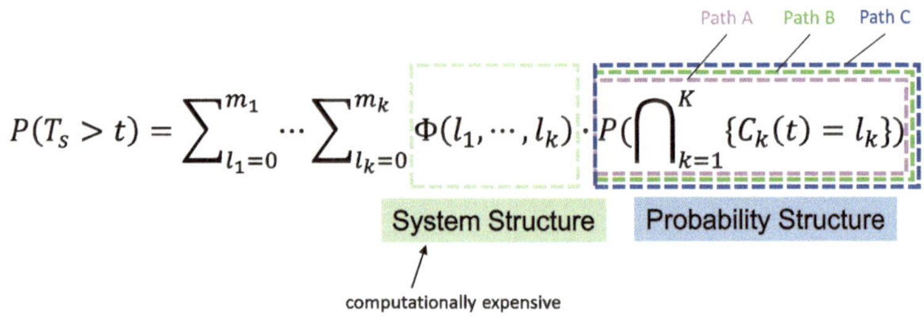

Fig. 12 Illustration of the advantageous properties of the concept of survival signature

vector specifies the state of $n = \sum_{k=1}^{K} n_k$ components in total, there are $\binom{n}{l}$ state vectors x with exactly l components refer to as S_l. In the case of $k \geq 2$, the survival survival signature summarizes the probability that a system is working as a function depending on the number of working components lk for each type $k = 1,...,K$. Assume the failure times within a component type to be *iid* or exchangeable. Consequently, all possible state vectors are equally likely to occur. Then, the survival signature is defined as

$$\Phi(l_1, l_2, \ldots, l_K) = \left[\prod_{k=1}^{K}\binom{n_k}{l_k}^{-1}\right] \sum_{x \in S_{l_1, l_2, ..., l_K}} \phi(x), \qquad (12)$$

where $\binom{n_k}{l_k}$ corresponds to the total number of state vectors xk of type k and $S_{l_1, l_2, ...l_k}$ denotes the set of all state vector of the entire system for which $l_k = \sum_{i=1}^{n_k} x_{k,i}$. Then the survival signature $(l_1, l_2, \ldots, l_K)$ characterizes the probability that a system is working given that exactly l of its components working for $l = 1,..., n$. Note that the survival signature depends only on the topological reliability of the system, independent of the time-dependent failure behavior of its components, namely, the probability structure. Note that it is differentiated between the concept of survival signature with its seperation property shown in Fig. 12 and the mathematical object survival signature itself shown in Eq. 12.

Further, assume the probability distribution for the failure times of type k to be known with $F_k(t)$, denoting the corresponding cumulative distribution function. Then,

$$P\left(\bigcap_{k=1}^{K}\{C_k(t) = l_k\}\right) = \prod_{k=1}^{K} P(C_k(t) = l_k)$$

$$= \prod_{k=1}^{K}\binom{n_k}{l_k}[F_k(t)]^{n_k - l_k}[1 - F_k(t)]^{l_k} \qquad (13)$$

describes the probability structure of the system, regardless of its topology. $C_k(t) \in$ $\{0, 1, \ldots, n_k\}$ represents the number of components of type k in a working state at time t.

Both Eqs. 12 and 13 form together the concept of survival signature illustrated in Fig. 12. The concept is integrated into the proposed framework, see phase three in Fig. 2, to leverage its salient beneficial properties for repeated model evaluations that are required for comprehensive MRO decision making.

6.3 Concept of Multi-state Survival Signature

While a binary-state consideration of systems and their components is state-of-the-art, further research on multistate systems with multi-state components is inevitable for a more realistic and comprehensive assessment of system reliability. In Eryilmaz and Tuncel (2016), Eryilmaz & Tuncel proposed a generalized concept of survival signature in the context of unrepairable homogeneous multi-state systems. In accordance with the approach presented in Coolen and Coolen-Maturi (2013), the survival function for multiple types with type $k = 1, \ldots, K$ can be derived as:

$$P\{T^{\geq J} > t\} = \sum_{i^1 \geq \cdots \geq i^J} \cdots \sum \Phi^{\geq J}\left(i_1^1, \ldots, i_k^j, \ldots, i_K^j\right)$$
$$\times P\left\{C_1^1(t) = i_1^1, \ldots, C_k^j(t) = i_k^j, \ldots, C_K^J(t) = i_K^J\right\}. \quad (14)$$

with maximum system and component level J and $T^{\geq J}$ that is the system failure time in state J. Thereby, $\Phi^{\geq J}(i_1^1, \ldots, i_k^j, \ldots, i_K^J)$ represents the j-th level survival signature for level J, i.e., the probability that the system is working in state J or above if i_k^j components are working for types $k = 1, \ldots, K$ and states $j = 1, \ldots,$ J with $i_k^{j-1} \geq i_k^j$. The total number of state vectors given i_k^j components of type k functioning in state j or above is

$$\upsilon_{n_1,\ldots,n_K}(i_1^1, \ldots, i_k^j, \ldots, i_K^J) = \prod_{j=1}^{J} \binom{n_1 - i_1^{j+1}}{i_1^j - i_1^{j+1}} \cdots \binom{n_K - i_K^{j+1}}{i_K^j - i_K^{j+1}}, \quad (15)$$

where $i_1^{J+1} = \ldots = i_K^{J+1} = 0$ and nk denotes the total number of components of type k. The j-th level survival signature for level J for multiple types is given as

Again the j-th level survival signature and the survival function with multiple components are derived similarly to Eqs. (16) and (14), respectively.

The approach proposed in Eryilmaz and Tuncel (2016) allows to compute the reliability of multi-state systems for J binary-state structure functions with components following a Markov degradation process with minor failures. Given the four

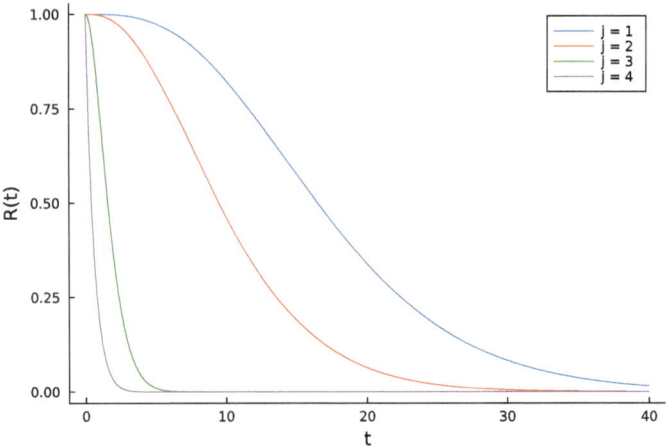

Fig. 13 j -th level survival functions for $j \in \{1,...,4\}$

binary-state structure functions established in Sect. 3.1 and the probability structure describing component degradation from state to state, the reliability of the multi-state axial compressor to be in one of four states can be evaluated. For illustrative purposes, the component degradation model was established as suggested in Eryilmaz and Tuncel (2016) with arbitrarily selected instantaneous degradation rates. Figure 13 shows the survival functions of the levels $j = 1, \ldots, 4$ of the multi-state axial compressor previously introduced. The combination of this multi-state system consideration with the developed analysis framework allows the assessment of the resilience of mutli-state systems with multi-state components.

7   Uncertainty Analysis

In reality, the information on a complex capital good and its behavior is subject to aleatoric or so-called irreducible uncertainties but typically also epistemic uncertainties or so-called imprecision. For instance, this is the case due to estimates of distribution parameters based on expert knowledge, measurement errors or a simple lack of data. In the context of the CRC 871, the influence of a regeneration measure on the survival behavior of a complex capital good might not be precisely known, i.e., the distribution parameters describing the failure behavior can only be estimated. Thus, the models and corresponding simulations are also governed by these uncertainties. However, for comprehensive decision making existing uncertainties need to be considered in analysis and therefore beneficial approaches to implement these are an important research topic, see Beer and Ferson (2012) and Beer et al. (2013). Consequently, the novel uncertainty analysis developed in this work constitutes the fourth and final phase of the analysis framework, see Fig. 2.

7.1 Imprecision and its Implementation via Fuzzy Probability

Figure 14 shows the concept of survival signature with an adaption of the probability structure via fuzzy probabilities, cf. Salomon et al. (2021b). Thus, the imprecision is propagated through the model. This allows the advantageous properties of this concept to be exploited while accounting for imprecision.

The result is not a sharp survival function, as seen in Fig. 15a, but an imprecise survival function with regions, as seen in Fig. 15b. The imprecise survival functions are computed on basis of stochastic input variables described via fuzzy probabilities. The reliability analysis under consideration of imprecision can be simplified by considering various discrete alpha-cuts with $\alpha \in [0, 1]$ as illustrated in Fig. 15b for $\alpha \in \{0, 0.2, 0.4, 0.6, 0.8\}$.

The survival function generated with $\alpha = 0$ represents the maximum level of uncertainty. For example, an expert specifies the parameter interval corresponding to an alpha level $\alpha = 0$ of the fuzzy probability as the maximum degree of uncertainty, i.e. the parameters will certainly not violate the interval limits. This might be the case in design and maintenance, if, e.g., only insufficient information on the installed components has been collected so far and only an educated guess of an expert is available. In contrast, an alpha-level of $\alpha = 1$ corresponds to a precise survival function that is typically not known. Note that this is only the case for triangular fuzzy probabilities as presented in Salomon et al. (2021b). However, there exist fuzzy probabilities that describe an interval of parameters for $\alpha = 1$ as well. Depending

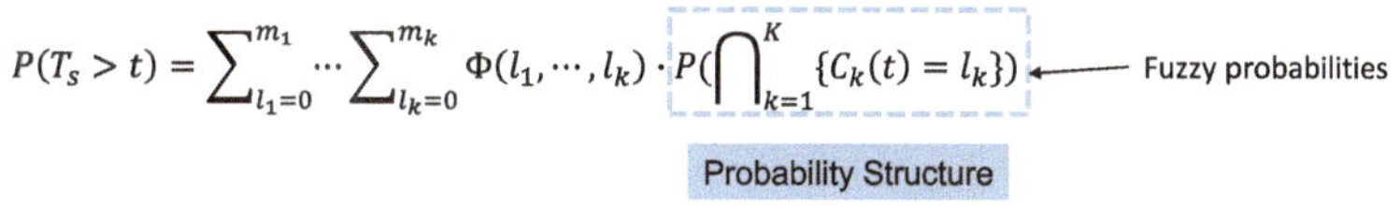

$$P(T_s > t) = \sum_{l_1=0}^{m_1} \cdots \sum_{l_k=0}^{m_k} \Phi(l_1, \cdots, l_k) \cdot P(\bigcap_{k=1}^{K} \{C_k(t) = l_k\}) \longleftarrow \text{Fuzzy probabilities}$$

Probability Structure

Fig. 14 Fuzzy probabilities included in the concept of survival signature

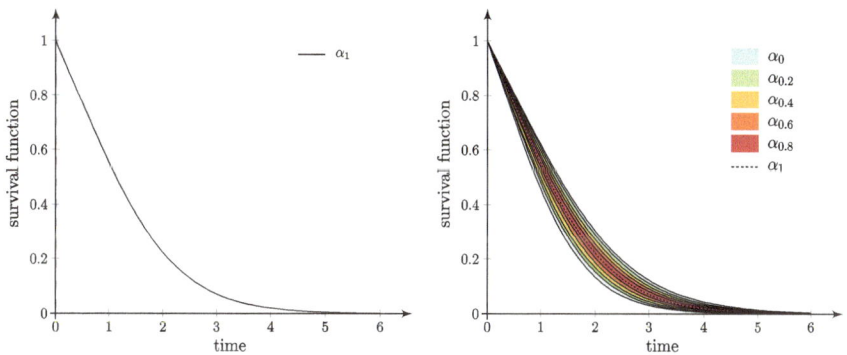

(a) Precise survival function that is equivalent to an alpha-cut $\alpha = 1$ of the fuzzy probability

(b) Imprecise survival function described via fuzzy probability with five alpha-cuts $\alpha \in \{0, 0.2, 0.4, 0.6, 0.8, 1\}$

Fig. 15 Precise and imprecise survival functions

on the budget, gathering precise information for each component type, e.g., via experimental campaigns, might not be feasible, impeding proper reliability analyses. In fact, a complete elimination of imprecision is in most cases neither necessary nor cost-efficient. Therefore, a method for determining a critical level of imprecision is crucial for cost-effective decision making that balances imprecision against the cost associated with reducing it. In Salomon et al. (2021b), Salomon et al. developed a comprehensive decision making procedure for uncertainty reduction. Integral parts are target values for the system reliability at certain points in time. If the current setting of parameters modeled via fuzzy probabilities fail to ensure these target reliabilites the imprecision in the fundamental data should be reduced. The procedure is cost-efficient, since it proceeds successively from $\alpha = 0$ up to $\alpha = 1$ until the reliability requirements are met.

7.2 Efficient Simulation Algorihtm Under Consideration of Imprecision

The consideration of both irreducible uncertainty and imprecision requires adequate treatment in systems analysis. A frequently implemented approach is a two-stage simulation, commonly known as a "double-loop" approach. Correspondingly, variables describing imprecision on parameters are sampled in an "outer loop" and variables representing irreducible uncertainty and depending on the imprecise parameters, as, e.g., failure time of components, are sampled in an "inner loop," cf. Hofer et al. (2002), or vice versa, in an "outer loop" aleatory variables are sampled and epistemic uncertainty is treated in the "inner loop", cf. Alvarez (2006). Clearly, for complex systems, this naive sampling approach leads to an extremely large sample size and consequently a high computational cost, see, e.g. Sarkar et al. (2015). Consequently, simulation approaches that increase computational efficiency and yield high accuracy at minimal sample size are desirable. Recently, the Non-Intrusive Stochastic Simulation (NISS), a promising approach to efficiently compute imprecise structural models with significantly reduced sample size was introduced in Wei et al. (2019). The method is divided into two basic approaches, Local Extended Monte Carlo Simulation (LEMCS) and Global Extended Monte Carlo Simulation (GEMCS), which provide different advantages in terms of accuracy and variation.

In Salomon et al. (2021b), a novel methodology was developed in the context of imprecise system reliability analysis by adapting the LEMCS and the GEMCS and combining them with the concept of survival signature. The imprecision of parameters is modeled via fuzzy probabilities. This imprecision is then propagated efficiently through the analysis framework by means of the new method as illustrated in Fig. 16. The complex amalgamation brings together the advantages of both the concept of survival signature and NISS concepts: a significant memory reduction of topological information and large efficiency benefits in repeated model evaluations, combined with a comprehensive consideration of uncertainties with only one

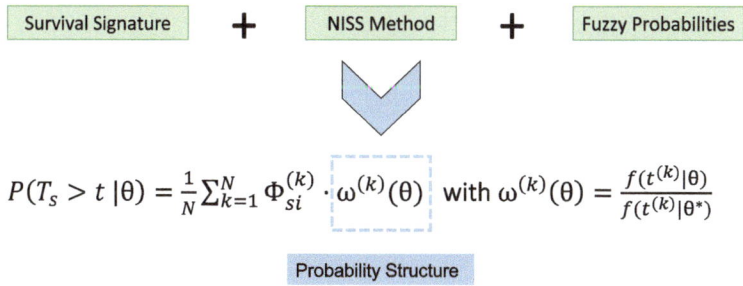

Fig. 16 NISS method and the LEMCS estimator

required stochastic simulation, thus drastically reducing the sample size. This combination leads to beneficial synergy effects, increasing the efficiency even more. The savings due to the new methodology regarding sampling effort compared with the naive double loop approach are illustrated in Fig. 17. Figure 17a and b The most attractive aspect of both the LEMCS and GEMCS algorithms is the fact that only a single stochastic simulation is necessary to account for the imprecisions. Therefore, the traditionally employed "double loop" simulation can be circumvented. In both LEMCS and GEMCS, the interval analysis and the stochastic analysis have been decoupled successfully, and the computational expense is mainly driven by the single stochastic simulation performed. Moreover, the stochastic analysis has been separated from the system topology by merging it with the survival signature, so that only one reliability analysis in terms of topology is required to generate the survival signature. In addition to these beneficial features of the survival signature, it is exactly the single stochastic simulation required that gives the proposed methodology its efficiency and differentiates it clearly from traditional approaches. Thanks to this approach the imprecise stochastic analysis used to estimate the bounds on the system survival function has been greatly simplified.

In Salomon et al. (2021b), the new methodology was demonstrated, among others, on the functional model of the axial compressir presented in Sect. 3 and shown in Fig. 4. The "double loop" approach is conducted with 5 000 samples (failure times) on the inner loop and 1 000 samples (model parameters) on the outer loop, i.e., a total of 5 000 000 samples. While even improving the quality in results, for both LEMCS and GEMCS, only one simulation was required with 100 000 generated samples (failure times for 100 000 different model parameters). Correspondingly, only 1/50th of the sample size compared to the "double loop" approach was required.

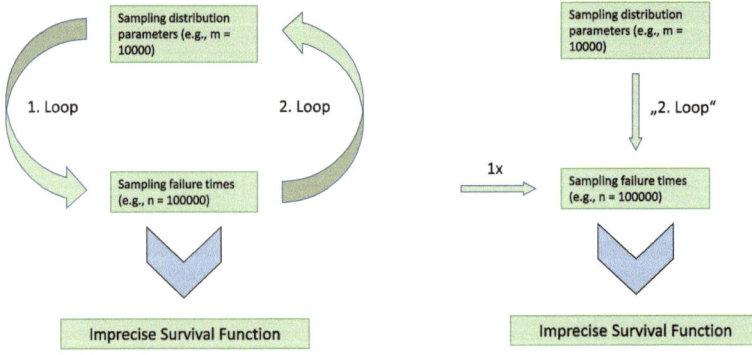

(a) Sampling procedure via the "double loop" approach   (b) Sampling procedure via novel developed uncertainty analysis

Fig. 17 Sampling via the "double loop" approach and the novel uncertainty analysis developed in this work

8 Conclusions

A decision making process has been developed that allows the identification of optimal trade-offs among numerous resilience-enhancing features/measures for complex capital goods of various types. During the period of the CRC 871 the approach as been applied to models of an axial compressor, of a flow network, of an arbitrary complex system, and of the Berlin metro network. The consideration of monetary and technical constraints into the decision making process is realized. The broad applicability of all developed methods is ensured, i.e., there are no limitations to a specific system type. A reduction in computational effort has been achieved, mainly due to the separation property of the survival signature, i.e., once the system structure has been computed, any possible characterization of the probabilistic part can be evaluated without the need to recompute the structure. The integration of uncertainties in the reliability analysis is enabled and the sample size is drastically reduced due to the adapted NISS methods requiring only a single stochastic simulation, avoiding the tedious "double loop" simulation traditionally applied.

It could be shown that functional models are a good and effective approach to represent physically complex systems. This approach was further developed to not only consider dependencies in the input parameters but also to include a time dependency of the sensitivities by means of importance indices. Still challenging is the merging of the several developed functional models of subsystems within the overall jet engine into an encompassing representative overall model. It is very computationally expensive due to the costly sensitivity analysis for complex systems with numerous input and output parameters. Further in-depth research on this topic is required to allow the generation of such an extensive model.

The comprehensive and encompassing analysis framework developed in this work, consisting of resilience analysis, consideration of technical and monetary constraints, reliability analysis, and uncertainty analysis, provides decision makers with an additional basis for decision making in MRO processes, enabling sophisticated decisions on an efficient background. Thereby, resilience applies to regeneration paths not exclusively as a part of the overall life cycle performance of the complex capital good, but as an important property for the regeneration path in itself. Clusters are identified of similar, equally acceptable regeneration paths to which, in the event of a problem, effortlessly can be switched, leading to a variety of resilient regeneration paths. Furthermore, strict attention was given that all developed phases within the analysis framework are applicable to complex capital goods of any kind and in every field of application, e.g., design process, optimization process etc.

The outlook that goes beyond the scope of this project is the combination of all presented approaches into a single encompassing methodology in order to be able to take even greater advantage of the excellent synergy effects between them.

Acknowledgements Funded by the Deutsche Forschungsgemeinschaft (DFG, German Research Foundation)—CRC 871/3—119193472.

References

Alvarez, D. A. (2006). On the calculation of the bounds of probability of events using infinite r andom sets. *International Journal of Approximate Reasoning*, 43(3):241–267.

Ayyub, B. M. (2014). Systems resilience for multihazard environments: Definition, m etrics, a nd v aluation for decision making. *Risk Analysis*, 34(2):340–355.

Ayyub, B. M. (2015). Practical resilience metrics for planning, design, and decision making. *ASCE-ASME Journal of Risk and Uncertainty in Engineering Systems, Part A: Civil Engineering*, 1(3):04015008.

Beer, M. and Ferson, S. (2012). Fuzzy probability in engineering analyses. In *Vulnerability, Uncertainty, and Risk: Analysis, Modeling, and Management*, pages 53–61.

Beer, M., Ferson, S., and Kreinovich, V. (2013). Imprecise probabilities in engineering analyses. *Mechanical Systems and Signal Processing*, 37(1):4–29.

Coolen, F. P. and Coolen-Maturi, T. (2013). Generalizing the signature to systems with multiple types of components. In *Complex Systems and Dependability*, pages 115–130. Springer.

Coolen, F. P. and Coolen-Maturi, T. (2016). The structure function for system reliability as predictive (imprecise) probability. *Reliability Engineering & System Safety*, 154:180–187.

Denkena, B., Nyhuis, P., Bergmann, B., Nu¨bel, N., and Lucht, T. (2019). Towards an autonomous maintenance, repair and overhaul process: Exemplary holistic data management approach for the regeneration of aero-engine blades. *Procedia Manufacturing*, 40:77–82.

Eryilmaz, S. and Tuncel, A. (2016). Generalizing the survival signature to unrepairable homogeneous multi-state systems. *Naval Research Logistics (NRL)*, 63(8):593–599.

Feinstein, Z., Rudloff, B., and Weber, S. (2017). Measures of Systemic Risk. *SIAM Journal on Financial Mathematics*, 8(1):672–708.

Henry, D. and Ramirez-Marquez, J. E. (2012). Generic metrics and quantitative approaches for system resilience as a function of time. *Reliability Engineering and System Safety*, 99:114–122.

Hofer, E., Kloos, M., Krzykacz-Hausmann, B., Peschke, J., and Woltereck, M. (2002). An approximate epistemic uncertainty analysis approach in the presence of epistemic and aleatory uncertainties. *Reliability Engineering & System Safety*, 77(3):229–238.

Jacques, J., Lavergne, C., and Devictor, N. (2006). Sensitivity analysis in presence of model uncertainty and correlated inputs. *Reliability Engineering & System Safety*, 91(10-11):1126–1134.

Keitel, H. and Dimmig-Osburg, A. (2010). Uncertainty and sensitivity analysis of creep models for uncorrelated and correlated input parameters. *Engineering Structures*, 32(11):3758–3767.

Kucherenko, S., Tarantola, S., and Annoni, P. (2012). Estimation of global sensitivity indices for models with dependent variables. *Computer physics communications*, 183(4):937–946.

Marelli, S., Lamas, C., Konakli, K., Mylonas, C., Wiederkehr, P., and Sudret, B. (2019). UQLab user manual—Sensitivity analysis. Technical report, Chair of Risk, Safety and Uncertainty Quantification, ETH Zurich, Switzerland. Report # UQLab-V1.3–106.

Mettas, A. (2000). Reliability allocation and optimization for complex systems. *Reliability and Maintainability Symposium, 2000. Proceedings. Annual*, pages 216–221.

Miro, S., Willeke, T., Broggi, M., Seume, J., and Beer, M. (2019). Reliability analysis of an axial compressor based on one-dimensional flow modeling and survival signature. *ASCE-ASME J Risk and Uncert in Engrg Sys Part B Mech Engrg*, 5(3).

Ouyang, M., Duen˜as-Osorio, L., and Min, X. (2012). A three-stage resilience analysis framework for urban infrastructure systems. *Structural Safety*, 36-37:23–31.

Salomon, J., Behrensdorf, J., Broggi, M., Weber, S., and Beer, M. (2019). Multidimensional resilience decision- making on a multistage high-speed axial compressor.

Salomon, J., Broggi, M., Kruse, S., Weber, S., and Beer, M. (2020). Resilience Decision-Making for Complex Systems. *ASCE-ASME J Risk and Uncert in Engrg Sys Part B Mech Engrg*, 6(2).

Salomon, J., Go¨ing, J., Lu¨ck, S., Broggi, M., Friedrichs, J., and Beer, M. (2021a). Sensitivity analysis of an aircraft engine model under consideration of dependent variables. In *Turbo Expo: Power for Land, Sea, and Air*, volume 84898, page V001T01A005. American Society of Mechanical Engineers.

Salomon, J., Winnewisser, N., Wei, P., Broggi, M., and Beer, M. (2021b). Efficient reliability analysis of complex systems in consideration of imprecision. *Reliability Engineering & System Safety*, 216:107972.

Sarkar, A., Guo, J., Siegmund, N., Apel, S., and Czarnecki, K. (2015). Cost-efficient sampling for performance prediction of configurable systems (T). In *2015 30th IEEE/ACM International Conference on Automated Software Engineering (ASE)*, pages 342–352. IEEE.

Sobol, I. M. (2001). Global sensitivity indices for nonlinear mathematical models and their monte carlo estimates. *Mathematics and computers in simulation*, 55(1–3):271–280.

Wei, P., Song, J., Bi, S., Broggi, M., Beer, M., Lu, Z., and Yue, Z. (2019). Non-intrusive stochastic analysis with parameterized imprecise probability models: I. Performance estimation. *Mechanical Systems and Signal Processing*, 124:349–368.

Interaction of Combined Module Variances and Influence on the Overall System Behaviour

Jan Goeing and Jens Friedrichs

Abstract Within the Collaborative Research Centre 871, geometrical variances caused by repair procedures and deterioration are evaluated for the turbomachinery of a high-bypass aircraft engine. Part of this evaluation is the investigation of the influence of isolated and combined geometric variances on the overall aircraft engine performance. For this purpose, a virtual twin of a research aircraft engine is developed in sub-project D6. This virtual aircraft engine is based on the Pseudo Bond Graph approach, which allows for transient manoeuvres and the effects of interactions to be simulated with a higher degree of accuracy compared to conventional methods. After validation of the model, a design of experiments is performed to analyse the sensitivities between the variances of modules and engine performance. Within the sensitivity analysis, it is shown that the evaluated steady-state and transient performances are mainly influenced by the high-pressure modules, especially by the mass flow and efficiency variances. Furthermore, it is shown that the sensitivities strongly depend on the operating points. However, significant interactions are found which can be attributed to both the high-pressure and low-pressure modules.

Keywords Aircraft engine · Performance simulation · Dynamic model · Gas path analysis · Sensitivity analysis

Nomenclature

α	Heat Transfer Coefficient
Δ	Deviation
$\dot{m}$	Mass Flow
η	Efficiency
Γ	Simplified Model

J. Goeing (✉) · J. Friedrichs
Institute of Jet Propulsion and Turbomachinery, Technische Universität Braunschweig,
Hermann-Blenk-Straße 37, 38108 Braunschweig, Germany
e-mail: j.goeing@ifas.tu-braunschweig.de
URL: https://www.tu-braunschweig.de

423

λ	Friction Coefficient
ω	Circumferential Speed
π	Pressure Ratio
ρ	Density
τ	Torque
Θ	Engine configuration
$q\cdot$	Specific heat flow
A	Cross section
BPR	Bypass Ratio
C	Compressor
c_v	Specific heat capacity
CRC	Collaborative research centre
DoE	Design of Experiments
E	Internal Energy
e	Effort, Internal energy
f	Fuel of Flow, Body force
GPA	Gas Path Analysis
HP	High-Pressure
h	Enthalpy
IAE	International Aero Engines
IPC	Booster
J	Moment of Inertia
L	Length
LP	Low-Pressure
MRO	Maintenance, Repair and Overhaul
$N1$	Rotational Speed of LP-System
$N2$	Rotational Speed of HP-System
OL	Operating Line
p	Pressure
Q	Heat Flow
R	Specific Gas Constant
s	Surface
S_y	First-Order Effect Kucherenko
S^T	Total-Effect Kucherenko
SF	Total-Effect Kucherenko
SFC	Scaling Factor
SL	Specific Fuel Consumption
EGT	Exhaust Gas Temperature
SM	Compressor Surge Margin
F	Surface Forces
T	Temperature, Turbine
t	Total or Time
V, v	Volume, Velocity

1 Introduction

Commercial aircraft engines have long service times of several decades (Weiner 2015). To maintain safe, reliable and efficient functionality, aircraft engines undergo extensive maintenance, repair and overhaul (MRO) activities during their lifetime.

More than 40% of the direct maintenance costs of an active aircraft are accounted for by the aircraft engines, which thus represent the most expensive and significant part (Markou and Cros 2021). Thus, a fundamental understanding of the influence of aircraft engine deterioration on performance has a significant economic viability for manufacturers, operators and maintainers. An effective and accurate regeneration or recovery of components can significantly reduce operating costs as well as the risk of failure.

The spectrum of maintenance work on aircraft engines is caused by wear and damage mechanisms. The mechanisms include, in particular, damage and contamination of the engine blading (erosion, corrosion and fouling) as well as deterioration of blade tip gaps (Müller 2013; Saravanamuttoo et al. 2001). In general, these local variances of the different modules have a direct impact on the overall performance, propulsion efficiency and safety margins of the aircraft engine. The condition monitoring of an aircraft engine requires performance-specific data as well as knowledge about the relationship between module deterioration and performance output (Fentaye et al. 2019; Spieler et al. 2008; Volponi 2014). To gain this knowledge, the non-linear behaviour of the aircraft engine has to be reproduced in the model and its ability to capture the influence and the interaction of combined module variances must not be neglected.

Therefore, the Collaborative Research Centre 871 "Regeneration of Complex Capital Goods" (CRC 871) develops a scientific basis for novel technologies and approaches to analyse, evaluate and determine the causes and effects of wear and to transfer the experience based approaches into knowledge-based approaches (Aschenbruck et al. 2014; Kellenbrink et al. 2022). Hence, these methods are exemplified using the V2500-A1, a mature high-bypass turbofan jet engine owned and operated by the Institute of Jet Propulsion and Turbomachinery (IFAS). In order to evaluate the degradation, a virtual process is developed in the CRC 871, which uses blades of the high-pressure turbine to determine the influence of geometrical variances on the performance output. This automated virtual process is represented in the flow chart in Fig. 1 (Goeing et al. 2022b).

Here, the geometry of the HPT blades is digitised (1), parameterised (2), and analysed (3) automatically. Furthermore, the influence on the local aerodynamics of the reconstructed blade is investigated using a computational fluid dynamics (CFD) simulation (4–6). The results of aerodynamic analyses are transferred into a simulation of the performance of the entire aircraft engine (7–9).

The main focus of this sub-project "Interaction of combined module variances and influence on the overall system behaviour" is on the investigation of sensitivities between module degradation and overall aircraft engine performance. Therefore, this sub-project is focused in particular on the blue-framed parts (7–9) in the virtual

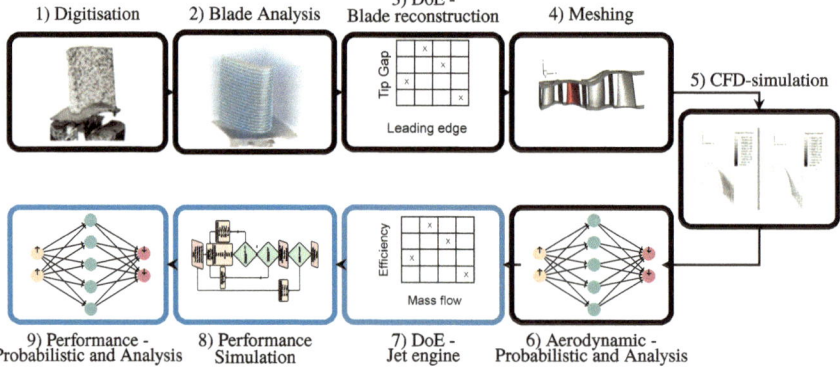

Fig. 1 Virtual performance evaluation process of the system demonstrator

assessment process of Fig. 1. In order to investigate these sensitivities between performance and module deterioration, a design of experiments (DoE) with a virtual twin of the research aircraft engine is carried out. The virtual twin is developed with the in-house performance analysis tool ASTOR (AircraftEngine Simulation for Transient Operation Research) to simulate the performance of the aircraft engine and validated experimentally and numerically. The module variances are integrated via performance maps of compressors and turbines in ASTOR. The impact of module variances on overall performance is evaluated by a variance-based sensitivity analysis that is able to evaluate the sensitivities and interactions of combined module variances.

Since the conventional 0D performance simulation approaches have fundamental limitations in dealing with the non-linear behaviour of the aircraft engine (Kurzke and Halliwell 2018; Fentaye et al. 2019), ASTOR is based on the quasi 1D consideration of the gas path, whereby volume effects as well as interactions in the gas path are taken into account.

Below, the developed and validated virtual twin and results of the sub-project D6 are briefly described and presented (for further details (Goeing et al. 2018, 2019a, b, 2020a, b, c, 2022a, b; Salomon et al. 2021; Lück et al. 2022).

2 Setup

In this section, the methods used to develop the virtual twin of the aircraft engine are presented. These are carried out on the basis of the IFAS research aircraft engine, the V2500-A1. The V2500 turbofan from IAE (International Aero Engines) usually powers the Airbus A320 medium-range aircraft. It is equipped with a low-pressure compressor (LPC), an intermediate-pressure compressor (IPC), a high-pressure compressor (HPC), a high-pressure turbine (HPT), a low-pressure turbine (LPT), and a common thrust nozzle.

2.1 ASTOR

The ASTOR model is able to simulate on- and off-design steady-state operations, as well as transient and highly dynamic engine responses, such as compressor surge. Furthermore, ASTOR is developed using Pseudo Bond Graph theory, which is a powerful unified approach to model interdisciplinary and dynamic systems. ASTOR is implemented in a Matlab/Simulink environment. The system of ordinary differential equations (ODE) is solved with an explicit multistep solver (Shampine et al. 1999) and a variable time step.

2.1.1 Dynamic Model

For the simulation of transient performance and dynamic phenomena, ASTOR considers main effects (dynamics of rotating machines) as well as secondary effects (heat transport, interactions and volume dynamics). For this system dynamic approach, the mass (1), momentum (1.2) and energy (1.3) conservation of fluids and solid bodies are considered on fixed control volumes:

$$\iiint_V \frac{\partial \rho}{\partial t} dV = -\iint_S \rho \cdot v_j \cdot n_j dS \tag{1}$$

$$\iiint_V \frac{\partial}{\partial t}(\rho \cdot v_i) \cdot dV + \iint_S \rho \cdot v_i \cdot v_j \cdot n_j \cdot dS = -\iint_S p \cdot n_i dS + \iiint_V \rho \cdot f_i dV \tag{2}$$

$$\iiint_V \frac{\partial}{\partial t}\left[\rho\left(e + \frac{v^2}{2}\right)\right]dV + \iiint_V \frac{\partial}{\partial x_j}\left[\rho\left(e + \frac{v^2}{2}\right)\right]dV = \iiint_V \frac{\partial}{\partial x_j}(\rho \cdot v_j)dV$$
$$+ \iiint_V \frac{\partial}{\partial x_l} \cdot \dot{q}dV \tag{3}$$

Based on the above integro-differential equation system, an ODE-system is spatially discretised (see Eqs. 4–8), which represents the quasi 1D gas path by finite control volumes. The turbomachinery component characteristics are included via the surface forces F_P and F_T and are equal to zero in a non-turbomachinery control volume. The pressure ratio π and efficiency η are imposed on the compressor and turbine volumes, while the burner efficiency and pressure losses are imposed on the burner volume. The energy and momentum contributions of gravitational forces have been neglected and the frictional shear stress tensor is replaced by the Fanning and Darcy friction coefficient λ. Thus, interaction effects within the gas path as well as with the boundaries of the control volume and within the control volume (inertia and capacity of the gas) can be taken into account

$$\frac{d}{dt}(\rho) = \frac{1}{V}(\dot{m}_{in} - \dot{m}_{out}) \tag{4}$$

$$\frac{d}{dt}(\dot{m}) = \frac{1}{L} \cdot \left(A_{in} \cdot p_{in} - A_{out} \cdot p_{out} + p \cdot (A_{in} - A_{out}) + \dot{m}_{in} \cdot v_{in} - \dot{m}_{out} \cdot v_{out} + F_{pt}\right)$$
$$- \rho \frac{\lambda}{D} \cdot \frac{v}{2} \cdot A \tag{5}$$

$$F_{pt} = \frac{A_{in} + A_{out}}{2} \cdot (p_{out} - p_{in}) + \dot{m}_{in} \cdot (v_{out} - v_{in}) \tag{6}$$

$$\frac{d}{dt}(\rho \cdot h_t - p_t) = \frac{1}{V}(\dot{m}_{in} \cdot h_{t,in} - \dot{m}_{out} \cdot h_{t,out} + F_{Tt} + \dot{m}_f \cdot h_f + \dot{Q}) \tag{7}$$

$$F_{Tt} = \dot{m}_{in} \cdot (h_{t,out} - h_{t,in}) \tag{8}$$

$$p = \rho \cdot R \cdot T \tag{9}$$

Equation 4–8 are first-order nonlinear state equations relating density, velocity and internal energy derivatives for the control volume. Equation 9 is the ideal gas equation.

Based on the energy conservation for solid bodies, the wall temperatures T_W of the blades, casing and disks are simulated using Eq. 10. The convective heat transfer is integrated into the ODE system with Eq. 11. For the heat transfer coefficient α, substitute models are created for the casing, blades and disks (see Stephan et al. (2019))

$$\frac{dT_w}{dt} = \frac{1}{V \cdot \rho \cdot c_v} \cdot (\dot{Q}_i - \dot{Q}_o) \tag{10}$$

$$\dot{Q} = \alpha \cdot A \cdot (T_{tx} - T_W) \tag{11}$$

Furthermore, the dynamic of rotating machines is included by the momentum equation for discretised solid bodies

$$\frac{dN}{dt} = \frac{1}{J \cdot 2\pi} \cdot (\tau_T - \tau_C) \tag{12}$$

Based on the thermal and mechanical stress, the radial expansion during a transient manoeuvre is also considered (Fiola 1993; Kypuros 2003).

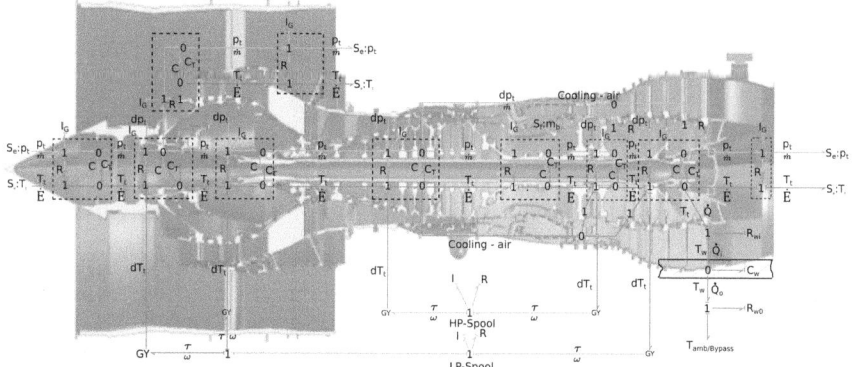

Fig. 2 CAD model and Pseudo Bond Graph of the V2500-A1

2.1.2 Boundary Conditions

Together with the sub-projects "Exhaust Jet Analysis", "Loss Behavior of Complex Surface Structures" and "Coupled geometric variances", the machine-specific performance maps are generated, e.g. through CFD studies. Apart from the performance maps of the miscellaneous modules, the geometry, material information as well as the initial steady-state operating point are required for the simulation of dynamic manoeuvres. Therefore, a full CAD-model based on the IFAS research V2500-A1 engine is designed (see background of Fig. 2). The mass of fan, HPC, HPT and LPT blades are weighed. The weight of the booster blades, the two spools and discs are approximated based on technical sketches. Mostly, titanium and nickel alloys are used as material for these components. The resulting moment of inertia J of the LP-system is 57 kgm^2 and that of the HP-system is 11 kgm^2.

In order to model the initial steady-state operating point and therefore the starting point for ASTOR, a global iterative 0D engine matching framework is developed. Matching in this context means iterating within the performance maps, turbine entry temperature, bypass ratio (BPR), the rotational speed N_1, until: (1) the turbine power output matches the compressor power, (2) the mass flow into all modules is the same (bypass, secondary air and fuel) and (3) the nozzle pressure is equal to the pressure downstream of the LPT (including friction). The convergence criterion E for all three equations is E < 10^{-6}.

2.1.3 Pseudo Bond Graph

The Pseudo Bond Graph theory is used to superimpose complex systems into a unified system-dynamic notation and categorises all quantities into efforts e and flows f. The advantage is that the complex compressible system of the gas turbine can be connected to other physical domains, which may be advantageous in view of future hybrid electric propulsion systems (Wahler et al. 2022). The Pseudo Bond Graph schematic is shown in Fig. 2. Here, efforts e (e.g. pressure p, temperature T or torque τ) and flows f (e.g. mass flow m, energy flow, heat flow or rotational speed) of the physical domains represented within a jet engine are layed out. The 0—and 1—junctions are applied to connect these efforts e and flows f of the miscellaneous domains. In the gas path, the conservation of momentum is solved at the 1—junction and the conservation of energy and mass, which take into account heat transfer and secondary air flow, at the 0—junction. Therefore, gas effects, which are based on inertia and capacity (mass storage and dynamic volumes), are directly included in the system of differential equations.

3   Results and Discussion

In this section, the results of sub-project D6 are discussed. First, ASTOR will be validated with the real IFAS research aircraft engine. Second, ASTOR is compared to conventional performance simulation methods to illustrate the improvement in performance prediction capabilities that can be achieved by using higher order approaches. Finally, the results of the DoE study and the sensitivity analysis are shown.

3.1   Validation

Numerous experiments were carried out to validate ASTOR. In Goeing et al. (2020b), ASTOR was shown to accurately predict the steady-state and transient performance of the IFAS experimental turbojet engine. However, the main focus of the validation study in this section is the comparison to the IFAS two-spool, high-bypass V2500-A1 engine. For this purpose, a complex pass-off test is conducted at the MTU-H-71 test bed. During the test run, the rotational speeds (N_1, N_2), thrust and temperatures behind the compressors (T_{t21}, T_{t25}, T_3) are recorded at a sampling rate of 10 Hz. In addition to the standard instrumentation, additional total pressure and total temperature probes are used in the HPC, behind the HPT (T_{t45}, p_{t45}) and behind the LPT (T_{t5}, p_{t5}) for a higher data acquisition. The test setup is depicted in Fig. 3a.

Furthermore, customised thrust curves are defined in order to obtain information at the steady-state operating points and to perform various transient manoeuvres. In Fig. 3b these thrust curves are shown. In general, four different thrust levels are approached (Band D/41 kN, C/72 kN, B/92 kN and A/101 kN) in this test campaign. Starting from the lowest level D, the different thrust levels are approached by accelerations. To reach steady-state operating conditions, the thrust levels are held for 2–3 min. After that, decelerations are carried out until the Band D is reached again. Finally, a slam acceleration and deceleration between Band D and Band A is performed.

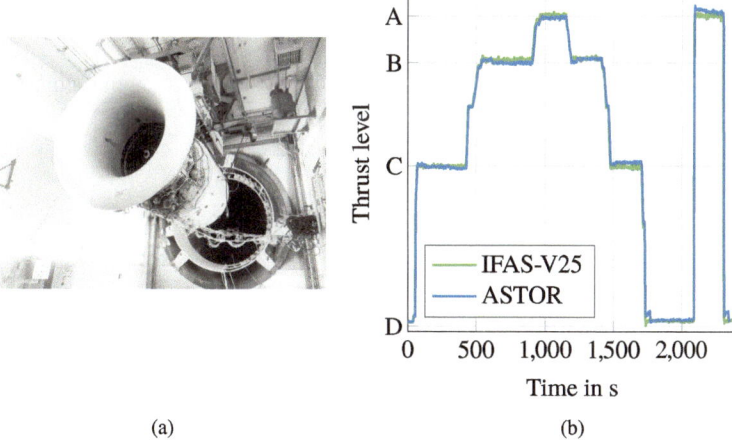

(a)   (b)

Fig. 3 **a** IFAS research aircraft engine in the MTU-H-71 test bed. **b** Test procedure and individual thrust curves

The steady-state performance data is then used to calibrate the virtual aircraft engine in ASTOR. Therefore, the performance maps of different turbomachinery components are modified iteratively until the deviation between measured and simulated steady-state data is below 5%. In the illustration 1.3b, the real engine data (green) is compared to the virtual engine data (blue). In general, the graphs are consistent with each other. The mean deviation during the whole manoeuvre is 1.1% with a 95% confidence interval of ±0.27%. During the fast acceleration between Band D and Band A, ASTOR overestimates the thrust by 2.5% compared to the steady-state operating point. This effect does not occur during the measurement.

In Fig. 4 the variations of temperatures T_{t45}, T_{t5} and the rotational speeds N_1, N_2 during the fast manoeuvre between Band D and Band A of the real and virtual aircraft engine are shown.

In general, the transient performance of the virtual twin matches the real aircraft engine. In the comparison of T_{t45} (see Fig. 4a), the mean deviation is $1.5 \pm 0.2\%$ during the acceleration and $1.7 \pm 0.3\%$ during the deceleration. In addition, the transient peak during the acceleration as well as the behaviour during the deceleration are reproduced by ASTOR (see magnification).

The mean deviation of T_{t5} (see Fig. 4b) is $1.2 \pm 0.2\%$ during the acceleration and $1.3 \pm 0.3\%$ during the deceleration. Similar to T_{t45}, the location and amplitude of the temperature peaks correspond to each other. Furthermore, the influence of the heat soakage, which dampens the temperature rise, is detected in both curves.

The reaction of the N_1 shaft is very sluggish due to the high moment of inertia, which means that there are no significant dynamic effects to the N_1 speed (see Fig. 4c). The largest deviation is 3% and is located at the steady-state operating point at Band A. The deviation can be traced back to the calibration of the performance maps.

In contrast to the N_2 speed where the steady-state deviation is between 0.3 and 0.7%. Furthermore, the mean deviation during acceleration is $0.6 \pm 0.2\%$ and during

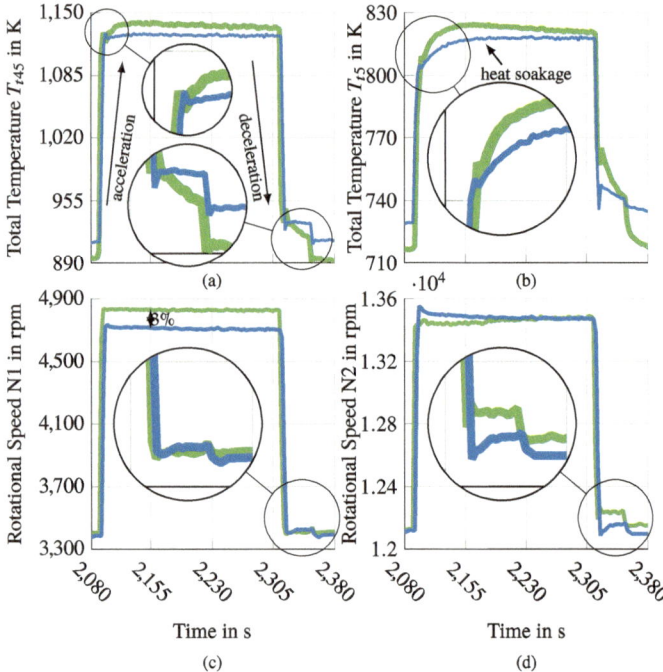

Fig. 4 Comparison of the transient performance of the real and virtual V2500-A1 turbofan during slam acceleration and deceleration

the deceleration $0.5 \pm 0.2\%$. However, the steady-state N2 at Band A is overestimated during the acceleration, which is not detected in the measurement. Finally, the deceleration is consistent except for an offset of 100 rpm.

3.2 Comparison of Performance Models

In this section, ASTOR is compared to a conventional engine matching (EM) approach, which is used in commercial performance tools (Kurzke and Halliwell 2018). This global engine matching (EM) method is implemented in Matlab to use same parameter and modelling errors. The EM approach iteratively determines the transient operating point by integrating the changes in rotational speeds dN/dt while matching mass flow rate inside the modules and the nozzle pressure and the pressure downstream of the LPT. On one hand, the most significant influence on the transient performance is exerted by the inertia of rotating machines as well as the convective heat transport. These effects can be taken into account by EM. On the other hand, through the ODE system, ASTOR allows a more accurate simulation of the interaction between the different effects (e.g. feedback between heat flow, gas path, material and system) as well as the influence of the volume dynamics.

To quantify the impact of the higher order approach, a slam acceleration and deceleration between idle and takeoff operating points is performed using ASTOR and the EM method. In Fig. 5 the mean deviation and 95% confidence interval between ASTOR and the EM are shown. Furthermore, the impact of interdependencies (green) due to the use of ODE and the impact of volume dynamics (yellow) are shown separately. In Fig. 5a, the temperature deviation of the full manoeuvre in each module is illustrated. The maximum deviation is observed downstream of the combustion chamber (2.8 ± 0.57%), followed by the HPT (2.6 ± 0.52%) and the LPT (2.2 ± 0.44%). Furthermore, it can be seen that the mean deviation increases particularly in high temperature ranges. In Fig. 5b, the mean deviation of the compressor surge margins is presented. These deviations have a magnitude of 5 to 7.5% with the maximum at the HPC. The impact of the volume dynamics is represented by the yellow area of the plot in both diagrams and is significantly lower compared to the total deviation (see Eq. 4–8). For the temperature, this deviation is up to 0.35% in the compressors, 0.7% in the turbines and 1.2% for the surge mar- gin. The remaining and thus evidently significant part of the deviation is based on the feedback between gas temperature T_t, wall temperature $T_{t,w}$, heat flow $\dot{Q}$ and their interaction with the entire system. These dynamics are inherently included in ASTOR, whereas they are neglected in the EM method.

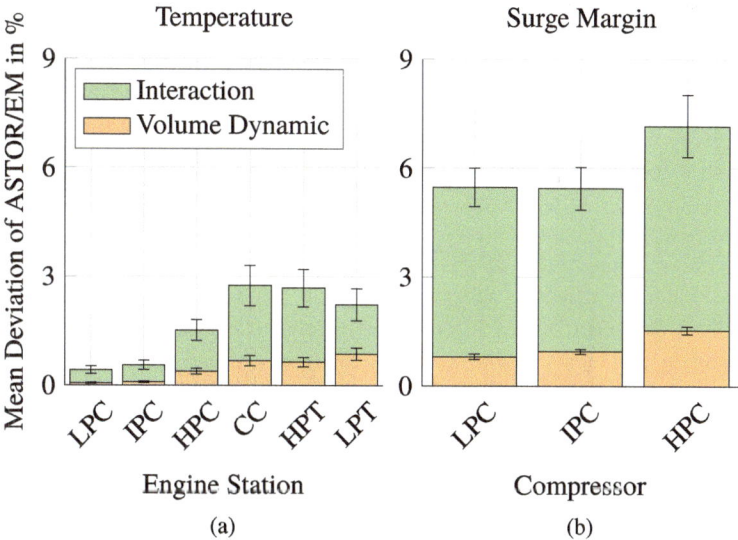

Fig. 5 Comparison between ASTOR and EM methods during a slam acceleration and deceleration. **a** Deviation in temperature. **b** Deviation in SM

3.3 Numerical Experiment

After validating the model and quantifying the improvement, a DoE is carried out to determine the interaction of combined module variances and its influence on the overall performance.

3.3.1 Latin Hypercube Sampling

For this purpose, the parameter space of the possible wear and tear and their combinations were first explored. Hence, a gas path analysis (GPA) is used to explain the performance-specific differences between a baseline (no specific V2500-A1 engine) and the deteriorated IFAS V2500-A1 engine. In the GPA, the individual characteristic curves of machine-specific performance maps of compressors and turbines are modified until the performance matches. The obtained scaling factors SF (see Eq. 13) thus represent the deterioration of the turbomachines

$$SF\pi = \frac{\pi}{\pi_{Ref,N_{corr}}}, SF\dot{m} = \frac{\dot{m}}{\dot{m}_{Ref,N_{corr}}}, SF\eta = \frac{\eta}{\eta_{Ref,N_{corr}}} \tag{13}$$

The performance of both aircraft engines is listed in Table 1. The table shows that the IFAS research aircraft engine requires 3.1% more fuel flow for the maximum thrust condition and the exhaust gas temperature T_{t5} has increased by 44 K. Moreover, the temperature T_{t5} of 850 K is 30 K above the EGT limit of 820 K, therefore the IFAS research aircraft engine is very well suited to define the limits of the parameter space for a deteriorated engine.

In addition to the thrust requirement F_N, the rotational speeds N_1 and N_2, the engine mass flow rate m, the fuel flow m_f, the temperatures T_{t25}, T_{t3}, T_{t45}, T_{t5}, as well as the pressure ratios of the compressors π_C are varied during the matching process. Since wear has a variety of effects on different areas of the performance maps, the GPA is performed at four thrust levels, based on the results of a pass-off test (see Fig. 3a). The scaling factors for the intermediate characteristic curves are interpolated. The resulting scaling factors for the higher and lower operating points of the turbomachines are shown in Table 2.

In general, the results show that the influence of the scaling factors increases with decreasing rotational speed. The mass flow rate, pressure ratio and efficiency of all compressors are reduced. Compared to the HPC, the performance of the LPC and IPC decreases less significantly. In general, the efficiencies of all compressors decrease less than the mass flow and pressure ratio. The deterioration of the turbines leads

Table 1 Corrected aircraft engine conditions at maximum thrust

Value	SFC	$pt5$	$Tt3$	$Tt5$	N_1	N_2
Unit	g/(kN·s)	kPa	K	K	rpm	rpm

Table 2 Parameter range and scaling factors SF for the characteristic lines at thrust level Band D and A

Module band	LPC		IPC		HPC		HPT		LPT	
	D	A	D	A	D	A	D	A	D	A
$N corr$	0.73	1	0.78	1	0.95	1	0.8	1.1	0.8	1
$\Delta m\cdot$	0.93	0.95	0.90	0.95	0.85	0.92	1.05	1.05	1.07	1.02
$\Delta \pi$	0.93	0.95	0.90	0.95	0.85	0.92	0.98	0.98	0.98	0.98
$\Delta \eta$	0.95	0.98	0.92	0.98	0.90	0.96	0.97	0.97	0.96	0.98

to a decrease of efficiency and pressure ratio while the mass flow increases. For the HPT, the same scaling factors are used as a consequence of the minor reduction of corrected HPT rotational speed between the considered operating points. In order to further analyses the performance and sensitivities, several simulations are carried out based on the parameter space from Table 2. The combinations were selected using Latin hypercube sampling (McKay et al. 2000) to efficiently sample the parameter space.

3.3.2 Performance Simulation

In order to analyse the interaction of combined module variances, the steady-state and transient performances are analysed in this section. For this purpose, the temperature load in the hot gas path and the surge margin of the HPC is investigated in more detail. Hence, different aircraft engine configurations Θ are first compared to each other:

As a representative example of a degraded component, the engine state from Table 2 is used.

In order to determine the influence of interactions, these results are compared to a simplified model Γ. As shown in a previous investigation (Goeing et al. 2020c), this model is able to represent the qualitative performance of combined deterioration, but not the interaction through system response. The combination of a deteriorated HP and LP system called Γ is calculated as below:

$$\Gamma = \frac{\Theta_4 \cdot \Theta_5}{\Theta_1} \tag{14}$$

In Fig. 6, the transient performance during a slam acceleration from the Band D to Band A, for the defined engine configurations, is shown. The first configuration Θ_1 is represented in blue, Θ_2 in orange, Θ_3 in green and Θ_6 in purple. The simplified combined engine Γ is shown in red.

In Fig. 6a, the EGT T_{t5} is plotted against rotational speed N_2. It can be clearly seen that the temperature T_{t5} increases with the level of deterioration at all operating points. Moreover, the distance ΔT between the transient temperature peak and

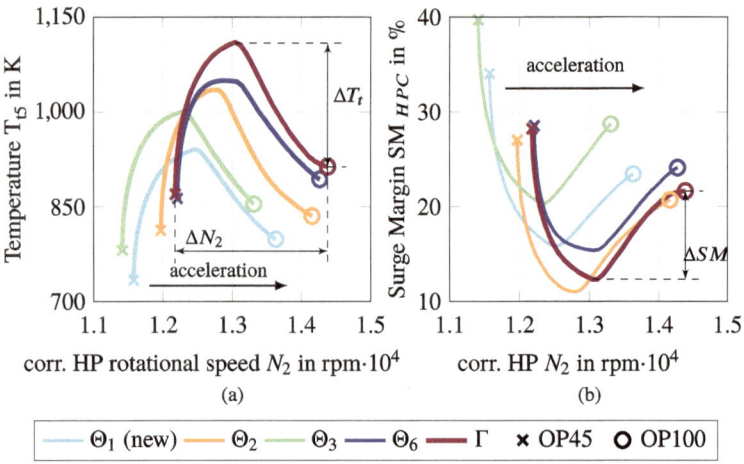

Fig. 6 Transient performance during slam acceleration from Band D to Band A of different aircraft engine configurations. **a** Exhaust Gas Temperature T_{t5}; **b** Surge margin SM. Annotations for definitions of ΔN, ΔSM and ΔT_t. Steady-state operating points highlighted by markers

the final EGT at Band A is influenced by deterioration. The ΔT_t for configuration Θ_2 is increased by 40% compared to the new engine. Configuration Θ_3 shows a nearly unchanged ΔT_t. Furthermore, the simplified combined engine overestimates the overload ΔT_t with 30% compared to Θ_6. This deviation is caused by interaction between the different module deterioration which is neglected in the simplified approach.

While the EGT increases with the degree of deterioration, the difference in rotational speeds between the two operating points has shown a different performance. On one hand the deterioration of HPT Θ_3 shifts the operating points to lower rotational speeds and decreases the operating range ΔN_2 compared to the new engine. On the other hand, configuration Θ_2 and Θ_6 have shown an increase in the rotational speed at the operating points and an increase in the operating range ΔN_2.

In the Fig. 6b the HPC surge margin is shown over the rotational speed N_2. Here, the surge margin SM is defined as below:

$$SM = \frac{\pi_{SL}}{\pi_{OP}} \cdot \frac{\dot{m}_{OP}}{\dot{m}_{SL}} - 1 \text{ with } \dot{m} = const. \tag{15}$$

In contrast to the EGT, the deterioration has positive and negative effects on the surge margin. The deterioration of the HPC in Θ_2 results in a shift of steady-state operation point and minimum surge margin to the lowest margin (0.12) among all configurations. In configuration Θ_3 the surge margin is shifted to higher values for all operating points. Configuration Θ_6 with the highest degree of deterioration produces a surge margin behaviour in between the configuration of isolated deteriorated modules. In addition, the configuration Θ_2 and Θ_3 shows an increase in ΔSM com- pared to the new engine. The configuration Θ_6 causes a decreased ΔSM but less than the configuration with single deterioration. ΔSM is overestimated by the simplified combined engine Γ of 7% compared to Θ_6.

3.3.3 Sensitivity Analysis

In order to evaluate the influence and interactions of combined module variances on the overall performance, a global variance-based sensitivity study is carried out (Saltelli et al. 2010). Therefore, the results of ASTOR are used to determine the first and total indices. These indices can be used to identify not only the direct effects of particular inputs on the variances of output variables (first order S_y), but also any potential effects due to interaction between the variances of the inputs and the outputs (total order S^T).

Here, the considered steady-state output variables are T_{t3}, T_{t5}, N_1, N_2, p_{t3} and specific fuel consumption (SFC). For the sake of simplicity, only the 5 highest values are considered. In Fig. 7 the first (dark) and total (bright) order indices are presented for Band D in green and Band A in red.

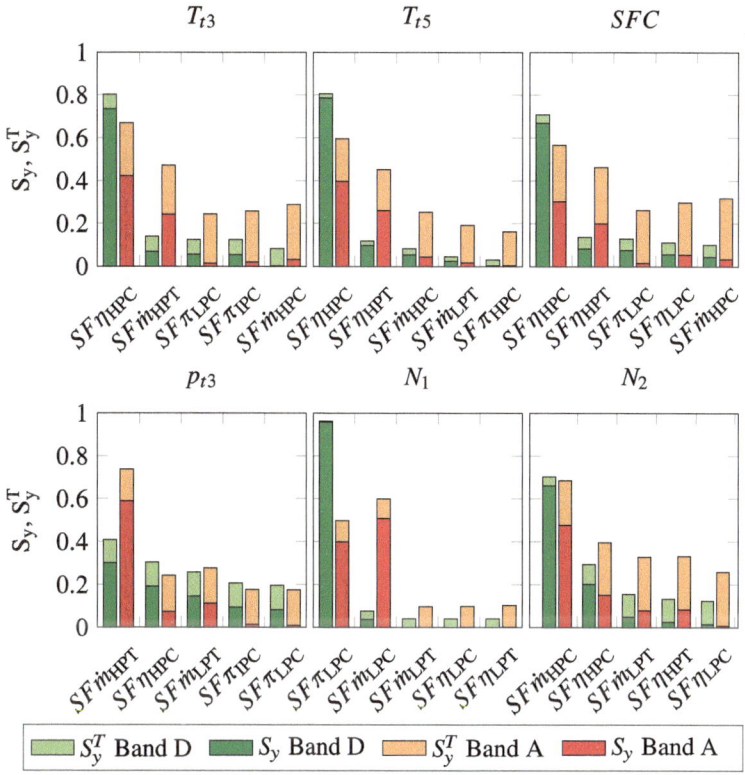

Fig. 7 First S_y and total S^T effect order Kucherenko indices for thrust level band D and band A

The main impact on the temperatures T_{t3} and T_{t5} is from the HPC efficiency at both operating points, which is quantified by a first index value of 0.42 and total index of 0.67. At the operating point Band D the sensitivity is increased to a first order index of 0.74 and total index of 0.8. The scaling factor with the second highest sensitivity for the temperature T_{t3} is the mass flow rate of the HPT and for T_{t5} it is the HPT efficiency. The temperature T_{t3} is influenced by the HPT mass flow as a consequence of a lower HPC pressure ratio. In addition, a higher temperature T_{t3} is caused by a lower efficiency of the compression. The decrease in HPT efficiency causes a higher temperature T_{t5}, which in terms results from a higher fuel flow to provide the required power to the compressor. Similarly, the decrease of HPC efficiency leads to an increase of the downstream temperature T_{t5}. Both types of deterioration cause an increase of the steady-state temperature T_{t5}, as described in Fig. 6. The SFC has almost the same sensitivities as the temperature T_{t5}. As a result of the decrease in HPT efficiency, a higher fuel flow is required to provide the power to the HPC. A deteriorated HPC requires more power input which causes an additional increase in fuel flow. This results in an further rise of the temperature T_{t5} at steady-state, as determined before. The pressure p_{t3} is most sensitive to the HPT mass flow rate. The first order index is determined to be 0.6 for Band A and 0.3 for Band D. The sensitivity of the HPC efficiency is the second highest. The HPC is shifted to lower pressure ratios due to the increased mass flow rate of the HPT. A reduced HPT efficiency provides less power to the HPC which causes a decrease in corrected mass flow and pressure ratio. The main effect on the low pressure spool speed N_1 is caused by the influence of the LPC mass flow and pressure ratio. All remaining scaling factors have no significant impact on N_1. The high pressure spool speed N_2 is mainly influenced by the mass flow of the HPC, which is shown in Fig.6. With the sensitivity analysis it can be determined that the increase of rotational speed is caused by the HPC mass flow. The HPC responds to a decrease in HPC mass flow with a shift of the steady-state operating points to higher rotational speeds. One reason for the decreasing sensitivity of the HPT and the increasing sensitivity of the HPC for Band D in comparison to Band A is due to the higher degree of deterioration of the HPC compared to the HPT at this operating point.

In Fig. 8 a global sensitivity analysis of the transient performance is presented. Therefore, the sensitivity of the temperature loads ΔT are determined. Furthermore, the impact of HPC surge margin and acceleration dN/dt are investigated. In general, all investigated temperature loads are mainly influenced by the HPC efficiency. The first order effect is between 0.7 and 0.8 for all ΔT. The remaining scaling factors show a minor impact on the temperature loads. However, the total order index of 0.2 indicates strong interaction effects. In the investigation of isolated deterioration in Fig. 6, it is shown that the HPC has a significant influence on the temperature load while the HPT has got a minor impact. The increased temperature load is caused by the effect of the deteriorated HPC on the acceleration of the high pressure spool speed. Due to the loss in HPC efficiency, a higher power input of the HPT is necessary. This results in a lower acceleration at the beginning of the transient manoeuvre. As a consequence of the lower acceleration of N_2, the temperature load increases. Because of the higher temperature load, the HPT provides more power, which allows a higher acceleration rate and thus causes the increase in ΔN_2.

The minimum surge margin of the HPC is influenced significantly by different scaling factors. The main impact is caused by the mass flow of the HPT, with a first order index of 0.3 and the HPC efficiency, whose first order index amounts to 0.2. The existence of interactions for the five scaling factors are indicated by the total order indices, which are significantly larger than the first order indices. Similar to the first order index, the total order indices indicate that the surge margin is mainly influenced by interactions that result from changes in HPT mass flow rate and HPC efficiency. With the results of the sensitivity analysis, it is observed that the increased HPT mass flow has a positive influence on the surge margin. The HPC operating point is shifted to lower pressure ratios due to an increase of the HPT mass flow, which results in a higher surge margin (see Fig. 6). In contrast to the HPT mass flow, the surge margin is decreased the most when the HPC efficiency is deteriorated, which is mentioned in the configuration Θ_2. This decreased surge margin is caused by the lower acceleration during the beginning of the transient maneuver. Mostly, the

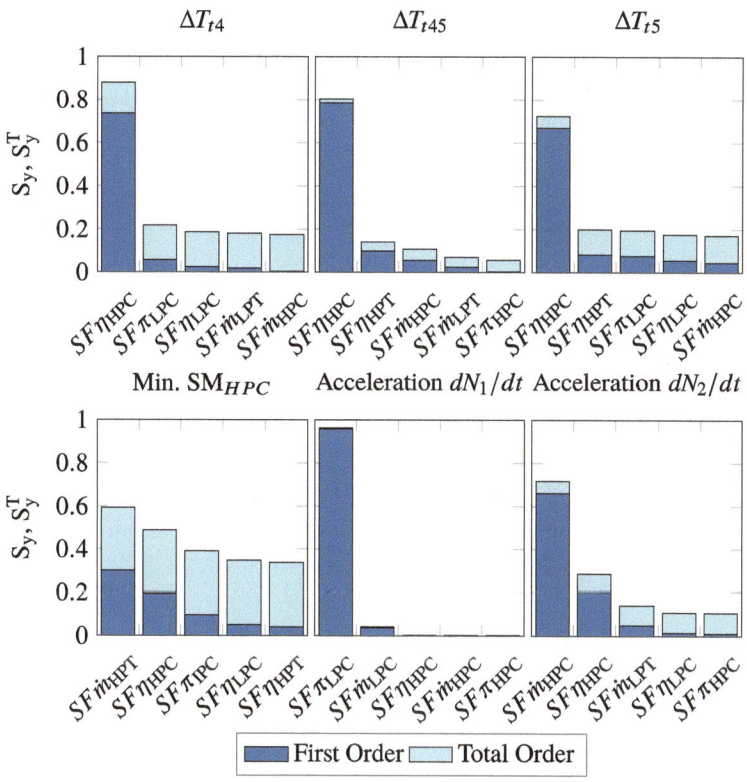

Fig. 8 First S_y and total S^T effect order Kucherenko indices for transient performance. ΔT_t according to definition in Fig. 6a

acceleration of the LP-spool is influenced by the pressure ratio of the LPC. The first order and the total order indices have reached a value of 0.95.

The rotational speed N2 is significantly influenced by the scaling factors compared to the low pressure spool speed. The acceleration of the spool is significantly influenced by the mass flow of the HPC. The first order index is calculated to be 0.65 and the total index to 0.7. Similar to the surge margin, the acceleration is also influenced by the efficiency of the HPC. However, the indices have shown that the influence of the HPC efficiency is less significant with a first order index of 0.25 and a total order index of 0.3. The sensitivity of the acceleration of the HP spool is shown in Fig. 6. The HPC rises the rotational speed between the two operating points. In conclusion it is observed that the investigated steady-state and transient performance are mostly influenced by HP-system.

4 Conclusions

A virtual twin of the IFAS V2500-A1 research aircraft engine was developed to investigate the interaction of combined module variances and their influence on the overall system performance in the sub-project D6 of the Collaborative Research Centre 871. For this task, the aircraft engine was transferred into a quasi 1D environment of the Pseudo Bond Graph notation in order to simulate the non-linear system dynamics in higher order. The virtual twin was validated with the IFAS research aircraft engine and the performance could be predicted with high accuracy (see Fig. 3a). In addition, ASTOR is compared with conventional performance analysis methods and it was found that the dynamic model leads to an improvement of the transient performance simulation capabilities by up to 8% in the surge margin during a slam acceleration (see Fig. 1.4), due to the volume dynamics and the interaction effects. Based on the validated virtual twin, a design of experiments was performed to investigate the sensitivities for steady-state and transient performance of combined module variances in a turbofan aircraft engine. The sub-project D6 showed that:

1. The qualitative and quantitative relationship between module variances and performance output strongly depends on the operating point (see Table 2 and Fig. 7).
2. The impact of a module with geometrical variances on the aircraft engine performance depends on the condition of the remaining modules (see Figs. 6 and 7).
3. The high-pressure system dominates the sensitivities in the steady-state and transient performance (see Figs. 7 and 8).
4. Strong interaction effects have been detected, especially due to the low-pressure system (see Figs. 6, 7 and 8).

These results can be used to improve detection and understanding of the sensitivities of the overall engine performance to the module variances, which is an essential field in production, maintenance, repair, and overhaul (MRO) as well as in the operation of an aircraft engine. The virtual twin and the results obtained from the sensitivity analysis are used in the virtual process from Fig. 1 to evaluate the scanned high-pressure turbine blades. In general, it is possible to integrate such a performance assessment analysis using the virtual twin for a condition-based decision making on component regeneration. This knowledge and the method of sensitivity analysis will become more important for modern aircraft engines due to more complex system architectures. In addition, recognising the interaction between low-pressure and high-pressure systems will become increasingly important with the progress towards ultra-high-bypass ratio aircraft engines.

Acknowledgements The authors gratefully acknowledge the funding as part of the Collaboration Research Centre 871 provided by the Deutsche Forschungsgemeinschaft (DFG, Ger- man Research Foundation)—SFB 871/3—119193472. Many thanks are also to Joern Staeding and Jonas Marx and MTU Maintenance Hannover GmbH for the continuous support.

References

Aschenbruck,J., Adamczuk, R., and Seume, J. R. (2014). Recent progress in turbine blade and compressor blisk regeneration. 22:256–262.

Fentaye, A. D., Baheta, A. T., Gilani, S. I., and Kyprianidis, K. G. (2019). A review on gas turbine gas-path diagnostics: State-of-the-art methods, challenges and opportunities. *Aerospace*, 6(7):83.

Fiola, R. (1993). *Berechnung des instationaeren Betriebsverhaltens von Gasturbinen unter besonderer Beruecksichtigung von Sekundaereffekten*. PhD thesis.

Goeing, J., Bode, C., Friedrichs, J., Seehausen, H., Herbst, F., and Seume, J. R. (2020a). Performance simulation of roughness induced module variations of a jet propulsion by using pseudo bond graph theory. In *Turbo Expo: Power for Land, Sea, and Air*, volume 84058, page V001T01A014. American Society of Mechanical Engineers.

Goeing, J., Hinz, L., Lueck, S., Bien, M., and Friedrichs, J. (2022a). Interaction of combined module variances and influence on the overall performance of a turbofan engine. *ISABE 2022 (accepted), No. 167*.

Goeing, J., Hogrefe, J., Lück, S., and Friedrichs, J. (2020b). Validation of a dynamic simulation approach for transient performance using the example of a turbojet engine. In *STAB/DGLR Symposium*, pages 559–568. Springer.

Goeing, J., Kellersmann, A., Bode, C., and Friedrichs, J. (2018). System dynamics of a single-shaft turbojet engine using pseudo bond graph. pages 427–436.

Goeing, J., Kellersmann, A., Bode, C., and Friedrichs, J. (2019a). Jet propulsion engine modelling using pseudo bond graph approach. In *Turbo Expo: Power for Land, Sea, and Air*, volume 58547, page V001T01A007. American Society of Mechanical Engineers.

Goeing, J., Lueck, S., Bode, C., and Friedrichs, J. (2019b). Simulation of the impact of a deteriorated high-pressure compressor on the performance of a tur- bofan engine using a pseudo bond graph modelling approach. *Global Power and Propulsion Society GPPS-BJ-2019-0160*.

Goeing, J., Seehausen, H., Pak, V., Lueck, S., Seume, J. R., and Friedrichs, J. (2020c). Influence of combined compressor and turbine deterioration on the overall performance of a jet engine using rans simulation and pseudo bond graph approach. *Journal of the Global Power and Propulsion Society*, 4:296–308.

Goeing, J., Seehausen, H., Stania, L., Nuebel, N., Salomon, J., Ignatidis, P., Dinkelacker, F., Beet, M., Denkena, B., Seume, R. J., and Friedrichs, J. (2022b). Interaction of combined module variances and influence on the overall performance of a turbofan engine.

Kellenbrink, C., Nübel, N., Schnabel, A., Gilge, P., Seume, J. R., Denkena, B., and Helber, S. (2022). A regeneration process chain with an integrated decision support system for individual regeneration processes based on a virtual twin. *International Journal of Production Research*, pages 1–22.

Kurzke, J. and Halliwell, I. (2018). *Propulsion and power: an exploration of gas turbine performance modeling*. Springer.

Kypuros, J. A. (2003). A reduced model for prediction of thermal and rotational effects on turbine tip clearance. 212226.

Lück, S., Wittmann, T., Goeing, J., Bode, C., and Friedrichs, J. (2022). Impact of condensation on the system performance of a fuel cell turbocharger. *Machines*, 10(1):59.

Markou, C. and Cros, G. (2021). Airline maintenance cost executive commentary: Fy2019 data.

McKay, M. D., Beckman, R. J., and Conover, W. J. (2000). A comparison of three methods for selecting values of input variables in the analysis of output from a computer code. *Technometrics*, 42(1):55–61.

Müller, M. H. (2013). Untersuchungen zum einfluss der betriebsbedingungen auf die schädigung und instandhaltung von turboluftstrahltriebwerken.

Salomon, J., Göing, J., Lück, S., Broggi, M., Friedrichs, J., and Beer, M. (2021). Sensitivity analysis of an aircraft engine model under consideration of dependent variables. In *Turbo Expo: Power for Land, Sea, and Air*, volume 84898, page V001T01A005. American Society of Mechanical Engineers.

Saltelli, A., Annoni, P., Azzini, I., Campolongo, F., Ratto, M., and Tarantola, S. (2010). Variance based sensitivity analysis of model output. design and estimator for the total sensitivity index. *Computer physics communications*, 181(2):259–270.

Saravanamuttoo, H. I., Rogers, G. F. C., and Cohen, H. (2001). *Gas turbine theory*. Pearson Education.

Shampine, L. F., Reichelt, M. W., and Kierzenka, J. A. (1999). Solving index-1 daes in matlab and simulink. *SIAM review*, 41(3):538–552.

Spieler, S., Staudacher, S., Fiola, R., Sahm, P., and Weißschuh, M. (2008). Probabilistic engine performance scatter and deterioration modeling. *Journal of engineering for gas turbines and power*, 130(4).

Stephan, P., Kabelac, S., Kind, M., Mewes, D., Schaber, K., and Wetzel, T. (2019). *VDI-Wärmeatlas: Fachlicher Träger VDI-Gesellschaft Verfahrenstechnik und Chemieingenieurwesen*. Springer-Verlag.

Volponi, A. J. (2014). Gas turbine engine health management: past, present, and future trends. *Journal of engineering for gas turbines and power*, 136(5).

Wahler, N., Radomsky, L., Hanisch, Lucas Goeing, J., Meyer, P., Mallwitz, R., Friedrichs, J., Henke, M., and Elham, A. (2022). An integrated framework for energy network modeling in hybrid-electric aircraft conceptual design. *AIAA 2022, No. 3688106*.

Weiner, M. (2015). Aeroreport: How an engine is maintained.

Process Chain for Condition-Based Regeneration

Nicolas Nuebel, Joerg R. Seume, and Berend Denkena

Abstract The regeneration of complex capital goods is dominated by manual labor and experience-based decisions about machining measures. The functional condition, which describes how well a part performs its task, is unknown before and after regeneration. Instead, the target geometries for the regeneration are based on continuity constraints and subjective assessments. In this study, a new type of process chain for condition- based regeneration is developed that selects the machining measures based on the achievable functional condition, the production effort as well as customer requirements. This allows a targeted and efficient decision on a regeneration process based on the customer requirements for the functional condition. A proof-of-concept process chain was successfully built to demonstrate the feasibility of this novel approach. It was shown that the concept is realizable with currently available technologies and that the functional assessment has minor influence on the regeneration time.

Keywords Digital twin · Repair · Overhaul · Regeneration · Process chain · Functional evaluation · High pressure turbine blade

N. Nuebel · B. Denkena (✉)
Institute of Production Engineering and Machine Tools, Leibniz University Hannover, An der Universitaet 2, 30823 Garbsen, Germany
e-mail: Denkena@ifw.uni-hannover.de
URL: https://www.tfd.uni-hannover.de/

J. R. Seume
Institute of Turbomachinery and Fluid Dynamics, Leibniz University Hannover, An der Universitaet 1, 30823 Garbsen, Germany
e-mail: Seume@tfd.uni-hannover.de
URL: https://www.ifw.uni-hannover.de/

445

1 Introduction

Currently, decisions about the repair measures in process chains for the regeneration of complex capital goods are often experience-based or dependent on geometric dimensions. Through manual machining with belt grinders or rotary tools, the skill and experience of the employee affect the result of the regeneration. Furthermore, the decisions are not always understandable, and the same good can be regenerated differently.

This results in different functional conditions and production efforts for each part, dependent on the executing employee (Tsong-Jye Ng et al. 2004).

One example of complex capital goods is aircraft engines. These engines are inspected and disassembled at fixed intervals, and damaged parts are repaired or replaced. Due to the high cost of new parts, regeneration is an economical alternative to replacement (Aschenbruck et al. 2014).

The typical regeneration process of aircraft engines starts with the disassembly, a cleaning process, and an inspection. Besides a visual inspection, computed tomography scans and x-ray inspection are used (Raj et al. 2000). Regulations then specify which damages are repairable. Generally, small dents can be removed by blend repair (N3Engine Overhaul Services 2011). For material deposition, soldering and welding are primarily used. In addition to the complete additive build-up of damaged areas of the blade, additional a patch of the same material can be welded on the blade (Eberlein 2007). Excess material is mostly machined manually either with rotary tools or with a belt grinder (Yilmaz et al. 2005). If the recontouring is automated, industrial robots in combination with a belt grinder or other machine tools are often used (Huang et al. 2002).

The automation of process chains for the regeneration of individual process steps is the focus of different research projects (Denkena et al. 2015; M'Saoubi et al. 2015). Besides an increase in productivity, quality, or repeatability, the emphasis is on substituting human workers with machines. Without automation and machine tools, or the optimization of machining processes themselves, an efficient regeneration process remains unachievable. The reason for this is that customers are primarily interested in the functional condition of the whole engine and not in compliance with geometric tolerances. Since the functional condition of the blade and its influence on the functional condition of the engine is unknown, an efficient and functional condition-oriented regeneration is not possible with the currently available technologies and process chains.

2 Objective

The objective of subproject S: "system demonstrator" of the collaborative research centre (CRC) 871, was to develop a process chain for condition-based regeneration and prove the feasibility using the example of high-pressure turbine blades. To achieve this, research results of nearly all subprojects of the CRC were integrated into something we called "system demonstrator". Besides the proof-of-concept of condition-based regeneration, the goal was to transfer the research results into the industry and give new impulses for future production systems.

In this new type of process chain, decisions about the machining measures are based on the customer's requirements on the functional condition, considering the production effort. The hypothesis is that through the integration of functional assessment and manufacturing simulations into an automated process chain it is possible to determine the feasible functional condition before the machining is executed. These high-fidelity functional and manufacturing simulations, which are computationally expensive and require extensive manual set-up efforts, are only used in product development.

In chapter "Multiscale Measurement of Blade Geometries with Robot-Supported, Laser-Positioned Multi-Sensor-Techniques", the structure, all components of the newly developed process chain, and the regeneration procedure are described. In chapter "Exhaust Jet Analysis", the condition-based approach is exemplified with one regeneration, and a summary of recommendations for industrial applications is provided.

3 Structure of the Condition-Based Process Chain

In the second funding period of CRC 871, a structure for a condition-based regeneration was developed (Fig. 1) (Aschenbruck et al. 2014). It consists of two layers, the real and virtual layers. The real layer includes measuring and inspection cells, as well as machine tools for automated machining. A virtual layer is included in the process chain to calculate the functional condition. Besides the functional assessment, manufacturing simulations and job scheduling are integrated through this layer. To connect both layers, a digital workpiece twin is used as the sole interface for the data.

At the beginning of the process chain, the blade is disassembled and inspected in the real layer. After a preliminary decision in which the repairability of the blade is determined, the functional condition of the operationally stressed blade is assessed.

Machining is divided into different paths. For each path, manufacturing simulations (process simulations) are used to calculate the changes in the geometry, the process times, and the manufacturing costs. In the functional assessment, these changes in the geometry are used to determine the achievable functional condition. In a subsequent evaluation, the achievable functional condition, process times, and process cost for all paths are compared with the customer's requirements. The path that best fits the customer's requirements is selected and executed in the real layer. Afterward, the engine is reassembled, and an acceptance test is carried out.

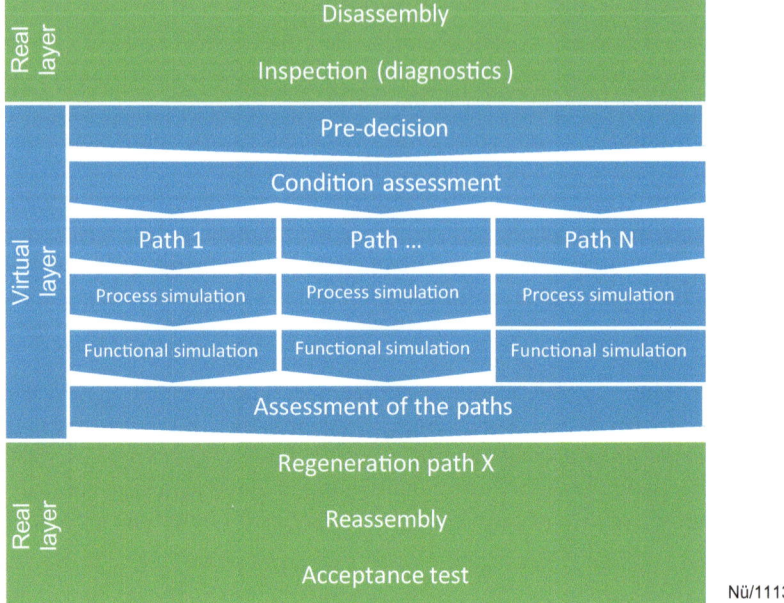

Fig. 1 Overview of the process chain developed by the CRC 871 (Kellenbrink et al. 2022)

3.1 Flexible Manufacturing System for the Condition-Based Regeneration

For the proof-of-concept of condition-based regeneration, a system that can integrate all of the regeneration necessary process technologies is built. The manufacturing processes are automated to better predict the results. This reduces the variances in machining times. Due to the high individuality of the operationally stressed components and the resulting variances, the machining times and necessary process steps vary for each workpiece. A fixed sequence of cells and machines is, therefore, not recommended. Also, some machines are used more than once in the regeneration process. For example, a milling center is used in preparation for welding and recontouring.

To achieve maximum flexibility in the sequence of processes, a flexible manufacturing system is chosen as a structure of the process chain for condition-based regeneration. For the handling of the workpiece, a combination of zero-point clamping modules in the machines with a workpiece carrier and a mobile handling system (MHS) is selected (Fig. 2). The MHS can transport up to seven workpiece carriers in the mobile storage and load/ unload the machines with a serial robot. It can move freely between the cells and needs no fence or barrier. Each process is executed in an individual cell, allowing the use of standard machines, e.g., milling machines or measuring cells. Expensive specialized machines can thus be avoided.

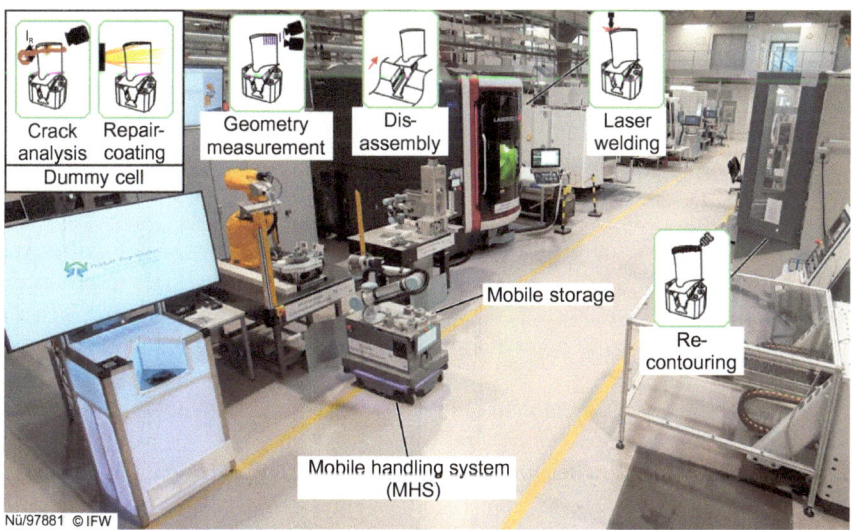

Fig. 2 Overview over the real layer (Kellenbrink et al. 2022)

In the disassembly cell, a piezo stack actuator is used to generate dynamic impacts on the mounted blade. This can reduce the force necessary compared to a static force (Mullo et al. 2019). For measuring the geometry of the blade, a robot-based fringe projection is used. Besides this fringe projection which measures the macroscopic geometry, other measuring methods are also included. For example, a low coherence interferometer (LCI) for surface roughness, or a bidirectional reflectance distribution function (BRDF) sensor to detect burns. The measuring data of all methods can be combined through hand-eye calibration (Betker et al. 2020).

High-frequency induction thermography is used to detect cracks and other defects under the coating. A robot moves an inductor to different positions next to the stationary blade and initiates an eddy current field. Cracks constrict the effective current field, which leads to a local heat-up. A thermography camera that can be moved on 4-axis detects these small heat-ups (Bruchwald et al. 2016). For recoating after the machining, a robot-based thermal spraying process is used. Through the combined application of nickel-based filler metal needed for filling small cracks and dents, and hot gas corrosion protective layer, the coating process chain can be shortened. Because of high logistical constraints, the crack analysis and the repair coating cannot be integrated directly into the system and are replaced with a dummy cell. Through this, the process can be integrated into the process chain without a dedicated cell.

To build up material, a 5-axis laser welding machine is selected because it allows for a targeted and controlled heat input. Additional material is fed into the focus point of the laser in the form of powder through a coaxial nozzle (Kaierle et al. 2017). The machining of the blade is carried out with a 5-axis milling machine. The blade preparation for welding is executed with a flat end mill and a 2.5-axis process.

The recontouring process to remove excess material after welding utilizes a ball end mill and a 5-axis process (Denkena et al. 2021a).

3.2 Digital Workpiece Twin—Universal Data Interface

To connect the real and virtual layers, a digital workpiece twin is used to store and transfer all data (Denkena et al. 2019). Although user interfaces in productive environments require individual configuration, for demonstration purposes a case in point was implemented to point out the potential of using a digital workpiece twin (Fig. 3).

The collection of different data is structured and organized through a hierarchic structure—comparable to file systems with folders and files. All elements can be enriched with arbitrary metadata, which in this case is transcribed by a list of pairs of descriptors and values. The grouping nodes can thus define collections and describe their shared attributes, while file nodes store concrete data, such as NC files, point clouds, pictures, or graphs.

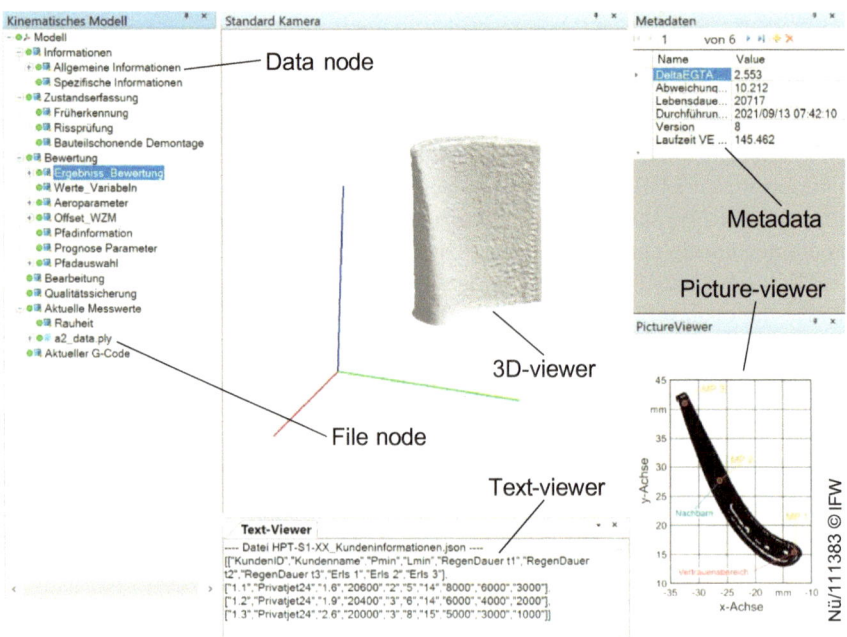

Fig. 3 Viewer of the digital workpiece twin (Denkena et al. 2021b)

3.3 Virtual Layer—Integration of Functional and Process Simulations

Direct integration of high-fidelity simulations into the process chain is not feasible due to the high setup effort, and long calculation times. Instead, the functional simulations are carried out outside the process chain based on a parameter space of a random sample of blades. The results are stored in look-up tables to reduce the calculation times.

In the virtual layer of the process chain, the point cloud of the operationally stressed blade is first read out of the digital workpiece twin and parameterized (Fig. 4). Besides the height of the blade, parameters such as the leading and trailing edge radius, chord length, stagger angle, and others are determined in 20 cutting planes (Stania and Seume 2022).

These parameters are transferred to the 'lifetime evaluation' where the look-up tables are used to calculate the remaining lifetime of the blade in flight cycles. The same procedure is carried out in the 'performance evaluation'. The 'exhaust gas temperature' (EGT) is selected to measure the performance. An increase in the EGT corresponds with the deterioration of the performance of the turbine blades.

Besides the evaluation, the parameters are also used in the 'Forecast'. This module first predicts the changes in the parameters through the machining for each path. Then, the parameters resulting from the machining are used with the performance and life evaluation to determine the achievable functional condition. The 'Job Scheduling', based on the achievable functional condition for each path in combination with the 'Customer Requirement', selects the path that best fits the customer (Kellenbrink et al. 2022).

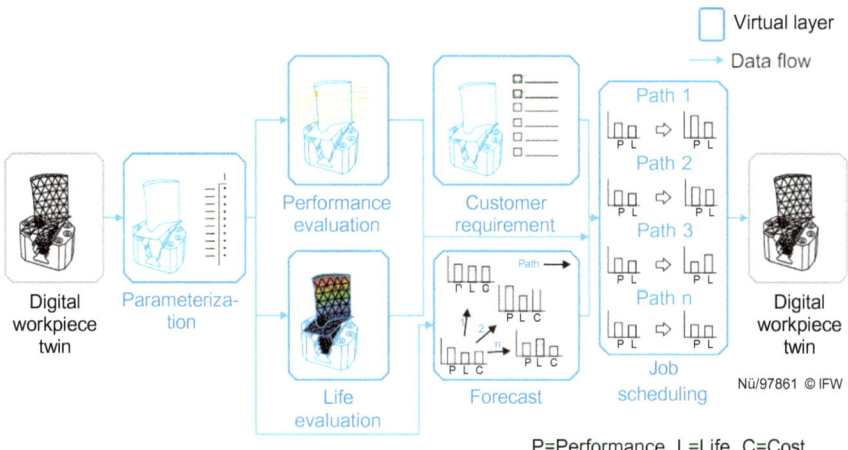

Fig. 4 Overview over the virtual layer (Kellenbrink et al. 2022)

3.4 Regeneration Process

The regeneration process is divided into three phases. First is the assessment and evaluation of the operational stressed blade. The second is machining. The third is quality control. In the assessment and evaluation, the first step is to disassemble the blade from the turbine disk and insert it into the workpiece carrier. Measured forces and used frequencies in the piezo stack actor are stored in the digital workpiece twin after the process is finished, as shown in Fig. 5.

If a crack is detected during crack analysis, the position and length are stored in the digital workpiece twin. After this inspection, the geometry measurement scans the blade in 13 different poses and creates a point cloud. Crack information and the point cloud are then used in the functional evaluation to predict the achievable functional condition for each path. Additionally, the process cells and times are also stored. The job scheduling reads this information out of the digital workpiece twin from all blades in the system. It optimizes the process sequence of all blades to minimize waiting times. Also, a path is selected for each blade that finished the functional evaluation. For this, the global profit, which is defined as the sum of the profit of all blades, is used.

To calculate the global profit, a heuristic approach is chosen that iterates through thousands of different work schedules. If no improvement between the iterations is achieved for 10 s time or the time limit of 45 s is expired, the algorithm stops, and the work schedule with the highest global profit is selected (Kellenbrink et al. 2022). For each blade, the selected path and all other information regarding the job schedules are stored in the digital workpiece twin.

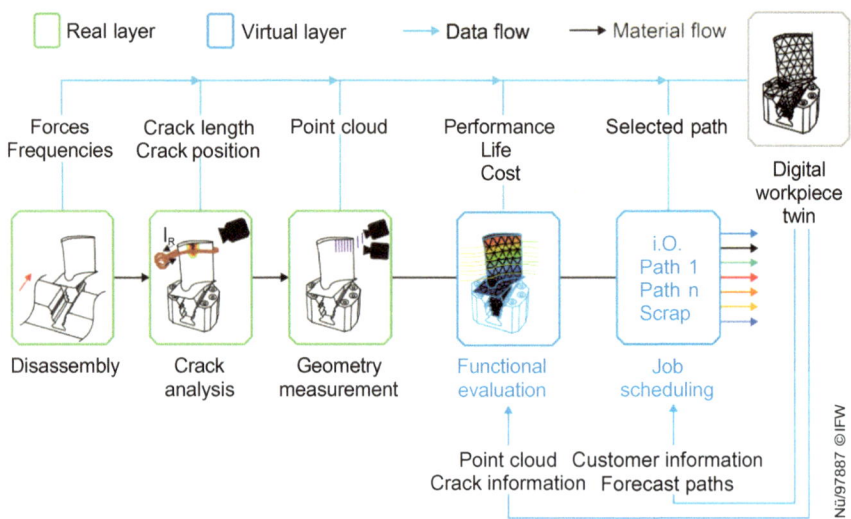

Fig. 5 Assessment and evaluation (Nübel and Denkena 2022)

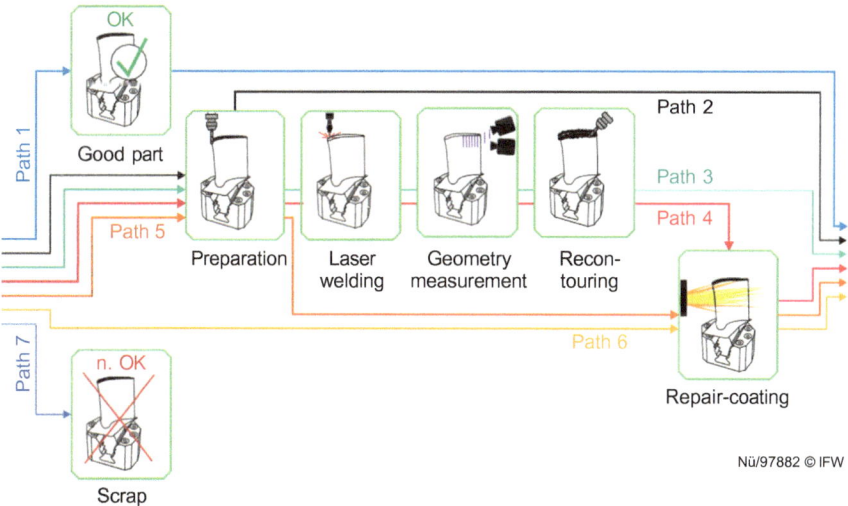

Fig. 6 Different paths for the machining (Kellenbrink et al. 2022)

The machining can be grouped into paths based on the machining technologies involved. Through a variation in the process parameters, for example the feed rate or the cutting depth, subvariants of a path can be created. Figure 6 shows the different paths available for selection.

If the blade has no critical damage, it can be reinstalled in path 1. If safety-relevant damages occur, this is no longer an option. In path 2, the tip is machined, and the blade is reinstalled afterward. This decreases performance and lifetime, but it is fast and cheap. Path 3 builds up a new tip on the machined surface. Because of the low accuracy of the laser welding process, recontouring is necessary to achieve the desired geometry. Path 4 adds a repair-coating process which can close small cracks or fill in dents. This process is also possible after the removal of the tip (path 5) or without machining (path 6). If the blade has damage that is not technologically or economically repairable, it can be scraped, and a new or refurbished blade will be installed instead in path 7.

After the machining, quality control ensures that only successfully repaired blades exit the process chain. The sequence of processes is shown in Fig. 7. After the blade is inspected for cracks and the regenerated geometry is measured, the functional evaluation calculates the after the machining achieved functional condition and compares this with the forecast. The difference is then used in a quality control loop to improve the forecast.

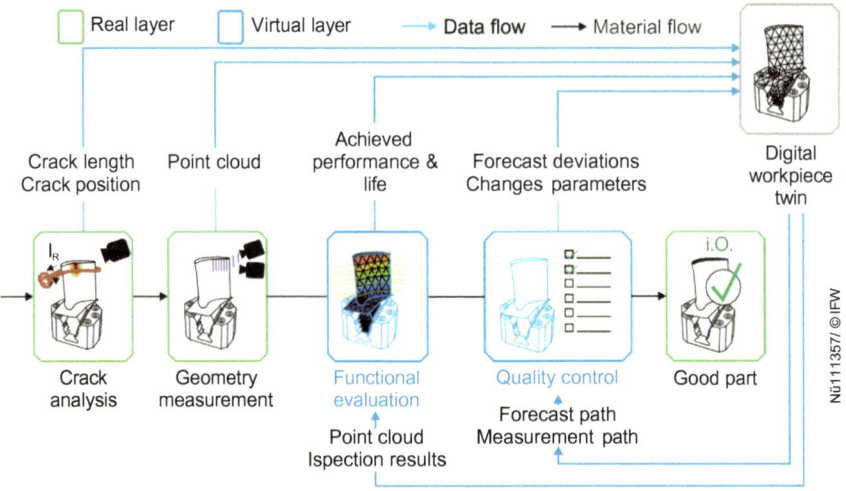

Fig. 7 Sequence of the quality control (Nübel and Denkena 2022)

4 Execution of the Condition-Based Regeneration

Having description of the process chain, the regeneration of a high-pressure turbine blade is used to exemplify the feasibility of the condition-based regeneration approach. For this, a customer with three different quality levels for the functional condition of the blade is modeled (Table 1). It also allows the option to consider individual customer priorities dependent on the business model, e.g., cargo or passenger flights.

The willingness to pay is modeled with a linear interpolation between three support points for each quality level. In addition, the willingness to pay is dependent on the duration of the regeneration. This is the time from the insertion of the blade into the process chain up to the reassembly. Before the first support point, the willingness is static. After the third point, the willingness drops directly to 0 € (Fig. 8). In this proof-of-concept, the performance, lifetime, and regeneration time are only considered for one blade. In an industrial application, these factors should be considered for the whole engine.

Table 1 Customer requirements for the functional condition

Name/ID	Quality level	Performance (EGT)	Lifetime
		K	Flight cycles
Customer 1.A	A	<2,5	>20.200
Customer 1.B	B	<3,5	>19.500
Customer 1.C	C	<5,0	>19.000

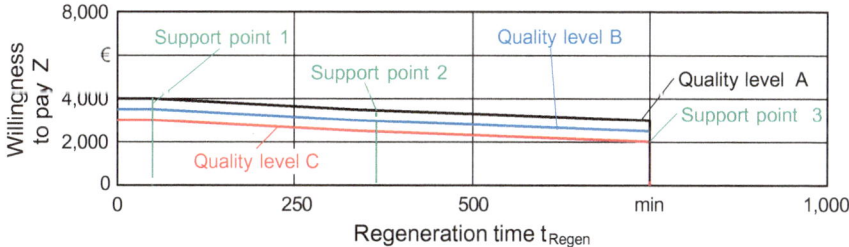

Fig. 8 Willingness to pay over the regeneration time (Nübel and Denkena 2022)

4.1 Assessment and Evaluation

After disassembly, a serial robot in the disassembly cell inserts the blade into the workpiece carrier, and the MHS transports it to the crack inspection. The results of the crack inspections are a thermography image of the blade (Fig. 9) and a table with all crack positions and lengths. Both the table and the image are stored in the digital workpiece twin.

After the inspection, the blade is scanned in the geometry measurement and parameterized. Before the parametrization can determine the parameters, the point cloud of the geometry measurement needs to be processed. Because of the assembly tolerances of the workpiece carrier and the low absolute accuracy of the geometry measurement, the position and orientation of the blade vary in this coordinate system. The result is low repeatability of the parametrization. To increase the repeatability, the measured blade is first aligned to a reference blade.

For this alignment, an iterative close point algorithm is used. To increase the speed and the quality of the algorithm, the point cloud of the blade is filtered using a boundary box and downsampled. In Fig. 10, the unfiltered and filtered point clouds are shown. Because of the high wear of the tip, the upper part of the blade is filtered out due to a negative influence on the alignment. Color information of the blade is deleted to reduce the size of the file. For the subsequent parametrization, a 3D-Model

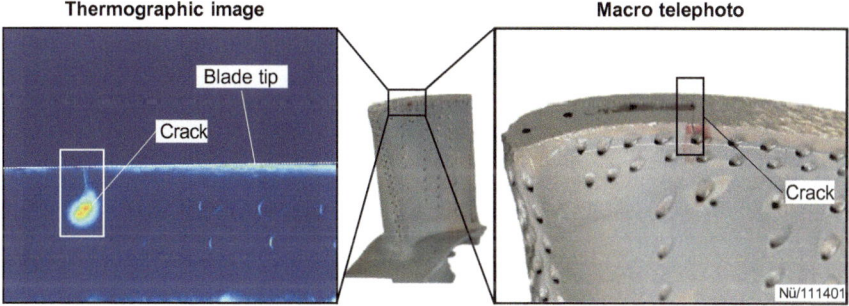

Fig. 9 Results of the crack inspection (Nübel and Denkena 2022)

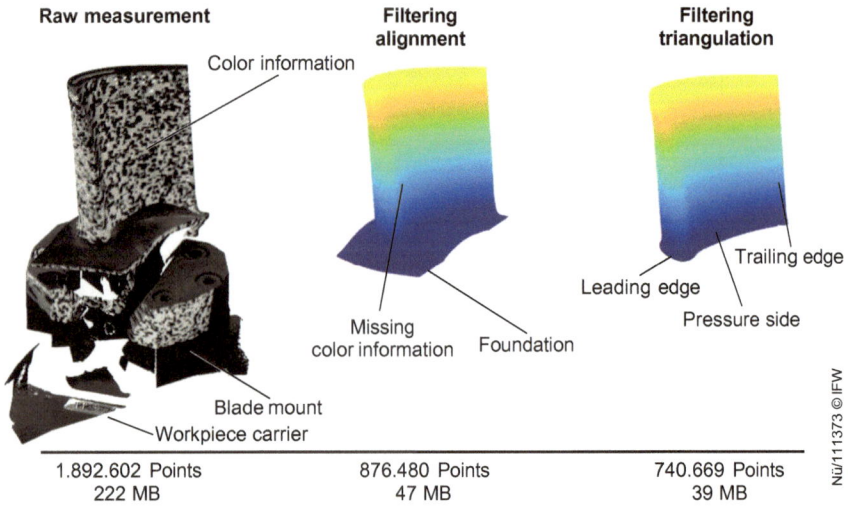

Fig. 10 Different filter steps of the point cloud (Nübel and Denkena 2022)

with surfaces is necessary instead of a point cloud. Therefore, the initial point cloud is moved using the calculated transformation matrix from the alignment. The moved raw measurement is filtered with a different boundary box to include the tip of the blade and exclude the foundation. This is because the foundation is not considered in the parameterization and leads to errors during the triangulation, which creates the 3D-Model with surfaces.

The height can be determined after the blade is aligned. Therefore, three measuring points are used, as shown in Fig. 11a. From these measuring points, an algorithm finds the closest 12 neighboring points. If the mean distance between the measuring points and the neighboring points is smaller than the illustrated confidence interval (red circle), the median z-height of all neighboring points is used to calculate the height of the measuring point. Then, the mean of all measuring points is used for the blade height. This height is subtracted from the channel height of the engine to get the tip clearance. This is the distance between the blade and the housing, which is an important parameter for functional evaluation.

For other parameters such as the leading and trailing edge radius, the chord length, or other aerodynamic parameters, 20 cutting planes are used. They are evenly distributed on the z-axis, as visualized in Fig. 11b. The repeatability of this parametrization is discussed in Table 2.

To increase the robustness of the system, the aligned point cloud is not directly triangulated for the parametrization, as this often leads to small holes or triangulation errors in the surfaces. Therefore, the aligned point cloud is filtered 600 µm above and below each cutting plane (z-axis). Every emerging point cloud (filtered cut) is triangulated independently using an envelope body. This body guarantees that all holes are closed, or small dents, chipped coating, or rough surfaces are filtered, which

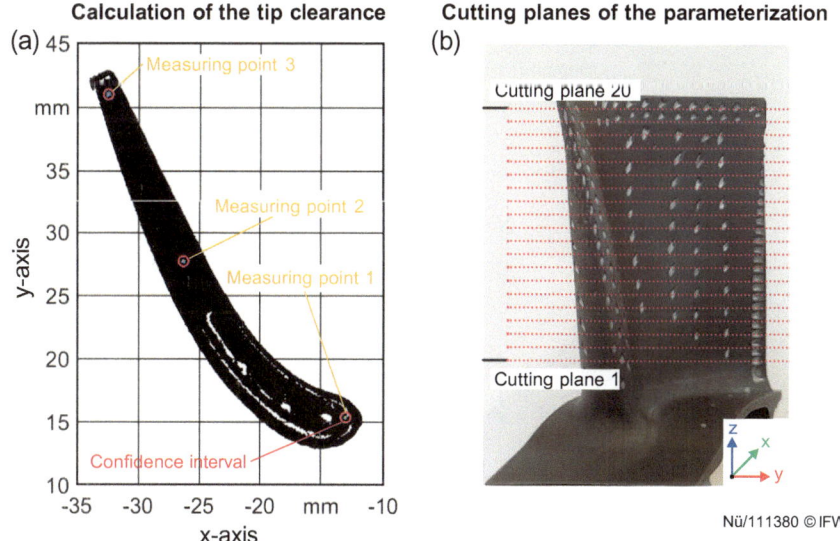

Fig. 11 Tip clearance calculation and cutting planes for the parametrization (Nübel and Denkena 2022)

Table 2 Range of the parameters

		Span leading edge radius	Span tip clearance
		μm	μm
RZPC	Unaligned	55	35
RZPC	Alignment	15	35
Reduction through alignment by		−54%	−0%
RWC	Unaligned	200	55
RWC	Alignment	35	30
Reduction through alignment by		−83%	−45%

increases the robustness of the system. The twenty filtered cuts around the cutting planes and the complete triangulated blade are shown in Fig. 12.

Especially very dark and dirty blades or blades with markings from felt pens would cause problems without this filter step. These surfaces do not reflect enough light, which leads to holes in the point cloud. The complete triangulated blade is saved in the digital workpiece twin, for example, for the control of the measurements by an employee.

The filtered cuts are then used in the parameterization. An algorithm forms the intersection of the filtered cut and a cutting plane to create a blade cut. With different geometric forms, for example, circles, triangles, or ellipses. The aerodynamic parameters are then determined (Fig. 13) (Stania and Seume 2022).

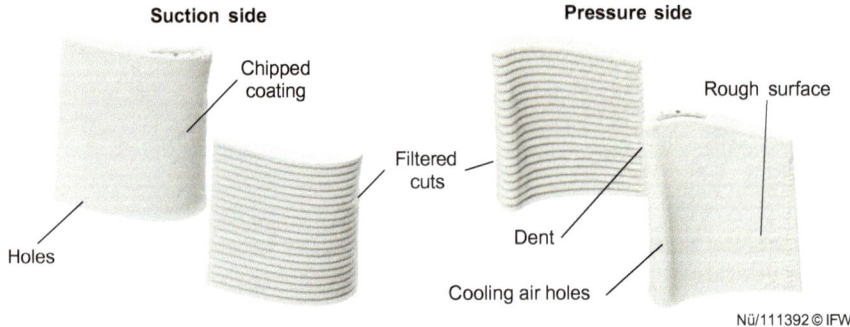

Fig. 12 Geometric models for the parameterization (Nübel and Denkena 2022)

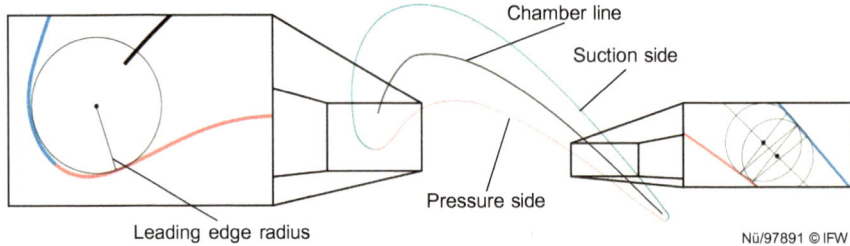

Fig. 13 Blade cut with geometric forms to determine different parameters (Nübel and Denkena 2022)

To evaluate the repeatability of the process, a blade is inserted into a workpiece carrier clamped in the zero-point clamping module. After one measurement, the workpiece carrier is lifted out of the clamping module and inserted again. This procedure is repeated five times while the blade remains in the workpiece carrier. The measurements are parameterized, with and without the alignment, and the range of the parameters is determined (RZPC). Additionally, the influence of the workpiece carrier on the parametrization is examined. For this purpose, the same procedure was used, with the exception that between measurements the blade was unclamped from the workpiece carrier and placed in another one (RWC). The results are shown in Table 2.

The alignment reduces the range of the parameters and thus the influence of the unavoidable measurement uncertainties, manufacturing, and assembly tolerances on the functional assessment. The achieved span is acceptable for the used measurement technologies but can be further reduced by using a coordinate measurement machine.

4.2   Functional Evaluation

For the calculation of the functional condition, FEM-simulations and CFD-simulations are used. These are normally only utilized in product development for an ideal geometry, requiring a manual setup and have long calculating times up to numerous weeks. To reduce the calculating and therefore waiting time in the process chain, these simulations are done in advance using a parameter model of the blade. The parameter space for this parametric simulation is derived from a random sample.

In the process chain, the parametrization-determined parameters are used in combination with the lookup table to get a quick assessment of the functional condition. Repeating calculations can also be avoided. If the parameters of a blade exceed the parameter space, it can be expanded outside the process chain.

The number of parameters considered to determine the functional state is arbitrary. The lookup table for the performance, for example, includes the leading-edge radius, tip clearance, and roughness. Between simulated support points, interpolation is performed to decrease the number of necessary simulations.

4.3   Forecast of the Achievable Functional Condition

For the selection of the regeneration measures, the functional condition of the operationally stressed blade is not relevant. Rather, the repair achievable one is. The forecast necessary for this is executed in a separate software module after the evaluation of the operationally stressed blade. First, each path is checked for its technical feasibility (Fig. 14). Afterward, process simulations predict the changes in the parameters for each path and the production effort. This can be divided into process times and cost. For each path with the changed parameters and the functional evaluation, the achievable functional condition is calculated and afterward stored in the digital workpiece twin.

For example, in Table 3 the forecast of a typical blade is shown. Paths 1–3 are not included since they are not able technically to regenerate all the damage. Due to a lack of space, all processes and process times have not been listed.

The job scheduling is executed after the forecast is finished. Because only one blade is in the system, it can directly calculate the duration of the regeneration without waiting times due to the limited capacity of the machines. The customer´s willingness to pay can be determined based on the achieved quality level of the path and the duration of the regeneration. If the costs are deducted from the willingness to pay, this results in a profit, as shown in Table 4. The path with the highest profit is selected, and a work plan is created. If more than one blade is in the system, the job scheduling selects the paths that lead to the maximum global profit. The selected path for each blade must not necessarily be the path with the highest profit due to the capacity limitations of the machines.

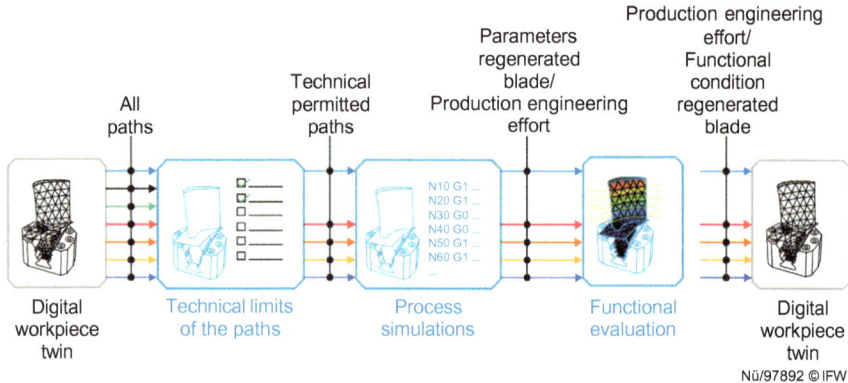

Fig. 14 Forecast of the achievable functional condition (Nübel and Denkena 2022)

Table 3 Forecast of the achievable functional condition and costs

	Lifetime	Performance	Cost
Path	Flight cycles	K	€
4	20,761	1.45	4,000
5	19,986	3.92	2,000
6	20,416	3.29	1,500
7	20,761	1.43	10,000

Table 4 Forecast of the achievable functional condition

	Quality level	Duration	Willingness to pay	Cost	Profit
Path		Min	€	€	€
4	A	647	3,123	4,000	−877
5	C	419	2,397	2,000	397
6	B	392	2,429	1,500	929
7	A	62	3,979	10,000	−6,021

4.4 Machining

For the typical blade, the job scheduling selects path 6 based on the possible profit, even if it only achieves quality level B. The machining in this path is carried out by a robot that applies the solder, the hot gas corrosion protective layer, and the thermal barrier coating via thermal spraying. The multilayer system is then processed in a high-vacuum furnace with a defined temperature profile. Due to the near-net-shape coating technology, machining the protruding solder is not necessary. This shortens the processing time significantly. In Fig. 15, the blade is shown before and after the coating process.

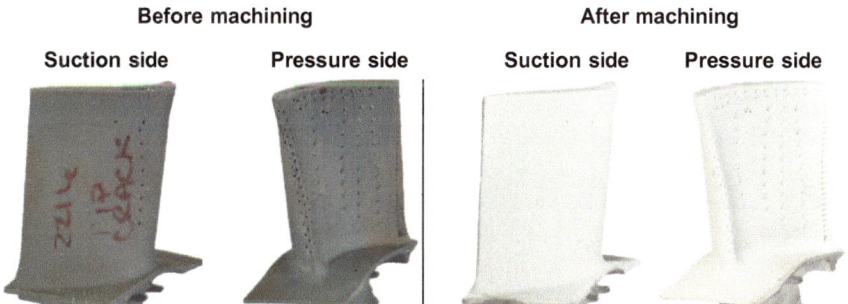

Fig. 15 Blade before and after machining (Nübel and Denkena 2022)

4.5 Quality Control

After machining, the blade is inspected for cracks, and the geometry is measured. Afterward, the achieved parameters and functional condition of the machined blade are determined with the functional evaluation. From the digital workpiece twin, the parameters from the forecast are read out, and the deviation between forecasted and the achieved parameters is calculated. This is then stored in a database as the forecast deviation. In the forecast of the next blade, the path-specific forecast errors for each parameter are subtracted from the forecast to improve on it.

In Fig. 16, this quality control loop is demonstrated for path 3. Besides the quality control loop for the forecast, a second quality control loop for the machining will be developed in other subprojects. The regeneration is finished if the blade has no defects and corresponds to the customer's requirements.

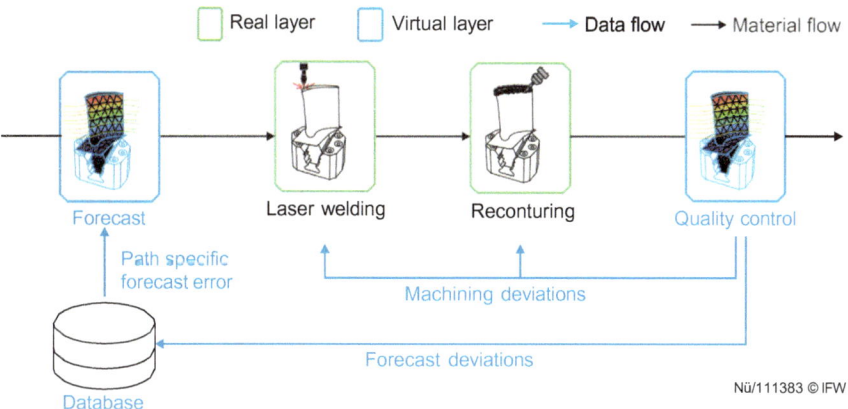

Fig. 16 Quality control loop (Nübel and Denkena 2022)

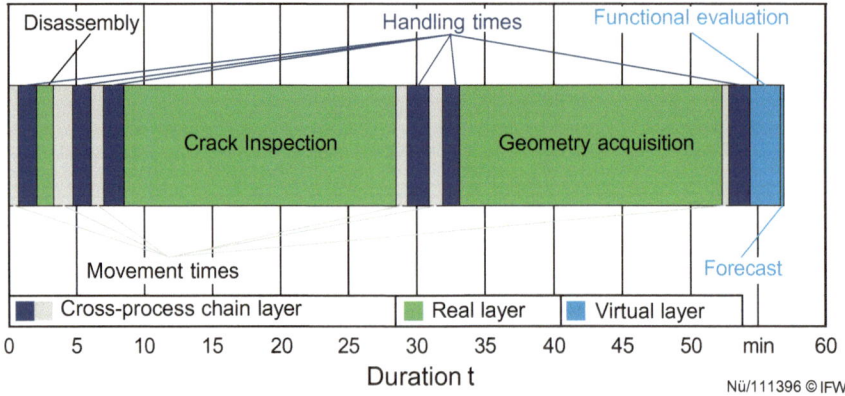

Fig. 17 Process times of the assessment and evaluation (Nübel and Denkena 2022)

4.6 Analysis of Process Times

In Fig. 17, the movement, handling, and process times of the assessment and evaluation are summarized. Because of the low absolute accuracy of the MHS, a time-intensive reference process is necessary, which increases the handling time. In comparison to the cells in the real layer or the movement and handling times, the functional evaluation and forecast have short process times. In an industrial implementation, the functional evaluation can be executed parallel to the handling, which can reduce the duration of the regeneration.

Figure 18 shows the proportion of the different layers for four different paths, each time with only one blade in the flexible manufacturing system. The new virtual layer, which allows condition-based regeneration, is, in all cases, under 5%. A greater share has the MHS, especially if many cells with short process times <5 min are included in a path. The regeneration time is dominated by the real layer.

It was demonstrated that integration of the virtual layer with the functional evaluation does not significantly increase the regeneration time. Through multiple blades in the system, the functional evaluation can be done parallel to the movement or handling times, which would further decrease the proportion of the regeneration time.

4.7 Lessons Learned from the Proof-Of-Concept

Parametrization of the point cloud in conjunction with lookup tables offers a robust method to quickly calculate the functional condition. Errors in the parametrization or the geometric measurement can be detected through fixed limits or a comparison

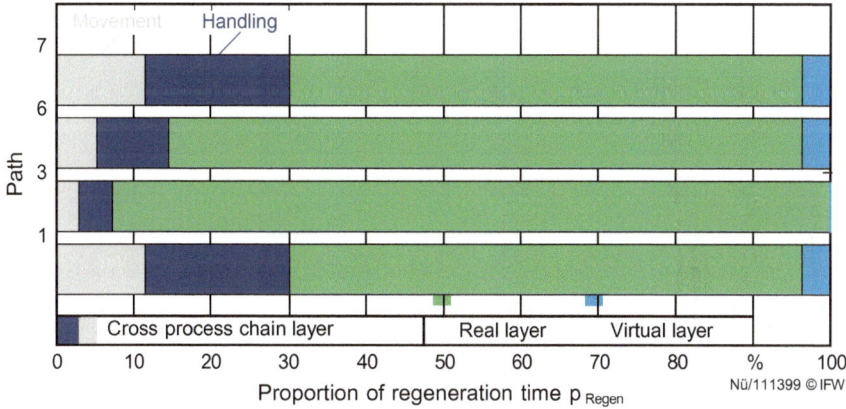

Fig. 18 Proportion of regeneration time of the different layers (Nübel and Denkena 2022)

between different cutting planes of parametrization. This is particularly necessary due to optical measurement technology on the blade's operational stress surfaces.

For the efficient utilization of the capacity of the cells with multiple blades in the system, the job scheduling needs to consider the movement time and the handling times. These can vary for each combination of start and finish points. In the process chain described, the movement and handling times were so constant that they only had to be measured once.

When using a digital workpiece twin as the sole interface for data, care must be taken to ensure that two programs do not access the twin at the same time. Especially job scheduling, which opens all twins to read out the customer information to create the work plan, triggers these access conflicts.

5 Conclusions

The process chain presented proves the feasibility of condition-based regeneration. By using a fully automated and digitized flexible manufacturing system in combination with the manufacturing and functional simulations, the selection of regeneration measures based on the achievable functional condition becomes possible. It was shown that integrating the simulations has only a minor influence on the regeneration time. Also, it is possible to build up such a process chain with existing technologies.

The flexible manufacturing system developed is scalable and easy to expand due to the mobile handling system (MHS). Since neither rails nor fences or other barriers are needed for the MHS, new technologies can easily be implemented through new process cells. If a machine is a bottleneck, a sister machine can be integrated. Even different workpieces, for example, different blade types, can be processed by the system simultaneously if they do not exceed the weight or size limitation of the MHS.

The regeneration of parts will become more important in the coming years because it requires less energy and materials compared to replacement. With the approach demonstrated, regeneration can be more efficient and economical through automation. Industrial implementation represents the next step for the approach presented.

Acknowledgements Funded by the Deutsche Forschungsgemeinschaft (DFG, German Research Foundation) SFB 871/3 119193472. Moreover, the authors would like to thank the other subprojects of CRC 871 for the substantial contribution of cells, software modules, and process knowledge.

References

Aschenbruck, J., Adamczuk, R., and Seume, J. R. (2014). Recent Progress in Turbine Blade and Compressor Blisk Regeneration. *Procedia CIRP*, *22*, 256–262. https://doi.org/10.1016/j.procir.2014.07.016

Betker, T., Quentin, L., Kästner, M., and Reithmeier, E. (Eds.) (2020). *3D Registration of Multiple Surface Measurements using Projected Random Patterns*. Optics and Photonics for Advanced Dimensional Metrology.

Bruchwald, O., Frackowiak, W., Reimche, W., and Maier, H. J. (2016). Applications of High Frequency Eddy Current Technology for Material Characterization of Thin Coatings. *Journal of Materials Science and Engineering a*, *6*(4). https://doi.org/10.17265/2161-6213/2016.7-8.001

Denkena, B., Bergmann, B., and Schumacher, T. (2021a). Anticipatory Online Compensation of Tool Deflection Using a Priori Information from Process Planning. *Journal of Manufacturing and Materials Processing*, *5*(3), 90.https://doi.org/10.3390/jmmp5030090

Denkena, B., Böß, V., Nespor, D., Flöter, F., and Rust, F. (2015). Engine Blade Regeneration: A Literature Review on Common Technologies in Terms of Machining. *The International Journal of Advanced Manufacturing Technology*, *81*(5/8), 917–924. https://doi.org/10.1007/s00170-015-7256-2

Denkena, B., Klemme, H., Nübel, N. (2021b). Bauteile simulationsgestützt aufarbeiten, Kostenkontrolle beim Instandsetzen. IT&Production, 10/2021, 32–33. https://www.it-production.com/produktentwicklung/kostenkontrolle-beim-instandsetzen

Denkena, B., Nyhuis, P., Bergmann, B., Nübel, N., and Lucht, T. (2019). Towards an Autonomous Maintenance, Repair and Overhaul Process. *Procedia Manufacturing*, *40*, 77–82. https://doi.org/10.1016/j.promfg.2020.02.014

Eberlein, A. (2007). Phases of High-Tech Repair Implementation. *18th International Symposium on Airbreathing Engines*, 1–8.

Huang, H., Gong, Z. M., Chen, X. Q., and Zhou, L. (2002). Robotic grinding and polishing for turbine-vane overhaul. *Journal of Materials Processing Technology*, *127*(2), 140–145.https://doi.org/10.1016/S0924-0136(02)00114-0

Kaierle, S., Overmeyer, L., Alfred, I., Rottwinkel, B., Hermsdorf, J., Wesling, V., and Weidlich, N. (2017). Single-Crystal Turbine Blade Tip Repair by Laser Cladding and Remelting. *CIRP Journal of Manufacturing Science and Technology*, *19*, 196–199.https://doi.org/10.1016/j.cirpj.2017.04.001

Kellenbrink, C., Nübel, N., Schnabel, A., Gilge, P., Seume, J. R., Denkena, B., and Helber, S. (2022). A Regeneration Process Chain with an Integrated Decision Support System for Individual Regeneration Processes Based on a Virtual Twin.

M'Saoubi, R., Axinte, D., Soo, S. L., Nobel, C., Attia, H., Kappmeyer, G., Engin, S., and Sim, W.-M. (2015). High Performance Cutting of Advanced Aerospace Alloys and Composite Materials. *CIRP Annals, 64*(2), 557–580.https://doi.org/10.1016/j.cirp.2015.05.002

Mullo, S. D., Pruna, E., Wolff, J., and Raatz, A. (2019). A vibration control for disassembly of turbine blades. *Procedia CIRP, 79*, 180–185.https://doi.org/10.1016/j.procir.2019.02.041

N3Engine Overhaul Services. (2011). *Reparatur und Überholung von Triebwerkskomponenten.* https://docplayer.org/111401854-Reparatur-und-ueberholung-von-triebwerkskompo nenten-repair-and-overhaul-of-aircraft-engine-components.html

Nübel, N. and Denkena, B. (2022). Digitaler Zwilling zur hochautomatisierten Einzelteilfertigung in der Reparatur. Dr.-Ing. Diss., Leibniz Universität Hannover.

Raj, B., Subramanian, C. V., and Jayakumar, T. (2000). *Non-destructive testing of welds.* Alpha Science.

Stania, L. and Seume, J. R (2022). Robust Probabilistic Analysis of Deterioration- Induced Aeroelasticity in an Axial Turbine. *Proceedings of ASME Turbo Expo 2022.*

Tsong-Jye Ng, B., Lin, W.-J., Chen, X., Gong, Z., and Zhang, J. (2004). Intelligent System for Turbine Blade Overhaul using Robust Profile Re-Construction Algorithm. In *ICARCV 2004 8th Control, Automation, Robotics and Vision Conference* (pp. 178–183). IEEE. https://doi.org/10.1109/ICARCV.2004.1468819

Yilmaz, O., Noble, D., Gindy, N., and Gao, J. (2005). A study of turbomachinery components machining and repairing methodologies. *Aircraft Engineering and Aerospace Technology, 77*(6), 455–466. https://doi.org/10.1108/00022660510628444

Target-Group Based Public Relations for the Collaborative Research Center 871

Claudia Schomaker, Gunnar Friege, Philipp Gilge, and Joerg R. Seume

Abstract This report describes the public relations work of the Collaborative Research Centre (CRC) 871 in its third and final funding period together with its partners. The public relations work aimed at two target groups: The expert audience and the general public with a focus on pupils. The expert audience includes research and industry, which were informed via symposiums, exhibitions and events. The general public, and in particular the target group of pupils, was informed about the contents of CRC 871 by events and newly developed interactive and multimedia learning materials. Furthermore, it is described how the public relations work was adapted to the restrictions of the Corona pandemic. For that adaptation, a new concept was developed and implemented in cooperation with the Faculty of Mechanical Engineering at Leibniz University Hannover, which primarily uses virtual media to make the contents of CRC 871 accessible to a broad public, even beyond the end of CRC 871.

Keywords Collaborative research centre · Public relations · Development of didactic materials

C. Schomaker
Institute of Special Education, Leibniz University Hannover, Schlosswender Straße 1, 30159 Hannover, Germany

G. Friege
Institute for Mathematics and Physics Didactics, Leibniz University Hannover, Welfengarten, 1A 30167 Hannover, Germany

P. Gilge · J. R. Seume (✉)
Institute of Turbomachinery and Fluid Dynamics, Leibniz University Hannover, An der Universitaet 1, 30823 Gabsen, Germany
e-mail: seume@tfd.uni-hannover.de
URL: https://www.tfd.uni-hannover.de

J. R. Seume et al. (eds.), *Regeneration of Complex Capital Goods*,
https://doi.org/10.1007/978-3-031-51395-4_23

1 Introduction

The purpose of public relations in CRC 871 was to inform the public about the obtained insights, newly developed technologies, and the concept of an automated and integral control of regeneration processes of complex capital goods. Therefore, the already established public relations activities have been further differentiated regarding the respective target group, intensified and consolidated. The public relations work in CRC 871 of today consists of numerous successful activities, measures and contacts which are based on the work in the first and the second funding periods of CRC 871. For example, the website www.sfb871.de was devised at the beginning of the first funding period together with an advertising agency. On this website information about the goals and the research priority as well as about events and publications is available. The content has been updated continually and has now, during the third funding period, been further developed and broadened, also because of the Covid pandemic. Additionally, campaigns and events also are a part of the public relation activities of CRC 871, in order to present the content to a wide public. To accomplish this different target groups have been specified: the expert audience, the general public and the potential young talents who are still in school.

The public relations work of CRC 871 has been developed with the Communications and Marketing of the Leibniz University Hannover (LUH). The department supports the CRC 871 when writing texts or articles for public relation activities. This has mainly been accomplished by one appointee who is in charge of supporting all the CRCs of LUH in their scientific communication and who has proven to be extremely helpful. Furthermore, the department's means of distributing press releases and event notifications have been used for the CRC 871 as well. Additionally, there was a close partnership with "Uni Transfer" for hosting events or participating in events. For example, the CRC 871 was part of the Hannover Messe in 2019 and "Die Nacht, die Wissen schafft" (science night) in 2018. Within the framework of the promotion of young talents the CRC 871 cooperated closely with the Student Advisory Services of LUH and the university equal opportunity office "ChancenVielfalt" in inviting pupils and teachers ("Schüler-Lehrer-Tag"), in visiting numerous schools with the "Leibniz-JuniorLab" and the "Mädchen-und-Technik-Kongress" (conference on girls in technology) (Sect. 4.1). Additionally, the newly deduced learning and teaching contents (Sect. 4.2) has been developed and put to test concertedly with the Institute of Special Education (IFS) Department "Sachunterricht und Inklusive Didaktik", the Institute for Didactics of Mathematics and Physics (IDMP), and Institute for Vocational Sciences in Metal Technology (IBM) of the LUH. Also, in order to develop a new concept of public relations in CRC 871 due to the Covid pandemic there was a collaboration with the dean's office of mechanical engineering (chapter "Non-destructive Characterization of Coating and Material Conditions of Heavily Stressed Turbine Components").

Within the framework of subproject Ö of CRC 871, the aim was to transfer the complex topic of CRC 871 into contexts close to everyday life, enabling primary and early secondary school pupils in the subjects of science and physics/natural sciences to become acquainted with the basic ideas of the CRC 871 and the working methods of scientists today. In the past 3½ project years, teaching units were developed and tested, evaluated and further developed with teachers in primary and lower secondary school lessons. In the area of physical education, the central results of the lesson development led to the conception of a picture book story and a related digital application (app), in order to connect the chosen everyday context of the CRC 871 topic ('repairing instead of throwing away') with the concrete working methods in the CRC 871 for pupils. For secondary level I, there is a tried and tested business game for 4–5 lessons and units on machines and troubleshooting as well as on repairing and upgrading headphones. In addition, there are materials on repairing zips for the primary grades and the beginning lessons at general education schools. The materials are compiled in loanable boxes for schools. The focus of the work so far has been on the development side (including a trial of the developed materials at schools in the city and region of Hannover). However, the supra-regional significance of the topic 'repair and recycling' as a social task and as a possible future occupational field has so far only been developed in rudimentary form for teaching in primary and secondary schools. Despite an increasing vocational orientation in schools, there are few concrete points of contact to the field of repair and recycling as a vocational field. At the same time, however, reference is made again and again to the existing shortage of skilled workers, so that an early linking of such topics is already necessary at primary level. The focus of the work so far has been on the development side (including a trial of the developed materials at schools in the city and region of Hanover). However, the supra-regional significance of the topic 'repair and recycling' as a social task and as a possible future occupational field has so far only been developed in rudimentary form for teaching in primary and secondary schools. Despite an increasing vocational orientation in schools, there are few concrete points of contact to the field of repair and recycling as a vocational field. At the same time, however, reference is made again and again to the existing shortage of skilled workers, so that an early linking of such topics is already necessary at primary level (Sect. 4.2).

2 Adaptation of Public Relations Because of the Consequences of the Corona-Pandemic

With the onset of the Corona pandemic and the associated restrictions on events, many of the planned public events could no longer be held. In 2020 and 2021, access restrictions to the university buildings meant that only very limited events could take place, which led to the cancellation of many of the events. Also, due to the discontinuation and conversion of school lessons, activities such as a school tour could not be implemented during this time. In order to continue public relations work under the conditions of the pandemic, new concepts were developed in CRC 871 in cooperation with the central institutions of LUH and the Faculty of Mechanical

Engineering. First of all, a dialogue took place on which activities were feasible under the conditions of the pandemic and could continue to be used meaningfully in the period after the pandemic. In addition, the actions were evaluated in terms of their acceptance and effect on the target group of young people. As a result, it was found that especially activities in the virtual space can be carried out safely and are well accepted by the target group. In addition, it was found that the formats planned so far for the virtual space (homepage, image films) were not sufficient to address the target group. Furthermore, the format of image films does not appear to attract the desired attention of young people and is rather perceived as an advertising measure that does not reflect reality.

With these findings, a new media concept was developed in close cooperation with the Faculty of Mechanical Engineering, which could be used by both the Faculty of Mechanical Engineering and the CRC 871 to continue public relations work under the conditions of the pandemic. The concept envisaged creating a broader media presence consisting of shorter videos with content on the research work in CRC 871, virtual tours of the experimental facilities and an expansion of the use of the CRC 871 homepage. The concept was supplemented by the content-related work in subproject "Ö" to create opportunities to present the content of CRC 871 in the area of schools and education. In addition, efforts were still made to hold in-person events. The specific measures that were implemented are the following:

- Expansion and maintenance of the homepage
- Creation of a channel on the video platform YouTube for the CRC 871
- Design of a virtual tour through the experimental facilities of CRC 871
- Creation of an infrastructure for creating short films together with the Faculty of Mechanical Engineering
- Generation of virtual content for the production of teaching material in the school sector
- Implementation of in-person events taking into account the restrictions of the Corona pandemic.

For the expansion and maintenance of the CRC 871 homepage, the existing structures were implemented in a new layout and the section containing the news and the description of the contents of the Collaborative Research Centre were specifically revised and their maintenance intensified. Furthermore, possibilities were established, for example, to integrate the virtual tours and the videos of the CRC 871's YouTube channel. By setting up our own channel on YouTube, several goals were achieved: On one hand, the overall visibility of the CRC 871 was increased by its presence on the YouTube video platform, and on the other hand, the very good infrastructure on the platform made it possible to avoid setting up redundant structures for putting the videos online and managing them. In addition, the videos on YouTube are very compatible with all kinds of end devices, so that the content can be shared and distributed quickly. In addition, the videos can be easily integrated on many popular online platforms and in the social media sector. This created the possibility to use the produced videos from CRC 871 efficiently and effectively for public relations. The videos could also be integrated in the area of the virtual tours and complement the possibility of visiting the experimental facilities in the CRC 871 in three dimensional spheres. The environment of the virtual tours also allows other media to be integrated, e.g., posters in PDF format. In order to be able to create these virtual environments and videos, the infrastructure had to be created first. For this purpose, the necessary material was compiled in cooperation with the Faculty of Mechanical Engineering, which already had experience in this field. Care was taken not to create redundancies with material that could already be contributed from the basic equipment of the faculty. In order to be able to produce videos and images spontaneously and extensively, a camera and lighting equipment were purchased and the staff of CRC 871 were introduced to the creation of media content in a seminar. This made it possible to create and implement high-quality content for public relations work even under pandemic conditions directly from the CRC 871.

This work was combined with the creation of new forms of communicating CRC 871 content to young pupils (as described in Sect. 4.2). For this purpose, among other things, interviews were conducted by pupils (2021) with the scientific staff and researchers from the CRC 871 during a visit of a class from the inclusive Otfried-Preußler-school. These interviews had been prepared beforehand in classroom sessions and then conducted at the Production Technology Centre at the Mechanical Engineering Campus in Garbsen in cooperation with the Faculty of Mechanical Engineering. The results were then integrated into an interactive app (see Sect. 4.2) and are also available on the CRC 871 YouTube channel. Furthermore, the described acquisitions and implementations of the concept were also used in the final colloquium, which was held as a hybrid event both in-person and online due to the pandemic. The lectures were recorded in English beforehand and in German during the event and made publicly available via the YouTube channel and links on the homepage. The same was done for the virtual tour.

3 Public Relations for the Professional Public and Industrial Stakeholders

Within the framework of CRC 871, a novel approach for the maintenance and repair of complex capital goods has been developed. Due to the broad interdisciplinary orientation, the developed methods and the findings of the individual subprojects are of interest to scientists and industrial research and development engineers from various disciplines. They have already been informed on the topic by articles in specialist journals and by presentations of the subprojects at scientific conferences. In addition, there has been and still is a constant professional exchange and knowledge transfer with various industrial companies on a bilateral basis. The CRC's novel approach uses methods that are classically only applied in product and manufacturing development, and transfers them to model-based or evidence-based decisions in Maintenance, Repair, and Overhaul (MRO) and scientifically extends them to the often more complex conditions of the stressed components and engines and their regeneration.

The CRC 871 was present at the Hannover Messe (Industrial exhibition in the city Hannover, Germany) 2019. There, the staff were able to present the CRC's research topics to a broad public as well as to decision-makers from politics and industry. Due to the pandemic, the Hannover Messe was cancelled in the following years and could therefore no longer be used for public relations. Instead, the overarching contents of CRC 871 were presented at international conferences, which had a clear technical reference to the regeneration of capital goods, but were not limited to only one discipline represented in CRC 871. For instance, the findings could be successfully presented at the international conference "Advanced Manufacturing and Repair for Gas Turbines" in 2019, 2020 and 2021. Also, as the only participant from the field of research, the CRC 871 and its findings were presented to a specialist audience at national conferences in Germany that explicitly address users, such as the "TÜV Süd Predictive Maintenance" conference in 2019. A keynote speech on system demonstrator from CRC 871 was also given at the Machine Innovations Conference 2020.

Furthermore, in order to inform the specialist public about new findings in the CRC, staff members have regularly presented the results of the subprojects at subject-specific international conferences. The publications are compiled on the CRC 871 homepage. The presentation of the general content and the subject-specific content of CRC 871 at conferences has also led to numerous direct contacts with companies and institutions, which have been used for professional exchange and/or to initiate cooperation and even transfer projects.

In addition to direct contact with potentially interested partners, general articles were also placed in non-specialist formats, i.e., in the "Uni Magazin" of LUH or an intranet article at the industrial partner MTU, which was also published in a publication in the associated "AEROREPORT" magazine.

4   Public Relations Within the Framework of the CRC 871 with Focus on Young People Promotion

4.1   Events and Affiliations

From September, 23th–28th 2018, a school tour called "LeibnizLAP" and which focussed on the CRC 871 was carried out in cooperation with the central university institution "uniKIK". Four schools in the Magdeburg area were visited and the contents of the CRC 871 were specifically presented to pupils. The school tour was aimed at 7th and 8th graders from grammar schools. The schools Dr. Frank Gymnasium, Albert-Einstein-Gymnasium Madgeburg, Käthe-Kollwitz-Gymnasium and Sportgymnasium Magdeburg took part. In order to convey the findings of the CRC 871 to the pupils, experiments and content were developed based on individual research projects. In particular, in addition to the theory on efficient regeneration of complex capital goods, parts of this knowledge could be directly experienced through an experiment in a wind tunnel.

On September 25th 2019, the third CRC 871 "Schüler-Lehrer-Tag" ("Pupil-Teacher-Day") was held under the name "Forschung macht Schule" ("Research goes to School") together with the central university institutions "UniKIK" and the Central Student Advisory Service of Leibniz University Hannover. The aim was, on one hand, to get high school students interested in technology in general, and in particular in topics of the CRC 871. On the other hand, teachers were shown examples of how they could possibly enrich mathematics, physics and/or chemistry lessons with current issues from research through simplified scientific experiments from the topics of CRC 871 and the simulation game (see also Sect. 4.2). The event lasted the whole day and the pupils were able to experience the contents of CRC 871 in groups and choose three out of six different available experiments. All experiments were inspired by questions that have also been worked on within the CRC 871. Therein, for instance experiments on lift in a simple wind tunnel could be carried out. Also, experiments on the subject of vibration were conducted on an airfoil and the influence of different air-gas compositions on combustion was investigated. The simulation game developed for work in schools (see Sect. 4.2) and a setup for optical measurement were also used. Experiences from previous funding periods were incorporated and results from network meetings of teachers and educationalists were used to further improve the experiments and materials of the "Schüler-Lehrer-Tag" in terms of content and didactics. A total of 73 students and 4 teachers from the three schools Gymnasium Stolzenau, Kooperative Gesamtschule Pattensen, and Berufsbildende Schulen II from Wolfsburg took part in the event. A direct evaluation by means of questionnaires at the end of the event, showed that the experiments improved the understanding of the contents of the CRC and generally increased the interest in a technical course of studies.

In order to further increase the visibility of the CRC 871 among female pupils, the CRC 871 participated in the Girls and Technology Day in 2019–2021. The Girls and Technology Day also involves female pupils working on scientific questions and thereby teaching them about research content.

4.2 Development of Didactic Materials for a Future-Oriented Science and Technology Teaching in Primary and Lower Secondary Schools

Within the framework of subproject Ö of CRC 871, the aim was to transfer the complex topic of CRC 871 into contexts close to everyday life, enabling primary and lower secondary school pupils in the subjects of science and physics/natural sciences to become acquainted with the basic ideas of the CRC 871 and the working methods of scientists today. In the past 3.5 project years, teaching units were developed and tested, evaluated and further developed with teachers in primary and lower secondary school lessons. In the area of physical education, the central results of the lesson development led to the conception of a picture book story and a related digital application (app), in order to connect the chosen everyday context of the CRC 871 topic ('repairing instead of throwing away', see Seume et al. 2019) with the concrete working methods in the CRC 871 for pupils.

For secondary level I, there is a tried and tested business game (Fig. 1) for 4–5 lessons and units on machines and troubleshooting as well as on repairing and upgrading headphones. In addition, there are materials on repairing zips for the primary grades and the beginning lessons at general education schools. The materials are compiled in loanable boxes for schools (Fig. 2). The focus of the work so far has been on the development side (including a trial of the developed materials at schools in the city and region of Hannover).

However, the supra-regional significance of the topic 'repair and recycling' as a social task and as a possible future occupational field has so far only been developed in rudimentary form for teaching in primary and secondary schools. Despite an

Fig. 1 Worksheet for the business game

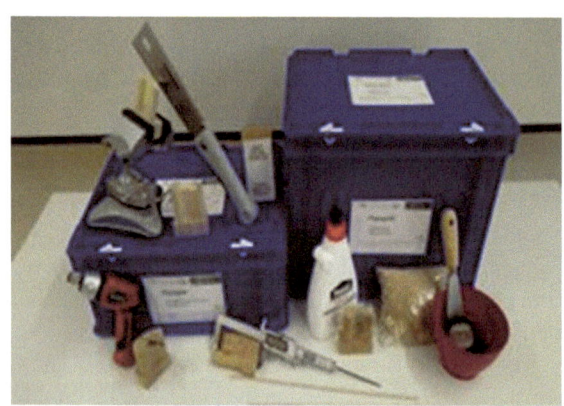

Fig. 2 Materials for the business game in boxes

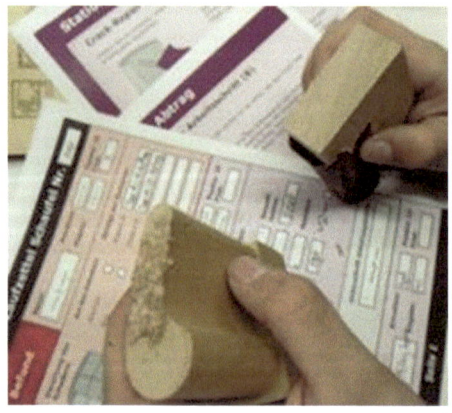

increasing vocational orientation in schools, there are few concrete points of contact to the field of repair and recycling as a vocational field. At the same time, however, reference is made again and again to the existing shortage of skilled workers, so that an early linking of such topics is already necessary at primary level. The focus of the work so far has been on the development side (including a trial of the developed materials at schools in the city and region of Hannover). However, the supra-regional significance of the topic 'repair and recycling' as a social task and as a possible future occupational field has so far only been developed in rudimentary form for teaching in primary and secondary schools. Despite an increasing vocational orientation in schools, there are few concrete points of contact to the field of repair and recycling as a vocational field. At the same time, however, reference is made again and again to the existing shortage of skilled workers, so that an early linking of such topics is already necessary at primary level.

4.2.1 Subject-Specificity Aims at Complexity

The task of science education is to support pupils in "understanding their natural, cultural, social and technical environment in a factual way, to open it up on this basis in an educationally effective way and to orientate themselves in it, to participate and to act" (GDSU 2013, 9).

In this way, the subject matter of the subject of physical education at the primary level and also of physical education at the secondary level aims at complexity. This means that educational content is to be selected in such a way that it

- "presuppose the recognition of the identity and integrity of person and thing,
- that they imply didactic processes of reciprocal development under the premises of uniqueness and exemplariness, connection to reality and phenomenological accessibility,
- that they demand developed knowledge and skills, problem- and science-orientation, relevance to the present and to the life-world, and
- that they enable pupils to gradually understand the nature of the world and to act responsibly in it: independently and self-determinedly, critically co-determining and ultimately also co-responsible in its implementation" (Lauterbach 2020, 152).

It is therefore necessary that complex societal issues find their way into lessons already at primary and secondary level. This demand is not new; the didactician Wolfgang Klafki already formulated so-called key problems in the 1980s, which should guide the selection of content in lessons. These key problems reflect social challenges of the present and presumably the future. In a 1985 publication, for example, he listed the peace issue, the environmental and ecological issue, socially produced inequality, the dangers and possibilities of the new technical control, information and communication media, as well as the subjectivity of the individual and the phenomenon of I-You relationships among the typical epochal issues of our time. However, the didactic approach to these questions has shown that complex issues are massively reduced in the classroom. Mehren and others state: "At present, learning groups often 'solve' the problems of the world in 45 minutes because the subject matter is oversimplified" (Mehren et al. 2014, 5). It is therefore necessary to develop didactic implementation possibilities that maintain the complexity of topics and at the same time enable students to access this complexity.

4.2.2 Repairing Instead of Throwing Away—Didactic Significance of a Complex Topic

Repairing technical things from everyday life is currently gaining increasing social significance. It is seen as a 'new social movement' and marked as the beginning of a developing 'repair society' (cf. Heckl 2013). Repairing as an activity that can be carried out by every citizen is thus "an expression of a growing technical maturity" (Krebs/Schabacher/Weber 2018, 10), which in particular also takes up the goals of sustainable management. According to Baier et al. (2016), this movement is characterised by an ethical interest that seeks to counter capitalist structures with "subsistence, participation, care and post-growth" (Krebs/Schabacher/Weber 2018, 10).

Furthermore, the representatives of the movement are united by the "commonality of a 'Do-it-together' (DIT) and the associated sharing of things" (ibid.) as well as the possibility of creating general access to repair knowledge, which is supported in particular by the internet.

However, the transformation process towards a sustainable society is an enormous challenge in which Education for Sustainable Development (ESD) plays a major role. The basis for sustainability-oriented values and norms as well as for corresponding social practices is a fundamental knowledge of sustainable strategies as well as the acquisition of competences that have the potential for their implementation.

The repair movement as part of the sustainability movement stands for a highly topical and at the same time ancient pattern that is still a natural part of social practice in many countries, but is being "reinvented" at the same time in western industrial nations. Repair is as old as technology itself. Even the tools of Stone Age people were repaired or used in other ways when they became blunt or a part of the blade broke off. Because the acquisition of necessary resources and the production of an object or tool required a high degree of energy, endurance and skill, the useful life of artefacts was extended in many ways until the 20th century. The introduction of mass production had the effect, among other things, of increasing the efficiency of production and consequently making products more affordable. The outsourcing of production to low wage countries accelerated the change to a throwaway society. The fact that modern societies are now rediscovering repair can be understood as a response to the consequences of an immoderate consumer society that lives beyond its means not only materially but also emotionally (cf. ibid.). For example, (online) trade has recently been criticised because many goods returned as returns were not resold but destroyed.

While repair work was still carried out by all social classes in preindustrial societies, repairing and the use of things has changed. "In rich mass-consumption societies, repairing is no longer a household strategy that encompasses the household goods, but rather a thing-specific devotion to the preservation of individual things that are considered worthy of repair, depending on their respective value" (ibid., 15). This is because, at the same time as the ownership of things has changed, things that have been sorted out can now be disposed of again with little effort (cf. ibid.). In addition, consumers today lack the possibilities to repair things because, for example, spare parts are not available or the purchase of a new appliance is far below the cost of repair. In poorer societies, on the other hand, the activity of repairing is considered very important because it allows goods that are difficult to obtain to be used for longer or to be kept as a resource ('means of exchange, spare parts store', cf. ibid., 19). Likewise, the continued use of used clothing and household appliances in particular is a way of dealing with things in everyday life in poorer regions of the world.

Although the 'repair' movement understands the activity of repairing in such a way that everyone should and can be enabled to carry it out, formal "technical knowledge about the construction and functioning of the objects to be repaired is important for repair, which is marked, among other things, by structured overviews of possible defects (so-called 'fault trees')" (cf. ibid., 25). (cf. ibid., 25). Added to this is longstanding experiential knowledge in handling these tools, which comes into play in the everyday practice of repairing and maintaining things: "This kind of knowledge is by no means merely intuitive, but is characterised by situational flexibility: the ability to choose between different knowledge resources—the intimate knowledge of different materials, construction methods and sensuously experienced error markers—in the interaction with the objects and work settings to be repaired" (ibid., 25).

Thus, different forms of mending can be distinguished, highlighting the respective intentions and goals of a repair and emphasising the social significance of this topic:

- "the practices of mending in relation to practices of replacement.
- the practices of waiting (with a view to the future of the object) in relation to the practices of repairing (with a view to the current malfunction, dysfunctionality of a thing)
- questions of material-material re-use (recycling) versus practices of re-use (repairing)
- Procedures of workarounds and repurposing as a question of 'actual' and 'inauthentic' repairs
- Relation of making and repairing" (ibid., 27).

The theme of 'repairing' is representative of processes of social change as a whole, which can be seen in particular in technical developments. These highlight the importance of technical issues for a future society (cf. Graube/Mammes 2016, Binder 2020). And so society's interest in technology and the associated changes is also high, with around 58% of the German population dealing with technical issues and considering them important. However, the majority of those surveyed take a critical view of the associated changes: almost half hope for an improvement in their personal quality of life (49.9%, acatech/TechnikRadar 2020), but many also fear that technical changes will lead to problems (26.7%, ibid.). At the same time, 45.5% trust that future technical solutions can help to overcome present-day problems such as hunger, poverty and climate change (ibid.). In order to be able to participate in social processes, it is important to deal with technical topics and challenges. At the same time, the way in which technical issues and topics occur in people's everyday lives makes it difficult to engage in a reflective debate: "Constructive mechanisms are hardly accessible to experience. Toys are mostly closed products and offer little opportunity for personal discovery; it is not possible to dismantle them without destroying them. Fully automated production processes are no longer visible to the end consumer and therefore cannot be traced. The increasing digitalisation that shapes people's everyday lives (e.g. driverless means of transport, self-service checkouts in supermarkets and public libraries, etc.) makes it difficult to understand technical processes. Due to the lack of transparency without educational processes, it is hardly possible to participate in social developments in a responsible manner and to develop individual evaluation skills" (Landwehr/Mammes/Murmann 2021, 7).

These contexts require the implementation of technical learning and education processes in a comprehensive sense and from the very beginning in order to enable children and young people to "use, understand, evaluate and assess technology and to use technical concepts and processes to read problems" (ibid.). The aim is to initiate technical literacy, which includes the acquisition of differentiated technical competences as well as the development of reflected attitudes and attitudes towards technical issues. Technical education also contributes to the development of the personality, because "whoever has solved a technical problem experiences himself as a successful agent, he not only gains knowledge about technology, but at the same time about himself, about his technology- specific productivity and creativity, his dexterity and strength" (Wiesmüller 2021, 33).

The following competencies were formulated both in the Perspective Framework for Teaching Science (GDSU 2013) as well as in other expert reports, which are to be exemplified in technical science teaching at the primary level:
"Central areas of knowledge are therefore

- Knowing technical inventions
- Recognising technical functional relationships
- Appropriate use of tools
- Producing objects themselves (planning, designing and executing)
- Analysis and modelling
- Observing, trying out, assembling, disassembling, recreating, constructing, reflecting
- Drawing processes of planning and construction
- Creating models from structured and unstructured material
- Transfer of acquired knowledge and experience
- Identifying the function and proper use of everyday objects by oneself
- Carry out maintenance and repair work

- Assessing and evaluating the quality of products
- Evaluate technology and technical innovations (human, social, economic, ecological aspects)" (Stiftung Haus der kleinen Forscher 2012, 10; cf. also GDSU 2013, 65f.).

These objectives are clearly supported by education policy, because an expansion of STEM education in general and technical education in particular is associated with the expectation that this will interest children and young people in these topics at an early and sustainable stage and thus also in future occupational fields in this area. In this way, the shortage of skilled workers caused by demographic change and the lack of interest in studying technology-oriented subjects is to be countered early and comprehensively (cf. MINT Action Plan BMBF 2019a). For example, the subject of computer science is rarely offered in the upper school and is not perceived as attractive by young people (only 1% of young people choose an advanced course in computer science, only 15% of whom are girls); the above-average number of first-year students in STEM subjects compared to other countries contrasts with a high number of dropouts (cf. acatech/MINT-Nachwuchsbarometer 2021). These aspects focus in particular on the implementation of gender-sensitive technical education. This is because girls still show significantly less interest in STEM-related issues and have significantly less confidence in dealing with them (with comparable performance in these school subjects, cf. ibid.). Thus, the aim is to address technical issues in class in a way that is detached from social stereotypes regarding technology, in order to place technology and gender in a relationship that is not presented and perceived as a given, but as a variable that is "constructed in social interactions" (Gilbert 2021, 72).

4.2.3 Technical Learning in Science and Physics Lessons/natural Sciences

Such a demand for technical learning can already be successfully implemented in primary school science lessons. Studies have shown (Mammes 2001, Möller 1991, Tenberge 2002, among others) that children's original interest in technical questions can be taken up in lessons in such a way that they are "able to solve [...] demanding technical problems" (Beinbrech 2014, 120). It is therefore possible to formulate technology-related problems for pupils, for which they develop concrete solutions by designing and building them using their own constructions, among other things. However, the competences to be developed in technical education do not only include the realisation of own constructions, also with the help of the professional use of certain techniques and tools, and their testing, but also the reflection on their functioning and quality (cf. Kosack et al. 2015). For example, many physics curricula include the treatment of technical machines and devices such as the electric motor or the lifting crane. Often, the focus is on understanding the principal mode of operation (e.g. motor as an all-current motor, function of individual components) of often elementary motors. The construction of motors from individual parts by pupils, the extension of the treatment to real motors and, if the motor is chosen appropriately, also the implementation of troubleshooting in the technical area in regular lessons (cf. e.g. Friege 2018), is also possible. Other examples concern the construction and use of modern sensor technology in everyday life, such as strain gauges or smoke detectors.

To date, however, studies on the implementation of technical content in school curricula or the consideration of 'technology' as an independent subject point to a "large gap between the social relevance of technical education and its reality in schools" (Binder 2020, 13). The content of technical education is integrated into the subjects of science and crafts at the primary level and into physics and science lessons at the secondary level, and is linked to other topics and subject-specific approaches. The objectives and questions of technical contexts can thus be developed from an interdisciplinary understanding. This offers the opportunity to consider questions such as 'teaching to assess the consequences of technology' in an interdisciplinary way and thus to reflect on technical progress from many perspectives (cf. Grunwald 2016, 34). However, this is also accompanied by the challenge of ensuring that technical issues are not regarded as mere appendages of other topics and that the independence of this content area is lost.

In view of the structural, curricular conditions for technical education, it is a challenge to integrate technical education content into primary school science lessons and secondary school science and technology lessons (cf. acatech/MINT-Nachwuchsbarometer 2021). Teachers are faced with the task of implementing STEM education that is fit for the future, in particular by linking technical topics and objectives to the con-tent areas and objectives of other subjects. Studies make it clear that technical content is thus given little consideration in educational processes. This has a lasting effect on the formation of interests and career choices of children and young people with regard to these topics. In an extensive empirical study, pupils were asked about the target areas of physics lessons. The results show that (a) the target area of occupation (how people work in certain physical/technical occupations) is clearly underrepresented compared to the target areas of society, science and everyday life, and (b) the interests of the pupils are clearly greater than the actual teaching on offer (cf. Häußler et al. 1998). In addition, in series such as 'The Big Bang Theory' or 'Breaking Bad', scientists (and also engineers) are portrayed as nerds and arouse considerable interest, especially among young people. Spitzer (2019), citing Kessels and Hannover (2002), notes: "The stereotypes conveyed, however, are not merely an image of scientists, they also shape decisions such as course choice or career choice of students" (Spitzer 2019, 4, after Kessels and Hannover 2002).

The ideas regarding possible occupational fields and areas of application are only rudimentarily developed; moreover, female pupils hardly trust themselves to contribute their skills and competences here, even if they show comparable performance in the STEM area as pupils. With regard to the associated necessary competences in the area of digital education, which are particularly important when dealing with technical issues, it is shown, for example, that approx. 33% of all pupils in the eighth grade show weak performance with regard to technology and information-related competences and only approx. 14% of high school graduates are able to systematically research information online and assess it (cf. acatech/MINT Nachwuchsbarometer 2021).

In order to effectively implement technical content in the design of teaching learning situations of STEM subjects, in addition to taking into account content that currently reflects the societal challenges in relation to technology, approaches are required that enable children and young people to learn working methods via adequate content that enable them to explore, understand, evaluate and assess technical contexts.

Based on the subject area and the objective of the technical sciences, the problem-oriented approach to the design of technical learning processes is of great importance here. Studies on the effectiveness of such teaching concepts also underline this, so that the selection of a problem that is interesting for boys and girls is considered an essential feature of good teaching design. This goes hand in hand with "enabling individual learning and problem-solving paths, providing suitable material for solving the problem and checking the associated partial solutions, [offering] help systems [and] conducting meta conversations as well as problem-solving conversations" (Beinbrech 2014, 124). International curricula and educational standards also focus on problem-solving learning in technical education and are oriented towards the concept of the 'design process' of the ITEA (2007, 237; cited in Kosack et al. 2015, 97). This is "a systematic problem-solving strategy used, given criteria and conditions, to find multiple possible solutions to solve a (technical) problem or to satisfy (technically shaped) needs or wants with the intention of narrowing down the number of solutions towards a final solution" (ibid.). In this context, Kosack et al. emphasise that the ability to solve problems in technical education should be combined with a focus on the creative abilities of students, which are indispensable for the development of technical solutions (cf. ibid., 119f.). Graube brings these demands on the design of technical learning processes together by focusing on the invention of technology as a central target dimension: "Inventing technology can open up a new perspective on technology. Children can recognise that technology is the finding of creative solutions to a technical problem or an end-means relationship and perceive themselves as inventors. The focus is on their own idea and their own solution or product [...]. This makes the variety of possible solutions visible to everyone, where there is no right or wrong, but only solutions that can be evaluated according to certain previously defined criteria. Functionality has priority here, but the creative idea, the appearance, the quality, the production effort, the effects on the environment, etc. can also be included in the evaluation" (Graube 2016, 42). Pupils should be given comprehensive access to technology through the interplay of inventing (constructing), discovering (reconstructing) and uncovering (deconstructing) technical contexts. The perspective of inventing is supplemented by the discovery of functional principles, technical processes, means- purpose relationships (What is it for?) and historical references (cf. ibid.). "The unmasking perspective requires the ability to take on a meta-level. [...] Questioning one's own inventions and the inventions of fellow pupils with a view to what could be better is primarily suitable for this" (ibid.).

The development of contents for technical education processes that have a high social relevance and make it possible to sensitise children and young people to these contents, to interest them and to teach them competences for the formation of technical literacy is an empty space in relation to STEM education. Learning environments need to be developed that take into account the principles of technical education in terms of content and access (hybrid teaching-learning environments with the integration of different for-mats) and thus also take up the goals of digital education contexts across the board. Digital education is understood here as the key to participation in a digital world: at work, as a consumer or as a citizen (cf. KMK 2016, BMBF 2019b). At the same time, new opportunities for and access to education through digitalisation can be used by creating new didactic means, dissemination channels and access to knowledge.

4.2.4   Project Development in the Subject of Science Education (Primary Level)

With a view to possible objectives and the choice of content in an inclusive subject teaching, which sees itself as a multi-perspective subject teaching, the claim to moti-vate and sensitise children for an enquiring attitude in order to adopt a questioning attitude towards the world (see Pech/Schomaker 2013) is of particular importance. Inclusive teaching that takes this principle of lesson design into account gives space to children's 'real' questions (see Schreier 1989), in which individual interests, an individual approach to the world and subjective meanings are reflected. According to the conceptual self-understanding of contemporary science education, children's motivation to link their individual experience with the world in which they live is the starting point for the planning of science education. Following Wagenschein, it is important to place the child and the object in an educational relationship: "With the child from the object, which is the object for the child" (Wagenschein 1990/2010: 11).

These intentions are taken up with the didactic use of (non-fiction) picture books, which depict a non-fictional reality and thus allow a concrete reference to empirically verifiable events.

In order to take up the previously outlined goals of technical education in physical education, a physical picture book (Fig. 3) was developed here, which is supple-mented by an app. The story of the picture book picks up on the topic of CRC 871 in the sense of 'repairing instead of throwing away'. A group of children prepares for a camping trip and discovers that the zip of the tent no longer works. Readers are encouraged to join the children in the story in discovering different ways to repair everyday objects. The story continues in an app accessible via a QR code in the book (Figs. 4 and 5).

The users are actively involved in the app, explanatory films, tutorials and inter-views conducted by students from the project with individual scientists from the CRC 871 connect the everyday context of the story with the concrete content of the CRC 871. It shows how the topic of 'repairing' is being researched by scientists at the PZH in Hanover. The book and app were developed together with teachers from

Fig. 3 Book: title page

Fig. 4 Scenes from the digital application 1

the primary school sector and unfortunately only tested in rudimentary form due to the pandemic (Schomaker 2018).

Fig. 5 Scene from the
application 2

4.2.5 Project Development for Physics/Science Lessons (Secondary Level I)

The development of materials and lessons for physics and science lessons in lower secondary education began with a simulation game on repairing an aircraft engine blade. This was based on tests at the CRC 871's teacher-pupil days and, in particular, a concept was developed for regular lessons and secondary school pupils. In the simulation game, the pupils deal with the analysis of the object to be repaired, develop repair paths and translate these into processes to be carried out. The process of a repair in the economy is reproduced by the pupils taking on different roles (e.g. business manager, mechanic, controller, ...). In addition, the students repair various defects on wooden aircraft turbine blades using different tools. The simulation game is designed for use in 8th-10th grades and has already been tested in elective courses at grammar schools. Other material boxes deal with the concept of machines (Figs. 6 and 7), the construction of everyday machines such as a kitchen mixer or electric knife, and the construction and repair of elementary machines with the help of technology kits. In addition, students can use other material boxes to focus on repairing zips (Fig. 8) and upgrading headphones by investigating the construction and troubleshooting, repairing or upgrading these items.

Fig. 6 Materials for the machine unit

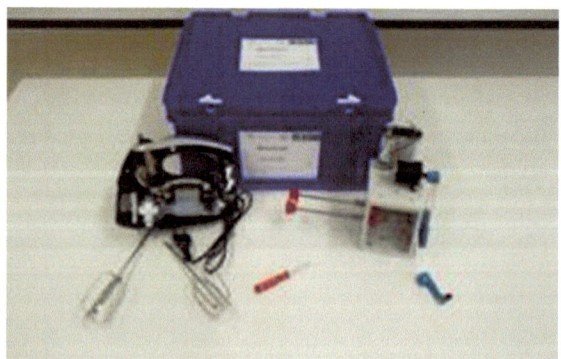

Fig. 7 Model of a kitchen blender as part of the machine unit

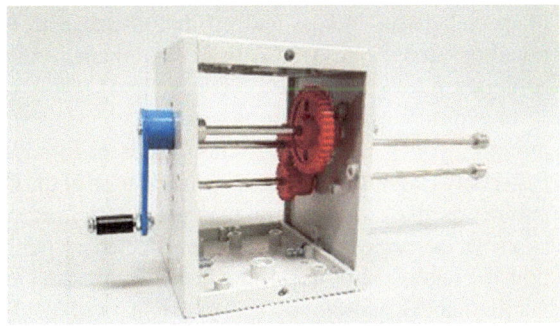

Fig. 8 Materials for the unit zippers

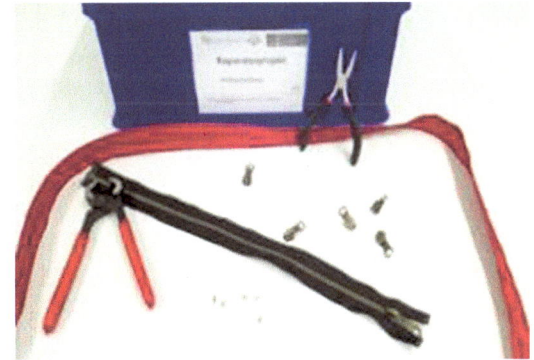

Within the framework of final theses at Bachelor's and Master's level), various content-related and structural questions regarding the handling and effectiveness of the materials created were investigated. The testing in the schools was massively impaired by the pandemic-related restrictions. Despite the great interest on the part of the teachers, it was hardly possible to carry out the project, as some of the pupils were taught online during the Corona period, the schools were not accessible to external persons for many months and school classes could not come to the university as part of an excursion. However, interested teachers will be able to use the developed materials for the duration of CRC 871.

5   Conclusions

The CRC 871's public relations work has created several opportunities to inform both the public and specific target groups about the contents of CRC 871. The expert public and industrial stakeholders were informed through a wide range of opportunities for participation in conferences and events. This includes, above all, participation in specialist conferences and exhibitions, such as the "Hannover Messe" international fair and the ASME "Advanced Manufacturing and Repair for Gas Turbines" conference.

In addition, events and new teaching concepts successfully addressed the target group of young people and schoolchildren. In addition to events held at Leibniz University Hannover, a direct communication of the CRC 871's content to the general public was also ensured through targeted activities, such as the creation of a learning app with interviews prepared and conducted by pupils with CRC 871 staff. In addition, the school subprojects have been incorporated into the publication of an issue of the journal "Grundschule Sachunterricht" with the focus on "Repairing" as well as articles in the journal "Unterricht Physik" with a focus on "Physics and Technology". The contents of the two subprojects will also be the subject of a one-week teacher training course in November 2022 and will be further developed in qualification work. It has been shown that the materials developed here are a didactically sensible way to address complex topics in lessons with young pupils.

By adapting public relations work to digital content, the restrictions caused by the Corona pandemic were successfully compensated. To this end, staff were empowered to create digital content for public relations by getting specific equipment and training. Together with the Faculty of Mechanical Engineering at Leibniz University Hannover, a concept was implemented that led to an increased online presence of the CRC 871. This made it possible, for example, to provide a virtual tour of the experimental facilities and to create videos of content and lectures from CRC 871, and share them to a broad public despite the restrictions imposed by the Corona pandemic. By integrating and expanding the homepage of the CRC 871 and freely accessible platforms (e.g. YouTube), this content can also be made available beyond the end of CRC 871.

Acknowledgements Funded by the Deutsche Forschungsgemeinschaft (DFG, German Research Foundation) SFB 871/3 119193472.

References

acatech (Hrsg.) (2020). Technik Radar 2020. Was die Deutschen über Technik denken. Schwerpunkt: Bioökonomie. München/Hamburg.

acatech (Hrsg.) (2021). MINT-Nachwuchsbarometer 2021. In: https://www.acatech.de/publikation/mint-nachwuchsbarometer-2021/ [letzter Zugriff: 20.10.21].

Baier, A., Hansing, T., Müller, C.,Werner, K. (2016). Die Welt reparieren: Eine Kunst des Zusammenmachens. In: dies. (Hrsg.): Die Welt reparieren. Open Source und Selbermachen als postkapitalistische Praxis. 1. Aufl. Bielefeld: transcript, 34–62.

Beinbrech, C. (2014). Technisches Lehren und Lernen. In: Hartinger, A., Lange, K. (Hrsg.): Sachunterricht. Didaktik für die Grundschule. Berlin: Cornelsen, 116–130.

Binder, M. (2020). Wie wäre es, technisch gebildet zu sein? Technische Bildung im Kontext Allgemeiner Bildung. Baltmannsweiler: Schneider.

Bundesministerium für Bildung und Forschung (BMBF) (2019a). Mit MINT in die Zukunft. DerMINT-AktionsplandesBMBF. In: https://www.bmbf.de/SharedDocs/Publikationen/de/bmbf/pdf/mit-mint-in-die-zu-kunft.pdf;jsessionid=29AE43960F6E690BAC583B40560 39FE3.live381?blob=publicationFile&v=2[letzter Zugriff: 20.10.21].

Bundesministerium für Bildung und Forschung (BMBF) (2019b). Digitale Zukunft: Lernen. Forschen. Wissen. Die Digitalstrategie des BMBF. In: https://www.bildung-for-schung.dig ital/digitalezukunft/shareddocs/Downloads/files/bmbf_digitalstrategie.pdf;jsessionid=4D6 62AFE692A88A6862FB5080CE57149.live092?blob=publicationFile&v=1[letzter   Zugriff: 20.10.21].

Finanzierungsantrag zum SFB 871 Regeneration komplexer Investitionsgüter 01.01.2018–31.12.2021. Unveröff. Antrag, Hannover: LUH.

Friege, G. (2018). Schülerinnen und Schüler bauen und testen einen Elektromotor – Eigenschaften erkunden – Fehler suchen, finden und beheben. Unterricht Physik, 29,163, S. 21–23.

Gesellschaft für Didaktik des Sachunterrichts (GDSU) (2013) (Hrsg.). Perspektivrahmen Sachunterricht. Vollständig überarbeitete und erweiterte Ausgabe. Bad Heilbrunn: Klinkhardt.

Gilbert, A.-F. (2021). Zum Verhältnis von Gender und Technik. Wege zu einer gendersensiblen Technischen Bildung. In: Müller, M., Schumann, S. (Hrsg.). Technische Bildung. Stimmen aus Forschung, Lehre und Praxis. Münster: Waxmann, 69–88.

Graube, G. (2016). Erfinden, Entdecken und Enttarnen: Didaktische Leitfragen für die Auseinandersetzung mit Basiskonzepten der Technik. In: Mammes, I. (Hrsg.): Technisches Lernen im Sachunterricht. Nationale und internationale Perspektiven. Baltmannsweiler: Schneider, 22–44.

Graube, G., Mammes, I. (2016). Gesellschaft im Wandel. Konsequenzen für die natur- und technikwissenschaftliche Bildung in der Schule. Bad Heilbrunn: Klinkhardt.

Grunwald, A. (2016). Gesellschaftliche Folgen des technischen Fortschritts: Anlass für interdisziplinä-re Erforschung und Gestaltung. In: Graube, G., Mammes, I. (Hrsg.): Gesellschaft im Wandel. Konsequenzen für natur- und technikwissenschaftliche Bildung in der Schule. Bad Heilbrunn: Klinkhardt, 24–37.

Häußler, P., Bünder, W., Duit, R., Gräber, W. & Mayer, J. (1998). Naturwissenschaftsdidaktische Forschung. Perspektiven für die Unterrichtspraxis. Kiel: IPN, S. 26–28.

Heckl, W. M. (2013). Die Kultur der Reparatur. München: Hanser.

Kessels, U. & Hannover, B. (2002). Die Auswirkungen von Stereotypen über Schulfächer auf Berufswahlabsichten. In: Spinath, B. (Hg.). Pädagogische Psychologie unter gewandelten gesellschaftlichen Bedingungen. Dokumentation des 5. Dortmunder Sympposiums für Pädagogische Psychologie, S. 53–67.

Kosack, W., Jeretin-Kopf, M., Wiesmüller, C. (2015). Zieldimensionen technischer Bildung im Elementar- und Primarbereich. In: Graube, G., Jeretin-Kopf, M., Kosack, W., Mammes, I., Renn, O., Wiesmüller, C. (Hrsg.): Wissenschaftliche Untersuchungen zur Arbeit der Stiftung, Haus der kleinen Forscher'. Bd. 7, Schaffhausen: Schubi, 30–156.

Krebs, S., Schabacher, G., Weber, H. (Hg.) (2018). Kulturen des Reparierens. Dinge - Wissen - Praktiken. Bielefeld: transcript (Edition Kulturwissenschaft, Band 133).

Kultusministerkonferenz (KMK) (2016). Bildung in der digitalen Welt. Strategie der Kultusminister-konferenz. In: https://www.kmk.org/fileadmin/pdf/PresseUndAktuelles/2018/Digitalstrategie_2017_mit_Weiterbildung.pdf [letzter Zugriff: 20.10.2021].

Landwehr, B., Mammes, I., Murmann, L. (Hrsg.) (2021). Technische Bildung im Sachunterricht der Grundschule. Elementar bildungsbedeutsam und dennoch vernachlässigt? Bad Heilbrunn: Klinkhardt.

Lauterbach, R. (2020). Bildungsinhalte bestimmen. In: Tänzer, S., Lauterbach, R., Blumberg, E., Grittner, F., Lange, J., Schomaker, C. (Hrsg.): Sachunterricht begründet planen. Das Prozessmodell Generativer Unterrichtsplanung Sachunterricht (GUS) und seine Grundlagen. 2. vollständig überarbeitete Auflage. Bad Heilbrunn: Klinkhardt,141–159.

Mammes, I. (2001). Förderung des Interesses an Technik. Eine Untersuchung zum Einfluss technischen Sachunterrichts auf die Verringerung von Geschlechterdifferenzen im technischen Interesse. Frank-furt/M.: Peter Lang.

Mehren, R., Rempfler, A., Ulrich-Riedhammer, E. M. (2014). Denken in komplexen Zusammen-hängen. Systemkompetenz als Schlüssel zur Steigerung derEigenkomplexität von Schülern. In: Praxis Geographie, H. 4, 4–8.

Möller, K. (1991). Handeln, Denken und Verstehen. Untersuchungen zum naturwissenschaftlich-technischen Sachunterricht in der Grundschule. Essen: Westarp Wissenschaften.

Pech, D., Schomaker, C. (2013): Inklusion und Sachunterrichtsdidaktik – Stand und Perspek-tiven. In: K.-E. Ackermann, O. Musenberg & J. Riegert (Hrsg.): Geistigbehindertenpädagogik!? Disziplin – Profession – Inklusion. Oberhausen: Athena, 341–359.

Schomaker, C. (2018). ‚Das große Rennen‘ – Kinder greifen aktuelle Fragestellungenaus der Forschung auf. In: Grundschule Sachunterricht, H. 78, 27–33.

Schreier, H. (1989): Ent-trivialisiert den Sachunterricht! In: Grundschule, 21, H. 3, 10–13. Wagenschein, M. (1990/2010): Kinder auf dem Wege zur Physik. Weinheim & Basel: Beltz.

Seume, J., Denkena, B., Hartmann, U. (2019). Reparieren statt Ersetzen. Über die Regenera-tion komplexer Investitionsgüter. In: Unimagazin. Forschungsmagazin der Leibniz Universität Hannover. Ausgabe 1/2, 2019, 56–59.

Spitzer, P. (2019). Klug, logisch denkend, aber unromantisch. Das Image von PhysikerInnen und ChemikerInnen. Plus Lucis, 4, 4–10.

Stiftung Haus der kleinen Forscher (2012). Technik – Bauen und Konstruieren. Hintergründe und Praxisideen für die Umsetzung in Hort und Grundschule.Berlin.

Tenberge, C. (2002). Persönlichkeitsentwicklung und Sachunterricht. Dissertation. Universität Münster.

Wiesmüller, C. (2021). Wirklich(e) technische Allgemeinbildung. In: Müller, M./Schumann, S. (Hrsg.): Technische Bildung. Stimmen aus Forschung, Lehre und Praxis. Münster: Waxmann, 25–39.